SECOND EDITION

Foundations of Earth Science

Frederick K. Lutgens

Edward J. Tarbuck

Illinois Central College

Illustrated by

Dennis Tasa

PRENTICE HALL
Upper Saddle River, New Jersey 07458

Library of Congress Cataloging-in-Publication Data

Lutgens, Frederick K.
    Foundations of earth science / Frederick K. Lutgens,
Edward J. Tarbuck; illustrated by Dennis Tasa. --2nd ed.
        p.   cm.
    Includes index.
    ISBN 0-13-914037-9
    1. Earth sciences.   I. Tarbuck, Edward J.   II. Title.
QE28.L96   1999
550—dc21   98–44447
                                                    CIP

Senior Editor: *Daniel Kaveney*
Editor in Chief: *Paul F. Corey*
Editorial Director: *Tim Bozik*
Assistant Vice President of Production and Manufacturing: *David W. Riccardi*
Executive Managing Editor: *Kathleen Schiaparelli*
Assistant Managing Editor: *Lisa Kinne*
Production Editor: *Edward Thomas*
Marketing Manager: *Leslie Cavaliere*
Creative Director: *Paula Maylahn*
Art Director: *Joseph Sengotta*
Art Manager: *Gus Vibal*
Photo Editors: *Lorinda Morris-Nantz* and *Melinda Reo*
Photo Researcher: *Clare Maxwell*
Copy Editor: *James Tully*
Assistant Editor: *Wendy Rivers*
Editorial Assistant: *Margaret Ziegler*
Production Assistant: *Nancy Bauer*
Marketing Assistant: *Rachele Triano*
Interior Design: *Amy Rosen*
Cover Design: *John Christiana*
Manufacturing Manager: *Trudy Pisciotti*
Manufacturing Buyer: *Benjamin Smith*
Text Composition: *Lido Graphics/ Molly Pike*
Cover Photo: *Mount Ritter and Banner Peak, in Ansel Adams Wilderness, Sierra Nevada, CA.*
    Photo © by Carr Clifton Photography.

© 1999, 1996 by Prentice-Hall, Inc.
Upper Saddle River, New Jersey 07458

10   9   8   7   6   5   4   3   2

ISBN 0-13-914037-9

Prentice-Hall International (UK) Limited, *London*
Prentice-Hall of Australia Pty. Limited, *Sydney*
Prentice-Hall Canada Inc., *Toronto*
Prentice-Hall Hispanoamericana, S.A., *Mexico*
Prentice-Hall of India Private Limited, *New Delhi*
Prentice-Hall of Japan, Inc., *Tokyo*
Pearson Education Asia Pte. Ltd., *Singapore*
Editora Prentice-Hall do Brasil, Ltda., *Rio de Janeiro*

Contents

Preface

Earth is a very small part of a vast universe, but it is our home. It provides the resources that support our modern society and the ingredients necessary to maintain life. Therefore, a knowledge and understanding of our planet is critical to our social well being and indeed, vital to our survival.

In recent years, media reports have made us increasingly aware of our place in the universe and the forces at work in our physical environment. We are also beginning to learn that human interactions with natural systems can upset delicate balances. News stories inform us of new discoveries in the solar system and beyond. Daily reports remind us of the destruction caused by hurricanes, earthquakes, floods, and landslides. We have been made aware of ozone depletion, potential global warming, and growing environmental concerns about the oceans. To comprehend, prepare for, and solve these and other concerns requires an awareness of how science is done and the scientific principles that influence our planet, its rocks, mountains, atmosphere, and oceans.

The Second Edition of *Foundations of Earth Science*, like its predecessor, is a college-level text designed for an introductory course in Earth science. It consists of seven units that emphasize broad and up-to-date coverage of basic topics and principles in geology, oceanography, meteorology, and astronomy. The book is intended to be a meaningful, nontechnical survey for undergraduate students with little background in science. Usually these students are taking an Earth science class to meet a portion of their college's or university's general requirements.

In addition to being informative and up-to-date, a major goal of *Foundations of Earth Science* is to meet the need of beginning students for a readable and user-friendly text, a book that is a highly usable "tool" for learning basic Earth science principles and concepts. To accomplish this goal we have incorporated the following features:

Readability. The language of this book is straightforward and *written to be understood*. Clear, readable discussions with a minimum of technical language are the rule. When new terms are introduced, they are placed in **boldface** and defined. The frequent headings and subheadings also help students follow discussions and identify the important ideas presented in each chapter.

Illustrations and Photographs. The Earth sciences are highly visual. Therefore, photographs and artwork are a very important part of an introductory book. *Foundations of Earth Science*, Second Edition contains hundreds of high-quality photographs that were carefully selected to aid understanding, add realism, and heighten the interest of the reader.

Many illustrations were revised and redesigned for the Second Edition so that ideas and concepts are presented even more clearly and realistically than before. The extensive art program was carried out by Dennis Tasa, a gifted artist and respected Earth science illustrator.

Focus on Learning. To assist student learning, every chapter opens with a series of questions. Each question alerts the reader to an important idea or concept in the chapter. When a chapter has been completed, four useful devices help students review. First, a helpful summary—**The Chapter in Review**—recaps all of the major points. Next is a checklist of **Key Terms** with page references. Learning the language of Earth science helps students learn the material. This is followed by **Questions for Review** that help students examine their knowledge of significant facts and ideas. Finally, a twenty-item chapter test—**Testing What You Have Learned**—wraps up the chapter-end review.

Acknowledgments

Writing a college textbook requires the talents and cooperation of many individuals. Working with Dennis Tasa, who is responsible for all of the outstanding illustrations, is always special for us. We not only value his outstanding artistic talents and imagination, but his friendship as well.

We are also grateful to Professor Ken Pinzke at Belleville Area College. In addition to his many helpful suggestions regarding the manuscript, Ken prepared the chapter-opening questions and the chapter-end summaries and tests that help make the book a usable tool for beginning students. Professor Pinzke is also responsible for preparing the *Student Study Guide* and *Laboratory Manual* that accompany the text.

Our students remain our most effective critics. Their comments and suggestions continue to help us maintain our focus on readability and understanding.

Special thanks goes to those colleagues who prepared in-depth reviews of the manuscript. Their critical comments and thoughtful input helped guide our work and clearly strengthened the text. We wish to thank:

Kim Bishop, *California State University, Los Angeles*
Barbara Callison, *San José State University*
Danny Childers, *Delaware County Community College*
Lindren L. Chyi, *University of Akron*
J. Warner Cribb, *Middle Tennessee State University*
Frank P. Florence, *Jefferson Community College*
Clayton D. Harris, *Middle Tennessee State University*
Sue Ellen Hirschfeld, *California State University, Hayward*
George C. Kelley, *Onondaga Community College*
Patricia D. Lee, *University of Hawaii*
John T. Leftwich, *Old Dominion University*
Kevin McCool, *Belleville Area College*
Ula L. Moody, *Florida Community College*
Carl Ojala, *Eastern Michigan University*

Roger A. Podewell, *Olive-Harvey College*
Paul Richards, *LDS Business College*
Peter B. Stifel, *University of Maryland*
Sabine F. Thomas, *University of Texas at San Antonio*
Anthony J. Vega, *Clarion University*

We also want to acknowledge the team of professionals at Prentice Hall. Thanks to Editor in Chief Paul Corey. We sincerely appreciate his continuing strong support for excellence and innovation. Thanks also to our editor Dan Kaveney. His strong communication skills and energetic style contributed greatly to the project. The production team, led by Ed Thomas, has done an outstanding job. They are true professionals with whom we are very fortunate to be associated.

Fred Lutgens
Ed Tarbuck

The Teaching and Learning Package

The authors and publisher have been pleased to work with a number of talented people to produce an excellent supplements package. This package includes the traditional supplements that students and professors have come to expect from authors and publishers, as well as some new kinds of supplements that involve electronic media.

For the Student

EarthShow CD-ROM

Each copy of *Foundations of Earth Science*, Second Edition comes with *EarthShow*, created by professional photographer and renowned geologic educator, Parvinder Sethi of Radford University. This provides students with a wide array of visual and audio resources for the study of Earth science. This technology has been extensively tested, and has proven to be a very effective study tool.

Internet Support

This site, specific to the text, contains numerous review exercises (from which students get immediate feedback), exercises to expand one's understanding of Earth science, and resources for further exploration. This Web site provides an excellent platform from which to start using the Internet for the study of Earth science. Please visit the site at **http://www.prenhall.com/lutgens**

Geosciences on the Internet: A Student's Guide

Written by Andrew T. Stull and Duane Griffin, this is a student's guide to the Internet and World Wide Web specific to geology. *Geosciences on the Internet* is available at no cost to qualified adopters of the text. Please contact your local Prentice Hall representative for details.

Study Guide

Written by experienced college educator Ken Pinzke in conjunction with the authors, the *Study Guide* complements the text by providing students with additional tools for learning Earth science more effectively. Each chapter of the *Study Guide* contains a chapter overview, learning objectives, a comprehensive review, key terms, a vocabulary review, and a practice test with corresponding answers. Professor Pinzke's publications have helped thousands of students master the Earth sciences.

For the Professor

Transparency Set

More than 125 full-color acetates of illustrations from the text are available free of charge to qualified adopters.

Slides

More than 150 slides of images taken from the text, many of which were taken by the authors, are also available to qualified adopters.

Presentation Manager

This user-friendly navigation software enables professors to custom build multimedia presentations.

Prentice Hall Presentation Manager 3.0 contains several hundred images from the text. The CD-ROM allows professors to organize material in whatever order they choose; preview resources by chapter; search the digital library by keyword; integrate material from their hard drive, a network, or the Internet; or edit lecture notes and annotate images with an overlay tool. This powerful presentation tool is available at no cost to qualified adopters of the text.

*The New York Times—Themes of the Times—*Changing Earth

This unique newspaper-format supplement features recent articles about geology from the pages of *The New York Times*. This supplement, available at no extra charge from your local Prentice Hall representative, encourages students to make connections between the classroom and the world around them.

Instructor's Resource Manual with Tests

In this manual Ken Pinzke gives new instructors the benefit of his years of teaching experience and experienced instructors a ready source of new ideas to complement their teaching style. The *Manual* contains a variety of lecture outlines, teaching tips, advice on how to integrate visual supplements, and many suggested test questions.

Test Item File

The *Test Item File* provides instructors with a wide variety of test questions.

PH Custom Test

Based on the powerful testing technology developed by Engineering Software Associates, Inc. (ESA), *Prentice Hall Custom Test* allows instructors to create and tailor exams to their own needs. With the on-line testing program, exams can also be administered online and data can then be automatically transferred for evaluation. The comprehensive desk reference guide is included along with online assistance.

For the Laboratory

Applications and Investigations in Earth Science

Written by Ed Tarbuck, Fred Lutgens, and Ken Pinzke, this full-color laboratory manual contains 22 exercises that provide students with hands-on experiences in geology, oceanography, meteorology, astronomy, and Earth science skills.

Introduction

A View of Earth

A view of Earth from space gives us a unique perspective of our planet (Figure I.1). At first, it may strike us that Earth is a fragile-appearing sphere surrounded by the blackness of space. In fact, it is just a speck of matter in a vast universe. As we look more closely at our planet from space, it becomes clear that Earth is much more than rock and soil. The swirling clouds suspended in the atmosphere and the vast global ocean are just as prominent as the continents.

From such a vantage point we can appreciate why Earth's physical environment is traditionally divided into three major parts: the solid earth; the water portion of our planet, the hydrosphere; and Earth's gaseous envelope, the atmosphere. However, our environment is highly integrated. It is not dominated by rock, water, or air alone. Rather, it is characterized by continuous interactions as air comes in contact with rock, rock with water, and water with air. Moreover, the biosphere, which is the totality of all plant and animal life on our planet, interacts with each of the three physical realms and is an equally integral part of Earth.

The interactions among Earth's four spheres are uncountable. Figure I.2 provides an easy-to-visualize example. The shoreline is an obvious meeting place for rock, water, and air. In this scene, ocean waves that were created by the drag of air moving across the water are breaking against the rocky shore. The force of the water can be powerful and the erosional work that is accomplished can be great.

Let us take a brief tour of Earth's four "spheres." Earth is sometimes called the *blue planet.* Water more than anything else makes Earth unique. The **hydrosphere** is a dynamic mass of water that is continually on the move, from the oceans to the atmosphere, to the land, and back to the ocean again. The global ocean is certainly the most prominent feature of the hydrosphere, blanketing nearly 71 percent of Earth's surface and accounting for about 97 percent of Earth's water. However, the hydrosphere also includes the fresh water found in streams, lakes, and glaciers, as well as that found underground.

Although these latter sources constitute just a tiny fraction of the total, they are much more important than their meager percentage indicates. In addition to providing the fresh water that is so vital to life on the continents, streams, glaciers, and groundwater are responsible for sculpturing and creating many of our planet's varied landforms.

Earth is surrounded by a life-giving gaseous envelope called the **atmosphere.** This thin blanket of air is an integral part of the planet. It not only provides the air that we breathe but also protects us from the Sun's intense heat and dangerous radiation. The energy exchanges that continually occur between the atmosphere and Earth's surface and between the atmosphere and space produce the effects we call weather.

If, like the Moon, Earth had no atmosphere, our planet would be lifeless because many of the processes and interactions that make Earth's surface such a dynamic place could not operate. Without weathering and erosion, the face of our planet might more closely resemble the lunar surface, which has not changed appreciably in nearly 3 billion years.

Figure I.1 Africa and Arabia are prominent in this image of Earth taken from Apollo 17. The tan cloud-free zones over the land coincide with major desert regions. The band of clouds across central Africa is associated with a much wetter climate that in places sustains tropical rain forests. The dark blue of the oceans and the swirling cloud patterns remind us of the importance of the oceans and the atmosphere. Antarctica, a continent covered by glacial ice, is visible at the South Pole. *(Courtesy of NASA-Science Source-Photo Researchers, Inc.)*

Figure I.2 The shoreline is one obvious meeting place for rock, water, and air. In this scene, ocean waves that were created by the force of moving air break against the rocky shore. The force of the water can be powerful and the erosional work that is accomplished can be great.*(Photo by H. Richard Johnston/Tony Stone Images)*

Lying beneath the atmosphere and the ocean is solid Earth. It is divided into three principal units: the dense *core;* the less dense *mantle;* and the *crust,* which is the light and very thin outer skin of Earth (Figure I.3). The crust is not a layer of uniform thickness. It is thinnest beneath the oceans and thickest where continents exist. Although the crust may seem insignificant when compared with the much thicker units of solid Earth, it was created by the same general processes that formed Earth's present structure. Thus, the crust is important in understanding the history and nature of our planet.

Much of our study of Earth focuses on the more accessible surface features. Fortunately, these features represent the outward expressions of Earth's dynamic interior. By examining the most prominent surface features and their global extent, we can obtain clues to the dynamic processes that shape our planet.

The fourth "sphere," the **biosphere,** includes all life on Earth and consists of the parts of the solid Earth, hydrosphere, and atmosphere in which living organisms can be found. Plants and animals depend on the physical environment for the basics of life.

However, organisms do more than just respond to their physical environment. Through countless interactions, life-forms help maintain and alter their physical environment. Without life, the makeup and nature of solid Earth, hydrosphere, and atmosphere would be very different.

The Earth Sciences

Earth Science is the name for all the sciences that collectively seek to understand Earth and its neighbors in space. It includes geology, oceanography, meteorology, and astronomy.

In this book, Units One through Four focus on the science of **geology,** a word that literally means "study of Earth." Geology is traditionally divided into two broad areas—physical and historical.

Physical geology examines the materials that make up Earth and seeks to understand the many processes that operate beneath and upon its surface. Earth is a dynamic, ever-changing planet. Forces within Earth create earthquakes, build mountains, and produce volcanic structures. At the surface, external processes break rock apart and sculpture a

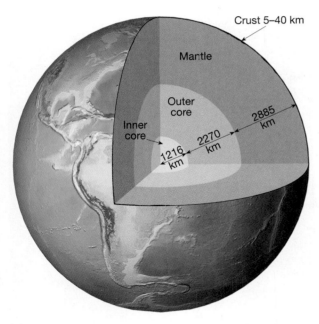

Figure I.3 View of Earth's layered structure. The inner core, outer core, and mantle are drawn to scale, but the thickness of the crust is exaggerated by about five times.

broad array of landforms. The erosional effects of water, wind, and ice result in diverse landscapes. Because rocks and minerals form in response to Earth's internal and external processes, their interpretation is basic to an understanding of our planet.

In contrast to physical geology, the aim of *historical geology* is to understand the origin of Earth and the development of the planet through its 4.6-billion-year history. It strives to establish an orderly chronological arrangement of the multitude of physical and biological changes that have occurred in the geologic past (Figure I.4).

Unit Five, *The Oceans,* is devoted to oceanography. **Oceanography** is actually not a separate and distinct science. Rather, it involves the application of all sciences in a comprehensive and interrelated study of the oceans in all their aspects and relationships. Oceanography integrates chemistry, physics, geology, and biology. It includes the study of the composition and movements of seawater, as well as coastal processes, seafloor topography, and marine life.

Unit Six, *The Atmosphere,* examines the mixture of gases that is held to the planet by gravity and thins rapidly with altitude. Acted on by the combined effects of Earth's motions and energy from the Sun, the formless and invisible atmosphere reacts by producing an infinite variety of weather, which, in turn, creates the basic pattern of global climates. **Meteorology** is the study of the atmosphere and the processes that produce weather and climate. Like oceanography, meteorology involves the application of other sciences in an integrated study of the thin layer of air that surrounds Earth.

Figure I.4 Partially exposed skeleton of an extinct sabretoothed cat. This lion-sized cat roamed the White River Badlands of South Dakota during the Oligocene epoch. The aim of historical geology is to understand the development of Earth and its life through time. Fossils are essential tools in that quest. *(Photo by T.A. Wiewandt/DRK Photo)*

Unit Seven, *Astronomy,* demonstrates that the study of Earth is not confined to investigations of its four interacting "spheres." The Earth sciences also attempt to relate our planet to the larger universe. Because Earth is related to all other objects in space, the science of **astronomy**—the study of the universe—is very useful in probing the origins of our own environment. Because we are so closely acquainted with the planet on which we live, it is easy to forget that Earth is just a tiny object in a vast universe. Indeed, Earth is subject to the same physical laws that govern the countless other objects that populate the expanses of space. Thus, to understand explanations of our planet's origin, it is useful to learn something about the other members of our solar system. Moreover, it is helpful to view the solar system as a part of the great assemblage of stars that comprise our galaxy, which, in turn, is but one of many galaxies.

Resources and Environmental Issues

Environment refers to everything that surrounds and influences an organism. Some of these things are biological and social, but others are nonliving. The factors in this latter category are collectively called our *physical environment.* The physical environment encompasses water, air, soil, and rock, as well as conditions such as temperature, humidity, and sunlight. The phenomena and processes studied by the Earth sciences are basic to an understanding of the physical environment. In this sense, most of Earth science may be characterized as environmental science.

However, when the term *environmental* is applied to Earth science today, it usually means relationships between people and the physical environment. Application of the Earth sciences is necessary to understand and solve problems that arise from these interactions.

Resources

Resources are an important environmental concern. Resources range from water and soil to metallic and nonmetallic minerals and energy. These materials are the very basis of modern civilization. The mineral and energy resources that are extracted from the crust are the raw materials from which the products used by society are made (Figure I.5).

Few people who live in highly industrialized nations realize the quantity of resources needed to

Figure I.5 As world population grows, demand for mineral and energy resources climbs. Earth scientists must deal with the search for additional supplies of traditional and alternative resources as well as the environmental impact of their extraction and use. *(Photo by Craig Aurness/WestLight)*

maintain their present standard of living. For example, the annual per capita consumption of metallic and nonmetallic mineral resources for the United States is nearly 10,000 kilograms (11 tons). This is each person's prorated share of the materials required by industry to provide the vast array of products modern society demands. Figures for other highly industrialized countries are comparable.

Resources are commonly divided into two broad categories. Some are classified as *renewable,* which means that they can be replenished over relatively short time spans. Common examples are plants and animals for food, natural fibers for clothing, and forest products for lumber and paper. Energy from flowing water, wind, and the Sun are also considered renewable.

By contrast, many other basic resources are classified as *nonrenewable.* Important metals such as iron, aluminum, and copper fall into this category, as do our most important fuels: oil, natural gas, and coal. Although these and other resources continue to form, the processes that create them are so slow that significant deposits take millions of years to accumulate. In essence, Earth contains fixed quantities of these substances. When the present supplies are mined or pumped from the ground, there will be no more. Although some nonrenewable resources, such as aluminum, can be used over and over again, others, such as oil, cannot be recycled.

How long will the remaining supplies of basic resources last? How long can we sustain the rising standard of living in today's industrial countries and still provide for the growing needs of developing regions? How much environmental deterioration are we willing to accept in pursuit of basic resources? Can alternatives be found? If we are to cope with an increasing demand and a growing world population, it is important that we have some understanding of our present and potential resources.

Environmental Problems

In addition to the quest for adequate mineral and energy resources, the Earth sciences must also deal with a broad array of other environmental problems.

Some are local, some are regional, and still others are global in extent. Serious difficulties face developed and developing nations alike. Urban air pollution, acid rain, ozone depletion, and global warming are just a few that pose significant threats (Figure I.6). Other problems involve the loss of fertile soils to erosion, the disposal of toxic wastes, and the contamination and depletion of water resources. The list continues to grow.

In addition to human-induced and human-accentuated problems, people must also cope with the many natural hazards posed by the physical environment (Figure I.7). Earthquakes, landslides, floods, and hurricanes are just four of the many risks. Others such as drought, although not as spectacular, are nevertheless equally important environmental concerns. In many cases, the threat of natural hazards is aggravated by increases in population as more people crowd into places where an impending danger exists or attempt to cultivate marginal lands that should not be farmed.

It is clear that as world population continues its rapid growth, pressures on the environment will increase as well. Therefore, an understanding of Earth is not only essential for the location and recovery of basic resources, but also for dealing with the human impact on the environment and minimizing

Figure I.6 Air pollution in downtown Los Angeles. Air quality problems affect many cities. Fuel combustion by motor vehicles and power plants provides a high proportion of the pollutants. Meteorological factors determine whether pollutants remain "trapped" in the city or are dispersed. *(Photo by Ted Spiegel/Black Star)*

Figure I.7 A helicopter rescues a man stranded on the roof of his Olivehurst, California, home after a levee broke along the Feather River. Huge floods occurred in California's Central Valley in January 1997. There are many Earth processes and phenomena that are hazardous to people, including volcanoes, earthquakes, landslides, floods, hurricanes, and tornadoes. *(Photo by John Trotter/The Sacramento Bee)*

the effects of natural hazards. Knowledge about our planet and how it works is necessary to our survival and well-being. Earth is the only suitable habitat we have, and its resources are limited.

The Nature of Scientific Inquiry

All science is based on the assumption that the *natural world behaves in a consistent and predictable manner.* The overall goal of science is to discover the underlying patterns in the natural world and then to use this knowledge to predict what will or will not happen, given certain facts or circumstances.

Collecting Facts

The development of new scientific knowledge involves some basic, logical processes that are universally accepted. To determine what is occurring in the natural world, scientists collect *facts* through observation and measurement (Figure I.8). These data are essential to science and serve as the springboard for the development of scientific theories.

Hypothesis

Once facts have been gathered and principles have been formulated to describe a natural phenomenon, investigators try to explain how or why things happen in the manner observed. They can do this by constructing a preliminary, untested explanation, which we call a scientific **hypothesis.** Often, several different hypotheses are advanced to explain the same facts and observations. Next, scientists think about what will occur or be observed if a hypothesis is correct and devise ways or methods to test the accuracy of predictions drawn from the hypothesis.

If a hypothesis cannot be tested, it is not scientifically useful, no matter how interesting it might seem. Testing usually involves making observations, developing models, and performing experiments. What if test results do not turn out as expected? One possibility is that there were errors in the observations or experiments. Of course, another possibility is that the hypothesis is not valid. Before rejecting the hypothesis, the tests may be repeated or new tests may be devised. The more tests the better.

Figure I.8 An automated weather station. The meteorologist is downloading the accumulated data into a laptop computer. *(Photo by David Parker/Science Photo Library/Photo Researchers, Inc.)*

The history of science is littered with discarded hypotheses. One of the best known is the idea that Earth was at the center of the universe, a proposal that was supported by the apparent daily motion of the Sun, Moon, and stars around Earth.

Theory

When a hypothesis has survived extensive scrutiny, and when competing hypotheses have been eliminated, a hypothesis may be elevated to the status of a scientific **theory.** In everyday language, we may say "that's only a theory." But a scientific theory is a well-tested and widely accepted view that scientists agree best explains certain observable facts. It is not enough for scientific theories to fit only the data that are already at hand. Theories must also fit additional observations that were not used to formulate them in the first place. Put another way, theories should have predictive power.

Scientific theories, like scientific hypotheses, are accepted only provisionally. It is always possible that a theory which has withstood previous testing may eventually be disproven. As theories survive more testing, they are regarded with higher levels of confidence. Theories that have withstood extensive testing, such as the theory of plate tectonics or the theory of evolution, are held with a very high degree of confidence.

Scientific Methods

The process just described, in which scientists gather facts through observations and formulate scientific hypotheses and theories, is termed the *scientific method.* Contrary to popular belief, the scientific method is not a standard recipe that scientists apply in a routine manner to unravel the secrets of our natural world. Rather, it is an endeavor that involves creativity and insight. Rutherford and Ahlgren put it this way: "Inventing hypotheses or theories to imagine how the world works and then figuring out how they can be put to the test of reality is as creative as writing poetry, composing music, or designing skyscrapers."*

There is not a fixed path that scientists always follow that leads unerringly to scientific knowledge. Nevertheless, many scientific investigations involve the following steps: (1) the collection of scientific facts through observation and measurement (Figure I.9); (2) the development of one or more working hypotheses to explain these facts; (3) development of observations and experiments to test the hypothesis; and (4) the acceptance, modification, or rejection of the hypothesis based on extensive testing.

Other scientific discoveries represent purely theoretical ideas, which stand up to extensive examination. Still other scientific advancements have been made when a totally unexpected happening occurred during an experiment. These serendipitous discoveries are more than pure luck, for as Louis Pasteur said, "In the field of observation, chance favors only the prepared mind."

---

*F. James Rutherford and Andrew Ahlgren, *Science for All Americans* (New York: Oxford University Press, 1990), p. 7.

Scientific knowledge is acquired through several avenues, so it might be best to describe the nature of scientific inquiry as the *methods* of science, rather than the scientific method.

Studying Earth Science

In this book, you will discover the results of centuries of scientific work. You will see the end product of millions of observations, thousands of hypotheses, and hundreds of theories. We have distilled all of this to give you a "briefing" in Earth science.

But realize that our knowledge of Earth is changing daily, as thousands of scientists worldwide make satellite observations, analyze drill cores from the sea floor, study earthquake waves, develop computer models to predict climate, examine the genetic codes of organisms, and discover new facts about our planet's long history. This new knowledge often updates hypotheses and theories. Expect to see many new discoveries and changes in scientific thinking in your lifetime.

Figure I.9 Geologist taking lava sample from a "skylight" in a lava tube. Near Kilauea Volcano, Hawaii. *(Photo by G. Brad Lewis/Tony Stone Images)*

UNIT ONE
Earth Materials

Bridalveil Falls in Yosemite Valley, Yosemite National Park, California.
Photo by Jack Dykinga & Associates

CHAPTER 1

Minerals

Building Blocks of Rocks

FOCUS ON LEARNING

To assist you in learning the important concepts
in this chapter, you will find it helpful to focus
on the following questions:

1. What are minerals and how are they different from rocks?
2. What are the smallest particles of matter? How do atoms bond together?
3. How do isotopes of the same element vary from each other and why are some isotopes radioactive?
4. What are some of the physical and chemical properties of minerals? How can these properties be used to distinguish one mineral from another?
5. What are the eight elements that make up most of Earth's continental crust?
6. What is the most abundant mineral group? What do all minerals within this group have in common?
7. When is the term *ore* used with reference to a mineral? What are the common ores of iron and lead?

Crystals of elbaite and albite from Brazil.
Photo by Jeff Scovil

Figure 1.1 Pyramid Peak, Moroon Bells, Colorado. *(Photo by Carr Clifton Photography)*

Earth's crust is the source of a wide variety of minerals, many of which are useful and essential to people. In fact, practically every manufactured product contains materials obtained from minerals. Most people are familiar with the common uses of basic metals, including aluminum in beverage cans, copper in electrical wiring, gold in jewelry, and silicon in computer chips. But fewer are aware that pencil "lead" does not contain lead metal but is really made of the soft black mineral called *graphite*. Baby powder ("talcum powder") is ground-up rock made of the mineral *talc*, with perfume added. Drill bits impregnated with pieces of *diamond* (a mineral) are used by dentists to drill through tooth enamel. The common mineral *quartz* is the main ingredient in ordinary glass and is the source of silicon for computer chips. And on and on. As the material demands of modern society grow, the need to locate additional supplies of useful minerals also grows, and becomes more challenging as the easily mined sources become depleted.

The economic uses of rocks and minerals are important to us. Also important to Earth scientists is that every process they study depends in some way on the properties of these basic Earth materials. Events such as volcanic eruptions, mountain building, weathering and erosion, and even earthquakes involve rocks and minerals (Figure 1.1). Consequently, a basic knowledge of Earth materials is essential to the understanding of all Earth science phenomena.

Minerals Versus Rocks

Many people think of rocks as nondescript, hard, often dirty objects. Minerals are considered by many to be part of the human diet ("vitamins and minerals"), or possibly rare ores or precious gems. However, these common perceptions are not entirely correct.

Figure 1.2 shows the relation between rocks and minerals. A **rock** can be defined simply as an aggregate of minerals. Minerals are the *building blocks* that make up rocks. In a rock, the minerals occur together as a *mixture*, so each mineral retains its distinctive properties, such as color, as you can see in Figure 1.2.

Although most rocks are made up of more than one mineral, certain minerals are commonly found by themselves in large, relatively pure quantities. In these instances, they are still considered to be rocks. A common example is the mineral *calcite*, which frequently

Granite
(Rock)

Quartz
(Mineral)

Hornblende
(Mineral)

Feldspar
(Mineral)

Figure 1.2 Rocks are aggregates of one or more minerals.

is the dominant constituent in widespread rock layers where it is given the name *limestone*.

Minerals, the building blocks of rocks, have a very precise definition. A **mineral** is a naturally occurring inorganic solid that possesses a definite chemical structure, which gives it a unique set of physical properties. Thus, for any Earth material to be considered a mineral, it must exhibit the following characteristics:

1. It must be naturally occurring.
2. It must be inorganic (was never alive).
3. It must be a solid.
4. It must possess a definite chemical structure.

When the term *mineral* is used by geologists, only those substances that fulfill these precise conditions are considered minerals. Consequently, synthetic diamonds, although chemically the same as natural diamonds, are human-made, and therefore are not considered minerals. Further, oil and natural gas, which are not solids, and which once were "living" (they are the remains of ancient organisms), are also not classified as minerals. An exception is material formed by the activity of animals, such as clam shells, which are considered minerals.

Composition and Structure of Minerals

Each of Earth's nearly 4000 minerals is a unique substance that is defined by its chemical composition and internal structure. What elements make up a particular mineral? In what pattern are its elements bonded together? And what important properties does this give each mineral?

At this point, let us briefly review the basic building blocks of minerals, the chemical **elements.** At present, 112 elements are known, although 20 of these have been produced only in the laboratory. Some minerals, such as gold and sulfur, are made entirely from one element. But most elements are not stable, and thus most minerals are a combination of two or more elements, joined to form a chemically stable compound.

To understand better how elements combine to form minerals, we must first consider the atom. The **atom** is the smallest particle of matter that has all the characteristics of an element. It is this extremely small particle that does the combining.

How Atoms are Constructed

Two simplified models of an atom are shown in Figure 1.3. Each atom has a central region, called the

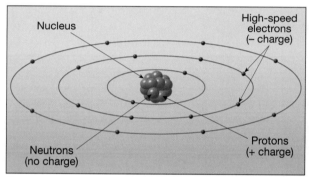

A.

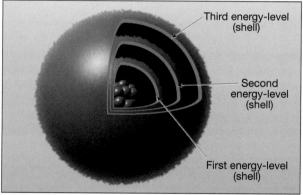

B.

Figure 1.3 Two models of the atom. **A.** This very simplified view of the atom has a central nucleus, consisting of protons and neutrons, encircled by high-speed electrons. **B.** Another model of the atom showing spherically shaped electron clouds (energy level shells). Note that these models are not drawn to scale. Electrons are minuscule in size compared to protons and neutrons, and the relative space between the nucleus and electron shells is much greater than illustrated.

nucleus, which contains very dense **protons** (particles with positive electrical charges) and equally dense **neutrons** (particles with neutral electrical charges). Orbiting the nucleus are **electrons,** which have negative electrical charges. For convenience, we often diagram atoms to show the electrons orbiting the nucleus, like the orderly orbiting of the planets around the Sun (Figure 1.3A). However, electrons move so rapidly that they create a sphere-shaped negative zone around the nucleus. Hence, a more realistic picture of the positions of electrons can be obtained by envisioning a cloud of negatively charged electrons surrounding the nucleus (Figure 1.3B).

It is also known that individual electrons are located at given distances from the nucleus in regions called **energy levels,** or **shells.** Three such shells are shown in Figure 1.3. As we shall see, an important fact about these shells is that each can have only a specific number of electrons.

The number of protons in an atom's nucleus determines its **atomic number** and name of the element. For example, all atoms with six protons are carbon atoms, all those with eight protons are oxygen atoms, and so forth. "Free" atoms (those not combined with other atoms) have the same number of electrons as protons, so the atomic number also equals the number of electrons surrounding the nucleus. Therefore, carbon has six electrons to match its six protons, and oxygen has eight electrons to match its eight protons. Neutrons have no charge, so the positive charge of the protons is exactly balanced by the negative charge of the electrons. Consequently, uncombined atoms are neutral electrically and have no overall electrical charge. Thus, the *element* is a large collection of electrically neutral atoms, all having the same atomic number.

How Atoms Bond Together

Elements combine with each other to form a wide variety of more complex compounds. A **compound** is composed of two or more elements bonded together in definite proportions. When the elements separate, the bonds are broken and the compound no longer exists; only its component elements do. Through experimentation, scientists have learned that the forces bonding atoms together are electrical. Further, it is known that chemical bonding results in a change in the arrangement of electrons in the shells of the bonded atoms.

When an atom combines chemically, it either gains, loses, or shares electrons with another atom.

An atom that gains electrons becomes negatively charged because it has more electrons than protons. Atoms that lose electrons become positively charged. Atoms that have an electrical charge because of a gain or loss of electrons are called **ions.** Simply stated, oppositely charged ions attract one another to produce a neutral chemical compound.

An example of chemical bonding involves sodium (Na) atoms and chlorine (Cl) atoms to produce sodium chloride molecules (NaCl). Sodium chloride is common table salt (Figure 1.4). When the sodium atom loses one electron it becomes a positive ion, and when the chlorine atom gains one electron it becomes a negative ion. These opposite charges act as an "electrical glue" that bonds the atoms together.

The properties of a chemical compound are dramatically different from the properties of the elements comprising it. For example, chlorine is a green, poisonous gas that is so toxic it was used as a weapon during World War I. Sodium is a soft, silvery metal that reacts vigorously with water and, if held in your hand, could burn it severely. Together, however, these atoms produce the compound sodium chloride (table salt), the familiar clear, crystalline solid that is essential for human life.

This example also illustrates an important difference between a rock and a mineral. When elements join to form a mineral, the constituent elements lose their physical properties. In a *rock,* on the other hand, which is a *mixture* of minerals, each mineral retains its own distinctive physical properties.

Isotopes and Radioactive Decay Subatomic particles are so incredibly small that a special unit was devised to express their mass. A proton or a neutron has a mass just slightly more than one *atomic mass unit*, whereas an electron is only about one two-thousandth of an atomic mass unit. Thus, although electrons play an active role in chemical reactions, they do not contribute significantly to the mass of an atom.

The **mass number** of an atom is simply the total of its neutrons and protons in the nucleus. Atoms of the same element always have the same number of protons, but commonly have varying numbers of neutrons. This means that an element can have more than one mass number. These variants of the same element are called **isotopes** of that element.

For example, carbon has two well-known isotopes, one having a mass number of 12 (carbon-12), the other a mass number of 14 (carbon-14). All atoms of the same element must have the same number of protons, and carbon always has six (atomic number = 6). Hence, carbon-12 must have six protons *plus six neutrons* to give it a mass number of 12, whereas carbon-14 must have six protons *plus eight neutrons* to give it a mass number of 14.

The *average* atomic mass of any random sample of carbon is much closer to 12 than 14, because carbon-12 is the more common isotope. This average is called *atomic weight.** In chemical behavior, all isotopes of the same element are nearly identical. To distinguish among them is like trying to differentiate identical twins, with one slightly heavier. Because isotopes of an element react the same chemically, different isotopes can become parts of the same mineral. For example, when calcite is formed from calcium, carbon, and oxygen, some of its carbon atoms are carbon-12 and some are carbon-14.

---

*The term *weight* is a misnomer that has resulted from long use. The correct term is *atomic mass.*

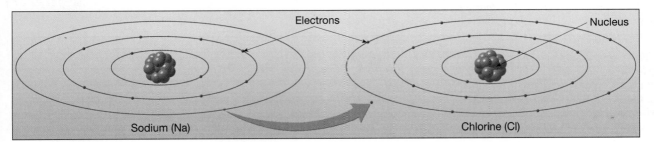

Figure 1.4 Chemical bonding of sodium and chlorine atoms to produce sodium chloride. Through the transfer of one electron in the outer shell of a sodium atom to the outer shell of a chlorine atom, the sodium becomes a positive ion and chlorine a negative ion.

Although the nuclei of most atoms are stable, many elements do have isotopes in which the nuclei are unstable. "Unstable" means that the isotopes disintegrate through a process called **radioactive decay** in which a decaying atom actively radiates energy and particles. Radioactive decay occurs when the forces that bind the nucleus are not strong enough to keep it together.

The rate at which unstable nuclei break apart (decay) is steady and measurable. This makes some isotopes useful "clocks" for dating the events of Earth history. A discussion of radioactive decay and its applications in dating past geologic events is found in Chapter 8.

Properties of Minerals

Minerals occur in a fascinating palette of colors, shapes, and lusters. They vary in hardness and may even have a distinctive taste or scent. We will now look at these properties, which are useful to help identify minerals.

Crystal Form

Crystal form is the external expression of a mineral's internal orderly arrangement of atoms. Generally, when a mineral forms without space restrictions, it will develop individual crystals with well-formed crystal faces. Figure 1.5A illustrates this for pyrite, an iron-and-sulfur mineral with cubic crystals. Figure 1.5B shows the distinctive hexagonal crystals of quartz that grow when space and time permit. However, most of the time, crystal growth is severely constrained. It is stunted because of competition for space, resulting in an intergrown mass of small, jammed crystals, none of which exhibit a crystal form. This is what happened to the minerals in the granite in Figure 1.2 (p. 13). Thus, most inorganic solid objects are composed of crystals, but they are not clearly visible to the unaided eye.

Luster

Luster is the appearance or quality of light reflected from the surface of a mineral. Minerals that have the appearance of metals, regardless of color, are said to have a *metallic luster,* like the pyrite crystals in Figure 1.5A. Minerals with a *nonmetallic luster* are described by various adjectives. These include vitreous (glassy, like the quartz crystals in Figure 1.5B), pearly, silky, resinous, and earthy (dull).

Color

Although **color** is an obvious feature of a mineral, it is often an unreliable diagnostic property. Slight impurities in the common mineral quartz, for

A.

B.

Figure 1.5 Crystal form is the external expression of a mineral's orderly internal structure. **A.** Pyrite, commonly known as "fool's gold," often forms cubic crystals that may contain parallel lines called striations. **B.** Quartz sample that exhibits well-developed hexagonal (six-sided) crystals with pyramidal-shaped ends. *(Photo by Breck P. Kent)*

example, give it a variety of colors, including pink, purple (amethyst), milky white, and even black.

Streak

Streak is the color of a mineral in its powdered form, which is a much more reliable indication of color. Streak is obtained by rubbing the mineral across a piece of hard, unglazed porcelain, termed a *streak plate*. Whereas the color of a mineral may vary from sample to sample, the streak usually does not, and is therefore the more consistent property. Streak can also help to distinguish minerals with metallic lusters from those having nonmetallic lusters. Metallic minerals generally have a dense, dark streak; minerals with nonmetallic lusters do not.

Hardness

One of the most useful diagnostic properties is **hardness,** a measure of the resistance of a mineral to abrasion or scratching. This property is determined by rubbing the mineral to be identified against another mineral of known hardness. One will scratch the other (unless they have the same hardness).

Geologists use a standard hardness scale, called the **Mohs scale.** It consists of ten minerals arranged in order from 10 (hardest) to 1 (softest), as shown in Table 1.1. Any mineral of unknown hardness can be rubbed against these to determine its hardness. In the field, other handy objects also work. For example, your fingernail has a hardness of 2.5, a copper penny 3, and a piece of glass 5.5. The mineral gypsum, which has a hardness of 2, can be easily scratched by your fingernail. On the other hand, the similar-looking mineral calcite, which has a hardness of 3, cannot be scratched by your fingernail. Calcite cannot scratch glass, because its hardness is less than 5.5. Quartz, the hardest of the common minerals at 7, will scratch a glass plate. Diamonds, hardest of all, scratch anything. You can see that by rubbing an unknown mineral against objects of known hardness, you can quickly narrow its hardness range.

Cleavage

In the crystal structure of a mineral, some bonds are weaker than others. These bonds are where a mineral will break when it is stressed. **Cleavage** is the tendency of a mineral to cleave, or break, along planes of weak bonding. Not all minerals have definite planes of weak bonding, but those that possess cleavage can be identified by the distinctive smooth surfaces that are produced when the mineral is broken.

The simplest type of cleavage is exhibited by minerals called micas (Figure 1.6). Because the micas have weak bonds in one direction, they cleave to form thin, flat sheets. Some minerals have several cleavage planes that produce smooth surfaces when broken, while others exhibit poor cleavage, and still others have no cleavage at all. When minerals break evenly in more than one direction, cleavage is described by the *number of planes* exhibited and the *angles at which they meet* (Figure 1.7).

Do not confuse *cleavage* with *crystal form!* When a mineral exhibits cleavage, it will break into pieces

---

Table 1.1 ■ Mohs scale of mineral hardness

Relative Scale		Mineral	Hardness of Some Common Objects
Hardest	10	Diamond	
	9	Corundum	
	8	Topaz	
	7	Quartz	
	6	Potassium Feldspar	
	5	Apatite	5.5 Glass, Pocketknife
	4	Fluorite	
	3	Calcite	3.5 Copper Penny
	2	Gypsum	2.5 Fingernail
Softest	1	Talc	

Figure 1.6 Sheet-type cleavage common to the micas. *(Photo by Chip Clark)*

that have the same geometry as each other. By contrast, the quartz crystals shown in Figure 1.5B do not have cleavage. If broken, they fracture into pieces that do not resemble each other or the original crystals.

Fracture

Minerals that do not exhibit cleavage when broken, such as quartz, are said to **fracture.** Even fracturing has variety: minerals that break into smooth curved surfaces like those seen in broken glass have a *conchoidal fracture* (Figure 1.8). Others break into splinters or fibers, like asbestos, but most minerals fracture irregularly.

Specific Gravity

Specific gravity compares the weight of a mineral to the weight of an equal volume of water. For example, if a cubic centimeter of a mineral weighs three times as much as a cubic centimeter of water, its specific gravity is 3. With a little practice, you can estimate the specific gravity of minerals by hefting them in your hand. For example, if a mineral feels as heavy as the common rocks you have handled, its specific gravity will probably be somewhere between 2.5 and 3. Some metallic minerals have a specific gravity noticeably greater than that of common rock-forming minerals. Galena, which is an ore of lead, has a specific gravity of roughly 7.5 (Figure 1.9). The specific gravity of pure 24-karat gold is a very hefty 20.

Other Properties of Minerals

In addition, some minerals can be recognized by other distinctive properties. For example, halite is ordinary salt, so it is quickly identified with your tongue. Thin sheets of mica will bend and elastically snap back. Gold is malleable, which means it can be easily hammered or shaped. Talc and graphite both have distinctive feels. Talc feels soapy. Graphite feels greasy (it is a principal ingredient in dry lubricants).

A few minerals, such as magnetite, have a high iron content and can be picked up with a magnet. Some varieties of magnetite (lodestone) are natural magnets and will attract small iron-based objects

Figure 1.7 Smooth surfaces produced when a mineral with cleavage is broken. The sample on the left (fluorite) exhibits four planes of cleavage (eight sides), whereas the other two samples exhibit three planes of cleavage (six sides). Also notice that the mineral in the center (halite) has cleavage planes that meet at 90-degree angles, whereas the mineral on the right (calcite) has cleavage planes that meet at 75-degree angles.

Figure 1.8 Conchoidal fracture. The smooth curved surfaces result when minerals break in a glasslike manner. *(Photo by E. J. Tarbuck)*

Figure 1.9 Galena is lead sulfide. Like other metallic ores, it has a relatively high specific gravity and therefore feels heavy when hefted.

such as pins and paper clips. Some minerals exhibit special optical properties. For example, when a transparent piece of calcite is placed over printed material, the letters appear doubled. This optical property is known as *double refraction*. In addition, the streak of many sulfur-bearing minerals smells like rotten eggs.

One very simple chemical test involves placing a drop of dilute hydrochloric acid from a dropper bottle on a freshly broken mineral surface. Certain minerals, called carbonates, will effervesce (fizz) with hydrochloric acid. This test is useful in identifying the mineral calcite, which is a common carbonate mineral.

In summary, a number of special physical and chemical properties are useful in identifying particular minerals. These include taste, smell, elasticity, malleability, feel, magnetism, double refraction, and chemical reaction to hydrochloric acid. Remember that every one of these properties depends on the composition (elements) of a mineral and its structure (how the atoms are arranged).

Mineral Groups

Nearly 4000 minerals have been named and about 40 to 50 new ones are identified each year. Fortunately, for students who are beginning to study minerals, no more than a few dozen are abundant! Collectively, these few make up most of the rocks of Earth's crust and as such,

are classified as the *rock-forming minerals*. It is also interesting to note that *only eight elements* make up the bulk of these minerals and represent over 98 percent (by weight) of the continental crust (Table 1.2).

The two most abundant elements are silicon and oxygen, comprising nearly three-fourths of Earth's continental crust. Silicon and oxygen combine to form the framework of the most common mineral group, the **silicates.** Perhaps the next most common mineral group is the carbonates (carbon plus oxygen plus other elements), of which calcite is the most prominent member. Other common rock-forming minerals include gypsum and halite.

Table 1.2 ■ Relative abundance of the most common elements in Earth's crust

Element	Approximate Percentage by Weight
Oxygen (O)	46.6
Silicon (S)	27.7
Aluminum (A)	8.1
Iron (Fe)	5.0
Calcium (Ca)	3.6
Sodium (Na)	2.8
Potassium (K)	2.6
Magnesium (Mg)	2.1
All others	1.7
Total	100

We will first discuss the most common mineral group, the silicates, and then consider other prominent mineral groups.

Rock-Forming Silicates

Each of the silicate minerals contains oxygen and silicon atoms. Except for a few silicate minerals such as quartz, most silicate minerals also contain one or more additional elements that join together to produce an electrically neutral compound. These elements give rise to the great variety of silicate minerals and their varied properties.

All silicates have the same fundamental building blocks, the **silicon-oxygen tetrahedron.** This structure consists of four oxygen atoms surrounding a much smaller silicon atom, as shown in Figure 1.10. Thus, a typical hand-size silicate mineral specimen contains millions of these silicon-oxygen tetrahedra, joined together in a variety of ways.

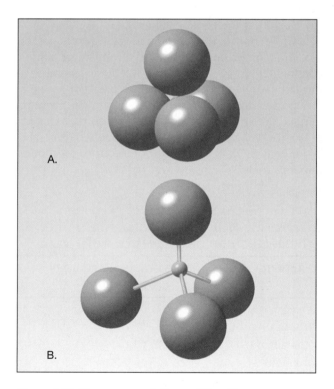

Figure 1.10 Two representations of the silicon-oxygen tetrahedron. **A.** The four large spheres represent oxygen atoms and the blue sphere represents a silicon atom. The spheres are drawn in proportion to the radii of the atoms. **B.** A model of the tetrahedron using rods to depict the bonds that connect the atoms.

In some minerals, the tetrahedra are joined into chains, sheets, or three-dimensional networks by sharing oxygen atoms (Figure 1.11). These larger silicate structures are then connected to one another by other elements. The primary elements that join silicate structures are iron (Fe), magnesium (Mg), potassium (K), sodium (Na), and calcium (Ca).

Major groups of silicate minerals and common examples are given in Figure 1.11. The feldspars are by far the most plentiful group, comprising over 50 percent of Earth's crust. Quartz, the second most abundant mineral in the continental crust, is the only common mineral made completely of silicon and oxygen.

Notice in Figure 1.11 that each mineral *group* has a particular silicate *structure*. A relationship exists between this internal structure of a mineral and the *cleavage* it exhibits. Because the silicon-oxygen bonds are strong, silicate minerals tend to cleave between the silicon-oxygen structures rather than across them. For example, the micas have a sheet structure and thus tend to cleave into flat plates (Figure 1.12). Quartz, which has equally strong silicon-oxygen bonds in all directions, has no cleavage, but fractures instead.

How do silicate minerals form? Most crystallize from molten rock as it cools. This cooling can occur at or near Earth's surface (low temperature and pressure) or at great depths (high temperature and pressure). The *environment* during crystallization and the *chemical composition of the molten rock* mainly determine which minerals are produced. For example, the silicate mineral olivine crystallizes at high temperatures (about 1200°C), whereas quartz crystallizes at much lower temperatures (about 700°C).

In addition, some silicate minerals form at Earth's surface from the weathered products of other silicate minerals. Still other silicate minerals are formed under the extreme pressures associated with mountain building. Each silicate mineral, therefore, has a structure and a chemical composition that *indicate the conditions under which it formed.* Thus, by carefully examining the mineral makeup of rocks, geologists can often determine the circumstances under which the rocks formed.

Important Nonsilicate Minerals

Nonsilicate minerals make up only about one-fourth of the continental crust, so they can be considered scarce compared to the silicates. However, many are important economically. Table 1.3 (p. 23) lists examples of nonsilicate minerals of economic value: oxides,

Mineral		Idealized Formula	Cleavage	Silicate Structure
Olivine		$(Mg, Fe)_2SiO_4$	None	Single tetrahedron 
Pyroxene group (Augite)		$(Mg,Fe)SiO_3$	Two planes at right angles	Single chains 
Amphibole group (Hornblende)		$Ca_2(Fe,Mg)_5Si_8O_{22}(OH)_2$	Two planes at 60° and 120°	Double chains 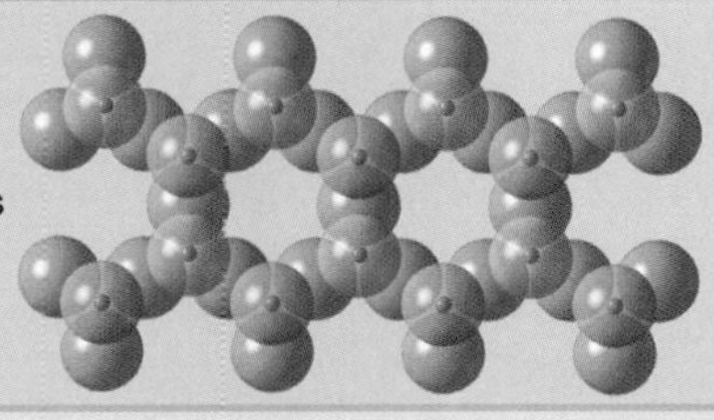
Micas	Biotite	$K(Mg,Fe)_3AlSi_3O_{10}(OH)_2$	One plane	Sheets 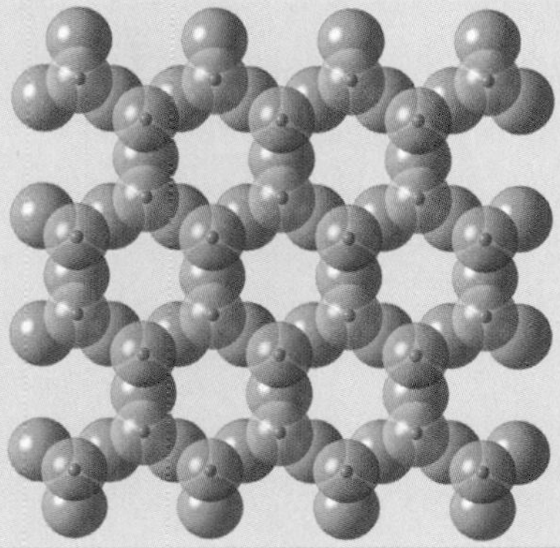
	Muscovite	$KAl_2(AlSi_3O_{10})(OH)_2$		
Feld-spars	Orthoclase	$KAlSi_3O_8$	Two planes at 90°	Three-dimensional networks Very complex structure
	Plagioclase	$(Ca,Na)AlSi_3O_8$		
Quartz		SiO_2	None	

Figure 1.11 Common silicate minerals. Note that the complexity of the silicate structure increases down the chart.

sulfides, sulfates, halides, and "native" elements. A major rock-forming group worldwide is the *carbonates,* which includes the mineral calcite ($CaCO_3$). This mineral is the major constituent in two well-known rocks: limestone and marble. Limestone is used commercially for road paving, building stone, and as the main ingredient in portland cement. Marble is used decoratively.

Two other nonsilicate minerals frequently found in sedimentary rocks are *halite* and *gypsum.* Both minerals are commonly found in thick layers, which are the last vestiges of ancient seas that long ago evaporated, leaving behind their salts (Figure 1.13).

Like limestone, halite and gypsum both are important nonmetallic resources. Halite is the mineral name for common table salt, sodium chloride. Gypsum is the mineral from which plaster and other similar building materials are composed.

In addition, a number of other minerals have economic value (Table 1.3). Ores of metals include hematite (iron), sphalerite (zinc), and galena (lead). Native (free-occurring, not in compounds) elements include gold, silver, copper, and carbon (diamonds and graphite). Other important minerals are fluorite, corundum, and sylvite.

Silicate minerals

Feldspar

Quartz

Muscovite

Hornblende

Non-silicate minerals

Calcite

Halite

Gypsum

Figure 1.12 Some common rock-forming minerals. Silicate minerals are common constituents of igneous rocks, while the nonsilicates along with quartz and clay are prominent minerals in many sedimentary rocks.

Mineral Resources

Mineral resources are Earth's storehouse of useful minerals that can be recovered for use. Resources include already identified deposits from which minerals can be extracted profitably, called **reserves,** as well as known deposits that are not yet recoverable under present economic conditions or technology. Deposits inferred to exist, but not yet discovered, are also considered mineral resources.

The term **ore** denotes useful metallic minerals that can be mined at a profit. In common usage, the term *ore* is also applied to some nonmetallic minerals, such as fluorite and sulfur. However, materials used for such purposes as building stone, road paving, abrasives, ceramics, and fertilizers are not usually called ores; rather, they are classified as *industrial rocks and minerals*.

Recall that more than 98 percent of Earth's crust is composed of only eight elements. Except for oxygen and silicon, all other elements make up a relatively small fraction (about one-fourth) of common crustal

Figure 1.13 Thick beds of halite (salt) are being drilled at an underground mine near Grand Saline, Texas.
(Used by permission of Morton International, Inc., Chicago, IL 60606)

Table 1.3 ■ Common nonsilicate mineral groups

Mineral Group	Name	Chemical Formula	Economic Use
Oxides	Hematite	Fe_2O_3	Ore of iron, pigment
	Magnetite	Fe_3O_4	Ore of iron
	Corundum	Al_2O_3	Gemstone, abrasive
	Ice	H_2O	Solid form of water
Sulfides	Galena	PbS	Ore of lead
	Sphalerite	ZnS	Ore of zinc
	Pyrite	FeS_2	Sulfuric acid production
	Chalcopyrite	$CuFeS_2$	Ore of copper
	Cinnabar	HgS	Ore of mercury
Sulfates	Gypsum	$CaSO_4 \cdot 2H_2O$	Plaster
	Anhydrite	$CaSO_4$	Plaster
	Barite	$BaSO_4$	Drilling mud
Native elements	Gold	Au	Trade, jewelry
	Copper	Cu	Electrical conductor
	Diamond	C	Gemstone, abrasive
	Sulfur	S	Sulfa drugs, chemicals
	Graphite	C	Pencil lead, dry lubricant
	Silver	Ag	Jewelry, photography
	Platinum	Pt	Catalyst
Halides	Halite	NaCl	Common salt
	Fluorite	CaF_2	Used in steel making
	Sylvite	KCl	Fertilizer
Carbonates	Calcite	$CaCO_3$	Portland cement, lime
	Dolomite	$CaMg(CO_3)_2$	Portland cement, lime

rocks (see Table 1.2, p. 19). Indeed, the natural concentrations of many elements are exceedingly small. A deposit containing only the average crustal percentage of a valuable element like gold is worthless because the cost of extracting it greatly exceeds the value of the material recovered.

To be considered of value, an element must be concentrated above the level of its average crustal abundance. For example, copper makes up about 0.0135 percent of the crust. However, for a material to be considered as copper ore, it must contain a concentration that is about 100 times this amount. Aluminum, on the other hand, represents 8.13 percent of the crust and must be concentrated to only about four times its average crustal percentage before it can be extracted profitably.

It is important to realize that economic changes may make a deposit profitable to extract, or lose its profitability. If demand for a metal increases and prices rise sufficiently, the status of a previously unprofitable deposit changes, and it becomes an ore. The status of unprofitable deposits may also change if a technological advance allows the material to be extracted at a lower cost than before.

Conversely, changing economic factors can turn a once-profitable ore deposit into an unprofitable deposit that can no longer be called an ore. This situation was illustrated recently at the copper mining operation located at Bingham Canyon, Utah, one of the largest open-pit mines on Earth (Figure 1.14). Mining was halted there in 1985 because outmoded equipment had driven the cost of extracting the copper beyond the current selling price. The owners responded by replacing an antiquated 1000-car railroad with conveyor belts and pipelines for transporting the ore and waste. These devices achieved a cost reduction of nearly 30 percent and returned this mining operation to profitability.

Figure 1.14 Aerial view of Bingham Canyon copper mine near Salt Lake City, Utah. This huge open-pit mine is about 4 kilometers across and 900 meters deep. Although the amount of copper in the rock is less than 1 percent, the huge volumes of material removed and processed each day (about 200,000 tons) yield significant quantities of metal. *(Photo by Michael Collier)*

The Chapter in Review

1. A *mineral* is a naturally occurring inorganic solid that possesses a definite chemical structure, which gives it a unique set of physical properties. Most *rocks* are aggregates composed of two or more minerals.

2. The building blocks of minerals are *elements*. An *atom* is the smallest particle of matter that still retains the characteristics of an element. Each atom has a *nucleus,* which contains *protons* and *neutrons.* Orbiting the nucleus of an atom are *electrons.* The number of protons in an atom's nucleus determines its *atomic number* and the name of the element. Atoms bond together to form a *compound* by either gaining, losing, or sharing electrons with another atom.

3. *Isotopes* are variants of the same element, but with a different *mass number* (the total number of neutrons plus protons found in an atom's nucleus). Some isotopes are unstable and disintegrate naturally through a process called *radioactive decay*.

4. The properties of minerals include: *crystal form, luster, color, streak, hardness, cleavage, fracture,* and *specific gravity*. In addition, a number of special physical and chemical properties (*taste, smell, elasticity, malleability, feel, magnetism, double refraction,* and *chemical reaction to hydrochloric acid*) are useful in identifying certain minerals. Each mineral has a unique set of properties that can be used for identification.

5. The eight most abundant elements found in Earth's continental crust (oxygen, silicon, aluminum, iron, calcium, sodium, potassium, and magnesium) make up the majority of minerals.

6. The most common mineral group is the *silicates*. All silicate minerals have the *silicon-oxygen tetrahedron* as their fundamental building block. In some silicate minerals the tetrahedra are joined in chains; in others, the tetrahedra are arranged into sheets, or three-dimensional networks. Each silicate mineral has a structure and a chemical composition that indicates the conditions under which it formed. The *nonsilicate* mineral groups include the *oxides* (e.g., magnetite, mined for iron), *sulfides* (e.g., sphalerite, mined for zinc), *sulfates* (e.g., gypsum, used in plaster and frequently found in sedimentary rocks), *native elements* (e.g., graphite, a dry lubricant), *halides* (e.g., halite, common salt and frequently found in sedimentary rocks), and *carbonates* (e.g., calcite, used in portland cement and a major constituent in two well-known rocks: limestone and marble).

7. The term *ore* is used to denote useful metallic minerals, such as hematite (mined for iron) and galena (mined for lead), that can be mined at a profit as well as some nonmetallic minerals, such as fluorite and sulfur, that contain useful substances.

Key Terms

atom (p. 14)

atomic number (p. 14)

cleavage (p. 17)

color (p. 16)

compound (p. 14)

crystal form (p. 16)

electron (p. 14)

element (p. 13)

energy levels (shells) (p. 14)

fracture (p. 18)

hardness (p. 17)

ions (p. 15)

isotope (p. 15)

luster (p. 16)

mass number (p. 15)

mineral (p. 13)

mineral resource (p. 22)

Mohs hardness scale (p. 17)

neutron (p. 14)

nucleus (p. 14)

ore (p. 22)

proton (p. 14)

radioactive decay (p. 16)

reserve (p. 22)

rock (p. 12)

silicates (p. 19)

silicon-oxygen tetrahedron (p. 20)

specific gravity (p. 18)

streak (p. 17)

Questions for Review

1. Define the term *rock*.
2. List the three main particles of an atom and explain how they differ from one another.
3. If the number of electrons in an atom is 35 and its mass number is 80, calculate the following:
   a. The number of protons.
   b. The atomic number.
   c. The number of neutrons.
4. What occurs in an atom to produce an ion?
5. What is an isotope?
6. Although all minerals have an orderly internal arrangement of atoms (crystalline structure), most mineral samples do not visibly demonstrate their crystal form. Why?
7. Why might it be difficult to identify a mineral by its color?
8. If you found a glassy-appearing mineral while rock hunting and had hopes that it was a diamond, what simple test might help you make a determination?
9. Table 1.3 (p. 23) lists a use for corundum as an abrasive. Explain why it makes a good abrasive in terms of Mohs hardness scale.
10. Gold has a specific gravity of almost 20. If a 25-liter pail of water weighs about 25 kilograms, how much would a 25-liter pail of gold weigh?
11. What are the two most common elements in Earth's crust?
12. What is the term used to describe the basic building block of all silicate minerals?
13. What are the two most common silicate minerals?
14. List three nonsilicate minerals that are commonly found in rocks.
15. Contrast a mineral *resource* and a mineral *reserve*.
16. What might cause a mineral deposit that had not been considered an ore to become reclassified as an ore?

Testing What You Have Learned

To test your knowledge of the material presented in this chapter, answer the following questions.

Multiple-Choice Questions

1. Most rock forming minerals are _____
   - **a.** oxides.
   - **b.** silicates.
   - **c.** carbonates.
   - **d.** sulfides.
   - **e.** sulfates.

2. Which of the following is NOT one of the eight most abundant elements found in Earth's continental crust?
   - **a.** carbon
   - **b.** potassium
   - **c.** oxygen
   - **d.** iron
   - **e.** silicon

3. A naturally occurring inorganic solid that possesses a definite chemical structure, which gives it a unique set of physical properties, is a _____
   - **a.** rock.
   - **b.** liquid.
   - **c.** glass.
   - **d.** mineral.
   - **e.** gas.

4. When an atom combines chemically with another atom, it either gains, loses, or shares _____
   - **a.** protons.
   - **b.** neutrons.
   - **c.** electrons.
   - **d.** compounds.
   - **e.** nuclei.

5. Variations of the same element, but with different mass numbers, are called _____
   - **a.** atoms.
   - **b.** minerals.
   - **c.** neutrons.
   - **d.** silicates.
   - **e.** isotopes.

6. Which one of the following is NOT a physical property of minerals?
   - **a.** hardness
   - **b.** cleavage
   - **c.** luster
   - **d.** streak
   - **e.** chemical composition

7. The fundamental building block of all silicate minerals is a tetrahedron composed of _____
   - **a.** carbon-oxygen.
   - **b.** silicon-oxygen.
   - **c.** iron-zinc.
   - **d.** iron-oxygen.
   - **e.** calcium-carbon.

8. Each silicate mineral group has its own particular silicate _____
   - **a.** structure.
   - **b.** luster.
   - **c.** color.
   - **d.** atoms.
   - **e.** taste.

9. The most plentiful silicate mineral group is the _____
   - **a.** micas.
   - **b.** sulfides.
   - **c.** feldspars.
   - **d.** pyroxenes.
   - **e.** amphiboles.

10. A metallic mineral that can be mined at a profit is called a(n) _____
    - **a.** ion.
    - **b.** proton.
    - **c.** rock.
    - **d.** isotope.
    - **e.** ore.

Fill-in Questions

11. Most _____ are aggregates of minerals.

12. Atomic particles called _____ have positive charges, whereas _____ have negative charges.

13. A _____ is a substance composed of two or more elements bonded together in definite proportions.

14. The external expression of the orderly internal arrangement of atoms in a mineral is called the _____.

15. The tendency of a mineral to break along planes of weak bonding is called _____.

True/False Questions

16. An atom is the smallest part of matter that still retains the characteristics of an element. _____

17. Protons orbit the nucleus of an atom. _____

18. The nuclei of some isotopes disintegrate naturally in a process called radioactivity. _____

19. One of the most reliable diagnostic physical properties of minerals is color. _____

20. The carbonates are an important nonsilicate rock-forming mineral group _____

Answers:

1. b; 2. a; 3. d; 4. c; 5. e; 6. e; 7. b; 8. a; 9. c; 10. e; 11. rocks; 12. protons, electrons; 13. compound; 14. crystal form; 15. cleavage; 16. T; 17. F; 18. T; 19. F; 20. T.

Web   Work

The following are informative and interesting World Wide Web sites that address topics related to those presented in the chapter:

- Minerals On the Internet
  http://www.swcp.com/~tasa/minlinks.html

- The Mineral Gallery
  http://www.galleries.com/

For direct access to these sites and other relevant Web locations, contact the *Foundations of Earth Science* WWW home page at http://www.prenhall.com/lutgens.

CHAPTER 2
Rocks:
Materials of the Lithosphere

FOCUS ON LEARNING

To assist you in learning the important concepts
in this chapter, you will find it helpful to focus
on the following questions:

1. What are the three groups of rocks and the geologic processes involved in the formation of each?
2. What two criteria are used to classify igneous rocks?
3. What are the two major types of weathering and the processes associated with each?

4. What are the names and environments of formation for some common detrital and chemical sedimentary rocks?
5. What are the names, textures, and environments of formation for some common metamorphic rocks?

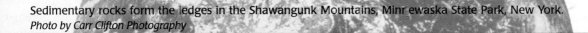

Sedimentary rocks form the ledges in the Shawangunk Mountains, Minnewaska State Park, New York.
Photo by Carr Clifton Photography

Figure 2.1 El Capitan, in Yosemite National Park, California. This mass of rock formed from molten material deep beneath Earth's surface. *(Photo by Jeff Gnass Photography)*

Why study rocks? You have already learned that rocks and minerals have great economic value. Furthermore, all Earth processes in some way depend on the properties of these basic materials. Events such as volcanic eruptions, mountain building, weathering, erosion, and even earthquakes involve rocks and minerals. Consequently, a basic knowledge of Earth materials is essential to understanding Earth phenomena.

Every rock contains clues about the environments in which it formed. For example, some rocks are composed entirely of small shell fragments. This tells Earth scientists that the particles making up the rock originated in a shallow marine environment. Other rocks contain clues that indicate they formed from a volcanic eruption, or deep in the Earth during mountain building (Figure 2.1). Thus, rocks contain a wealth of information about events that have occurred over Earth's long history.

We divide rocks into three groups, based on their mode of origin. The groups are igneous, sedimentary, and metamorphic. Before examining each group, we will view the *rock cycle*, which depicts the interrelationships among these rock groups.

The Rock Cycle

The **rock cycle** (Figure 2.2) shows important relations among the three rock types. By studying the rock cycle, we may ascertain the origin of each rock type. We also may gain insight into the role of geologic processes in transforming one rock type into another.

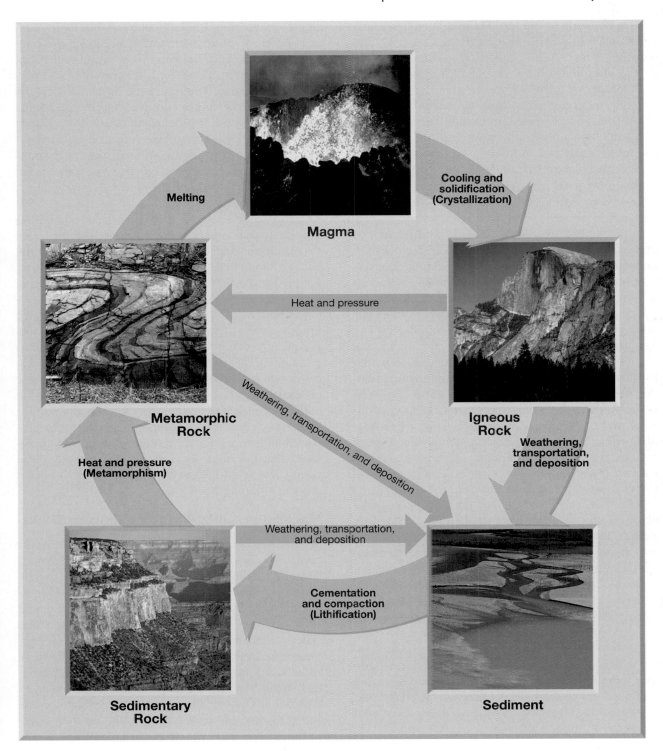

Melting

Magma

Cooling and
solidification
(Crystallization)

Heat and pressure

Weathering, transportation, and deposition

Metamorphic
Rock

Igneous
Rock

Weathering,
transportation,
and deposition

Heat and pressure
(Metamorphism)

Weathering, transportation,
and deposition

Cementation
and compaction
(Lithification)

Sedimentary
Rock

Sediment

Figure 2.2 The rock cycle illustrates the interrelationships among the three basic rock groups and the various geologic processes that act to transform one rock type into another. *(Photos by E. J. Tarbuck)*

The concept of the rock cycle essentially is an outline of physical geology. The cycle was initially proposed in the late 1700s by James Hutton, a founding father of modern geology. The version shown in Figure 2.2 uses arrows to indicate chemical and physical processes, and photos to represent Earth materials.

Let us begin at the top. **Magma** is molten material found inside Earth. Eventually magma cools and solidifies. This process, called **crystallization**, may occur either beneath the surface or, following a volcanic eruption, at the surface. In either situation, the resulting rocks are called **igneous rocks**.

When igneous rocks are exposed at the surface of Earth, they undergo **weathering**, in which the day-in and day-out influences of the atmosphere slowly disintegrate and decompose the rock. The resulting material is picked up, transported, and deposited by various *erosional agents*—gravity, running water, glaciers, wind, or waves.

Once this material, called **sediment**, is deposited, usually as horizontal beds in the ocean, it undergoes **lithification**, a term meaning "conversion into rock." Sediment becomes lithified when it is compacted by the weight of overlying layers, or when it is cemented as percolating groundwater fills pore spaces in the sediment with mineral matter. The result is **sedimentary rock**.

If sedimentary rock becomes buried deep within Earth or involved in the dynamics of mountain building, it is subjected to great pressures and heat. The sedimentary rock will react to this changing environment and turn into the third rock type, **metamorphic rock**. When metamorphic rock is subjected to still greater heat and pressure, it melts to create magma, completing the rock cycle. This magma will eventually solidify as igneous rock, and so on.

The full cycle just described does not always take place completely. "Shortcuts" in the cycle happen, as indicated by dashed lines in Figure 2.2. Igneous rock, for example, rather than being exposed to weathering and erosion, may be subjected to the heat and pressure far below the surface and change to metamorphic rock. Conversely, metamorphic and sedimentary rocks may become exposed at the surface and eroded into new materials for sedimentary rock.

As you study the remaining pages of this chapter, which discuss details of each of the three basic rock types, remember the rock cycle. Although rocks may seem to be unchanging masses, the rock cycle shows that they are not. The changes, however, take time—great amounts of time.

Igneous Rocks: "Formed by Fire"

In our discussion of the rock cycle, we pointed out that igneous rocks form as magma cools and crystallizes. This molten rock, which originates at depths as great as 200 kilometers within Earth, consists primarily of the elements found in silicate minerals (silicon and oxygen, plus aluminum, iron, calcium, sodium, potassium, magnesium, and others). Magma also contains some gases, particularly water vapor, which are confined within the magma by the surrounding rocks. Because the magma body is less dense than the surrounding rocks, it works its way toward the surface over time spans of thousands to millions of years. On occasion, the magma breaks through the surface, producing a volcanic eruption.

The spectacular explosions that sometimes accompany an eruption are produced by the gases that escape as the confining pressure lessens near the surface. Sometimes blockage of the volcanic vent, coupled with surface-water seepage into the magma chamber, can produce catastrophic explosions.

Along with ejected rock fragments, a volcanic eruption often generates extensive lava flows. **Lava** is similar to magma, except that most of the gaseous component has escaped (Figure 2.3). The rocks that result when lava solidifies are classified as **volcanic** or **extrusive**, because they are *extruded* onto the surface. The magma not able to reach the surface eventually crystallizes at depth. Igneous rocks produced in this manner are termed **plutonic**, or **intrusive**, because they *intrude* existing rocks. These deep rocks would never be observed were it not for erosion stripping away the overlying rock.

Magma Crystallizes to Form Igneous Rocks

Magma is basically a very hot, thick fluid, but it also contains solids and gases. The solids are mineral crystals. The liquid portion of a magma body is composed of ions that move about freely. However, as magma cools, the random movements of the ions slow, and the ions begin to arrange themselves into orderly patterns. This process is called *crystallization*. Usually all of the molten material does not solidify at the same time. Rather, as it cools, numerous small crystals develop. In a systematic fashion, ions are added to these centers of crystal growth. When the crystals grow large enough for their edges to meet, their growth ceases for lack of space, and crystallization continues elsewhere. Eventually, all of the liquid is transformed into a solid mass of interlocking crystals.

The rate of cooling strongly influences crystal size. If a magma cools very slowly, relatively few centers of crystal growth develop. Slow cooling also

Figure 2.3 Lava of basaltic composition flowing from a lava tube, Hawaii Volcanoes National Park. *(Photo by J. D. Griggs, U.S. Geological Survey, U. S. Department of the Interior)*

allows ions to migrate over relatively great distances. Consequently, *slow cooling results in the formation of large crystals*. On the other hand, if cooling occurs quite rapidly, the ions lose their motion and quickly combine. This results in a large number of tiny crystals that all compete for the available ions. Therefore, the outcome of rapid cooling is the formation of a solid mass of *small intergrown crystals*.

Thus, if a geologist discovers rock containing large, finger-sized crystals, it means the rock cooled very slowly. But if the crystals can be seen only with a microscope, the geologist knows that the magma cooled very quickly.

If the molten material is quenched almost instantly, there is not sufficient time for the ions to arrange themselves into a crystalline network at all. Therefore, solids produced in this manner consist of randomly distributed ions. Such rocks are called *glass* and are quite similar to ordinary manufactured glass. "Instant" quenching occurs during violent volcanic eruptions that produce tiny shards of glass called *volcanic ash*.

Classifying Igneous Rocks

In addition to the rate of cooling, the composition of a magma and the amount of dissolved gases influence crystallization. Because magmas differ in each of these aspects, the physical appearance and mineral composition of igneous rocks vary widely. Nevertheless, it is possible to classify igneous rocks based on their **texture** and *mineral constituents*. We will now look at both features.

Igneous Textures **Texture** describes the overall appearance of an igneous rock, based on the *size* and *arrangement* of its interlocking crystals. Texture is a very important characteristic, because it reveals a great deal about the environment in which the rock formed. You learned that rapid cooling produces small crystals, whereas very slow cooling produces much larger crystals. As you might expect, the rate of cooling is quite slow in magma chambers lying deep within the crust, whereas a thin layer of lava extruded upon Earth's surface may chill "rock solid" in a matter of hours. Small molten blobs ejected into the air during a violent eruption can solidify almost instantly.

Igneous rocks that form rapidly at the surface or as small masses within the upper crust have a very **fine-grained texture**, with the individual crystals too small to be seen with the unaided eye (Figure 2.4A). Common in many fine-grained igneous rocks are voids left by gas bubbles trapped as the magma solidifies. These openings are called *vesicles* and are limited to the fast-cooling upper portion of lava flows (Figure 2.5).

When large masses of magma solidify far below the surface, they form igneous rocks that exhibit a **coarse-grained texture**. These coarse-grained rocks have the appearance of a mass of intergrown crystals, which are roughly equal in size and large enough that the individual minerals can be identified with the unaided eye. Granite is a classic example (Figure 2.4B).

A large mass of magma located at depth may require tens of thousands, even millions, of years to solidify. Because all materials within a magma do not crystallize at the same rate or at the same time during cooling, it is possible for some crystals to become quite large before others even start to form. If magma that already contains some large crystals suddenly erupts at the surface, the remaining molten portion of the lava would cool quickly. The resulting rock, which has large crystals embedded in a matrix of smaller crystals, is said to have a **porphyritic texture** (Figure 2.5).

← 2 cm →

A. Rhyolite

Close up

← 2 cm →

B. Granite

Close up

Figure 2.4 **A**. Rhyolite exhibits a fine-grained texture. **B.** Granite is a common igneous rock that has a coarse-grained texture. *(Photos by E. J. Tarbuck)*

During some volcanic eruptions, molten rock is ejected into the atmosphere, where it is quenched very quickly. Rapid cooling of this type may generate rock with a **glassy texture**. As was indicated, glass results when the ions do not have sufficient time to unite into an orderly crystalline structure. *Obsidian*, a common type of natural glass, is similar in appearance to a dark chunk of manufactured glass (Figure 2.7A). Another volcanic rock that often

exhibits a glassy texture is *pumice*. Usually found with obsidian, pumice forms when large amounts of gas escape through lava to generate a gray, frothy mass (Figure 2.7B). In some samples, the vesicles are quite noticeable, whereas in others, the pumice resembles fine shards of intertwined glass. Because of the large volume of air-filled voids, many samples of pumice will float in water.

Mineral Composition The mineral makeup of an igneous rock depends on the chemical composition of the magma from which it crystallizes. Such a large

Figure 2.5 Scoria is a volcanic rock that exhibits a vesicular texture. Vesicles form as gas bubbles escape near the top of a lava flow.

← 5 cm →

Figure 2.6 Andesite porphyry. Notice the two distinctively different sizes of crystals. *(Photo by E. J. Tarbuck)*

← 2 cm →

A.

B.

Figure 2.7 Igneous rocks that exhibit a glassy texture. **A**. Obsidian, a glassy volcanic rock. **B**. Pumice, a glassy rock containing numerous tiny voids.

variety of igneous rocks exists that you might assume an equally large variety of magmas must also exist. However, various eruptive stages of the same volcano often extrude lavas exhibiting somewhat different mineral compositions, particularly if long periods separate the eruptions. Such evidence led geologists to examine the possibility that a single magma might produce rocks of varying mineral content.

N. L. Bowen studied this idea in the first quarter of the twentieth century. Bowen discovered that, as magma cools in the laboratory, certain minerals crystallize first, at very high temperatures (top of Figure 2.8). At successively lower temperatures, other minerals crystallize.

Bowen also demonstrated that if a mineral remains in the melt after crystallization, it will react

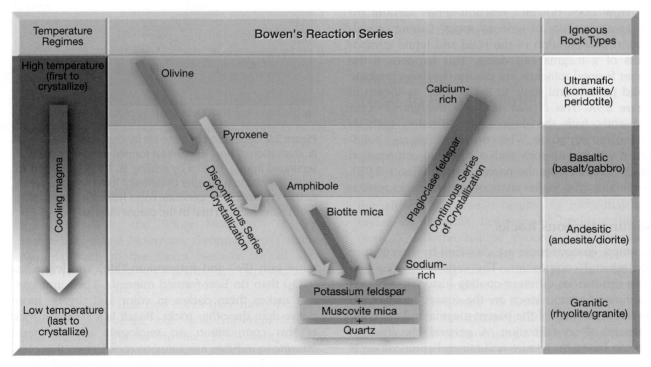

Figure 2.8 Bowen's reaction series shows the sequence in which minerals crystallize from a magma. Compare this figure to the mineral composition of the rock groups in Table 1.3. Note that each rock group consists of minerals that crystallize at the same time.

Note that gradations exist (Table 2.1). For example, an abundant extrusive igneous rock called

response, this rock mass will gradually bring it is once again in equilibrium, or balance, with its new environment. Such transformation of rock is what we call **weathering**.

In the following sections we will discuss the two kinds of weathering—mechanical and chemical. Mechanical weathering is the physical breaking up of rocks. Chemical weathering actually alters what a rock is, changing it into a different substance. Although we will consider these two processes separately, keep in mind that they usually work simultaneously in nature. Furthermore, the activities of erosional agents—wind, water, and glaciers—which transport weathered rock particles, are important. As these mobile agents move rock debris, they relentlessly disintegrate it further.

Mechanical Weathering of Rocks

When a rock undergoes **mechanical weathering**, it is broken into smaller and smaller pieces. Each piece retains the characteristics of the original material. The end result is many small pieces from a single large one. Figure 2.10 shows that breaking a rock into smaller pieces increases the surface area available for chemical attack. An example is adding sugar to water. In this situation, a chunk of rock candy will dissolve much more slowly than will an equal volume of sugar granules because of the vast difference in surface area. Hence, by breaking rocks into smaller pieces, mechanical weathering increases the amount of surface area available for chemical weathering.

In nature, three important physical processes break rocks into smaller fragments: frost wedging, expansion resulting from unloading, and biological activity.

Frost Wedging Alternate freezing and thawing of water is one of the most important processes of mechanical weathering. Water has the unique property of expanding about 9 percent when it freezes. This increase in volume occurs because, as ice forms, the water molecules arrange themselves into a very open crystalline structure. As a result, when water freezes, it expands and exerts a tremendous outward force. Here is everyday proof: water in a car's cooling system will freeze in winter, expanding and cracking the engine block. This is why antifreeze is added; it lowers the temperature at which the solution freezes.

In nature, water works its way into every crack or void in rock and, upon freezing, expands and enlarges the opening. After many freeze-thaw cycles, the rock is broken into pieces. This process is appropriately called *frost wedging* (Figure 2.11). Frost wedging is most pronounced in mountainous regions in the middle latitudes where a daily freeze-thaw cycle often exists. Here, sections of rock are wedged loose and may tumble into large piles called *talus slopes* that often form at the base of steep rock outcrops (Figure 2.11).

Unloading When large masses of igneous rock are exposed by erosion, entire slabs begin to break loose, like layers of an onion. This *sheeting* is thought to occur because of the great reduction in pressure when the overlying rock is eroded away. Accompanying the unloading, the outer layers expand more than the rock below, and thus separate from the rock body. Granite is particularly prone to sheeting.

Continued weathering eventually causes the slabs to separate and spall, causing *exfoliation domes*. Excellent examples of exfoliation domes include Stone Mountain, Georgia, and Liberty Cap and Half Dome in Yosemite National Park (Figure 2.12).

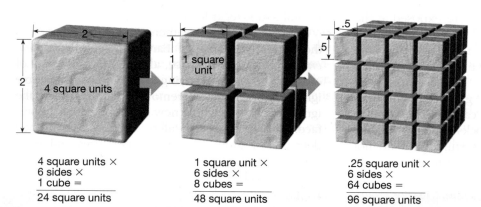

4 square units ×
6 sides ×
1 cube =
_____
24 square units

1 square unit ×
6 sides ×
8 cubes =
_____
48 square units

.25 square unit ×
6 sides ×
64 cubes =
_____
96 square units

Figure 2.10 Chemical weathering can occur only to those portions of a rock that are exposed to the elements. Mechanical weathering breaks rock into smaller and smaller pieces, thereby increasing the surface area available for chemical attack.

Frost wedging

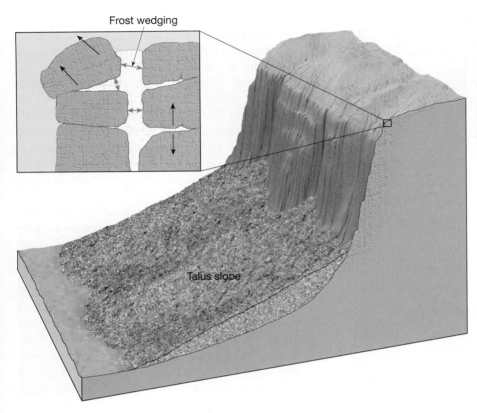

Talus slope

Figure 2.11 Frost wedging. As water freezes it expands to 109 percent of its original volume, exerting a force great enough to break rock. When frost wedging occurs in a setting such as this, the broken rock fragments fall to the base of the cliff and create a cone-shaped accumulation known as talus.

Biological Activity Weathering is also accomplished by the activities of organisms, including plants, burrowing animals, and humans. Plant roots in search of water grow into fractures, and as the roots grow they wedge the rock apart (Figure 2.13). Burrowing animals further break down the rock by moving fresh material to the surface, where physical and chemical processes can more effectively attack it.

Chemical Weathering of Rocks

Chemical weathering alters the internal structure of minerals by removing and/or adding elements. During this transformation, the original rock is altered into substances that are stable in the surface environment.

Water is the most important agent of chemical weathering. Oxygen dissolved in water will *oxidize* some materials. For example, when an iron nail is found in the soil, it will have a coating of rust (iron oxide), and if the time of exposure has been long, the nail will be so weak that it can be broken as easily as a toothpick. When rocks containing iron-rich minerals (such as hornblende) oxidize, a yellow-to-reddish brown rust will appear on the surface.

Carbon dioxide (CO_2) dissolved in water (H_2O) forms carbonic acid (H_2CO_3). This is the same weak acid produced when soft drinks are carbonated. Rain dissolves some carbon dioxide as it falls through the atmosphere, so normal rainwater is mildly acidic. Water in the soil also dissolves carbon dioxide released by decaying organic matter. The result is that acidic water is everywhere on Earth's surface.

Figure 2.12 Summit of Half Dome, an exfoliation dome in Yosemite National Park. *(Photo by Stephen Trimble)*

Figure 2.13 Root wedging widens fractures in rocks and aids the process of mechanical weathering. *(Photo by E. J. Tarbuck)*

mineral grains, or be carried to the ocean, where microscopic animals will build silica shells from it.

Quartz, the other main component of granite, is very resistant to chemical weathering. Because it is durable, quartz remains substantially unaltered when attacked by weak acid. As granite weathers, the feldspar crystals become dull and slowly turn to clay, releasing the once-interlocked quartz grains, which still retain their fresh, glassy appearance. Although some quartz remains in the soil, much is transported to the sea and other sites, where it becomes sandy beaches and sand dunes.

To summarize, *the chemical weathering of granite produces clay minerals along with potassium ions and silica, which enters into solution.* In addition, durable quartz grains are freed.

Table 2.2 lists the weathered products of some of the most common silicate minerals. Remember that silicate minerals make up most of Earth's crust and are composed primarily of just eight elements (Table 1.2, p. 19). When chemically weathered, the silicate minerals yield sodium, calcium, potassium, and magnesium ions. These may be used by plants or removed by groundwater. The element iron combines with oxygen to produce iron oxide compounds that give soil a reddish-brown or yellowish color. The three remaining elements, aluminum, silicon, and oxygen, join with water to produce clay minerals that become an important part of the soil. Ultimately, the products of weathering form the raw materials for building sedimentary rocks, which we consider next.

How does rock decompose when attacked by carbonic acid? Consider the weathering of the common igneous rock, granite. Recall that granite is composed mainly of quartz and potassium feldspar. As the weak acid slowly reacts with crystals of potassium feldspar, potassium ions are displaced. *This destroys the mineral's crystalline structure.*

The most abundant products of the chemical breakdown of feldspar are clay minerals. Because clay minerals are the end product of chemical weathering, they are very stable under surface conditions. Consequently, clay minerals make up a high percentage of the inorganic material in soils.

In addition to the formation of clay minerals, some silica (SiO_2) is dissolved from the feldspar structure and is carried away by groundwater. The dissolved silica will eventually precipitate to produce a hard, dense sedimentary rock (chert), or fill pore spaces between

Table 2.2 ■ Products of weathering

Original Mineral	Weathers to Produce	Released into Solution
Quartz	Quartz grains	Silica (SiO_2)
Feldspar	Clay minerals	Silica (SiO_2) Ions of potassium, sodium, and calcium
Hornblende	Clay minerals Iron minerals (limonite and hematite)	Silica (SiO_2) Ions of calcium and magnesium
Olivine	Iron minerals (limonite and hematite)	Silica (SiO_2) Ions of magnesium

Figure 2.14 Sedimentary rocks exposed in Canyonlands National Park, Utah. About 75 percent of all rock outcrops on the continents are sedimentary rocks. *(Photo by Jeff Gnass Photography)*

Sedimentary Rocks: Compacted and Cemented Sediment

Recall the rock cycle, which shows the origin of sedimentary rocks. Weathering begins the process. Next, gravity and erosional agents (running water, wind, waves, and glacial ice) remove the products of weathering and carry them to a new location where they are deposited. Usually the particles are broken down further during this transport phase. Following deposition, this **sediment** may become lithified, or "turned to rock." Commonly, **compaction** and **cementation** transform the sediment into solid sedimentary rock.

The word *sedimentary* indicates the nature of these rocks, for it is derived from the Latin *sedimentum*, which means "settling," a reference to a solid material settling out of a fluid. Most sediment is deposited in this fashion. Weathered debris is constantly being swept from bedrock and carried away by water, ice, or wind. Eventually the material is deposited in lakes, river valleys, seas, and countless other places. The particles in a desert sand dune, the

mud on the floor of a swamp, the gravels in a stream bed, and even household dust are examples of sediment produced by this never-ending process.

The weathering of bedrock and the transport and deposition of the weathering products are continuous. Therefore, sediment is found almost everywhere. As piles of sediment accumulate, the materials near the bottom are compacted by the weight of the overlying layers. Over long periods, these sediments are cemented together by mineral matter deposited from water in the spaces between particles. This forms solid sedimentary rock.

Geologists estimate that sedimentary rocks account for only about 5 percent (by volume) of Earth's outer 16 kilometers (10 miles). However, the importance of this group of rocks is far greater than this percentage implies. If you sampled the rocks exposed at Earth's surface, you would find that the great majority are sedimentary (Figure 2.14). Indeed, about 75 percent of all rock outcrops on the continents are sedimentary. Therefore, we can think of sedimentary rocks as comprising a relatively thin and

41

somewhat discontinuous layer in the uppermost portion of the crust. This makes sense because sediment accumulates at the surface.

It is from sedimentary rocks that geologists reconstruct many details of Earth's history. Because sediments are deposited in a variety of different settings at the surface, the rock layers that they eventually form hold many clues to past surface environments. They may also exhibit characteristics that allow geologists to decipher information about the method and distance of sediment transport. Furthermore, it is sedimentary rocks that contains fossils, which are vital evidence in the study of the geologic past.

Finally, many sedimentary rocks are very important economically. Coal, for example, is a sedimentary rock that presently is the source for over half the electric power generated in the United States. Other major energy resources (petroleum and natural gas) occur in pores within sedimentary rocks. Other sedimentary rocks are major sources of iron, aluminum, manganese, fertilizer, and sand and gravel.

Classifying Sedimentary Rocks

Materials accumulating as sediment have two principal sources. First, sediments may originate as solid particles from weathered rocks, such as the igneous rocks earlier described. These particles are called *detritus*, and the sedimentary rocks that they form are called **detrital sedimentary rocks**.

The second major source of sediment is soluble material produced largely by chemical weathering. When these dissolved substances are precipitated back as solids, they are called *chemical sediment*, and they form **chemical sedimentary rocks**. We will now look at detrital and chemical sedimentary rocks.

Detrital Sedimentary Rocks Though a wide variety of minerals and rock fragments may be found in detrital rocks, clay minerals and quartz dominate. As you learned earlier, clay minerals are the most abundant product of the chemical weathering of silicate minerals, especially the feldspars. Quartz, on the other hand, is abundant because it is extremely durable and very resistant to chemical weathering. Thus, when igneous rocks such as granite are weathered, individual quartz grains are set free.

Geologists use particle size to distinguish among detrital sedimentary rocks. Table 2.3 presents the four size categories for particles making up detrital rocks. When gravel-sized particles predominate, the rock is called *conglomerate* if the sediment is rounded

Table 2.3 ■ Particle size classification for detrital rocks

Size Range (millimeters)	Particle Name	Common Sediment Name	Detrital Rock
>256	Boulder		
64-256	Cobble	Gravel	Conglomerate
4-64	Pebble		or breccia
2-4	Granule		
1/16-2	Sand	Sand	Sandstone
1/256-1/16	Silt	Mud	Shale or
<1/256	Clay		mudstone

(Figure 2.15A) and *breccia* if the pieces are angular (Figure 2.15B). Angular fragments indicate that the particles were not transported very far from their source prior to deposition, and so have not had corners and rough edges abraded. *Sandstone* is the name given rocks when sand-sized grains prevail (Figure 2.15C). *Shale*, the most common sedimentary rock, is made of very fine-grained sediment (Figure 2.15D). *Siltstone*, another rather fine-grained rock, is sometimes difficult to differentiate from rocks such as shale, which are composed of even smaller clay-sized sediment.

Particle size is not only a convenient method of dividing detrital rocks; the sizes of the component grains also provide useful information about the environment in which the sediment was deposited. Currents of water or air sort the particles by size. The stronger the current, the larger the particle size carried. Gravels, for example, are moved by swiftly flowing rivers, rockslides, and glaciers. Less energy is required to transport sand; thus, it is common in windblown dunes, river deposits, and beaches. Because silts and clays settle very slowly, accumulations of these materials are generally associated with the quiet waters of a lake, lagoon, swamp, or marine environment.

Although detrital sedimentary rocks are classified by particle size, in certain cases the mineral composition is also part of naming a rock. For example, most sandstones are predominantly quartz-rich, and they are often referred to as quartz sandstone. In addition, rocks consisting of detrital sediments are rarely composed of grains of just one size. Consequently, a rock containing quantities of both sand and silt can be correctly classified as sandy siltstone or silty sandstone, depending on which particle size dominates.

A.

B.

C.

D.

Figure 2.15 Common detrital sedimentary rocks. **A.** Conglomerate (rounded particles). **B.** Breccia (angular particles). **C.** Sandstone. **D.** Shale with plant fossil. *(Photos by E. J. Tarbuck)*

Chemical Sedimentary Rocks In contrast to detrital rocks, which form from the solid products of weathering, chemical sediments are derived from material that is carried in solution to lakes and seas. This material does not remain dissolved in the water indefinitely. When conditions are right, it precipitates to form chemical sediments. This precipitation may occur directly as the result of physical processes, or indirectly through life processes of water-dwelling organisms. Sediment formed in this second way has a *biochemical* origin.

An example of a deposit resulting from physical processes is the salt left behind as a body of salt water evaporates. In contrast, many water-dwelling animals and plants extract dissolved mineral matter to form shells and other hard parts. After the organisms die, their skeletons may accumulate on the floor of a lake or ocean.

Limestone is the most abundant chemical sedimentary rock. It is composed chiefly of the mineral calcite ($CaCO_3$). Ninety percent of limestone is biochemical sediment. The rest precipitates directly from seawater.

One easily identified biochemical limestone is *coquina*, a coarse rock composed of loosely cemented shells and shell fragments (Figure 2.16).

← 5 cm →

Figure 2.16 This rock, called coquina, consists of shell fragments; therefore, it has a biochemical origin.

Another less obvious but familiar example is *chalk*, a soft, porous rock made up almost entirely of the hard parts of microscopic organisms that are no larger than the head of a pin (Figure 2.17).

Inorganic limestones form when chemical changes or high water temperatures increase the concentration of calcium carbonate to the point that it precipitates. *Travertine*, the type of limestone that decorates caverns, is one example. Groundwater is the source of travertine that is deposited in caves. As water drops reach the air in a cavern, some of the carbon dioxide dissolved in the water escapes, causing calcium carbonate to precipitate.

Dissolved silica (SiO_2) precipitates to form varieties of microcrystalline quartz. Minerals composed of microcrystalline quartz include chert (light color), flint (dark), jasper (red), and agate (banded). These chemical sedimentary rocks may have either an inorganic or biochemical origin, but the mode of origin is usually difficult to determine.

Very often, evaporation causes minerals to precipitate from water. Such minerals include halite, the chief component of *rock salt*, and gypsum, the main ingredient of *rock gypsum*. Both materials have significant commercial importance. Halite is familiar to everyone as the common salt used in cooking and seasoning foods. Of course, it has many other uses and has been considered important enough that people have sought, traded, and fought over it for much of human history. Gypsum is the basic ingredient of plaster of Paris. This material is used most extensively in the construction industry for "drywall" and plaster.

In the geologic past, many areas that are now dry land were covered by shallow arms of the sea that had only narrow connections to the open ocean. Under these conditions, water continually moved into the bay to replace water lost by evaporation. Eventually the waters of the bay became saturated and salt deposition began. Today, these arms of the sea are gone, and the remaining deposits are called **evaporites**.

Figure 2.17 White Chalk Cliffs, East Sussex, England. *(Photo by Kevin Schafer/Tony Stone Images)*

Figure 2.18 Bonneville Salt Flats, Utah. *(Photo by Scott T. Smith)*

Lignite and bituminous coals are sedimentary rocks, but anthracite is a metamorphic rock. Anthracite forms when sedimentary layers are subjected to the folding and deformation associated with mountain building.

In summary, we divide sedimentary rocks into two major groups: detrital and chemical. The main criterion for classifying detrital rocks is particle size, whereas chemical rocks are distinguished by their mineral composition. The categories presented here are more rigid than is the actual state of nature. Many detrital sedimentary rocks are a mixture of more than one particle size. Furthermore, many sedimentary rocks classified as chemical also contain at least small quantities of detrital sediment, and practically all detrital rocks are cemented with material that was originally dissolved in water.

On a smaller scale, evaporite deposits can be seen in such places as Death Valley, California. Here, following rains or periods of snowmelt in the mountains, streams flow from surrounding mountains into an enclosed basin. As the water evaporates, *salt flats* form from dissolved materials left behind as a white crust on the ground (Figure 2.18).

Coal is quite different from other chemical sedimentary rocks. Unlike other rocks in this category, which are calcite- or silica-rich, coal is made mostly of organic matter. Close examination of a piece of coal under a microscope or magnifying glass often reveals plant structures such as leaves, bark, and wood that have been chemically altered but are still identifiable. This supports the conclusion that coal is the end product of the burial of large amounts of plant material over extended periods.

The initial stage in coal formation is the accumulation of large quantities of plant remains. However, special conditions are required for such accumulations, because dead plants normally decompose when exposed to the atmosphere. An ideal environment that allows for the buildup of plant material is a swamp. Because stagnant swamp water is oxygen-deficient, complete decay (oxidation) of the plant material is not possible. At various times during Earth history, such environments have been common. Coal undergoes successive stages of formation. With each successive stage, higher temperatures and pressures drive off impurities and volatiles, as shown in Figure 2.19.

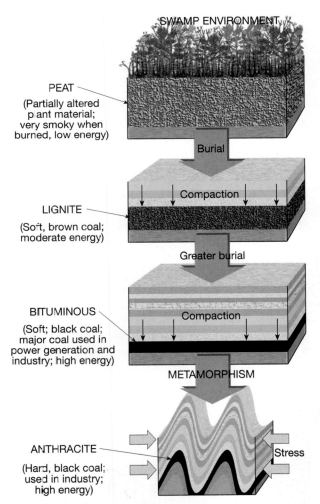

Figure 2.19 Successive stages in the formation of coal.

Lithification of Sediment

Lithification refers to the processes by which sediments are transformed into solid sedimentary rocks. One of the most common processes is *compaction*. As sediments accumulate through time, the weight of overlying material compresses the deeper sediments. As the grains are pressed closer and closer, pore space is greatly reduced. For example, when clays are buried beneath several thousand meters of material, the volume of the clay may be reduced as much as 40 percent. Compaction is most significant in fine-grained sedimentary rocks such as shale, because sand and other coarse sediments compress little.

Cementation is another important means by which sediments are converted to sedimentary rock. The cementing materials are carried in solution by water percolating through the pore spaces between particles. Through time, the cement precipitates onto the sediment grains, fills the open spaces, and joins the particles. Calcite, silica, and iron oxide are the most common cements. Identification of the cementing material is simple. Calcite cement will effervesce (fizz) with dilute hydrochloric acid. Silica is the hardest cement and thus produces the hardest sedimentary rocks. When a sedimentary rock has an orange or red color, this usually means iron oxide is present.

Features of Sedimentary Rocks

Sedimentary rocks are particularly important evidence of Earth's long history. These rocks form at Earth's surface, and as layer upon layer of sediment accumulates, each records the nature of the environment at the time the sediment was deposited. These layers called **strata,** or **beds,** are the *single most characteristic feature of sedimentary rocks* (see Figure 2.14, p. 41). The thickness of beds range from microscopically thin to tens of meters thick. Separating the strata are *bedding planes*, flat surfaces along which rocks tend to separate or break. Generally, each bedding plane marks the end of one episode of sedimentation and the beginning of another.

Sedimentary rocks provide geologists with evidence for deciphering past environments. A conglomerate, for example, indicates a high-energy environment, such as a rushing stream, where only the coarse materials can settle out. By contrast, black shale and coal are associated with a low-energy, organic-rich environment such as a swamp or lagoon. Other features found in some sedimentary rocks also give clues to past environments (Figure 2.20).

Fossils, the traces or remains of prehistoric life, are perhaps the most important inclusions found in some sedimentary rock (Figure 2.21). Knowing the nature of the life-forms that existed at a particular

Figure 2.20 A. Ripple marks preserved in sedimentary rocks may indicate a beach or stream channel environment. *(Photo by Stephen Trimble)* **B**. Mud cracks form when wet mud or clay dries and shrinks, perhaps signifying a tidal flat or desert basin. *(Photo by David Muench Photography, Inc.)*

A.

B.

Figure 2.21 Fossil of a trilobite, an ancient marine organism, preserved in shale. Fossils are important tools used to determine past environmental conditions. They are also valuable time indicators.

time may help to answer many questions about the environment. Was it land or ocean? A lake or swamp? Was the climate hot or cold, rainy or dry? Was the ocean water shallow or deep, turbid or clear? Furthermore, fossils are important time indicators and play a key role in matching up rocks from different places that are the same age. Fossils are important tools used in interpreting the geologic past and will be examined in some detail in Chapter 8.

Metamorphic Rocks: Changed in Form

Metamorphism literally means to "change form." Thus, metamorphism changes existing rocks in size, shape, texture, and even the minerals they contain. Metamorphic rocks can form from igneous, sedimentary, or even from other metamorphic rocks. The agents of change include heat, pressure, and chemically active fluids.

Metamorphism varies broadly in degree. In some instances, rocks are only slightly changed. For example, under *low-grade metamorphism*, the common sedimentary rock shale becomes the metamorphic rock called *slate*. Although slate is more compact than shale, hand samples of these rocks are sometimes difficult to distinguish. In other cases, the transformation is so complete that the identity of the original rock cannot be determined. In such *high-grade metamorphism*, features like bedding planes, fossils, and vesicles that may have existed in the parent rock are completely obliterated. Further, when

rocks at depth are subjected to differential forces, they flow and bend into intricate folds (Figure 2.22).

In the most extreme metamorphic environments, the temperatures approach those at which rocks melt. However, during metamorphism the deformed material must remain essentially solid, for once complete melting occurs, we have re-entered the realm of igneous activity.

Metamorphism takes place when rock is subjected to conditions unlike those in which it originally formed. As a result, the rock becomes unstable and gradually changes until a state of equilibrium with the new environment is reached. The changes occur at the elevated temperatures and pressures of the region extending from a few kilometers below Earth's surface to the crust–mantle boundary. Because the formation of metamorphic rocks is completely hidden from view (which is not the case for many sedimentary and some igneous rocks), metamorphism is one of the most difficult processes for geologists to study.

Metamorphism most often occurs in one of two settings. First, during mountain building, great quantities of rock are subjected to the intense stresses and high temperatures associated with large-scale deformation. The end result may be extensive areas of metamorphic rocks that are said to have undergone **regional metamorphism**. The greatest volume of metamorphic rock is produced in this fashion. Second, when rock is in contact with, or near, a mass

Figure 2.22 Deformed metamorphic rocks exposed in a road cut in the Eastern Highland of Connecticut. Imagine the tremendous force required to fold rock in this manner. *(Photo by Phil Dombrowski)*

of magma, **contact metamorphism** takes place. In this circumstance, the changes are caused primarily by the high temperatures of the molten material that "bake" the surrounding rock.

Extensive areas of metamorphic rocks are exposed on every continent. Metamorphic rocks are an important component of many mountain belts, where they make up a large portion of a mountain's crystalline core. Even the stable continental interiors, which are generally covered by sedimentary rocks, are underlain by metamorphic basement rocks. In all of these settings, the metamorphic rocks are usually highly deformed and intruded by igneous masses. Indeed, significant parts of Earth's continental crust are composed of metamorphic and associated igneous rocks.

Three Metamorphic Agents

As stated, the agents of metamorphism include *heat, pressure,* and *chemically active fluids.* During metamorphism, rocks are often subjected to all three agents simultaneously. However, the degree of metamorphism and the contribution of each agent vary greatly from one environment to another. In low-grade metamorphism, rocks are subjected to temperatures and pressures only slightly greater than those associated with the lithification of sediments. High-grade metamorphism, on the other hand, involves extreme conditions closer to those at which rocks melt.

In addition, the minerals making up the parent rock determine, to a large extent, the degree to which various metamorphic agents will cause change. For example, when an intruding igneous mass enters a rock unit, hot, ion-rich fluids circulate through the host rock. If the host rock is quartz sandstone, very little alteration may take place. On the other hand, if the host rock is limestone, the impact of these fluids can be dramatic and the effects of metamorphism may extend for several kilometers from the igneous mass.

Heat is perhaps the most important agent of metamorphism. Heat provides the energy to drive chemical reactions that recrystallize minerals. Rocks formed near Earth's surface may be subjected to intense heat when they are intruded by molten material rising from below. Because temperature increases with depth, rocks that originate in a surface environment may also be subjected to extreme temperatures if they are subsequently buried deep within Earth. When buried to a depth of only a few

kilometers, certain minerals, such as clay, become unstable and recrystallize into minerals that are stable in this more intense environment. Other minerals, particularly those in crystalline igneous rocks, are stable at relatively high temperatures and pressures and therefore require burial to 20 kilometers or more before metamorphism will occur.

Pressure, like temperature, also increases with depth. Buried rocks are subjected to the force exerted by the load above. This confining pressure is analogous to air pressure, where the force is applied equally in all directions. In addition to the pressure exerted by the load of material above, rocks are also subjected to differential *stresses* during mountain building. In this situation the force is unequal in different directions, and the material is squeezed as if it had been placed in a vise. Rock located at great depth is quite warm and behaves plastically during deformation. This accounts for its ability to flow and bend into intricate folds (Figure 2.22).

Chemically active fluids also influence the metamorphic process. Most common is the water contained in the pore spaces of virtually every rock. Water that surrounds the crystals acts as a catalyst by aiding the migration of ions. In some instances, the minerals recrystallize into more stable crystal forms. In other cases, ion exchange among minerals results in the formation of completely different minerals.

Metamorphism Changes Texture

The degree of metamorphism is reflected in the rock's texture and mineralogy. When rocks are subjected to low-grade metamorphism, they become more compact and thus more dense. A common example is the metamorphic rock slate, which forms when shale is subjected to temperatures and pressures only slightly greater than those associated with the compaction that lithifies sediment. In this case, differential stress causes the microscopic clay minerals in shale to align into the more compact arrangement found in slate. Under more extreme conditions, stress causes certain minerals to recrystallize.

In general, recrystallization encourages the growth of larger crystals. Consequently, many metamorphic rocks consist of visible crystals, much like coarse-grained igneous rocks. The crystals of some minerals will recrystallize with a preferred orientation, essentially perpendicular to the direction of the compressional force. The resulting mineral alignment usu-

A.

B.

Figure 2.23 Under the pressures of metamorphism, some mineral grains become reoriented and aligned at right angles to the stress. The resulting orientation of mineral grains gives the rock a foliated (layered) texture. If the coarse-grained igneous rock (granite) on the left underwent intense metamorphism, it could end up closely resembling the metamorphic rock on the right (gneiss). *(Photos by E. J. Tarbuck)*

ally gives the rock a layered or banded appearance termed **foliated texture** (Figure 2.23). Simply, foliation results whenever the minerals of a rock are brought into parallel alignment.

Not all metamorphic rocks have a foliated texture. Such rocks are said to exhibit a **nonfoliated texture**. Metamorphic rocks consisting of only one mineral that forms equidimensional crystals are, as a rule, not visibly foliated. For example, limestone, if pure, consists of a single mineral, calcite. When a fine-grained limestone is metamorphosed, the small calcite crystals combine to form larger interlocking crystals. The resulting rock resembles a coarse-grained igneous rock. This nonfoliated metamorphic equivalent of limestone is called *marble*.

In some environments, new materials are actually introduced during the metamorphic process. For example, host rock adjacent to a large intruding magma body may be altered by *hydrothermal* (hot water) *solutions* released during the latter stages of crystallization. Hydrothermal solutions are rich in ions, including those of valuable metals. Many metallic ore deposits are formed by the precipitation of minerals from hydrothermal solutions.

Classifying Metamorphic Rocks

To review, metamorphic processes cause many changes in existing rocks, including increased density, growth of larger crystals, foliation (reorientation of the mineral grains into a layered or banded appearance), and the transformation of low-temperature minerals into high-temperature minerals. Further, the introduction of ions generates new minerals, some of which are economically important.

Here is a brief look at common rocks produced by metamorphic processes.

Foliated Rocks *Slate* is a very fine-grained foliated rock composed of minute mica flakes (Figure 2.24). The most noteworthy characteristic of slate is its excellent rock cleavage, meaning that it splits easily into flat slabs. This property has made slate a most useful rock for roof and floor tile, chalkboards, and billiard tables. Slate is most often generated by the low-grade metamorphism of shale, although less frequently it forms from the metamorphism of volcanic ash. Slate can be almost any color, depending on its mineral constituents. Black slate contains organic material; red

Figure 2.24 Slate, a common metamorphic rock produced by the low-grade metamorphism of shale. *(Photo by E. J. Tarbuck)*

slate gets its color from iron oxide; and green slate is usually composed of chlorite, a micalike mineral.

Schists are strongly foliated rocks, formed by regional metamorphism. They are "platy" and can be readily split into thin flakes or slabs. Like slate, the parent material from which many schists originate is shale, but in the case of schist, the metamorphism is more intense.

The term *schist* describes the *texture* of a rock regardless of composition. For example, schists composed primarily of muscovite and biotite are called *mica schists* (Figure 2.25).

Gneiss (pronounced "nice") is the term applied to banded metamorphic rocks that contain mostly elongated and granular, as opposed to platy, minerals (see Figure 2.23B). The most common minerals in gneisses are quartz and feldspar, with lesser amounts of muscovite, biotite, and hornblende. Gneisses exhibit strong segregation of light and dark silicates, giving them a characteristic banded texture. While in a plastic state, these banded gneisses can be deformed into intricate folds.

Nonfoliated Rocks   *Marble* is a coarse, crystalline rock whose parent rock is limestone (Figure 2.26). Marble is composed of large interlocking calcite crystals, which form from the recrystallization of smaller grains in the parent rock.

Because of its color and relative softness (hardness of only 3 on the Mohs scale) marble is a popular building stone. White marble is particularly prized as a stone from which to carve monuments and statues, such as the famous statue of David by Michelangelo. Often the limestone from which marble forms contains impurities that color the marble. Thus, marble can be pink, gray, green, or even black.

Quartzite is a very hard metamorphic rock most often formed from quartz sandstone. Under moderate-to-high-grade metamorphism, the quartz grains in sandstone fuse. Pure quartzite is white, but iron oxide may produce reddish or pinkish stains and dark minerals may impart a gray color.

Figure 2.25 Mica shist, a common metamorphic rock composed of shiny mica flakes. *(Photo by E. J. Tarbuck)*

Figure 2.26 Marble, a crystalline rock formed by the metamorphism of limestone. *(Photo by E. J. Tarbuck)*

The Chapter in Review

1. The three rock groups are igneous, sedimentary, and metamorphic. *Igneous rock* forms from *magma* that cools and solidifies in a process called *crystallization*. *Sedimentary rock* forms from the *lithification* of *sediment*. *Metamorphic rock* forms from rock that has been subjected to great pressure and heat in a process called *metamorphism*.

2. Igneous rocks are classified by their *texture* and *mineral composition*.

3. The rate of cooling of magma greatly influences the size of mineral crystals in igneous rock, and thus its texture. The four basic igneous rock textures are (1) *fine-grained*, (2) *coarse-grained*, (3) *porphyritic*, and (4) *glassy*.

4. The mineral makeup of an igneous rock is ultimately determined by the chemical composition of the magma from which it crystallized. N. L. Bowen showed that as magma cools, minerals crystallize in an orderly fashion at different temperatures. *Crystal settling* can change the composition of magma and cause more than one rock type to form from a common parent magma.

5. *Weathering* is the response of surface materials to a changing environment. *Mechanical weathering*, the physical disintegration of material into smaller fragments, is accomplished by *frost wedging*, expansion resulting from *unloading*, and *biological activity*. *Chemical weathering* involves processes by which the internal structures of minerals are altered by the removal and/or addition of elements. It occurs when materials are *oxidized* or *react with acid*, such as carbonic acid.

6. *Detrital sediments* originate as solid particles derived from weathering and are transported. *Chemical sediments* are soluble materials produced largely by chemical weathering that are precipitated by either inorganic or organic processes. *Detrital sedimentary rocks*, which are classified by particle size, contain a variety of mineral and rock fragments, with clay minerals and quartz the chief constituents. *Chemical sedimentary rocks* often contain the products of biological processes, or mineral crystals that form as water evaporates and minerals precipitate. *Lithification* refers to the processes by which sediments are transformed into solid sedimentary rocks.

7. Common detrital sedimentary rocks include *shale* (the most common sedimentary rock), *sandstone*, and *conglomerate*. The most abundant chemical sedimentary rock is *limestone*, consisting chiefly of the mineral calcite. *Rock gypsum* and *rock salt* are chemical rocks that form as water evaporates.

8. Some features of sedimentary rocks that are often used in the interpretation of Earth history and past environments include *strata* or *beds* (the single most characteristic feature), *bedding planes*, and *fossils*.

9. Two types of metamorphism are (1) *regional metamorphism* and (2) *contact metamorphism*. The agents of metamorphism include *heat*, *pressure*, and *chemically active fluids*. Heat is perhaps the most important because it provides the energy to drive the reactions that result in the *recrystallization* of minerals. Metamorphic processes cause many changes in rocks, including *increased density*, growth of *larger mineral crystals, reorientation of the mineral grains* into a layered or banded appearance known as *foliation*, and the formation of *new minerals*.

10. Some common metamorphic rocks with a *foliated texture* include *slate*, *schist*, and *gneiss*. Metamorphic rocks with a *nonfoliated texture* include *marble* and *quartzite*.

Key Terms

chemical sedimentary rock (p. 42)

chemical weathering (p. 39)

coarse-grained texture (p. 33)

contact metamorphism (p. 48)

crystallization (p. 32)

detrital sedimentary rock (p. 42)

evaporite deposit (p. 44)

extrusive (volcanic) (p. 32)

fine-grained texture (p. 33)

foliated texture (p. 49)

glassy texture (p. 34)

igneous rock (p. 32)

intrusive (plutonic) (p. 32)

lava (p. 32)

lithification (p. 32)

magma (p. 32)

mechanical weathering (p. 38)

metamorphic rock (p. 32)

nonfoliated texture (p. 49)

porphyritic texture (p. 33)

regional metamorphism (p. 48)

rock cycle (p. 30)

sediment (p. 32)

sedimentary rock (p. 32)

strata (beds) (p. 46)

texture (p. 33)

weathering (p. 32)

Questions for Review

1. Explain the statement "One rock is the raw material for another" using the rock cycle.
2. If a lava flow at Earth's surface had a basaltic composition, what rock type would the flow likely be (see Table 2.1, p. 37)? What igneous rock would form from the same magma if it did not reach the surface but instead crystallized at great depth?
3. What does a porphyritic texture indicate about the history of an igneous rock?
4. How are granite and rhyolite different? The same? (see Table 2.1, p. 37.)
5. Relate the classification of igneous rocks to Bowen's reaction series.
6. If two identical rocks were weathered, one mechanically and the other chemically, how would the products of weathering for the two rocks differ?
7. How does mechanical weathering add to the effectiveness of chemical weathering?
8. How is carbonic acid formed in nature? What are the products when this acid reacts with potassium feldspar?
9. What minerals are most common in detrital sedimentary rocks? Why are these minerals so abundant?
10. What is the primary basis for distinguishing among various detrital sedimentary rocks?
11. Distinguish between the two categories of chemical sedimentary rocks.
12. What are evaporite deposits? Name a rock that is an evaporite.
13. Compaction is an important lithification process with which sediment size?
14. What is probably the single most characteristic feature of sedimentary rocks?
15. What is metamorphism? What are the *agents of change*?
16. Distinguish between regional and contact metamorphism.
17. What feature would easily distinguish schist and gneiss from quartzite and marble?
18. In what ways do metamorphic rocks differ from the igneous and sedimentary rocks from which they formed?

Testing What You Have Learned

Multiple-Choice Questions

1. Molten material found within Earth is called _____
   a. sediment.   c.   magma. e.   none of the above.
   b. foliation.   **d.** pyroxene.
2. The term used to describe the overall appearance of an igneous rock based on the size and arrangement of its interlocking crystals is _____
   a. texture.   **c.** foliation   **e.** mineral content.
   b. color.   **d.** hardness.
3. The term used to describe the composition of an igneous rock rich in the elements magnesium and iron is _____
   a. granitic.   **c.** glass.   **e.** porphyritic.
   b. fema.   **d.** basaltic.
4. Which of the following is NOT a process of mechanical weathering?
   a. frost wedging   **c.** oxidation   **e.** both c. and d.
   b. unloading   **d.** biological activity

5. Which one of the following is a common product of chemical weathering?
   a. clay   c. basalt   e. olivine
   b. slate   d. granite

6. Sedimentary rocks are _____
   a. foliated.   c. layered, or bedded.   e. volcanic
   b. porphyritic.   d. plutonic.

7. Which one of the following is the primary basis for distinguishing among various detrital sedimentary rocks?
   a. color   c. particle size   e. mineral composition
   b. hardness   d. crystal form

8. Limestone, the most abundant chemical sedimentary rock, is made of the mineral _____
   a. gypsum.   c. clay.   e. calcite.
   b. halite.   d. quartz.

9. Which of the following is NOT an agent of metamorphism?
   a. pressure   c. lithification   e. both a. and c.
   b. heat   d. chemical fluids

10. Which one of the following pairs of parent rocks and their possible metamorphic equivalent is NOT correct?
    a. shale-slate   c. sandstone-marble   e. limestone-marble
    b. shale-schist   d. sandstone-quartzite

Fill-In Questions

11. _____ is magma at Earth's surface, except that most of the _____ component has escaped.

12. An igneous rock that has large crystals embedded in a matrix of smaller crystals is said to have a _____ texture.

13. The two major types of weathering are _____ and _____ weathering.

14. A detrital sedimentary rock with gravel-sized particles is called a _____ if the sediment is rounded and _____ if the pieces are angular.

15. The greatest volume of metamorphic rock is produced during _____ metamorphism; however, when a rock is adjacent to, or near, a mass of magma, _____ metamorphism takes place.

True/False Questions

16. Rocks are divided into three groups based on their different mineral compositions. _____

17. N. L. Bowen demonstrated that minerals crystallize from magma in a systematic fashion. _____

18. The term *lithification* means "conversion into rock." _____

19. Water is the most important agent of chemical weathering. _____

20. Minerals in a foliated metamorphic rock are aligned essentially parallel to the compressional force that produced the rock. _____

Answers:

Web Work

The following are informative and interesting World Wide Web sites that address topics related to those presented in the chapter:

- Major Rock Groups and Rock Cycle
  http://duke.usask.ca/~reeves/prog/geoe118/geoe118.010.html

- Rocks and Minerals Reference
  http://luigi.calpoly.edu/StudWeb95/rocks/rocks.html

For direct access to these sites and other relevant Web locations, contact the *Foundations of Earth Science* WWW home page at http://www.prenhall.com/lutgens.

UNIT TWO
Earth's External Processes

Weathered sandstone, Paria Canyon, Vermillion Cliffs Wilderness.
Photo by Jack Dykinga and Associates

CHAPTER 3
Landscapes Fashioned By Water

FOCUS ON LEARNING

To assist you in learning the important concepts
in this chapter, you will find it helpful to focus
on the following questions:

1. What is the geological process called *mass wasting*?

2. What is the water cycle? What is the source of energy that powers the cycle?

3. What are the factors that determine the velocity of water in a stream?

4. How does base level influence a stream's ability to erode?

5. The work of a stream includes what three processes?

6. What are the two general types of stream valleys and some features that are associated with each?

7. What common drainage patterns do streams produce?

8. What is the importance of groundwater as a resource and as a geological agent?

9. What is groundwater, and how does it move?

10. What are some common natural phenomena associated with groundwater?

11. What are some environmental problems associated with groundwater?

North Fork of the White River, Colorado.
Photo by Carr Clifton Photography

Earth is a dynamic planet. Volcanic and other internal forces elevate the land while opposing external processes continually wear it down. The Sun and gravity drive external processes occurring at Earth's surface. Rock is disintegrated and decomposed, moved to lower elevations by gravity, and carried away by water, wind, or ice. In this manner the physical landscape is sculptured (Figure 3.1).

Mass Wasting

Earth's surface is never perfectly flat but instead consists of slopes. Some are steep and precipitous; others are moderate or gentle. Some are long and gradual; others are short and abrupt. Some slopes are mantled with soil and covered by vegetation; others consist of barren rock and rubble. Their form and variety are great.

Although most slopes appear to be stable and unchanging, they are not static features, because the force of gravity causes material to move downslope. At one extreme, the movement may be gradual and practically imperceptible. At the other extreme, it may consist of a thundering landslide or rockfall.

Landslides, such as the one pictured in Figure 3.2, are spectacular examples of a common geologic process called mass wasting. **Mass wasting** is the downslope movement of rock and soil under the direct influence of gravity. It is distinct from the erosional processes because mass wasting does not require a transporting medium. There is a broad array of mass wasting processes. Four are illustrated in Figure 3.3.

Although gravity is the controlling force of mass wasting, other factors can play an important part in bringing about the downslope movement of material. Water is one of these factors. When the open spaces in sediment become saturated with water, the cohesion between particles is destroyed. This allows gravity to more easily set the material in motion.

Oversteepening of slopes is another factor that triggers mass movements. Loose, undisturbed particles assume a stable slope called the *angle of repose*, the steepest angle at which material remains stable. If the angle is increased, the rock debris will adjust by moving downslope. Many situations in nature exist where this takes place. A stream undercutting a valley wall and waves pounding against the base of a cliff are two familiar examples. Furthermore, through their activities, people often create oversteepened and unstable slopes that become prime sites for slides, slumps, and flows.

Figure 3.1 The Colorado River in Marble Canyon, Grand Canyon National Park, Arizona. Slopes here, as elsewhere, are places where materials are continually moving from higher to lower elevations. Weathering begins the process by attacking the solid rock exposed at the surface. Next, gravity moves the weathered debris downslope. This step, termed mass wasting, may range from a slow and gradual creep to a thundering landslide. Eventually, the material that was once high up the slope reaches the stream at the bottom. The moving water then transports the debris away. *(Photo by Tom Bean/DRK Photo)*

Figure 3.2 This rockslide near Lakeside, Montana, closed the highway for two days. Fortunately, no one was injured when the mass of rock gave way. *(Photo by AP/Wide World Photos)*

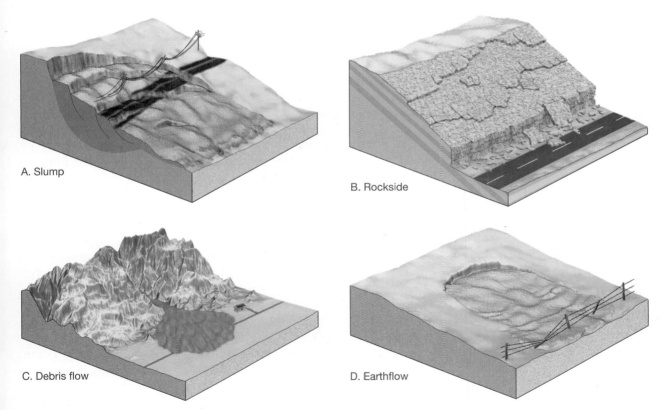

A. Slump

B. Rockside

C. Debris flow

D. Earthflow

Figure 3.3 The four processes illustrated here are all considered to be relatively rapid forms of mass wasting. Because materials in slumps (**A**) and rockslides (**B**) move along well-defined surfaces, they are said to move by sliding. By contrast, when material moves downslope as a viscous fluid, the movement is described as a flow. Debris flow (**C**) and earthflow (**D**) advance downslope in this manner.

Mass wasting is the step that follows weathering in the evolution of most landforms. Once weathering weakens and breaks rock apart, mass wasting transfers the debris downslope, where a stream, acting as a conveyor belt, usually carries it away. Although there may be many intermediate stops along the way, the sediment is eventually transported to its ultimate destination, the sea. It is the combined effects of mass wasting and running water that produce stream valleys, which are the most common and conspicuous landforms at Earth's surface and the focus of the next part of this chapter. If streams alone were responsible for creating the valleys in which they flow, valleys would be very narrow features. However, the fact that most river valleys are much wider than they are deep is a strong indication of the significance of gravity and mass wasting processes in supplying material to streams.

The Water Cycle

All the rivers run into the sea; yet the sea is not full; unto the place from whence the rivers come, thither they return again.

Ecclesiastes 1:7

As the perceptive writer of Ecclesiastes indicated, water is continually on the move, from the ocean to the land and back again in an endless cycle. The remainder of this chapter deals with that part of the water cycle that returns water to the sea. Some water travels quickly via a rushing stream, and some moves more slowly below the surface. We shall examine the factors that influence the distribution and movement of water, and we shall look at how water sculptures the landscape. The Grand Canyon, Niagara Falls, Old Faithful, and Mammoth Cave all owe their existence to the action of water on its way to the sea.

The amount of water on Earth is immense, an estimated 1.36 billion cubic kilometers (326 million cubic miles). Of this total, the vast bulk—97.2 percent—is part of the world ocean. Ice sheets and glaciers account for another 2.15 percent, leaving only 0.65 percent to be divided among lakes, streams, subsurface water, and the atmosphere (Figure 3.4). Although the percentage of Earth's total water found in each of the latter sources is just a small fraction of the total inventory, the absolute quantities are great.

The water found in each of the reservoirs depicted in Figure 3.4 does not remain in these

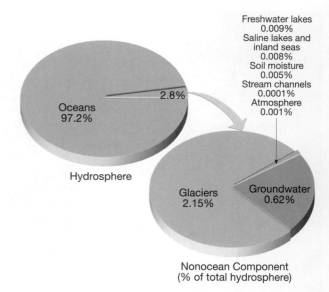

Figure 3.4 Distribution of Earth's water.

places indefinitely. Water can readily change from one state of matter (solid, liquid, or gas) to another at the temperatures and pressures that occur at Earth's surface. Therefore, water is constantly moving among the hydrosphere, the atmosphere, the solid Earth, and the biosphere. This unending circulation of Earth's water supply is called the **water cycle**. It is a gigantic worldwide system powered by energy from the Sun in which the atmosphere provides the vital link between the oceans and continents (Figure 3.5).

Water evaporates into the atmosphere from the ocean and to a much lesser extent from the continents. Winds transport this moisture-laden air, often great distances, until conditions cause the moisture to condense into clouds, and precipitation to fall. The precipitation that falls into the ocean has completed its cycle and is ready to begin another. The water that falls on the continents, however, must make its way back to the ocean.

What happens to precipitation once it has fallen on land? A portion of the water soaks into the ground, slowly moving downward, then laterally, finally seeping into lakes, streams, or directly into the ocean. When the rate of rainfall exceeds Earth's ability to absorb it, the surplus water flows over the surface into lakes and streams. Much of the water that soaks in or runs off eventually returns to the atmosphere because of evaporation from the soil, lakes, and streams. Also, some of the water that infiltrates the

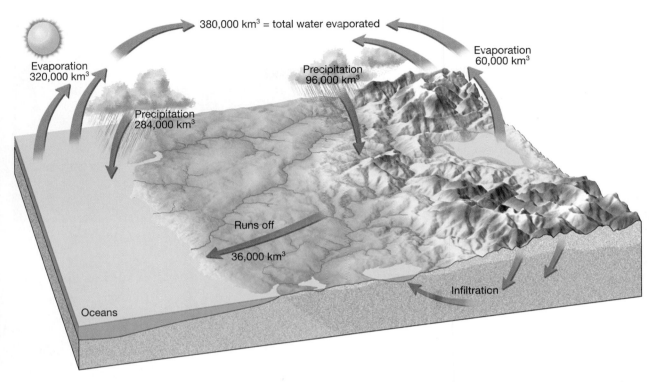

380,000 km³ = total water evaporated

Evaporation
60,000 km³

Evaporation
320,000 km³

Precipitation
96,000 km³

Precipitation
284,000 km³

Runs off

36,000 km³

Infiltration

Oceans

Figure 3.5 Earth's water balance. Each year, solar energy evaporates about 320,000 cubic kilometers of water from the oceans, while evaporation from the land (including lakes and streams) contributes 60,000 cubic kilometers of water. Of this total of 380,000 cubic kilometers of water, about 284,000 cubic kilometers fall back to the ocean, and the remaining 96,000 cubic kilometers fall on the land. Of the 96,000 cubic kilometers, only 60,000 cubic kilometers of water evaporate from the land, leaving 36,000 cubic kilometers of water to erode the land during the journey back to the oceans.

ground surface is absorbed by plants, which then release it into the atmosphere. This process is called **transpiration**. Each year a field of crops may transpire the equivalent of a water layer 60 centimeters (2 feet) deep over the entire field. The same area of trees may pump twice this amount into the atmosphere.

When precipitation falls at high elevations or high latitudes, where temperatures are cold, the water may not immediately soak in, run off, or evaporate. Instead, it may become part of a snowfield or a glacier. In this way, glaciers store large quantities of water on land. If present-day glaciers were to melt and release all their water, sea level would rise by several dozen meters. This would submerge many heavily populated coastal areas. As we shall see in Chapter 4, over the past 2 million years, huge ice sheets have formed and melted on several occasions, each time affecting the balance of the water cycle.

A diagram of Earth's water cycle is shown in Figure 3.5. The amount of water vapor in the air at any one time is just a tiny fraction of Earth's total water supply. But the *absolute* quantities that are cycled through the atmosphere over a 1-year period are immense—some 380,000 cubic kilometers— enough to cover Earth's entire surface to a depth of about 1 meter (39 inches).

It is important to know that the water cycle is *balanced*. Because the total amount of water vapor in the atmosphere remains about the same, the average annual precipitation worldwide must be equal to the quantity of water evaporated. Therefore, the average annual precipitation worldwide must equal the quantity of water evaporated. However, for all of the continents taken together, precipitation exceeds evaporation. Conversely, over the oceans, evaporation exceeds precipitation. Because the level of the world ocean is not dropping, the system must be in balance. In Figure 3.5, the 36,000 cubic kilometers of water that annually runs off from the land to the ocean causes enormous erosion. In fact, this immense volume of moving water is *the single most important agent sculpturing Earth's land surface.*

To summarize, the water cycle is the continuous movement of water from the oceans to the atmosphere, from the atmosphere to the land, and from the land back to the sea. The land-back-to-the-sea step is the primary action that wears down Earth's land surface. In this chapter, we will first observe the work of water running over the surface, including floods, erosion, and the formation of valleys. Then we will look underground at the slow labors of groundwater as it forms springs and caverns, and provides drinking water on its long migration to the sea.

Running Water

Running water is of great importance to people. We depend on rivers for energy, transportation, and irrigation. Their fertile floodplains have been favored sites for agriculture and industry ever since the dawn of civilization. As the dominant agent of erosion, running water has shaped much of our physical environment.

Although people have always depended on running water, its source eluded them for centuries. It was not until the sixteenth century that they first realized streams were supplied by runoff and underground water, which ultimately had their sources as rain and snow.

Streamflow

Water makes its way to the sea under the influence of gravity. The time required for the journey depends on the velocity of the stream. Velocity is the distance that water travels in a unit of time. Water in some sluggish streams travels at less than 0.8 kilometer (0.5 mile) per hour, whereas water in a few rapid streams reaches speeds as high as 32 kilometers (20 miles) per hour. Velocities are measured at gauging stations (Figure 3.6). Along straight stretches, the highest velocities are near the center of the channel just below the surface, where friction is lowest. But when a stream curves, its zone of maximum speed shifts toward its outer bank (Figure 3.6).

The ability of a stream to erode and transport materials depends on its velocity. Even slight variations in velocity can lead to significant changes in how much sediment can be transported by the water. Several factors determine the velocity of a stream, including (1) gradient; (2) shape, size, and roughness of the channel; and (3) discharge.

Gradient is the slope of a stream channel expressed as the vertical drop of a stream over a specified distance. For example, portions of the lower Mississippi River have very low gradients of 10

A.

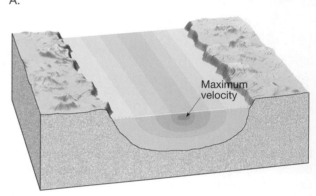

B.

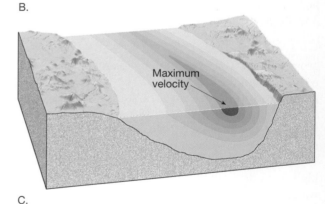

C.

Figure 3.6 A. Continuous records of stage and discharge are collected by the U.S. Geological Survey at more than 7000 gauging stations in the United States. Average velocities are determined by using measurements from several spots across the stream. This station is on the Rio Grande south of Taos, New Mexico. *(Photo by E.J. Tarbuck)* **B**. Along straight stretches, stream velocity is highest at the center of the channel. **C**. When a stream curves, its zone of maximum speed shifts toward the outer bank.

centimeters per kilometer or less. By contrast, some mountain stream channels decrease in elevation at a rate of more than 40 meters per kilometer, or a gradient 400 times steeper than the lower Mississippi

Figure 3.7 Rapids are common where the gradient is steep and the channel is rough and irregular. *(Photo by Tom Bean)*

(Figure 3.7). Gradient varies not only between streams, but over a stream's length. The steeper the gradient, the more energy available for streamflow. If two streams were identical in every respect except gradient, the stream with the higher gradient would obviously have the greater velocity.

A stream's channel is a conduit that guides the flow of water, but the water encounters friction as it flows. The shape, size, and roughness of the channel affect the amount of friction. Larger channels have more efficient flow because a smaller proportion of water is in contact with the channel. A smooth channel promotes a more uniform flow, whereas an irregular channel filled with boulders creates enough turbulence to slow the stream significantly.

The **discharge** of a stream is the volume of water flowing past a certain point in a given unit of time. This is usually measured in cubic meters per second or cubic feet per second. The largest river in North America, the Mississippi, discharges an average of 17,300 cubic meters (611,000 cubic feet) per second. Although this is a huge quantity of water, it is nevertheless dwarfed by the mighty Amazon in South America, the world's largest river. Draining a rainy region that is nearly three-fourths the size of the conterminous United States, the Amazon discharges 12 times more water than the Mississippi!

The discharge of an individual stream is far from constant. This is true because of such variables as rainfall and snowmelt. If discharge changes, the factors noted earlier adjust. Indeed, measurements show that when the amount of water in a stream increases, the width, depth, and velocity all increase. To handle the additional water, the stream will increase the size of its channel by widening and deepening it. The larger the channel, the less friction, and the more swiftly the water will flow.

Changes from Upstream to Downstream

One useful way of studying a stream is to examine its *profile*. A profile is simply a cross-sectional view of a stream from its source area (called the *head*) to its *mouth*, the point downstream where the river ends by emptying into another water body. By examining Figure 3.8, you can see that the most obvious feature of a typical profile is a constantly decreasing gradient from the head to the mouth. Although many local irregularities may exist, the overall profile is a smooth curve.

The profile shows that the gradient decreases downstream. To see how other factors change in a downstream direction, observations and measurements must be made. When data are collected from several gauging stations along a river, they show that

Head

Mouth

Steep gradient

Longitudinal profile

Gentle gradient

Ocean

Figure 3.8 A longitudinal profile is a cross section along the length of a stream. Note the concave-upward curve of the profile, with a steeper gradient upstream and a gentler gradient downstream. Moving downstream from the head, the discharge of most streams increases because tributaries and groundwater contribute water to the main channel.

discharge increases toward the mouth. This should come as no surprise because, as we move downstream, more and more tributaries contribute water to the main channel (Figure 3.8). Furthermore, in most humid regions, rainwater that soaks into the ground continually migrates into streams. Thus, as you move downstream, the stream's width, depth, and velocity change in response to the increased volume of water carried by the stream.

The observed increase in velocity that occurs downstream is primarily attributable to the greater efficiency of the larger channel in a downstream direction. In the headwaters region where the gradient is steep, the water must flow in a relatively small and often boulder-strewn channel. The small channel and rough bed create great friction and inhibit movement by scattering water in all directions. However, downstream, the material on the bed of the stream becomes much smaller, offering less resistance to flow, and the width and depth of the channel increase to accommodate the greater discharge. These factors, especially the wider and deeper channel, permit the water to flow more rapidly.

In summary, you have seen an inverse relationship between gradient and discharge. Where gradient is steep, discharge is small, and where discharge is great, gradient is small. Stated another way, a stream can maintain a higher velocity near its mouth even though it has a lower gradient than upstream because of the greater discharge, larger channel, and smoother bed.

Base Level

Streams cannot endlessly erode their channels deeper and deeper. There is a lower limit to how deep a stream can erode, and that limit is the stream's base level. **Base level** is the lowest point to which a stream can erode its channel. Two general types of base level exist. Sea level is considered the *ultimate base level*, because it is the lowest level to which stream erosion could lower the land. *Temporary*, or *local, base levels* include lakes, resistant layers of rock, and main streams that act as base level for their tributaries.

For example, when a stream enters a lake, its velocity quickly approaches zero and its ability to erode decreases. Thus, the lake prevents the stream from eroding below its level at any point upstream from the lake. However, because the outlet of the lake can cut downward and drain the lake, the lake is only a temporary hindrance to the stream's ability

to downcut its channel. In a similar manner, a layer of resistant rock acts as a temporary base level. Until the hard rock is eroded away, it will limit the amount of downcutting upstream.

Any change in base level will cause a corresponding readjustment of stream activities. When a dam is built along a stream, the reservoir that forms behind it raises the base level of the stream (Figure 3.9). Upstream from the dam the stream gradient is reduced, lowering its velocity and, hence, its sediment-transporting ability. The stream, now having too little energy to transport all of its load, will deposit sediment. This builds up its channel. This process continues until the stream again has a gradient sufficient to carry its load.

If, on the other hand, the base level were lowered, either by an uplift of the land or by a drop in the base level, the stream would readjust. The stream, now above base level, would have excess energy and would downcut its channel to establish a balance with its new base level. Erosion would progress most rapidly near the mouth, then work upstream until the stream profile was adjusted along its full length.

Work of Streams

The work of streams includes erosion, transportation, and deposition. These activities go on simultaneously in all stream channels, even though they are presented individually here.

Erosion

Erosion is the removal of rock and soil. Much of the material carried by streams reaches them through underground water, overland flow, and mass wasting. Streams also create some of their load by eroding their own channels. If a channel is made up of bedrock, most of the erosion is accomplished by the abrasive action of water armed with sediment, a process analogous to sandblasting. Pebbles caught in swirling eddies act like cutting tools and bore circular "potholes" into the channel floor. In channels made of loose material, considerable lifting and removal can be accomplished by the impact of water alone.

Transportation

Streams transport their load of sediment in three ways: (1) in solution (**dissolved load**); (2) in suspension (**suspended load**); and (3) scooting or rolling along the bottom (**bed load**).

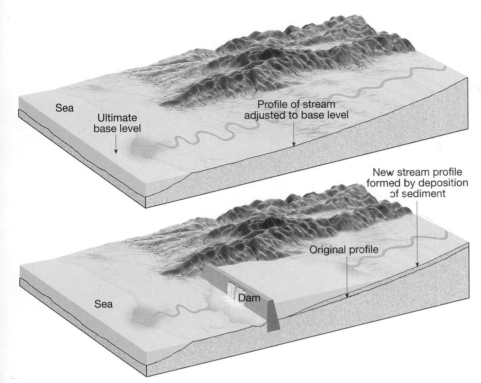

Figure 3.9 When a dam is built and a reservoir forms, the stream's base level is raised. This reduces the stream's velocity and leads to deposition and a reduction of the gradient upstream from the reservoir.

Most of the *dissolved load* is contributed by groundwater. In addition, some may be acquired as the stream dissolves material in its channel. Usually the dissolved load is expressed as parts of dissolved material per million parts of water (parts per million, or ppm). Although some rivers may have a dissolved load of 1000 ppm or more, the average amount for the world's rivers is estimated at 115 to 120 ppm. Almost 4 billion metric tons of dissolved mineral matter are supplied to the oceans each year by streams.

Most streams (but not all) carry the largest part of their load in *suspension*. The visible cloud of sediment suspended in the water is the most obvious portion of a stream's load. Usually only sand, silt, and clay can be carried this way, but during a flood larger particles are transported as well. Also, during a flood, the total quantity of material carried in suspension increases dramatically, as can be verified by anyone whose home has been a site for the deposition of this material.

A portion of a stream's load of solid material consists of sediment that is too large to be carried in suspension. These coarser particles move along the bottom of the stream and constitute the *bed load*. Unlike the suspended and dissolved loads, which are constantly in motion, the bed load is in motion only intermittently, when the force of the water is sufficient

to move the larger particles. In terms of the erosional work accomplished by a downcutting stream, the grinding action of the bed load is of great importance.

Each year, the Mississippi River transports about 750 million tons of material to the Gulf of Mexico. Of this total, it is estimated that approximately 500 million tons are carried in suspension, 200 million tons in solution, and the remaining 50 million tons as bed load; however, such proportions vary widely from stream to stream.

Streams vary in their ability to carry a load. Their ability is determined by two criteria. First, the **competence** of a stream measures the maximum size of particles it is capable of transporting. The stream's velocity determines its competence. If the velocity of a stream doubles, its competence increases four times; if the velocity triples, its competence increases nine times; and so forth. This explains how large boulders that seem immovable can be transported during a flood, which greatly increases a stream's velocity.

Second, the **capacity** of a stream is the maximum load it can carry. The capacity of a stream is directly related to its discharge. The greater the volume of water flowing in a stream, the greater is its capacity for hauling sediment.

It should now be clear why the greatest erosion and transportation of sediment occur during a flood.

Figure 3.10 The suspended load is clearly visible because it gives the flooding river a "muddy" appearance. During floods both capacity and competency increase. Therefore, the greatest erosion and sediment transport occur during these high-water periods. Here we see the sediment-filled floodwaters of Kenya's Mara River. *(Photo by Tim Davis/Tony Stone Images)*

The increase in discharge results in a greater capacity and the increase in velocity results in greater competence. With rising velocity the water becomes more turbulent, and larger and larger particles are set in motion. In just a few days, or perhaps a few hours, a flooding stream can erode and transport more sediment than during months of normal flow (Figure 3.10).

Deposition

Whenever a stream slows down, the situation reverses. As velocity falls, competence decreases and sediment begins to drop out, largest particles first. Each particle size has a *critical settling velocity*. As streamflow drops below the critical settling velocity of a certain particle size, sediment in that category begins to settle out. Thus, stream transport provides a mechanism by which solid particles of various sizes are separated. This process, called **sorting**, explains why particles of similar size are deposited together.

The well-sorted material typically deposited by a stream is called **alluvium**, the general term for any stream-deposited sediment. Many different depositional features are composed of alluvium. Some occur within stream channels, some occur on the valley floor adjacent to the channel, and some exist at the mouth of the stream.

Deltas When a stream enters the relatively still waters of an ocean or lake, its velocity drops abruptly, and the resulting deposits form a **delta**. As the delta grows, the gradient of the river continually lessens. This causes the stream to seek a shorter route to base level. Frequently the main channel divides into several smaller ones called **distributaries**. These shifting channels act in an opposite way from tributaries, *distributing* water and sediment instead of contributing it. Rather than carrying water into the main channel, distributaries carry water away from the main channel (Figure 3.11).

Many large rivers have deltas extending over thousands of square kilometers. The delta of the Mississippi River is one example. It resulted from the accumulation of huge quantities of sediment derived from the vast region drained by the river and its tributaries. Today, New Orleans rests where there was ocean less than 5000 years ago. Figure 3.12 shows that portion of the Mississippi delta that has been built

Figure 3.11 Satellite view of the Mississippi delta. Distributaries are clearly visible on the right side of the image. For the past 500 years or so, the main flow of the river has been along its present course, extending southeast from New Orleans. During that span, the delta advanced into the Gulf of Mexico at a rate of about 10 kilometers (6 miles) per century. *(Photo courtesy of NASA Headquarters)*

over the past 5000 to 6000 years. As shown, the delta is actually a series of seven coalescing subdeltas. Each formed when the river left its existing channel in favor of a shorter, more direct path to the Gulf of Mexico. The individual subdeltas interfinger and partially cover one another to produce a very complex structure. The present subdelta, called a *bird-foot* delta because of the configuration of its distributaries, has been built by the Mississippi in the last 500 years.

Natural Levees   Some rivers occupy valleys with broad, flat floors and build **natural levees** that parallel their channels on both banks (Figure 3.13). Natural levees are built by successive floods over many years. When a stream overflows its banks, its velocity immediately diminishes, leaving sediment deposited in strips bordering the channel. As the water spreads out over the valley, a lesser amount of fine sediment is deposited over the valley floor. This uneven distribution of material produces the very gentle slope of the natural levee.

The natural levees of the lower Mississippi rise 6 meters (20 feet) above the valley floor. The area behind the levee is characteristically poorly drained for the obvious reason that water cannot flow up the levee and into the river. Marshes called **back-swamps** result. A tributary stream that cannot enter a river because its levees block the way often has to flow parallel to the river until it can breach the levee. Such streams are called **yazoo tributaries** after the Yazoo River, which parallels the Mississippi for over 300 kilometers.

Artificial Levees   *Artificial Levees* are earthen mounds built on the banks of a river to increase the volume of water the channel can hold. Artificial levees are usually easy to distinguish from natural levees because their slopes are much steeper. When a river is confined by levees during periods of high water, it deposits material in its channel as the discharge diminishes. This is sediment that otherwise would have been dropped on the floodplain. Thus, each time there

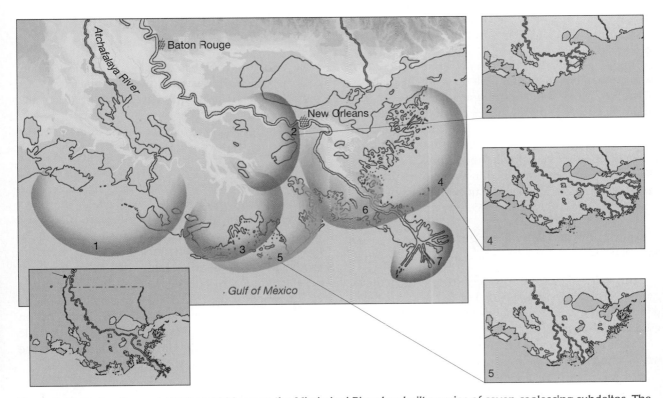

Figure 3.12   During the past 5000 to 6000 years, the Mississippi River has built a series of seven coalescing subdeltas. The numbers indicate the order in which the subdeltas were deposited. The present bird-foot delta (number 7) represents the activity of the past 500 years. Without ongoing human efforts, the present course will shift and follow the path of the Atchafalaya River. The inset on left shows the point where the Mississippi may someday break through (arrow) and the shorter path it would take to the Gulf of Mexico. *(After C. R. Kolb and J. R. Van Lopik)*

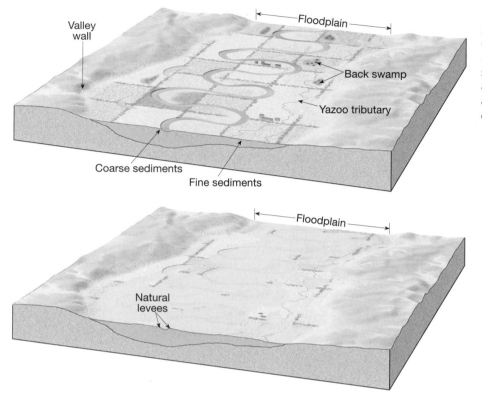

Figure 3.13 Natural levees are gently sloping deposits that are created by repeated floods. Because the ground next to the stream channel is higher then the adjacent floodplain, back swamps and yazoo tributaries may develop.

is a high flow, deposits are left on the river bed and the bottom of the channel is built up. With the buildup of the bed, less water is required to overflow the original levee. As a result, the height of the levee may have to be raised periodically to protect the floodplain. Moreover, many artificial levees are not built to withstand periods of extreme flooding. For example, levee failures were numerous in the Midwest during the summer of 1993, when the upper Mississippi and many of its tributaries experienced record floods.

Stream Valleys

Stream valleys can be divided into two general types. Narrow V-shaped valleys and wide valleys with flat floors exist as the ideal forms, with many gradations between.

Most stream valleys are much broader at the top than is the width of their channel at the bottom. This would not be the case if the only agent responsible for eroding valleys were the streams flowing through them. The sides of most valleys are shaped by a combination of weathering, overland flow, and mass wasting. In some arid regions, where downcutting is rapid and weathering is slow, and in places where rock is particularly resistant, narrow valleys may not be V-shaped but rather may have nearly vertical walls.

Narrow Valleys

The Yellowstone River provides an excellent example of a narrow valley (Figure 3.14). A narrow V-shaped valley indicates that the primary work of a stream has been downcutting toward base level. The most prominent features of a narrow valley are *rapids* and *waterfalls*. Both occur where the stream profile drops rapidly, a situation usually caused by variations in the erodibility of the bedrock into which a stream channel is cutting. Resistant beds create rapids by acting as a temporary base level upstream while allowing downcutting to continue downstream. Once erosion has eliminated the resistant rock, the stream profile smooths out again. Waterfalls are places where the stream profile makes a vertical drop.

Wide Valleys

Once a stream has cut its channel closer to base level, downward erosion becomes less dominant. At

Figure 3.14 V-shaped valley of the Yellowstone River. The rapids and waterfalls indicate that the river is vigorously downcutting. *(Photo by Art Wolfe, Inc.)*

sideways and slightly downstream. The sideways movement occurs because the maximum velocity of the stream shifts toward the outside of the bend, causing erosion of the outer bank called the *cut bank* (Figure 3.16). At the same time, the reduced current at the inside of the meander results in the deposition of coarse sediment, especially sand, called a *point bar*. Thus, by eroding its outer bank and depositing material along its inner bank, a stream moves sideways without changing its channel size.

Because of the slope of the channel, erosion is more effective on the downstream side of a meander. Therefore, in addition to migrating laterally, the bends also gradually migrate down the valley. Sometimes the downstream migration of a meander is slowed when it reaches a more resistant portion of

this point more of the stream's energy is directed from side to side. The result is a widening of the valley as the river cuts away first at one bank and then at the other (Figure 3.15).

Floodplains The side-to-side cutting of a stream eventually produces a flat valley floor, or **floodplain**. It is appropriately named because the river is confined to its channel, except during flood stage, when it overflows its banks and inundates the floodplain.

When a river erodes laterally, creating a floodplain as just described, it is called an *erosional floodplain*. Floodplains can also be depositional in nature. *Depositional floodplains* are produced by a major fluctuation in conditions, such as a change in base level. The floodplain in California's Yosemite Valley is one such feature, and was produced when a glacier gouged the former stream valley deeper by about 300 meters (1000 feet). After the glacial ice melted, the stream readjusted to its former base level by refilling the valley with alluvium.

Meanders Streams that flow on floodplains move in sweeping bends called **meanders** (Figure 3.15). Meanders continually change position by eroding

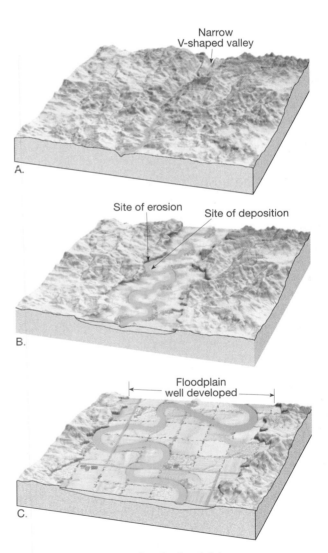

Figure 3.15 Stream eroding its floodplain.

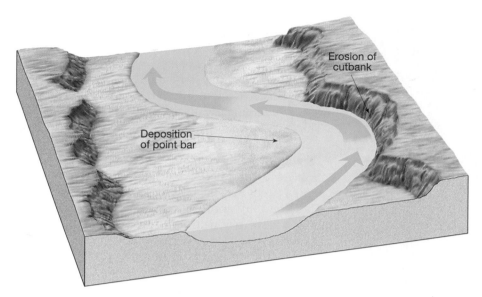

Figure 3.16 Lateral movement of meanders. By eroding its outer bank and depositing material on the inside of the bend, a stream is able to shift its channel.

the floodplain. This allows the next meander upstream to overtake it. Gradually the neck of land between the meanders is narrowed until the river breaks through the narrow neck of land to the next loop (Figure 3.17). The new, shorter channel segment is called a **cutoff** and, because of its shape, the abandoned bend is called an **oxbow lake**.

Artficial Cutoffs One method of flood control is to straighten a channel by creating *artificial cutoffs.* The idea is that by shortening the stream, the gradient and, hence, the velocity are increased. By increasing velocity, the larger discharge associated with flooding can be dispersed more rapidly.

Since the early 1930s, the Army Corps of Engineers has created many artificial cutoffs on the Mississippi for the purpose of increasing the channel efficiency and reducing the threat of flooding. In all, the river has been shortened more than 240 kilometers (150 miles). The program has been somewhat successful in reducing the height of the river in flood. Because the river's tendency to meander still exists, preventing the river from returning to its previous condition is an ongoing challenge.

Drainage Basins and Patterns

Every stream, no matter how large or small, has a **drainage basin** (Figure 3.18). This is the land area that contributes water to the stream. The drainage basin of one stream is separated from the drainage basin of another by an imaginary line called a **divide**. Divides range in scale from a ridge separating two small gullies on a hillside to a *continental divide* which splits continents into enormous drainage basins. For example, the continental divide that runs somewhat north-south through the Rocky Mountains separates the drainage that flows west to the Pacific Ocean from that which flows to the Gulf of Mexico. Although divides separate the drainage of two streams, if both streams are tributaries of the same river, they are both part of the river's drainage system.

Drainage systems are networks of streams that together form distinctive patterns. The nature of a drainage pattern can vary greatly from one type of terrain to another, primarily in response to the kinds of rock on which the streams developed or the structural pattern of faults and folds.

Certainly the most commonly encountered drainage pattern is the **dendritic pattern** (Figure 3.19A). This pattern of irregularly branching tributary streams resembles the branching pattern of a deciduous tree. In fact, the word *dendritic* means "tree-like." The dendritic pattern forms where the underlying material is relatively uniform. Because the surface material is essentially uniform in its resistance to erosion, it does not control the pattern of streamflow. Rather, the pattern is determined chiefly by the direction of slope of the land.

When streams diverge from a central area like spokes from the hub of a wheel, the pattern is said to be **radial** (Figure 3.19B). This pattern typically develops on isolated volcanic cones and domal uplifts.

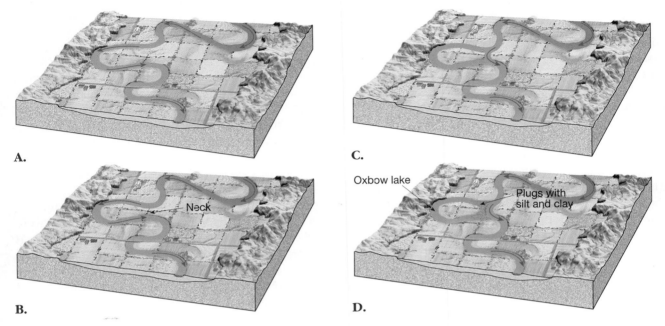

A.

B.

C.

D.

Neck

Oxbow lake

Plugs with silt and clay

Figure 3.17 Formation of a cutoff and oxbow lake.

Figure 3.19C illustrates a **rectangular pattern,** in which many right-angle bends can be seen. This pattern develops when the bedrock is crisscrossed by a series of joints and/or faults. Because these structures are eroded more easily than unbroken rock, their geometric pattern guides the directions of valleys.

Figure 3.19D illustrates a **trellis drainage pattern**, a rectangular pattern in which tributary streams are nearly parallel to one another and have the appearance of a garden trellis. This pattern forms in areas underlain by alternating bands of resistant and less resistant rock.

Figure 3.18 A *drainage basin* is the land area drained by a stream and its tributaries. *Divides* are the boundaries separating drainage basins.

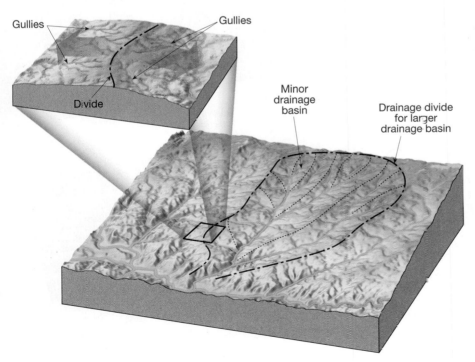

Gullies

Gullies

Divide

Minor drainage basin

Drainage divide for larger drainage basin

Stages of Valley Development

Two centuries ago, many people believed Earth to be only a few thousand years old, and that rainwater flowed down valleys that had been there since the beginning of time. After much careful observation an early geologist named James Hutton believed otherwise and proposed that streams were responsible for cutting the valleys in which they flowed. Later geologic work substantiated Hutton's proposal and further indicated that the development of stream valleys occurs in a somewhat predictable fashion. To learn about the evolution of a valley, it is helpful to divide its development into three stages: youth, maturity, and old age.

As long as the stream is downcutting, it is considered *youthful*. Rapids, an occasional waterfall, and a narrow *V*-shaped valley are all visible signs of vigorous downcutting (see Figure 3.14). Other characteristics of youth include a steep gradient, little or no floodplain, and a relatively straight course without meanders (Figure 3.20A).

When a stream reaches *maturity*, downward erosion diminishes and lateral erosion dominates. Thus, the mature stream begins to create a floodplain and meander upon it (Figure 3.20B, C). In contrast to the gradient of a youthful stream, the gradient of a mature stream is much lower and the profile is much smoother because all rapids and waterfalls have been eliminated.

A stream enters *old age* after it has cut its floodplain several times wider than its *meander belt*, which is the width of the meander (Figure 3.20D). When this stage is reached, the stream is rarely near the valley walls; hence, it ceases to significantly enlarge the floodplain. Thus, the primary work of a river in an old-age valley is the reworking of unconsolidated floodplain deposits. Because this task is easier than cutting bedrock, a stream in an old-age valley shifts more rapidly than a stream in a mature valley. Natural levees are also common features of old-age valleys and, when present, are accompanied by backswamps and yazoo tributaries.

Thus far we have assumed that the base level of a stream remains constant as a river very gradually

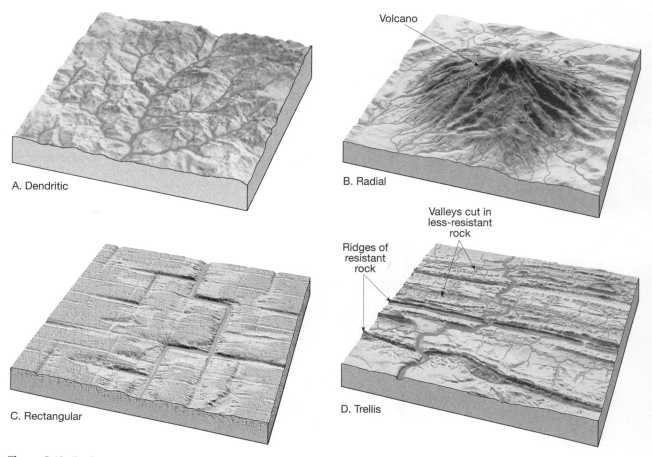

Figure 3.19 Drainage patterns. **A**. Dendritic. **B**. Radial. **C**. Rectangular. **D**. Trellis.

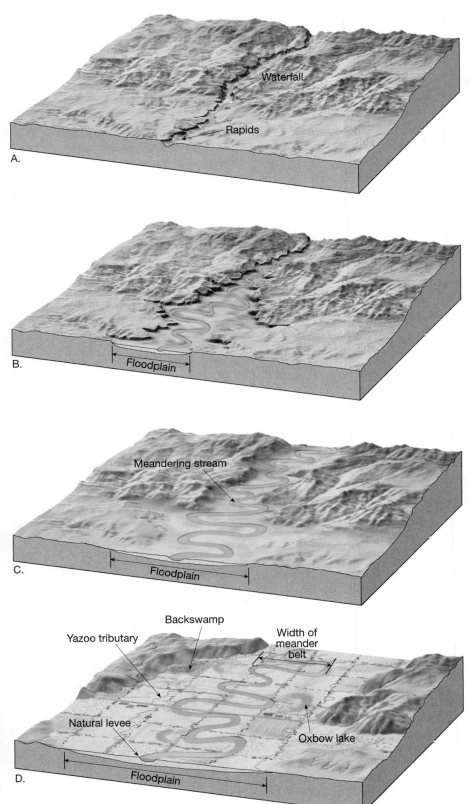

A.

B.

C.

D.

Figure 3.20 Stages of valley development. **A**. Youth. The youthful stage is characterized by down-cutting and a V-shaped valley. **B**. and **C**. Maturity. Once a stream has sufficiently lowered its gradient, it begins to erode laterally, producing a wide valley. **D**. Old age. After the valley has been cut several times wider than the width of the meander belt, it has entered old age. *(After Ward's Natural Science Establishment, Inc., Rochester, N.Y.)*

changes from youth to old age. On many occasions, however, the land is uplifted or the base level is lowered. The effect of uplifting on a youthful stream is to increase its gradient and accelerate its rate of downcutting. However, uplifting of a mature stream would cause it to abandon lateral erosion and revert to downcutting. Rivers of this type are said to be **rejuvenated** ("made young again"). The meanders stay in the same place but become deeper, and are called **incised meanders** (Figure 3.21). Mature streams may eventually readjust to uplift by cutting a new floodplain at a level below the old one. The remnants of the old higher floodplain often remain as flat surfaces called *terraces*.

Two additional points concerning valley development should be made. First, the time required for a stream to reach any given stage depends on several factors, including the erosive ability of the stream, the nature of the material through which the stream must cut, and the stream's height above base level. Consequently, a stream that starts out very near base level and has to cut only through unconsolidated sediments may reach maturity in a matter of a few hundred years. On the other hand, the Colorado River, where it is actively cutting the Grand Canyon, has retained its youthful nature for an estimated 5

million to 6 million years. Second, individual portions of a stream reach each stage at different times. Often the lower reaches of a stream attain old age while the headwaters are still youthful in character.

Water Beneath the Surface

Groundwater, water that has soaked into the ground, is one of our most important and widely available resources. Yet people's perceptions of groundwater are often unclear and incorrect. The reason is that groundwater is hidden from view except in caves and mines, and the impressions people gain from these subsurface openings are often misleading. Observations on the land surface give an impression that Earth is "solid." This view is not changed very much when we enter a cave and see water flowing in a channel that appears to have been cut into solid rock. Because of such observations many people believe that groundwater occurs only in underground "rivers." But actual rivers underground are extremely rare.

In reality, most of the subsurface environment is not "solid" at all but contains a huge volume of openings that exist as spaces between grains of soil and sediment and occur as narrow joints and fractures in bedrock. It is in these tiny openings that groundwater collects and moves.

The Importance of Groundwater

The importance of groundwater can be demonstrated by comparing its volume with the quantity of water in other parts of the hydrosphere. Of all the world's water, only about six-tenths of 1 percent occurs underground. Nevertheless, the amount of water stored in the rocks and sediments beneath Earth's surface is vast. When the oceans are excluded and only sources of freshwater are considered, the significance of groundwater becomes more apparent. Table 3.1 contains estimates of the distribution of freshwater in the hydrosphere. Clearly, the largest volume occurs as glacial ice. Second in rank is groundwater, with slightly more than 14 percent of the total. However, when ice is excluded and just liquid water is considered, more than 94 percent is groundwater. Without question, *groundwater represents the largest reservoir of freshwater that is readily available to humans*. Its value in terms of economics and human well-being is incalculable.

In many parts of the world, wells and springs provide the water needed not only for great numbers of people, but also for crops, livestock, and industry.

Figure 3.21 A close-up view of incised meanders of the Colorado River in Canyonlands National Park, Utah. The river began downcutting because of the uplift of the Colorado Plateau. *(Photo by Michael Collier)*

Table 3.1 ■ Freshwater of the hydrosphere			
Parts of the Hydrosphere	Volume of Freshwater (km³)	(ml³)	Share of Total Volume of Freshwater (percent)
Ice sheets and glaciers	24,000,000	5,800,000	84.945
Groundwater	4,000,000	960,000	14.158
Lakes and reservoirs	155,000	37,000	0.549
Soil moisture	83,000	20,000	0.294
Water vapor in the atmosphere	14,000	3,400	0.049
River water	1,200	300	0.004
Total	28,253,200	6,820,700	100.000

Source: U.S. Geological Survey Water Supply Paper 2220, 1987.

In the United States, groundwater is the source of about 40 percent of the water used for all purposes except hydroelectric power generation and power plant cooling. Groundwater provides drinking water for more than 50 percent of the population, as well as 40 percent of the water for irrigation and 26 percent of industry's needs. In some areas, however, overuse of this basic resource has caused serious problems, including streamflow depletion, land subsidence, and increased pumping costs. In addition, groundwater contamination due to human activities is a real and growing threat in many places.

Groundwater's Geological Roles

Geologically, groundwater is important as an erosional agent. The dissolving action of groundwater slowly removes rock, allowing surface depressions known as sinkholes to form as well as creating subterranean caverns (Figure 3.22). Groundwater is also an equalizer of streamflow. Much of the water that flows in rivers is not direct runoff from rain and snowmelt. Rather, a large percentage of precipitation soaks in and then moves slowly underground to stream channels. Groundwater is thus a form of storage that sustains streams during periods when rain does not fall. When we see water flowing in a river during a dry period, it is water from rain that fell at some earlier time and was stored underground.

Distribution and Movement of Groundwater

When rain falls, the water may run off immediately, evaporate, be taken up by plants and transpired back into the air, or soak into the ground. This last path is the primary source of practically all subsurface water. The amount of water that takes each of these paths can vary considerably. For any location, the steepness of the slope, the nature of the surface material, the intensity of the rainfall, and the type and amount of vegetation are all influential factors. Heavy rains that fall on steep slopes underlain by impervious materials will obviously result in a high percentage of the water running off. Conversely, a gentle, steady rain that falls on more gradual slopes consisting of materials more easily penetrated by water would result in a much larger percentage of the water soaking into the ground.

Distribution

Some of the water that soaks in does not travel far, because it is held by molecular attraction as a surface film on soil particles. A portion of this moisture evaporates back into the atmosphere. Much of the remainder is used by plants between rains. But water that is not held near the surface penetrates downward until it reaches a zone where all the open spaces in sediment and rock are completely filled with water. This is the **zone of saturation**. Water within it is called **groundwater**. The upper limit of this zone is known as the **water table**. The area above the water table where the soil, sediment, and rock are not saturated is called the **zone of aeration** (Figure 3.23). The open spaces here are filled mainly with air.

The water table is rarely level as we might expect a table to be. Instead, its shape is usually a subdued replica of the surface, reaching its highest elevations beneath hills and decreasing in height toward valleys (Figure 3.23). When you see a wetland (swamp), it

Figure 3.22 The dissolving action of groundwater creates caverns. Calsbad Caverns National Park, New Mexico. *(Photo by Harold Hoffman/Photo Researchers, Inc.)*

tells you that the water table is right at the surface. Lakes and streams generally occupy areas where the land surface is below the water table.

Although many factors contribute to the irregular surface of the water table, the most important cause is that groundwater moves very slowly. For this reason, the water tends to "pile up" beneath hills. If rainfall were to cease completely, these water "hills" would slowly subside and gradually approach the level of the valleys. However, new supplies of

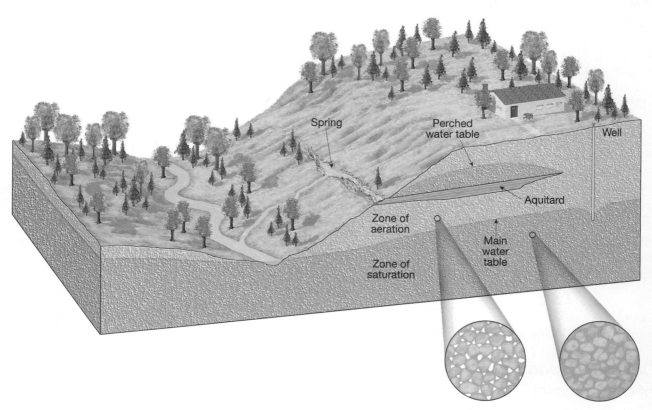

Figure 3.23 This diagram illustrates the relative positions of many features associated with subsurface water.

76

rainwater are usually added often enough to prevent this. Nevertheless, in times of extended drought, the water table may drop enough to fall below otherwise productive wells.

Movement

Groundwater exists in the pore spaces and fractures in rock and sediment. Consequently, the movement of most groundwater is exceedingly slow, from pore to pore.

The nature of subsurface materials strongly influences the rate of groundwater movement and the amount of groundwater that can be stored. Two factors are especially important—porosity and permeability.

Porosity The quantity of groundwater that can be stored depends on the **porosity** of the material, which is the percentage of the total volume of rock or sediment that consists of pore spaces. Voids most often are spaces between sedimentary particles, but also common are joints and faults, and cavities formed by the dissolving of soluble rocks such as limestone.

Permeability Porosity alone cannot measure a material's capacity to yield groundwater. Rock or sediment may be very porous yet still not allow water to move through it. The pores must be *connected* to allow water to move, and they must be *large enough* to allow movement. Thus, the **permeability** of a material, its ability to transmit water, is also very important.

Groundwater moves by twisting and turning through small interconnected openings. The smaller the pore spaces, the slower the groundwater moves. Clay exemplifies this circumstance. Although the ability of clay to store water is often high, its pore spaces are so small that water is unable to move through it. Thus, we say that clay is *impermeable*.

Aquitards and Aquifers Impermeable layers that hinder or prevent water movement are termed **aquitards**. Clay is a good example. On the other hand, larger particles, such as sand or gravel, have larger pore spaces. Therefore, the water can move with relative ease. Such permeable rock strata or sediments that transmit groundwater freely are called **aquifers** ("water carriers"). Sands and gravels are common examples. Aquifers are important because they are the water-bearing layers that well drillers seek. A good aquifer can provide water for numerous wells, homes, cities, and even for irrigation and industries.

Springs

Springs have aroused the curiosity and wonder of people for thousands of years. The fact that springs were (and to some people still are) rather mysterious phenomena is not difficult to understand, for here is water flowing freely from the ground in all kinds of weather in seemingly inexhaustible supply but with no obvious source. Today we know that the source of springs is water from the zone of saturation and that the ultimate source of this water is precipitation.

Whenever the water table intersects Earth's surface, a natural outflow of groundwater results, which we call a **spring**. Many circumstances create springs. Springs such as the one pictured in Figure 3.24 form when an aquitard blocks the downward movement of groundwater and forces it to move laterally. Where the permeable bed (aquifer) outcrops, a spring or several of springs result.

Another situation that can produce a spring is illustrated in Figure 3.23. Here an aquitard is situated above the main water table. As water percolates downward, a portion of it is intercepted by the aquitard, thereby creating a localized zone of saturation and a *perched water table*.

Springs, however, are not confined to places where a perched water table creates a flow at the surface. Many geological situations lead to the formation of springs.

Figure 3.24 Minnie Miller Springs, Thousand Springs Preserve, along the Snake River near Hagerman, Idaho. *(Photo by William H. Mullins/Photo Researchers, Inc.)*

Hot Springs and Geysers

By definition, the water in **hot springs** is 6° to 9°C (10° to 15°F) warmer than the mean annual air temperature for the localities where they occur. In the United States alone, there are well over 1000 such springs.

Temperatures in deep mines and oil wells usually rise with an increasing depth an average of about 2°C per 100 meters (1°F per 100 feet). Therefore, when groundwater circulates at great depths, it becomes heated. If it rises to the surface, the water may emerge as a hot spring. The water of some hot springs in the eastern United States is heated in this manner. However, the great majority (over 95 percent) of the hot springs (and geysers) in the United States are found in the West. The reason for such a distribution is that the source of heat for most hot springs is cooling igneous rock, and it is in the West that igneous activity has occurred most recently.

Geysers are intermittent hot springs or fountains in which columns of water are ejected with great force at various intervals, often rising 30 to 60 meters (100 to 200 feet) into the air. After the jet of water ceases, a column of steam rushes out, usually with a thunderous roar. Perhaps the most famous geyser in the world is Old Faithful in Yellowstone National Park, which erupts about once each hour (Figure 3.25). Geysers are also found in other parts of the world, notably New Zealand and Iceland. In fact, the Icelandic word *geysa*, to gush, gives us the name *geyser*.

Geysers occur where extensive underground chambers exist within hot igneous rocks. As relatively cool groundwater enters the chambers, it is heated by the surrounding rock. At the bottom of the chamber, the water is under great pressure because of the weight of the overlying water. This great pressure prevents the water from boiling at the normal surface temperature of 100°C (212°F). For example, at the bottom of a 300-meter (1000-foot) chamber, water must attain a temperature of nearly 230°C (450°F) before it will boil. The heating causes the water to expand, with the result that some is forced out at the surface. This loss of water reduces the pressure on the remaining water in the chamber. The reduced pressure lowers the boiling point and a portion of the water deep within the chamber quickly turns to steam and causes the geyser to erupt. Following the eruption, cool groundwater again seeps into the chamber and the cycle begins anew.

Figure 3.25 A wintertime eruption of Old Faithful in Yellowstone National Park. One of the world's most famous geysers, it emits as much as 45,000 liters (12,000 gallons) of hot water and stream about once each hour. *(Photo by Dallas and John Heaton/WestLight)*

Wells

The most common method for removing groundwater is the **well**, a hole bored into the zone of saturation (see Figure 3.23). Wells serve as small reservoirs into which groundwater migrates and from which it can be pumped to the surface. The use of wells dates back many centuries and continues to be an important method of obtaining water. By far the single greatest use of this water in the United States is irrigation for agriculture. More than 65 percent of the groundwater used each year is for this purpose. Industrial uses rank a distant second, followed by the amount used in city water systems and rural homes.

The water table level may fluctuate considerably during the course of a year, dropping during dry seasons and rising following periods of precipitation. Therefore, to ensure a continuous supply of water, a well must penetrate below the water table. Whenever water is withdrawn from a well, the water table around the well is lowered. This effect, termed **drawdown**, decreases with increasing distance from the well. The result is a depression in the water table, roughly conical in shape, known as a **cone of**

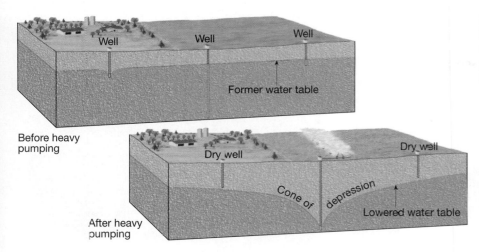

Figure 3.26 A cone of depression in the water table often forms around a pumping well. If heavy pumping lowers the water table, some wells may be left dry.

depression (Figure 3.26). For most small domestic wells, the cone of depression is negligible. However, when wells are used for irrigation or for industrial purposes, the withdrawal of water can be great enough to create a very wide and steep cone of depression that may substantially lower the water table in an area and cause nearby shallow wells to become dry. Figure 3.26 illustrates this situation.

Artesian Wells

In most wells, water cannot rise on its own. If water is first encountered at 30 meters depth, it remains at that level, fluctuating perhaps a meter or two with seasonal wet and dry periods. However, in some wells, water rises, sometimes overflowing at the surface.

The term **artesian** is applied to any situation in which groundwater rises in a well above the level where it was initially encountered. For such a situation to occur, two conditions must exist (Figure 3.27): (1) water must be confined to an aquifer that is inclined so that one end is exposed at the surface, where it can receive water; and (2) aquitards both above and below the aquifer, must be present to prevent the water from escaping. When such a layer is tapped, the pressure created by the weight of the water above will force the water to rise. If there were no friction, the water in the well would rise to the level of the water at the top of the aquifer. However, friction reduces the height of this pressure surface. The greater the distance from the recharge area (area where water enters the inclined aquifer), the greater the friction and the smaller the rise of water.

In Figure 3.27, Well 1 is a *nonflowing artesian well*, because at this location the pressure surface is

below ground level. When the pressure surface is above the ground and a well is drilled into the aquifer, a *flowing artesian well* is created (Well 2, Figure 3.27).

Artesian systems act as conduits, transmitting water from remote areas of recharge great distances to the points of discharge. In this manner, water that fell in central Wisconsin years ago is now taken from the ground and used by communities many kilometers away in Illinois. In South Dakota, such a system brings water from the Black Hills in the west, eastward across the state.

On a different scale, city water systems may be considered as examples of artificial artesian systems. The water tower, into which water is pumped, may be considered the area of recharge, the pipes the confined aquifer, and the faucets in homes the flowing artesian wells.

Environmental Problems of Groundwater

As with many of our valuable natural resources, groundwater is being exploited at an increasing rate. In some areas, overuse threatens the groundwater supply. In other places, groundwater withdrawal has caused the ground and everything resting upon it to sink. Still other localities are concerned with the possible contamination of their groundwater supply.

Treating Groundwater as a Nonrenewable Resource

For many people, groundwater appears to be an endlessly renewable resource, for it is continually replenished by rainfall and melting snow. But in

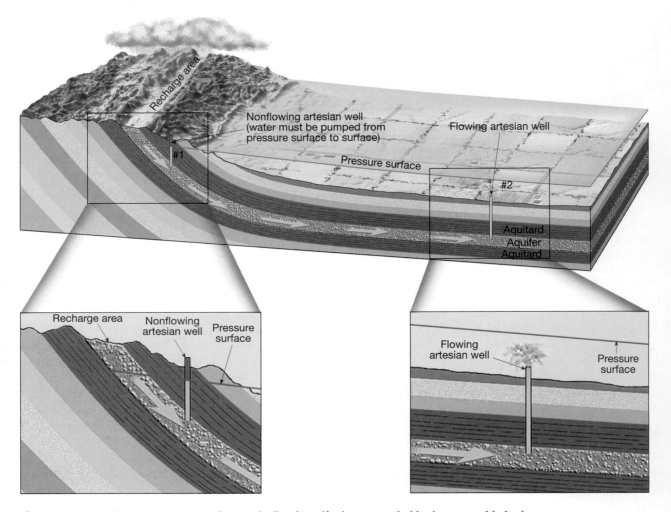

Figure 3.27 Artesian systems occur when an inclined aquifer is surrounded by impermeable beds.

some regions, groundwater has been and continues to be treated as a *nonrenewable* resource. Where this occurs, the amount of water available to recharge the aquifer is significantly less than the amount being withdrawn. The High Plains, a relatively dry region that extends from South Dakota to western Texas, provides one example. Here an extensive agricultural economy is largely dependent on irrigation. As a result, there are an estimated 168,000 wells being used to irrigate more than 65,000 square kilometers (16 million acres) of land. In the southern part of this region, which includes the Texas panhandle, the natural recharge of the aquifer is very slow and the problem of declining groundwater levels is acute. In fact, in years of average or below-average precipitation, recharge is negligible because all or nearly all of the meager rainfall is returned to the atmosphere by evaporation and transpiration.

Therefore, where intense irrigation has been practiced for an extended period, depletion of groundwater can be severe. Declines in groundwater levels at rates as great as 1 meter per year have led to an overall drop in the water table of between 15 and 60 meters (50 and 200 feet) in some areas. Under these circumstances, it can be said that the groundwater is literally being "mined." Even if pumping were to cease immediately, it could take hundreds or thousands of years for the groundwater to be fully replenished.

Land Subsidence Caused by Groundwater Withdrawal

As you shall see later in this chapter, surface subsidence can result from natural processes related to groundwater. However, the ground may also sink when water is pumped from wells faster than natural recharge processes can replace it. This effect is particularly pronounced in areas underlain by thick layers of loose sediments. As water is withdrawn, the weight of the overburden packs the sediment grains more tightly together and the ground subsides.

Many areas can be used to illustrate such land subsidence. A classic example in the United States occurred in the San Joaquin Valley of California (Figure 3.28). This important agricultural region relies heavily on irrigation. Land subsidence due to groundwater withdrawal began in the valley in the mid-1920s and locally exceeded 8 meters (28 feet) by 1970. Then, because of the importation of surface water and a decrease in groundwater pumping, water levels in the aquifer recovered and subsidence ceased.

However, during a drought in 1976–1977, heavy groundwater pumping led to renewed subsidence. This time, water levels dropped at a much faster rate than during the previous period because of the reduced storage capacity caused by earlier compaction of material in the aquifer. In all, more than 13,400

square kilometers (5200 square miles) of irrigable land, one-half the entire valley, were affected by subsidence. Damage to structures, including highways, bridges, water lines, and wells, was extensive. Because subsidence changed the gradients of some streams, flooding also became a costly problem. Many other examples of land subsidence due to groundwater pumping occur in the United States and elsewhere in the world.

Groundwater Contamination

The pollution of groundwater is a serious matter, particularly in areas where aquifers provide a large part of the water supply. One common source of groundwater pollution is sewage emanating from an ever-increasing number of septic tanks. Other sources are inadequate or broken sewer systems, and farm wastes.

If sewage water, which is contaminated with bacteria, enters the groundwater system, it may become purified through natural processes. The harmful bacteria may be mechanically filtered by the sediment through which the water percolates, destroyed by chemical oxidation, and/or assimilated by other organisms. For purification to occur, however, the aquifer must be of the correct composition. For example, extremely permeable aquifers (such as highly fractured crystalline rock, coarse gravel, or cavernous limestone) have such large openings that contaminated groundwater may travel long distances without being cleansed. In this case, the water flows too rapidly and is not in contact with the surrounding material long enough for purification to occur. This is the problem at Well 1 in Figure 3.29A.

Conversely, when the aquifer consists of sand or permeable sandstone, the water can sometimes be purified after traveling only a few dozen meters through it. The openings between sand grains are large enough to permit water movement, yet the movement of the water is slow enough to allow ample time for its purification (Well 2, Figure 3.29B).

Because groundwater movement is usually slow, polluted water may go undetected for a long time. In fact, most contamination is discovered only after drinking water has been affected and people become ill. By this time, the volume of polluted water could be quite large, and even if the source of contamination is removed immediately, the problem is not solved. Although the sources of groundwater contamination are numerous, the solutions are relatively few.

Once the source of the problem has been identified and eliminated, the most common practice is simply to abandon the water supply and allow the

Figure 3.28 The shaded area on the map shows California's San Joaquin Valley. The marks on the utility pole in the photo indicate the level of the surrounding land in preceding years. Between 1925 and 1975, this part of the San Joaquin Valley subsided almost 9 meters (30 feet) because of the withdrawal of groundwater and the resulting compaction of sediments. *(Photo courtesy of U.S. Geological Survey, U.S. Department of the Interior.)*

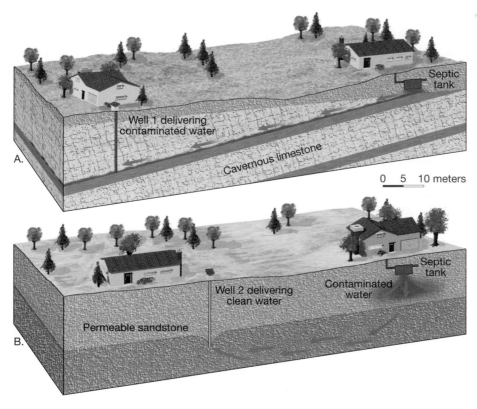

Figure 3.29 A. Although the contaminated water has traveled more than 100 meters before reaching Well 1, the water moves too rapidly through the cavernous limestone to be purified. **B**. As the discharge from the septic tank percolates through the permeable sandstone, it is purified in a relatively short distance.

pollutants to be flushed away gradually. This is the least costly and easiest solution, but the aquifer must remain unused for many years. To accelerate this process, polluted water is sometimes pumped out and treated. Following removal of the tainted water, the aquifer is allowed to recharge naturally or, in some cases, the treated water or other fresh water is pumped back in. This process is costly, time consuming, and it may be risky because there is no way to be certain that all of the contamination has been removed. Clearly, the most effective solution to groundwater contamination is prevention.

The Geologic Work of Groundwater

Groundwater dissolves rock. This fact is key to understanding how caverns and sinkholes form. Because soluble rocks, especially limestone, underlie millions of square kilometers of Earth's surface, it is here that groundwater carries on its important role as an erosional agent. Limestone is nearly insoluble in pure water, but is quite easily dissolved by water containing small quantities of carbonic acid, and most groundwater contains this acid. It forms   because rainwater readily dissolves carbon dioxide from the air and from decaying plants. Therefore, when groundwater comes in contact with limestone, the carbonic acid reacts with calcite (calcium carbonate) in the rocks to form calcium bicarbonate, a soluble material that is then carried away in solution.

Caverns

Among the most spectacular results of groundwater's erosional handiwork are limestone **caverns**. In the United States alone about 17,000 caves have been discovered. Although most are relatively small, some have spectacular dimensions. Carlsbad Caverns in southeastern New Mexico and Mammoth Cave in Kentucky are famous examples. One chamber in Carlsbad Caverns has an area equivalent to 14 football fields and enough height to accommodate the U.S. Capitol Building. At Mammoth Cave, the total length of interconnected caverns extends for more than 540 kilometers (340 miles).

Most caverns are created at or just below the water table in the zone of saturation. Here acidic groundwater follows lines of weakness in the rock, such as joints and bedding planes. As time passes, the dissolving process slowly creates cavities and gradually enlarges them into caverns. Material that is dissolved by the groundwater is eventually discharged into streams and carried to the ocean.

Certainly the features that arouse the greatest curiosity for most cavern visitors are the stone formations that give some caverns a wonderland appearance. These are not erosional features, like the caverns in which they reside, but depositional features. They are created by the seemingly endless dripping of water over great spans of time. The calcite that is left behind produces the limestone we call travertine. These cave deposits, however, are also commonly called *dripstone*, an obvious reference to their mode of origin.

Although the formation of caverns takes place in the zone of saturation, the deposition of dripstone is not possible until the caverns are above the water table, in the zone of aeration. This commonly occurs as nearby streams cut their valleys deeper, lowering the water table as the elevation of the rivers drops. As soon as the chamber is filled with air, the conditions are right for the decoration phase of cavern building to begin.

Of the various dripstone features found in caverns, perhaps the most familiar are **stalactites**. These icicle-like pendants hang from the ceiling of the cavern and form where water seeps through cracks above. When water reaches air in the cave, some of the dissolved carbon dioxide escapes from the drop and calcite begins to precipitate. Deposition occurs as a ring around the edge of the water drop. As drop after drop follows, each leaves an infinitesimal trace of calcite behind, and a hollow limestone tube is created. Water then moves through the tube, remains suspended momentarily at the end, contributes a tiny ring of calcite, and falls to the cavern floor. The stalactite just described is appropriately called a *soda straw* (Figure 3.30A). Often the hollow tube of the soda straw becomes plugged or its supply of water increases. In either case, the water is forced to flow and deposit along the outside of the tube. As deposition continues, the stalactite takes on the more common conical shape.

Formations that develop on the floor of a cavern and reach upward toward the ceiling are called **stalagmites** (Figure 3.30B). The water supplying the calcite for stalagmite growth falls from the ceiling and splatters over the surface. As a result, stalagmites do not have a central tube and are usually more massive in appearance and more rounded on their upper ends than are stalactites. Given enough time, a downward-growing stalactite and an upward-growing stalagmite may join to form a *column*.

Karst Topography

Many areas of the world have landscapes that, to a large extent, have been shaped by the dissolving power of groundwater. Such areas are said to exhibit **karst topography**, named for the **Krs** region in the border area between Slovenia (formerly a part of

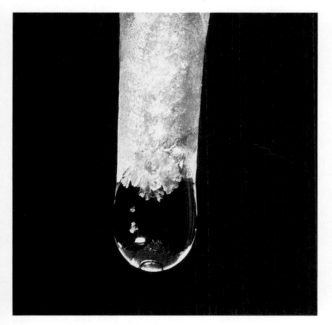

A.

B.

Figure 3.30   A. A "live" solitary soda straw stalactite. *(Photo by Clifford Stroud, National Park Service)* **B**. Stalagmites grow upward from the floor of a cavern. *(Photo by Ballman, Wider, Downey/K.R. Downey Photography)*

Yugoslavia) and Italy where such topography is strikingly developed. In the United States, karst landscapes occur in many areas that are underlain by limestone, including portions of Kentucky, Tennessee, Alabama, southern Indiana, and central and northern Florida. Generally, arid and semiarid areas do not develop karst topography because there is insufficient groundwater. When solution features exist in such regions, they are likely to be remnants of a time when rainfall was more abundant.

Figure 3.31 This small sinkhole formed suddenly in 1991 when the roof of a cavern collapsed, destroying this home in Frostproof, Florida. *(Photo by* St. Petersburg Times/*Liaison Agency, Inc.)*

Karst areas typically have irregular terrain punctuated with many depressions called **sinkholes** or, simply, **sinks**. In the limestone areas of Florida, Kentucky, and southern Indiana, there are literally tens of thousands of these depressions varying in depth from just a meter or two to a maximum of more than 50 meters (Figure 3.31).

Sinkholes commonly form in one of two ways. Some develop gradually over many years without any physical disturbance to the rock. In these situations, the limestone immediately below the soil is dissolved by downward-seeping rainwater that is freshly charged with carbon dioxide. These depressions are usually not deep and are characterized by relatively gentle slopes. By contrast, sinkholes can also form suddenly and without warning when the roof of a cavern collapses under its own weight. Typically, the depressions created in this manner are steep-sided and deep. When they form in populous areas, they may represent a serious geologic hazard (Figure 3.31).

In addition to a surface pockmarked by sinkholes, karst regions characteristically show a striking lack of surface drainage (streams). Following a rainfall, runoff is quickly funneled below ground through sinks. It then flows through caverns until it finally reaches the water table. Where streams do exist at the surface, their paths are usually short. The names of such streams often give a clue to their fate. In the Mammoth Cave area of Kentucky, for example, there is Sinking Creek, Little Sinking Creek, and Sinking Branch. Some sinkholes become plugged with clay and debris, creating small lakes or ponds.

The Chapter in Review

1. *Mass wasting* is the downslope movement of rock and soil under the direct influence of gravity. *Gravity* is the controlling force of mass wasting, but *water often influences mass wasting* by saturating the pore spaces and destroying the cohesion between particles. *Oversteepening* of slopes can trigger mass wasting.

2. The *water cycle* describes the continuous interchange of water among the oceans, atmosphere, and continents. Powered by energy from the Sun, it is a global system in which the atmosphere provides the link between the oceans and continents. The processes involved in the water cycle include: *precipitation, evaporation, infiltration* (the movement of water into rocks or soil through cracks and pore spaces), *runoff* (water that flows over the land, rather than infiltrating into the ground), and *transpiration* (the release of water vapor to the atmosphere by plants).

3. The factors that determine a stream's *velocity* are *gradient*

(slope of the stream channel), *shape, size,* and *roughness* of the channel, and the stream's *discharge* (amount of water passing a given point per unit of time). Most often, the gradient and roughness of a stream decrease downstream; whereas width, depth, discharge, and velocity increase.

4. The two general types of *base level* (the lowest point to which a stream may erode its channel) are (a) *ultimate base level,* and (b) *temporary,* or *local base level.* Any change in base level will cause a stream to adjust and establish a new balance. Lowering base level will cause a stream to erode, while raising base level results in deposition of material in the channel.

5. The work of a stream includes *erosion, transportation* (as *dissolved load, suspended load,* and *bed load*), and whenever a stream's velocity decreases, *deposition.*

6. Although many gradations exist, the two general types of stream valleys are (a) *narrow V-shaped valleys* and (b) *wide valleys with flat floors*. Because the dominant activity is downcutting toward base level, narrow valleys often contain *waterfalls* and *rapids*. When a stream has cut its channel closer to base level, its energy is directed from side to side and erosion produces a flat valley floor, or *floodplain*. Streams that flow on floodplains often move in sweeping bends called *meanders*. Widespread meandering may result in shorter channel segments, called *cutoffs*, and abandoned bends, called *oxbow lakes*.

7. Common *drainage patterns* produced by streams include (a) *dendritic*, (b) *radial*, (c) *rectangular*, and, (d) *trellis*.

8. As a resource, *groundwater* represents the largest reservoir of freshwater that is readily available to humans. Geologically, the dissolving action of groundwater produces caves and sinkholes. Groundwater is also an equalizer of streamflow.

9. Groundwater is that water which occupies the pore spaces in sediment and rock in a zone beneath the surface, called the *zone of saturation*. The upper limit of this zone is the *water table*. The *zone of aeration* is above the water table where the soil, sediment, and rock are not saturated. Groundwater generally moves within the zone of saturation.

The quantity of water that can be stored depends on the *porosity* (the volume of open spaces) of the material. The *permeability* (the ability to transmit a fluid through interconnected pore spaces) of a material is a key factor affecting the movement of groundwater.

10. Phenomena that involve groundwater include springs, wells, caverns, and sinkholes. *Springs* occur whenever the water table intersects the land surface and a natural flow of groundwater results. *Wells*, openings bored into the zone of saturation, withdraw groundwater and create roughly conical depressions in the water table, known as *cones of depression*. *Artesian wells* occur when water rises above the level where it was initially encountered. Most *caverns* form in limestone at or below the water table when acidic groundwater dissolves rock. *Karst topography* exhibits an irregular terrain punctuated with many depressions, called *sinkholes*.

11. Some of the current environmental problems involving groundwater include (a) *overuse* by intense irrigation, (b) *land subsidence* caused by groundwater withdrawal, and (c) *contamination*.

Key Terms

alluvium (p. 66)

aquitards (p. 77)

aquifer (p. 77)

artesian well (p. 79)

backswamp (p. 67)

base level (p. 64)

bed load (p. 64)

capacity (p. 65)

cavern (p. 82)

competence (p. 65)

cone of depression (p. 78)

cutoff (p. 70)

delta (p. 66)

dendritic pattern (p. 70)

discharge (p. 63)

dissolved load (p. 64)

distributary (p. 66)

divide (p. 70)

drainage basin (p. 70)

drawdown (p. 78)

floodplain (p. 69)

geyser (p. 78)

gradient (p. 62)

groundwater (p. 75)

hot spring (p. 78)

incised meander (p. 74)

karst topography (p. 83)

mass wasting (p. 58)

meander (p. 69)

natural levee (p. 67)

oxbow lake (p. 70)

permeability (p. 77)

porosity (p. 77)

radial pattern (p. 71)

rectangular pattern (p. 71)

rejuvenation (p. 74)

sinkhole (sink) (p. 84)

sorting (p. 66)

spring (p. 77)

stalactite (p. 83)

stalagmite (p. 83)

suspended load (p. 64)

transpiration (p. 61)

trellis drainage pattern (p. 71)

water cycle (p. 60)

water table (p. 75)

well (p. 78)

yazoo tributary (p. 67)

zone of aeration (p. 75)

zone of saturation (p. 75)

Questions for Review

1. Describe the role of mass wasting in the formation of most landforms.

2. Describe the movement of water through the water cycle. Once precipitation has fallen on land, what paths might it take?

3. A stream starts out 2000 meters above sea level and travels 250 kilometers to the ocean. What is its average gradient in meters per kilometer?

4. Suppose that the stream mentioned in Question 3 developed extensive meanders so that its course was lengthened to 500 kilometers. Calculate its new gradient. How does meandering affect gradient?

5. When the discharge of a stream increases, what happens to the stream's velocity?

6. Define *base level*. Name the main river in your area. For what streams does it act as base level? What is the base level for the Mississippi River? The Missouri River?

7. In what three ways does a stream transport its load?

8. If you collect a jar of water from a stream, what part of its load will settle to the bottom of the jar? What portion will remain in the water?

9. Differentiate between competency and capacity.

10. Why must the height of many artificial levees be increased periodically?

11. What is an artificial cutoff? What is its purpose?

12. What is a divide?

13. Each of the following statements refers to a particular drainage pattern. Identify the pattern.

    a. Streams diverge from a central high area such as a volcano.

    b. Streams form a branching, "treelike" pattern.

    c. A pattern that develops when bedrock is crisscrossed by joints and faults.

14. Rivers often are used as political boundaries. Do mature and old-age streams make good political boundaries? Explain.

15. What percentage of freshwater is groundwater? (See Table 3.1) If glacial ice is excluded and only liquid freshwater is considered, about what percentage is groundwater?

16. Geologically, groundwater is important as an erosional agent. Name another significant geological role of groundwater.

17. Define groundwater and relate it to the water table.

18. How do porosity and permeability differ?

19. What is the source of heat for most hot springs and geysers? How is this reflected in the distribution of these features?

20. What is meant by the term *artesian*? Under what circumstances do artesian wells form?

21. What problem is associated with the pumping of groundwater for irrigation in the southern part of the High Plains?

22. Briefly explain what happened in the San Joaquin Valley of California as the result of excessive groundwater withdrawal.

23. Which aquifer would be most effective in purifying polluted groundwater: one composed mainly of coarse gravel, one consisting of sand, or one in cavernous limestone?

24. List two conditions required for the development of karst topography.

25. Differentiate between stalactites and stalagmites. How do these features form?

Testing What You Have Learned

Multiple-Choice Questions

1. Which one of the following is NOT a form of mass wasting?

    a. rockslide   **c.** earthflow   **e.** slump

    b. mudflow   **d.** deflation

2. Which one of the following is NOT a process included in the water cycle?

    a. precipitation   **c.** evaporation   **e.** runoff

    b. erosion   **d.** infiltration

3. The single most important agent sculpturing Earth's land surface is _____

    a. running water.   **c.** wind.   **e.** groundwater.

    b. ice sheets.   **d.** ocean waves.

4. The ability of a stream to erode is directly related to its _____

    a. load.   **c.** velocity.   **e.** temperature.

    b. depth.   **d.** width.

5. The lowest level to which a stream can erode its channel is the stream's _____

    a. meander.   **c.** gradient.   **e.** base level.

    b. discharge.   **d.** divide.

6. Which of the following is NOT a way a stream transports its load of sediment?

    a. abrasion   **c.** solution   **e.** both a. and c.

    b. suspension   **d.** rolling

7. When a stream slows down, its competence is reduced and sediment begins to be _____
   - **a.** eroded.
   - **b.** transported.
   - **c.** suspended.
   - **d.** deposited.
   - **e.** abraded.

8. Which one of the following features would least likely be found in a wide valley?
   - **a.** waterfall
   - **b.** floodplain
   - **c.** oxbow lake
   - **d.** meander
   - **e.** cutoff

9. The upper limit of the zone of saturation is called the _____
   - **a.** aquitard.
   - **b.** water table.
   - **c.** aquifer.
   - **d.** drawdown.
   - **e.** zone of aeration.

10. Landscapes that have been shaped by the dissolving power of groundwater are said to exhibit this type of topography.
    - **a.** alluvial
    - **b.** fluvial
    - **c.** karst
    - **d.** eolian
    - **e.** glacial

Fill-In Questions

11. _____ is the controlling force of all forms of mass wasting.

12. The unending circulation of Earth's water supply is called the _____.

13. A stream's _____ is the slope of its channel expressed as the vertical drop of the stream over a specified distance.

14. The drainage basin of one stream is separated from the drainage basin of another by an imaginary line called a _____.

15. Without question, _____ represents the largest reservoir of freshwater that is readily available to humans.

True/False Questions

16. The combined effects of mass wasting and running water produce stream valleys. _____

17. A lake prevents a stream from eroding below its level at any point upstream from the lake. _____

18. Most streams, but not all, carry the largest part of their load in solution. _____

19. As the velocity of a stream decreases, its competence is reduced and sediment begins to be deposited, smallest particles first. _____

20. A steady rain that falls on a gentle slope composed of permeable materials results in a high percentage of infiltration. _____

Answers:

1. d; 2. b; 3. a; 4. c; 5. e; 6. a; 7. d; 8. a; 9. b; 10. c; 11. Gravity; 12. water cycle; 13. gradient; 14. divide; 15. groundwater; 16. T; 17. T; 18. F; 19. F; 20. T.

Web Work

The following are informative and interesting World Wide Web sites that address topics related to those presented in this chapter:

- Mammoth Cave National Park
  http://www.nps.gov/maca/macahome.htm

- Water Resources of the United States
  http://h2o.usgs.gov/

For direct access to these sites and other relevant Web locations, contact the *Foundations of Earth Science* WWW home page at http://www.prenhall.com/lutgens.

CHAPTER 4
Glacial and Arid Landscapes

FOCUS ON LEARNING

To assist you in learning the important concepts
in this chapter, you will find it helpful to focus
on the following questions:

1. What is a glacier? Where on Earth are the two general types of glaciers located today?

2. How do glaciers move and what are the various processes of glacial erosion?

3. What materials make up the features created by glacial deposition? What are the most widespread features?

4. What is some evidence for the Ice Age? What indirect effects did Ice Age glaciers have on the land and sea?

5. What are the causes of deserts in both the lower and middle latitudes?

6. What are the roles of water and wind in arid climates?

7. How have many of the landscapes in the dry Basin and Range region of the United States evolved?

8. What are the ways that wind erodes?

9. What are some depositional features produced by wind?

Sand dunes are sometimes striking features in deserts, but they only represent a small percentage of the total desert area. These dunes are in California's Death Valley. *Photo by Carr Clifton Photography*

Today, glaciers cover nearly 10 percent of Earth's land surface; however, in the recent geologic past ice sheets were three times more extensive, covering vast areas with ice thousands of meters thick. Many regions still bear the mark of these glaciers. The first part of this chapter examines glaciers and the erosional and depositional features they create. The second part is devoted to dry lands and the geologic work of wind. Because desert and near-desert conditions prevail over an area as large as that affected by the massive glaciers of the Ice Age, the nature of such landscapes is indeed worth investigating.

Types of Glaciers

Many present-day landscapes were modified by the widespread glaciers of the most recent Ice Age and still strongly reflect the handiwork of ice. The basic features of such diverse places as the Alps, Cape Cod, and Yosemite Valley were fashioned by now vanished masses of glacial ice. Moreover, Long Island, the Great Lakes, and the fiords of Norway and Alaska all owe their existence to glaciers. Glaciers, of course, are not just a phenomenon of the geologic past. As we shall see, they are still sculpturing and depositing in many regions today.

Recall the water cycle in Chapter 3: water evaporates from the oceans into the atmosphere, precipitates on land, and flows in rivers and underground back to the sea. However, when precipitation falls at high elevations or high latitudes, the water may not immediately make its way toward the sea. Instead, it

may become part of a glacier. Although the ice will eventually melt, allowing the water to continue its path to the sea, water can be stored as glacial ice for tens, hundreds or even thousands of years.

A **glacier** is a thick ice mass that forms over hundreds or thousands of years. It originates on land from the accumulation, compaction, and recrystallization of snow. A glacier appears to be motionless, but it is not—glaciers move very slowly. Like running water, groundwater, wind, and waves, glaciers are dynamic erosional agents that accumulate, transport, and deposit sediment. Although glaciers are found in many parts of the world today, most are located in remote areas, either near Earth's poles or in high mountains.

Literally thousands of relatively small glaciers exist in lofty mountain areas, where they usually follow valleys originally occupied by streams. Unlike the rivers that previously flowed in these valleys, the glaciers advance slowly, perhaps only a few centimeters per day. Because of their location, these moving ice masses are termed **valley glaciers** or **alpine glaciers** (Figure 4.1). Each glacier is a stream of ice, bounded by precipitous rock walls, that flows downvalley from a snow accumulation center near its head. Like rivers, valley glaciers can be long or short, wide or narrow, single or with branching tributaries. Generally the widths of alpine glaciers are small compared to their lengths. In length, some extend for just a fraction of a kilometer, whereas others go on for many tens of kilometers. The west branch of the Hubbard Glacier, for example, runs

Figure 4.1 An active valley glacier in Denali National Park, Alaska. Note how it flows down the valley like a river of ice. Also notice the rock debris it has eroded and is transporting from the mountains. *(Photo by Michael Collier)*

through 112 kilometers of mountainous terrain in Alaska and Canada's Yukon Territory.

In contrast to valley glaciers, **ice sheets** exist on a much larger scale. These enormous masses flow out in all directions from one or more centers and completely obscure all but the highest areas of underlying terrain. Although many ice sheets have existed in the past, just two achieve this status at present. In the Northern Hemisphere, Greenland is covered by an imposing ice sheet averaging nearly 1500 meters thick. It occupies 1.7 million square kilometers, or about 80 percent of this large island. In the Southern Hemisphere, the huge Antarctic Ice Sheet attains a maximum thickness of nearly 4300 meters and covers an area of more than 13.9 million square kilometers. Because of the proportions of these huge features, they often are called *continental ice sheets*. Indeed, the combined areas of present-day continental ice sheets represent almost 10 percent of Earth's land area.

How Glaciers Move

The movement of glacial ice is generally referred to as *flow*. The fact that glacial movement is described in this way seems paradoxical—how can a solid flow? Glacial ice flows in two ways. One mechanism involves plastic movement within the ice. Ice behaves as a brittle solid until the pressure upon it is equivalent to the weight of about 50 meters (165 feet) of ice. Once that load is surpassed, ice behaves as a plastic material and flow begins. A second and often equally important mechanism of glacial movement consists of the entire ice mass slipping along the ground. The lowest portions of most glaciers probably move by this sliding process.

The uppermost 50 meters of a glacier is appropriately referred to as the *zone of fracture*. Because there is not enough overlying ice to cause plastic flow, this upper part of the glacier consists of brittle ice. Consequently, the ice in this zone is carried along piggyback style by the ice below. When the glacier moves over irregular terrain, the zone of fracture is subjected to tension, with cracks called **crevasses** resulting (Figure 4.2). These gaping cracks, which often make travel across glaciers dangerous, may extend to depths of 50 meters (165 feet). Below this depth, plastic flow seals them off.

Observing and Measuring Movement

Unlike streamflow, glacial movement is not obvious. If we could watch a valley glacier move, we would see that, like the water in a river, all of the ice does not move downstream at the same rate. Flow is greatest in the center of the glacier because of the drag created by the walls and floor of the valley.

Early in the nineteenth century, the first experiments involving the movement of glaciers were designed and carried out in the Alps. Markers were placed in a straight line across an alpine glacier. The position of the line was marked on the valley walls so that if the ice moved, the change in position could be detected. Periodically the positions of the markers were noted, revealing the movement just described. Although most glaciers move too slowly for direct visual detection, the experiments succeeded in demonstrating that movement nevertheless occurs. The experiment illustrated in Figure 4.3 was carried out at Switzerland's Rhone Glacier later in the nineteenth century. It not only traced the movement of markers within the ice, but also mapped the position of the glacier's terminus.

How rapidly does glacial ice move? Average rates vary considerably from one glacier to another. Some move so slowly that trees and other vegetation may become well established in the debris that accumulates on the glacier's surface. Others advance up to several meters each day. Movement of some glaciers is characterized by periods of extremely rapid advance followed by periods during which movement is practically nonexistent.

Accumulation and Wastage

Snow is the raw material from which glacial ice originates. Therefore, glaciers form in areas where more snow falls in winter than can melt during the summer. Glaciers are constantly gaining and losing ice. Snow accumulation and ice formation occur in the **zone of accumulation** (Figure 4.2). Here the addition of snow thickens the glacier and promotes movement. Beyond this area of ice formation is the **zone of wastage**. Here there is a net loss to the glacier as all of the snow from the previous winter melts, as does some of the glacial ice (Figure 4.2).

In addition to melting, glaciers also waste as large pieces of ice break off the front of a glacier in a process called *calving*. Calving creates *icebergs* where glaciers reach the sea (Figure 4.4). Because icebergs are just slightly less dense than seawater, they float very low in the water, with more than 80 percent of their mass submerged. The margins of the Greenland Ice Sheet produce thousands of icebergs each year. Many drift southward and find their

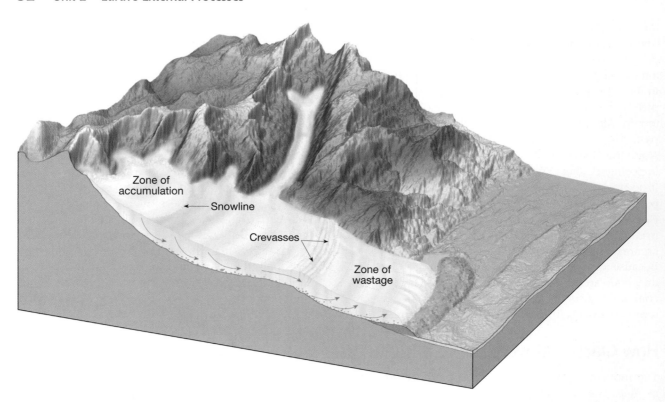

Figure 4.2 The snowline separates the zone of accumulation and the zone of wastage. Above the snowline, more snow falls each winter than melts each summer. Below the snowline, the snow from the previous winter completely melts, as does some of the underlying ice. Whether the margin of a glacier advances, retreats, or remains stationary depends on the balance or lack of balance between accumulation and wastage. When a glacier moves across irregular terrain, *crevasses* form in the brittle portion.

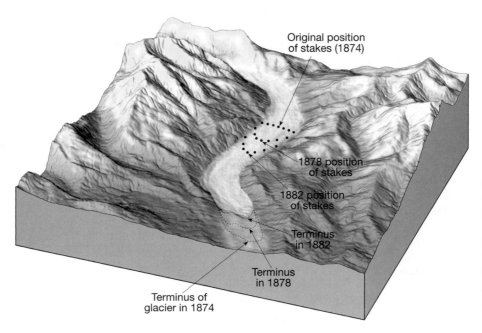

Figure 4.3 Ice movement and changes in the terminus at Rhone Glacier, Switzerland. In this classic study of a valley glacier, the movement of stakes clearly showed that ice along the side of the glacier moves slowest. Also notice that even though the ice front was retreating, the ice within the glacier was advancing.

Figure 4.4 Icebergs are created when large pieces break off the front of a glacier after it reaches a water body. These icebergs are in the Weddell Sea off the coast of Antarctica. Only about 20 percent of an iceberg protrudes above the water line. *(Photo by Frans Lanting/Minden Pictures)*

way into the North Atlantic, where they are a hazard to navigation.

Whether the margin of a glacier is advancing, retreating, or remaining stationary depends on the *budget of the glacier*. That is, it depends on the balance or lack of balance between accumulation on the one hand and wastage on the other. If ice accumulation exceeds wastage, the glacial front advances until the two factors balance. At this point, the terminus of the glacier becomes stationary. At a later time, when wastage exceeds accumulation, the ice front will retreat until a balance is again reached.

Whether the margin of a glacier is advancing, retreating, or stationary, the ice within the glacier continues to flow forward. In the case of a receding glacier, the ice simply does not flow forward rapidly enough to offset wastage. This point is illustrated in Figure 4.3. While the line of stakes within the Rhone Glacier continued to move downvalley, the terminus of the glacier slowly retreated upvalley.

Glacial Erosion

Glaciers erode tremendous volumes of rock. For anyone who has observed the terminus of an alpine glacier, the evidence of its erosive force is plain. You can witness firsthand the melting ice unlocking rock material of all sizes from huge boulders to tiny grit. All signs lead to the conclusion that the ice has scraped, scoured, and torn rock debris from the floor and walls of the valley and carried it downvalley.

Indeed, as a transporter of sediment, ice has no equal, because once the debris is acquired by the ice, it cannot settle out as does the load carried by a stream or by the wind. Consequently, glaciers can carry huge blocks that no other erosional agent could possibly budge. Although glaciers are of limited importance as an erosional agent today, many landscapes that were modified by the widespread glaciers of the recent Ice Age still reflect to a high degree the work of ice.

How Glaciers Erode

Glaciers erode land primarily in two ways. First, as a glacier flows over a fractured bedrock surface, it loosens and lifts blocks of rock, incorporates them into the ice, and carries them off. This process, known as **plucking**, occurs when meltwater penetrates the cracks and joints along the rock floor of the glacier and refreezes. As the water expands, it exerts tremendous leverage that pries the rock loose. In this manner sediment of all sizes becomes part of the glacier's load.

The second major erosional process is *abrasion*. As the ice with its load of rock fragments moves along, it acts as a giant rasp or file and grinds the surface below as well as the rocks within the ice. The pulverized rock produced by the glacial "grist mill" is appropriately called **rock flour**. So much rock flour may be produced that meltwater streams leaving a glacier often have the grayish appearance of skimmed milk—visible evidence of the grinding power of the ice.

When the embedded material includes large fragments, long scratches and grooves called **glacial striations** may be gouged into the bedrock passed over by the glacier (Figure 4.5). These linear

Figure 4.5 Glacial striations are scratches and grooves in bedrock caused by glacial abrasion (ice "armed" with sediment acting like sandpaper). Crater Lake National Park, Oregon. *(Photo by Jeff Gnass Photography)*

scratches on the bedrock surface provide clues to the direction of glacial movement. By mapping the striations over large areas, glacial flow patterns can often be reconstructed. On the other hand, not all abrasive action produces striations. The rock surfaces over which the glacier moves may become highly polished by the ice and its load of finer particles.

Although the erosional accomplishments of ice sheets can be tremendous, landforms carved by these huge ice masses usually do not inspire the same awe as do the erosional features created by valley glaciers. In regions where the erosional effects of ice sheets are significant, glacially scoured surfaces and subdued terrain are the rule. By contrast, in mountainous areas, erosion by valley glaciers produces many truly spectacular features. Much of the rugged mountain scenery so celebrated for its majestic beauty is the product of erosion by valley glaciers.

Features of Valley Glaciers

Take a moment to study Figure 4.6, which shows a mountain setting before, during, and after glaciation. You will refer to this often in the following discussion.

Prior to glaciation, alpine valleys are characteristically V-shaped because streams are well above base level and are therefore downcutting (Figure 4.6A). However, in mountainous regions that have been glaciated, the valleys are no longer narrow. As a glacier moves down a valley once occupied by a stream, the ice modifies it in three ways: The glacier widens, deepens, and straightens the valley, so that what was once a narrow V-shaped valley is transformed into a U-shaped **glacial trough** (Figures 4.6C and 4.7).

Because the amount of glacial erosion depends in part on the thickness of the ice, the main or trunk glacier cuts its valley deeper than can the smaller tributary glaciers. Thus, after the ice has receded, the valleys of tributary glaciers are left standing above the main trough and are termed **hanging valleys**. Rivers flowing through hanging valleys may produce spectacular waterfalls, such as those in Yosemite National Park, California (Figure 4.8).

At the head of a glacial valley is a characteristic and often imposing feature associated with an alpine glacier—a **cirque**. As Figures 4.6 and 4.9 illustrate, these hollowed-out, bowl-shaped depressions have precipitous walls on three sides but are open on the downvalley side. The cirque represents the focal point of the glacier's source—that is, the area of snow accumulation and ice formation. Cirques begin as irregularities in the mountainside that are subsequently enlarged by frost wedging and plucking along the sides and bottom of the glacier. The glacier, in turn, acts as a conveyor belt that carries the debris away. After the glacier has melted away, the cirque basin is often occupied by a small lake.

The Alps, Northern Rockies, and many other mountain landscapes carved by valley glaciers reveal more than glacial troughs and cirques. In addition, sinuous, sharp-edged ridges called **arêtes** and sharp, pyramid-like peaks called **horns** project above the surroundings (Figure 4.6C). Both features can originate from the same basic process—the enlargement of cirques produced by plucking and frost action. Cirques around a single high mountain create the spires of rock called horns. As the cirques enlarge and converge, an isolated horn is produced. The most famous example is the Matterhorn in the Swiss Alps (Figure 4.10).

94

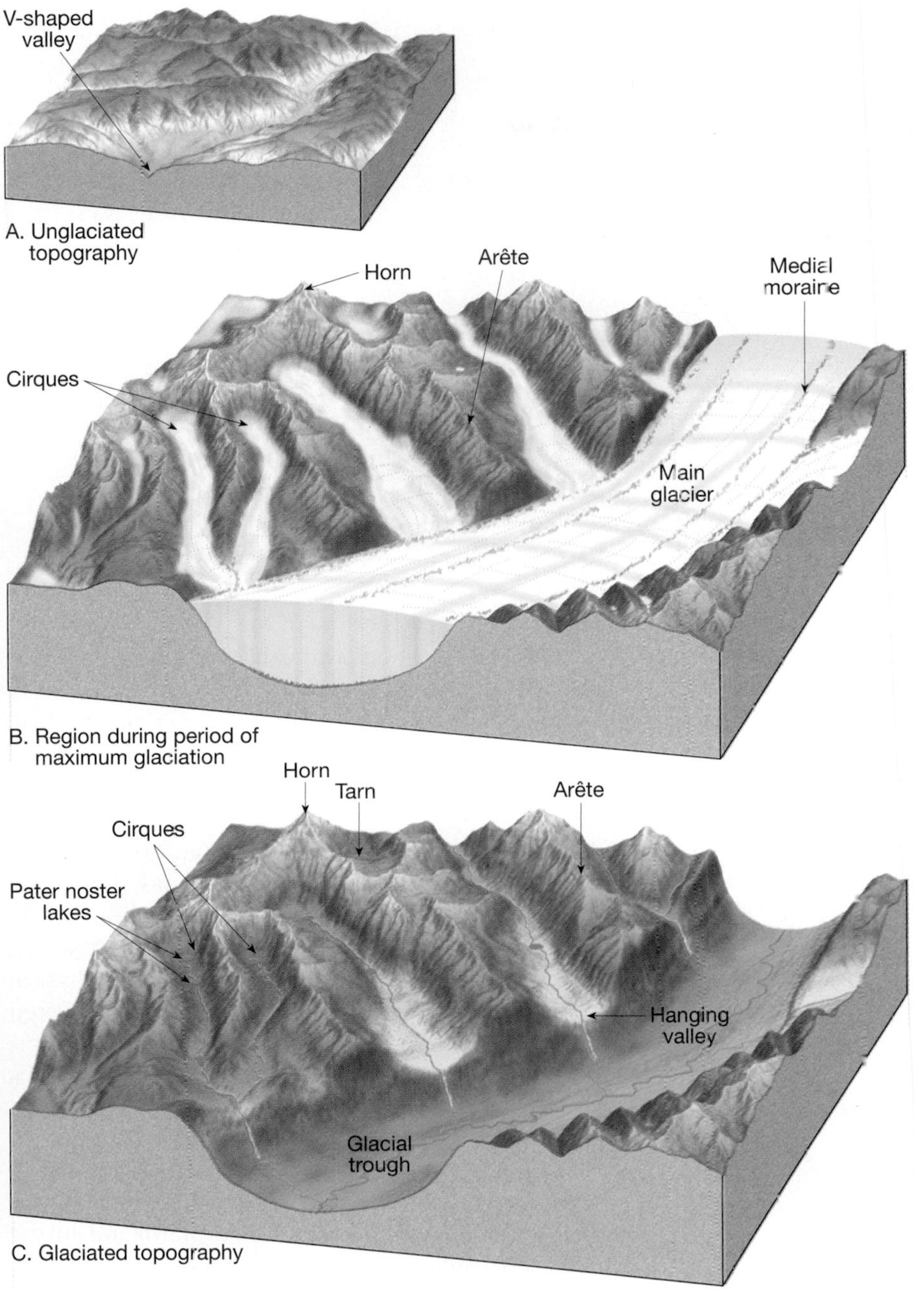

Figure 4.6 Erosional landforms created by alpine glaciers.

Arêtes can form in a similar manner except that the cirques are not clustered around a point but rather exist on opposite sides of a divide. As the cirques grow, the divide separating them is reduced to a very narrow, knifelike partition. An arête may also be created in another way. When two glaciers occupy parallel valleys, an arête can form when the divide separating the moving tongues of ice is progressively narrowed as the glaciers scour and widen their valleys.

Fiords are deep, often spectacular, steep-sided inlets of the sea that exist in many high-latitude areas of the world where mountains are adjacent to the ocean (Figure 4.11). Norway, British Columbia, Greenland, New Zealand, Chile, and Alaska all have coastlines characterized by fiords. They are glacial troughs that became submerged as the ice left the valley and sea level rose following the Ice Age. The depth of some fiords exceeds 1000 meters (3300 feet).

Figure 4.7 Prior to glaciation, a mountain valley is typically narrow and V-shaped. During glaciation, a valley glacier widens, deepens, and straightens the valley, creating a U-shaped glacial trough. This valley is in Glacier National Park, Montana. *(Photo by John Montagne)*

However, the great depths of these flooded troughs are only partly explained by the post-Ice Age rise in sea level. Unlike the situation governing the downward erosional work of rivers, sea level does not act as base level for glaciers. As a consequence, glaciers are capable of eroding their beds far below the surface of the sea. For example, a valley glacier 300 meters (1000 feet) thick can carve its valley floor more than 250 meters (800 feet) below sea level before downward erosion ceases and the ice begins to float.

Glacial Deposits

Glaciers pick up and transport a huge load of debris as they slowly advance across the land. Ultimately these materials are deposited when the ice melts. In regions where glacial sediment is deposited, it can play a truly significant role in forming the physical landscape. For example, in many areas once covered by the ice sheets of the recent Ice Age, the bedrock is rarely exposed because glacial deposits that are tens or even hundreds of meters thick completely mantle the terrain. The general effect of these deposits is to level the topography. Indeed, many of today's familiar rural landscapes—rocky pastures in New England, wheat fields in the Dakotas, rolling farmland in the Midwest—resulted directly from glacial deposition.

Types of Glacial Drift

Long before the theory of an extensive Ice Age was proposed, much of the soil and rock debris covering portions of Europe was recognized as coming from elsewhere. At the time, these foreign materials were believed to have been "drifted" into their present positions by floating ice during an ancient flood. As a consequence, the term *drift* was applied to this sediment. Although rooted in a concept that was not

Figure 4.8 Bridalveil Falls in Yosemite National Park cascades from a hanging valley into the glacial trough below. *(Photo by Marc Muench/David Muench Photography, Inc.)*

Figure 4.9 Aerial view of bowl-shaped depressions called cirques in the Uinta Range, Utah. *(Photo by John S. Shelton)*

correct, this term was so well established by the time the true glacial origin of the debris became widely recognized that it remained in the glacial vocabulary. Today, **drift** is an all-embracing term for sediments of glacial origin, no matter how, where, or in what form they were depositied.

Glacial drift is divided into two distinct types: (1) materials deposited directly by the glacier, which are known as *till*, and (2) sediments laid down by glacial meltwater, called *stratified drift*. Here is the difference: **till** is deposited as glacial ice melts and drops its load of rock fragments. Unlike moving water and wind, ice cannot sort the sediment it carries; there-

Figure 4.10 Horns are sharp, pyramid-like peaks that are fashioned by alpine glaciers. This example is the famous Matterhorn in the Swiss Alps. *(Photo by E. J. Tarbuck)*

Figure 4.11 Like other fiords, Tracy Arm, Alaska, is a drowned glacial trough. *(Photo by Tom Bean)*

Figure 4.12 Glacial till is an unsorted mixture of many sediment sizes. *(Photo by E. J. Tarbuck)*

fore, deposits of till are characteristically unsorted mixtures of many particle sizes (Figure 4.12). **Stratified drift** is sorted according to the size and weight of the fragments. Because ice is not capable of such sorting activity, these sediments are not deposited directly by the glacier. Rather, they reflect the sorting action of glacial meltwater.

Some deposits of stratified drift are made by streams coming directly from the glacier. Other stratified deposits involve sediment that was originally laid down as till and later picked up, transported, and redeposited by meltwater beyond the margin of the ice. Accumulations of stratified drift often consist largely of sand and gravel, because the meltwater is not capable of moving larger material and because the finer rock flour remains suspended and is commonly carried far from the glacier. An indication that stratified drift consists primarily of sand and gravel can be seen in many areas where these deposits are actively mined as aggregate for road work and other construction projects.

When boulders are found in the till or lying free on the surface, they are called **glacial erratics** if they are different from the bedrock below. Of course, this means that they must have been derived from a source outside the area where they are found (Figure 4.13). Although the locality of origin for most erratics is unknown, the origin of some can be determined. Therefore, by studying glacial erratics as well as the mineral composition of the till, geologists can sometimes trace the path of a lobe of ice. In portions of New England as well as other areas, erratics may be seen dotting pastures and farm fields. In some places, these rocks were cleared from fields and piled to make fences and walls.

Moraines, Outwash Plains, and Kettles

Perhaps the most widespread features created by glacial deposition are *moraines*, which are simply layers or ridges of till. Several types of moraines are identified; some are common only to mountain valleys, and others are associated with areas affected by either ice sheets or valley glaciers. Lateral and medial moraines fall in the first category, whereas end moraines and ground moraines are in the second.

The sides of a valley glacier accumulate large quantities of debris from the valley walls. When the glacier wastes away, these materials are left as ridges, called **lateral moraines**, along the sides of the valley (Figure 4.14). **Medial moraines** are formed when two valley glaciers coalesce to form a single ice stream. The till that was once carried along the edges of each glacier joins to form a single dark stripe of debris within the newly enlarged glacier. The creation of these dark stripes within the ice stream is one obvious proof that glacial ice moves, because the medial moraine could not form if the ice did not flow downvalley (Figure 4.14).

End moraines, as the name implies, form at the terminus of a glacier. Here, while the ice front is

Figure 4.13 Land cleared of glacial erratics, which were then piled atop one another to build this stone wall near West Bend, Wisconsin. *(Photo by Tom Bean)*

Figure 4.14 Lateral moraines form from the accumulation of debris along the sides of a valley glacier. Medial moraines form when the lateral moraines of merging valley glaciers join. Medial moraines could not form if the ice did not advance downvalley. Therefore, these dark stripes are proof that glacial ice moves. Kaskawulsh Glacier, Kluane National Park, Yukon, Canada. *(Photo by Theo Allofs/Tony Stone Images)*

stationary (because accumulation and wastage are equal), the glacier continues to carry in and deposit large quantities of rock debris, creating a ridge of till tens to hundreds of meters high. The end moraine marking the farthest advance of the glacier is called the *terminal moraine*, and those moraines that formed as the ice front periodically became stationary during retreat are termed *recessional moraines*. As the glacier recedes, a layer of till is laid down, forming a gently undulating surface of **ground moraine**. Ground moraine has a leveling effect, filling in low spots and clogging old stream channels, often leading to a disruption of drainage.

At the same time that an end moraine is forming, meltwater emerges from the ice in rapidly moving streams. Often they are choked with suspended material and carry a substantial bed load. As the water leaves the glacier, it rapidly loses velocity and much of its bed load is dropped. In this way a broad, ramplike surface of stratified drift is built adjacent to the downstream edge of most end moraines. When the feature is formed in association with an ice sheet, it is termed an **outwash plain**, and when it is confined to a mountain valley, it is commonly referred to as a **valley train**. Figure 4.15 shows an outwash plain and other common depositional features.

Often end moraines, outwash plains, and valley trains are pockmarked with basins or depressions known as **kettles** (Figure 4.15). Kettles form when blocks of stagnant ice become buried in drift and

melt, leaving pits in the glacial sediment. Most kettles do not exceed 2 kilometers in diameter and the typical depth of most kettles is less than 10 meters (33 feet). Water often fills the depression and forms a pond or lake. One well-known example is Walden Pond near Concord, Massachusetts. It is here that Henry David Thoreau lived alone for two years in the 1840s and about which he wrote *Walden*, his classic of American literature.

Drumlins, Eskers, and Kames

Drumlins are streamlined asymmetrical hills composed of till (Figure 4.15). They range in height from 15 to 60 meters (50 to 200 feet) and average 0.4 to 0.8 kilometer (0.25–0.50 mile) in length. The steep side of the hill faces the direction from which the ice advanced, while the gentler slope points in the direction the ice moved. Drumlins are not found singly, but rather occur in clusters, called *drumlin fields*. One such cluster, east of Rochester, New York, is estimated to contain about 10,000 drumlins. Their streamlined shape indicates that they were molded in the zone of flow within an active glacier. It is thought that drumlins originate when glaciers advance over previously deposited drift and reshape the material.

In some areas that were once occupied by glaciers, sinuous ridges composed largely of sand and gravel may be found. These ridges, called **eskers**, are deposits made by streams flowing in tunnels

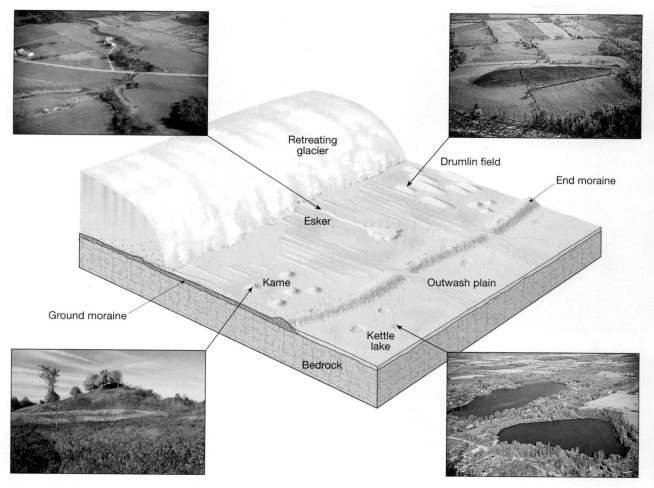

Figure 4.15 This hypothetical area illustrates many common depositional landforms. *(Photo A, C, and D, by Richard Jacobs/JLM Visuals) (Photo B courtesy of Ward's Natural Science Establishment, Inc.)*

beneath the ice, near the terminus of a glacier (Figure 4.15). They may be several meters high and extend for many kilometers. In some areas they are mined for sand and gravel, and for this reason, eskers are disappearing in some localities.

Kames are steep-sided hills that, like eskers, are composed largely of stratified drift (Figure 4.15). Kames originate when glacial meltwater washes sediment into openings and depressions in the stagnant wasting terminus of a glacier. When the ice eventually melts away, the stratified drift is left behind as mounds or hills.

Glaciers of the Past

At various points in the preceding pages, we mentioned the Ice Age, a time when ice sheets and alpine glaciers were far more extensive than they are today. There was a time when the most popular

explanation for drift was that the material had been drifted in by means of icebergs or perhaps simply swept across the landscape by a catastrophic flood. However, during the nineteenth century, field investigations by many scientists provided convincing proof that an extensive Ice Age was responsible for these deposits, and for many other features.

By the beginning of the twentieth century, geologists had largely determined the extent of Ice Age glaciation. Further, they discovered that many glaciated regions had not one layer of drift, but several. Close examination of these older deposits showed well-developed zones of chemical weathering and soil formation as well as the remains of plants that require warm temperatures. The evidence was clear: there had not been just one glacial advance but several, each separated by extended periods when climates were as warm or warmer than at present. The Ice Age had not simply been a time when the ice advanced over the land, lingered for a while, and then receded. Rather, the period was a very complex event characterized by a number of advances and withdrawals of glacial ice.

The glacial record on land is punctuated by many erosional gaps. This makes it difficult to reconstruct clearly the episodes of the Ice Age. But sediment on the ocean floor provides an uninterrupted record of climate cycles for this period. Studies of cores drilled from these seafloor sediments show that glacial/inter-glacial cycles have occurred about every 100,000 years. About 20 such cycles of cooling and warming were identified for the span we call the Ice Age.

During the Ice Age, ice left its imprint on almost 30 percent of Earth's land area, including about 10 million square kilometers of North America, 5 million square kilometers of Europe, and 4 million square kilometers of Siberia (Figure 4.16). The amount of glacial ice in the Northern Hemisphere was roughly twice that of the Southern Hemisphere. The primary reason is that the Southern Hemisphere has little land in the middle latitudes and therefore the southern polar ice could not spread far beyond the margins of Antarctica. By contrast, North America and Eurasia provided great expanses of land for the spread of ice sheets.

Today we know that the Ice Age began between 2 million and 3 million years ago. This means that most of the major glacial episodes occurred during a division of the geologic time scale called the **Pleistocene epoch**. Although the Pleistocene is commonly used as a synonym for the Ice Age, this epoch does not encompass it all. The Antarctic Ice Sheet, for example, formed at least 14 million years ago, and, in fact, might be much older.

Glaciers have not been ever-present features throughout Earth's long history. In fact, for most of geologic time, glaciers have been absent. Evidence does indicate that, in addition to the Pleistocene epoch, there were at least three earlier periods of glacial activity: 2 billion, 600 million, and 250 million years ago. However, the most recent period of glaciation is of greatest interest, because the features of many present-day landscapes are a reflection of the work of Pleistocene glaciers.

Figure 4.16 Maximum extent of glaciation in the Northern Hemisphere during the Ice Age.

Some Indirect Effects of Ice Age Glaciers

In addition to the massive erosional and depositional work carried on by Pleistocene glaciers, the ice sheets had other, sometimes profound, effects on the landscape. For example, as the ice advanced and retreated, animals and plants were forced to migrate. This led to stresses that some organisms could not tolerate. Furthermore, many present-day stream courses bear little resemblance to their preglacial routes. The Missouri River once flowed northward toward Hudson Bay in Canada. The Mississippi River followed a path through central Illinois, and the head of the Ohio River reached only as far as Indiana. Some rivers that today carry only a trickle of water but occupy broad channels are a testament to the fact that they once carried torrents of glacial meltwater.

In areas that were centers of ice accumulation, such as Scandinavia and northern Canada, the land has been slowly rising for the past several thousand years. The land had downwarped under the tremendous weight of 3-kilometer-thick masses of ice. Following the removal of this immense load, the crust has been adjusting by gradually rebounding upward ever since.

A far-reaching effect of the Ice Age was the worldwide change in sea level that accompanied each advance and retreat of the ice sheets. The snow that nourishes glaciers ultimately comes from moisture evaporated from the oceans. Therefore, when the ice sheets increased in size, sea level fell and the shoreline moved seaward (Figure 4.17). Estimates suggest that sea level was as much as 100 meters (330 feet) lower than today. Consequently, the Atlantic Coast of the United States was located more than 100 kilometers (60 miles) to the east of New York City. Moreover, France and Britain were joined where the English Channel is today. Alaska and Siberia were connected across the Bering Strait, and Southeast Asia was tied by dry land to the islands of Indonesia.

The formation and growth of ice sheets was an obvious response to significant changes in climate. But the existence of the glaciers themselves triggered climatic changes in the regions beyond their margins. In arid and semiarid areas on all continents, temperatures were lowered, which meant evaporation rates were also lowered. At the same time, precipitation was moderate. This cooler, wetter climate resulted in the formation of many lakes called **pluvial lakes** (from the Latin term *pluvia* meaning "rain"). In North America, pluvial lakes were concentrated in the vast Basin and Range region of Nevada and Utah. Although most are now gone, there are a few remnants, the largest being Utah's Great Salt Lake.

Deserts

Climate has a strong influence on the nature and intensity of Earth's external processes. This was clearly demonstrated in the preceding section on glaciers and the Ice Age. Another excellent example of the strong link between climate and geology is seen when we examine the development of arid landscapes.

The word *desert* literally means "deserted" or "unoccupied." For many dry regions this is a very appropriate description, although where water is

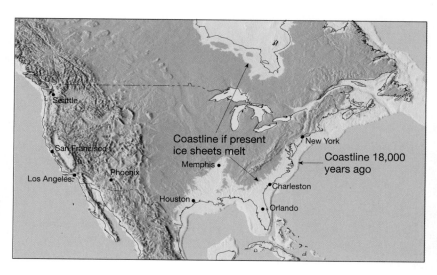

Figure 4.17 This map of a portion of North America shows the present-day coastline compared to the coastline that existed during the last ice-age maximum (18,000 years ago) and the coastline that would exist if present ice sheets in Greenland and Antarctica melted. *(After R. H. Dott, Jr., and R. L. Battan, Evolution of the Earth. New York: McGraw-Hill, 1971. Reprinted by permission of the publisher.)*

available in deserts, life thrives. Nevertheless, the world's dry regions are among the least familiar land areas on Earth outside of the polar realm.

Desert landscapes frequently appear stark. Their profiles are not softened by a carpet of soil and abundant plant life. Instead, barren rocky outcrops with steep, angular slopes are common. At some places the rocks are tinted orange and red. At others they are gray and brown and streaked with black. For many visitors desert scenery exhibits a striking beauty; to others, the terrain seems bleak. No matter which feeling is elicited, it is clear that deserts are very different from the more humid places where most people live.

As you shall see, arid regions are not dominated by a single geologic process. Rather, the effects of tectonic forces, running water, and wind are all apparent. Because these processes combine in different ways from place to place, the appearance of desert landscapes varies a great deal as well (Figure 4.18).

Distribution and Causes of Dry Lands

The dry regions of the world encompass about 42 million square kilometers, nearly 30 percent of Earth's land surface. No other climatic group covers so large a land area. Within these water-deficient regions, two climatic types are commonly recognized: **desert**, or arid, and **steppe**, or semiarid. The

Figure 4.18 Desert plants surrounded by snow-covered ground in dry southern Utah. As this scene demonstrates, deserts are not necessarily hot, lifeless, dune-covered expanses. *(Photo by Stephen Trimble)*

two categories have many features in common; their differences are primarily a matter of degree. The steppe is a marginal and more humid variant of the desert and represents a transition zone that surrounds the desert and separates it from bordering humid climates. The world map showing the distribution of desert and steppe regions reveals that dry lands are concentrated in the subtropics and in the middle latitudes (Figure 4.19).

Deserts in places such as Africa, Arabia, and Australia are primarily the result of the prevailing global distribution of air pressure and winds. Coinciding with dry regions in the lower latitudes are zones of high air pressure known as the *subtropical highs*. These pressure systems are characterized by subsiding air currents. When air sinks, it is compressed and warmed. Such conditions are just the opposite of what is needed to produce clouds and precipitation. Consequently, these regions are known for their clear skies, sunshine, and ongoing drought.

Middle-latitude deserts and steppes exist principally because they are sheltered in the deep interiors of large landmasses. They are far removed from the ocean, which is the ultimate source of moisture for cloud formation and precipitation. In addition, the presence of high mountains across the paths of prevailing winds further acts to separate these areas from water-bearing, maritime air masses. In North America, the Coast Ranges, Sierra Nevada, and Cascades are the foremost mountain barriers to moisture from the Pacific (Figure 4.20). In their rainshadow lies the dry and expansive Basin and Range region of the American West.

Middle-latitude deserts provide an example of how mountain-building processes affect climate. Without mountains, wetter climates would prevail where dry regions exist today.

The Role of Water in Arid Climates

Permanent streams are normal in humid regions, but practically all desert streambeds are dry most of the time (Figure 4.21A). Deserts have **ephemeral streams**, which means that they carry water only in response to specific episodes of rainfall. A typical ephemeral stream may flow only a few days or perhaps just a few hours during the year. In some years, the channel might carry no water at all.

This fact is obvious even to the casual traveler who notices the numerous bridges with no streams beneath them, or numerous dips in the road where

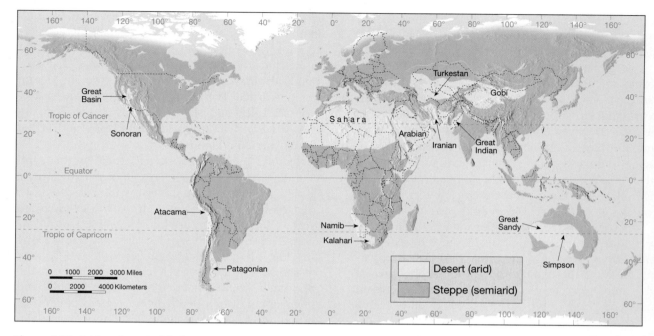

Figure 4.19 Arid and semiarid climates cover about 30 percent of Earth's land area. No other climatic group covers so large an area.

dry channels cross. However, when the rare heavy showers do come, so much rain falls in such a short time that it cannot all soak in. Because desert vegetative cover is sparse, runoff is largely unhindered and therefore rapid, often creating flash floods along valley floors (Figure 4.21B). These floods are quite unlike floods in humid regions. A flood on a river like the Mississippi may take several days to reach its crest and then subside. But desert floods arrive suddenly and subside quickly. Because much surface

material in a desert is not anchored by vegetation, the amount of erosional work that occurs during a single short-lived rain event is impressive.

In the dry western United States, different names are used for ephemeral streams, including *wash* and *arroyo*. In other parts of the world, a dry desert stream may be a *wadi* (Arabia and North Africa), a *donga* (South America), or a *nullah* (India).

Humid regions are notable for their well-developed drainage systems. But in arid regions, streams

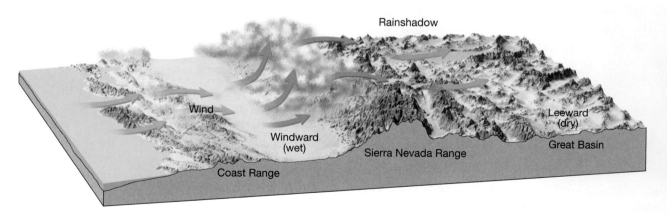

Figure 4.20 Mountains contribute to the formation of middle-latitude dry regions by creating a rainshadow. As moving air meets a mountain barrier, it is forced to rise. Clouds and precipitation on the windward side result. Air descending the leeward side is much drier. The mountains effectively cut the leeward side off from the source of moisture, creating a rainshadow.

A.

B.

Figure 4.21 A. Most of the time, desert stream channels are dry. **B**. This photo shows an ephemeral stream shortly after a heavy shower. Although such floods are short-lived, large amounts of erosion occur. *(Photos by E. J. Tarbuck)*

usually lack an extensive system of tributaries. In fact, a basic characteristic of desert streams is that they are small and die out before reaching the sea. Because the water table is usually far below the surface, few desert streams can draw upon it as streams do in humid regions. Without a steady supply of water, the combination of evaporation and soaking in soon depletes the stream.

The few permanent streams that do cross arid regions, such as the Colorado and Nile Rivers, originate *outside* the desert, often in mountains where water is more plentiful. Here the supply of water must be great to compensate for the losses that occur as the stream crosses the desert. For example, after the Nile leaves its headwaters in the lakes and mountains of central Africa, it traverses almost 3000 kilometers (1800 miles) of the Sahara Desert *without a single tributary.*

It should be emphasized that *running water, although infrequent, nevertheless is responsible for most of the erosional work in deserts.* This is contrary to a common belief that wind is the most important erosional agent sculpting desert landscapes. Although wind erosion is indeed more significant in dry areas than elsewhere, most desert landforms are carved by running water. The main role of wind, as you shall see later in this chapter, is in the transportation and deposition of sediment, which creates and shapes the ridges and mounds we call dunes.

Basin and Range: The Evolution of a Desert Landscape

Because arid regions typically lack permanent streams, they are characterized as having **interior drainage**. This means that they have a discontinuous pattern of intermittent streams that do not flow out of the desert to the ocean. In the United States, the dry Basin and Range region provides an excellent example. The region includes southern Oregon, all of Nevada, western Utah, southeastern California, southern Arizona, and southern New Mexico (Figure 4.22). The name Basin and Range is an apt description for this almost 800,000-square-kilometer region, as it is characterized by more than 200 relatively small mountain ranges that rise 900 to 1500 meters above the basins that separate them.

In this region, as in others like it around the world, most erosion occurs without reference to the ocean (ultimate base level), because the interior drainage never reaches the sea. Even where permanent streams flow to the ocean, few tributaries exist, and thus only a narrow strip of land adjacent to the stream has sea level as its ultimate level of land reduction.

The block models in Figure 4.22 depict how the landscape has evolved in the Basin and Range region. During and following uplift of the mountains, running water begins carving the elevated mass and depositing large quantities of debris in the basin. In

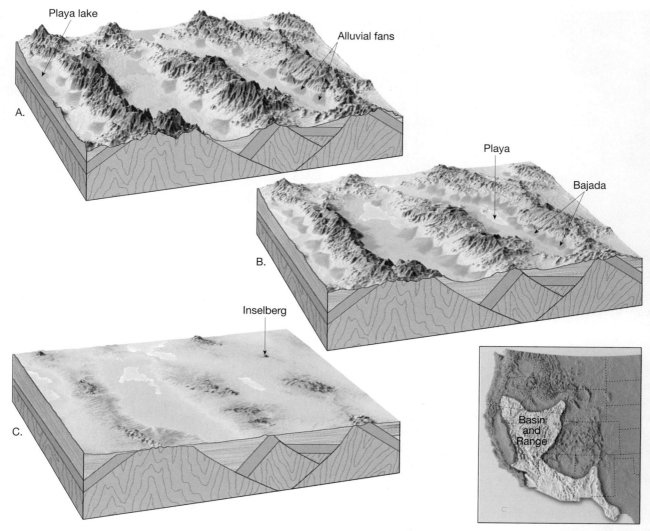

Figure 4.22 Stages of landscape evolution in a mountainous desert such as the Basin and Range region of the West. As erosion of the mountains and deposition in the basins continue, relief diminishes. **A.** Early stage. **B.** Middle stage **C.** Late stage.

this early stage, relief (difference in elevation between high and low points in an area) is greatest. As erosion lowers the mountains and sediment fills the basins, elevation differences diminish.

When the occasional torrents of water produced by sporadic rains move down the mountain canyons, they are heavily loaded with sediment. Emerging from the confines of the canyon, the runoff spreads over the gentler slopes at the base of the mountains and quickly loses velocity. Consequently, most of its sediment load is dumped within a short distance. The result is a cone of debris known as an **alluvial fan** at the mouth of a canyon (Figure 4.23). Over the years, a fan enlarges, eventually coalescing with fans from

adjacent canyons to produce an apron of sediment (*bajada*) along the mountain front.

On the rare occasions of abundant rainfall, streams may flow across the alluvial fans to the center of the basin, converting the basin floor into a shallow **playa lake**. Playa lakes last only a few days or weeks, before evaporation and infiltration remove the water. The dry, flat lake bed that remains is termed a *playa*.

Playas occasionally become encrusted with salts that are left behind when the water in which they were dissolved evaporates. These precipitated salts may be uncommon. A case in point is the sodium borate (better known as borax) mined from ancient playa lake deposits in Death Valley, California.

Figure 4.23 Alluvial fans develop where the gradient of a stream changes abruptly from steep to flat. Such a situation exists in Death Valley, California, where streams emerge from the mountains into a flat basin. As a result, Death Valley has many large alluvial fans. *(Photo by Michael Collier)*

With the ongoing erosion of the mountain mass and the accompanying sedimentation, the local relief continues to diminish. Eventually nearly the entire mountain mass is gone. Thus, by the late stages of erosion, the mountain areas are reduced to a few large bedrock knobs projecting above the sediment-filled basin.

Each of the stages of landscape evolution in an arid climate depicted in Figure 4.22 can be observed in the Basin and Range region. Recently uplifted mountains in an early stage of erosion are found in southern Oregon and northern Nevada. Death Valley, California, and southern Nevada fit into the more advanced middle stage, while the late stage, with its inselbergs and extensive pediments, can be seen in southern Arizona.

Wind Erosion

Compared to running water and glaciers, wind is a relatively insignificant erosional agent. Recall that even in deserts most erosion is performed by intermittent running water, not by the wind. Wind erosion is more effective in arid lands than in humid areas because in humid places moisture binds particles together and vegetation anchors the soil. For wind to be an effective erosional force, dryness and scanty vegetation are important prerequisites. When such circumstances exist, wind may pick up, transport, and deposit great quantities of fine sediment. During the 1930s, parts of the Great Plains experienced vast dust storms (Figure 4.24). The plowing under of the natural vegetative cover for farming, followed by severe drought, exposed the land to wind erosion and led to the area being labeled the Dust Bowl.

Moving air, like moving water, is turbulent and able to pick up loose debris and transport it to other locations. Just as in a stream, the velocity of wind increases with height above the surface. Also like a

Figure 4.24 Dust blackens the sky on May 21, 1937, near Elkhart, Kansas. It was because of storms like this that portions of the Great Plains were called the "Dust Bowl" in the 1930s. *(Photo reproduced from the collection of the Library of Congress)*

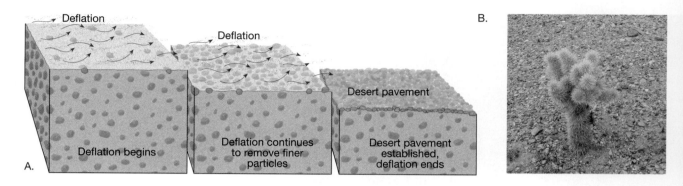

Figure 4.25 A. Formation of desert pavement. Coarse particles gradually become concentrated into a tightly packed layer as deflation lowers the surface by removing sand and silt. **B**. If left undisturbed, desert pavement will protect the surface from further deflation. *(Photo by David Muench Photography)*

stream, wind transports fine particles in suspension while heavier ones are carried as bed load. However, the transport of sediment by wind differs from that by running water in two significant ways. First, wind has a low density compared to water; thus, it is not capable of picking up and transporting coarse materials. Second, because wind is not confined to channels, it can spread over large areas, as well as high into the atmosphere.

One way that wind erodes is by **deflation**, the lifting and removal of loose material. Wind can suspend only fine sediment such as clay and silt. Larger grains of sand are rolled or skipped along the surface (a process called *saltation)* and comprise the bed load. Particles larger than sand are usually not transported by wind. Deflation sometimes is difficult to notice because the entire surface is being lowered at the same time, but it can be significant. In portions of the 1930s Dust Bowl, the land was lowered by as much as 1 meter in only a few years.

The most noticeable result of deflation in some places is shallow depressions called **blowouts**. In the Great Plains region, from Texas north to Montana, thousands of blowouts can be seen. They range from small dimples less than 1 meter deep and 3 meters wide to depressions that are over 45 meters deep and several kilometers across.

In portions of many deserts, the surface is characterized by a layer of coarse pebbles and cobbles that are too large to be moved by the wind. This stony veneer, called **desert pavement**, is created as deflation lowers the surface by removing sand and silt until eventually only a continuous cover of

coarse particles remains (Figure 4.25). Once desert pavement becomes established, a process that might take hundreds of years, the surface is effectively protected from further deflation if left undisturbed. However, as the layer is only one or two stones thick, the passage of vehicles or animals can dislodge the pavement and expose the fine-grained material below. If this happens, the surface is once again subject to deflation.

Like glaciers and streams, wind erodes in part by *abrasion.* In dry regions as well as along some beaches, windblown sand will cut and polish exposed rock surfaces. However, abrasion is often given too much credit. Such features as balanced rocks that stand high atop narrow pedestals and intricate detailing on tall pinnacles are not the results of abrasion. Sand seldom travels more than a meter above the surface, so the wind's sandblasting effect is obviously limited in vertical extent. But in areas prone to such activity, telephone poles have actually been cut through near their bases. For this reason, collars are often fitted on the poles to protect them from being "sawed" down.

Wind Deposits

Although wind is relatively unimportant in producing erosional landforms, there are significant depositional landforms created by the wind in some regions. Accumulations of windblown sediment are particularly conspicuous in the world's dry lands and along many sandy coasts. Wind deposits are of two distinctive types: (1) extensive blankets of silt, called

Figure 4.26 A vertical loess bluff near the Mississippi River in southern Illinois. *(Photo by James E. Patterson)*

loess, that once were carried in suspension, and (2) mounds and ridges of sand from the wind's bed load, which we call *dunes*.

Loess

In some parts of the world the surface topography is mantled with deposits of windblown silt, called **loess**. Over thousands of years, dust storms deposited this material. When loess is breached by streams or road cuts, it tends to maintain vertical cliffs and lacks any visible layers, as you can see in Figure 4.26.

The distribution of loess worldwide indicates two primary sources for this sediment: deserts and glacial deposits of stratified drift. The thickest and most extensive loess deposits occur in western and northern China. They were blown there from the extensive desert basins of central Asia. Accumulations of 30 meters are not uncommon, and thicknesses of more than 100 meters have been measured. It is this fine, buff-colored sediment that gives the Yellow River (Huang Ho) its name.

In the United States, deposits of loess are significant in many areas, including South Dakota, Nebraska, Iowa, Missouri, and Illinois as well as portions of the Columbia Plateau in the Pacific Northwest. Unlike the deposits in China, which originated in deserts, the loess in the United States and Europe is an indirect product of glaciation. Its source is deposits of stratified drift. During the retreat of the ice sheets, many river valleys were choked with glacial sediment. Strong winds sweeping across the barren floodplains picked up the finer sediment and dropped it as a blanket on areas adjacent to the valleys.

Figure 4.27 Large sand dunes in Africa's Namib Desert. *(Photo by Art Wolfe/Tony Stone Images)*

Figure 4.28 Cross beds are an obvious characteristic of the Navajo Sandstone, here exposed in Zion National Park, Utah. When dunes are buried and become part of the sedimentary record, the cross-bedded structure is preserved. *(Photo by Martin G. Miller)*

Sand Dunes

Like running water, wind releases its load of sediment when its velocity falls and the energy available for transport diminishes. Thus, sand begins to accumulate wherever an obstruction across the path of the wind slows its movement. Unlike deposits of loess, which form blanketlike layers over broad areas, winds commonly deposit sand in mounds or ridges called **dunes** (Figure 4.27).

As moving air encounters an object, such as a clump of vegetation or a rock, the wind sweeps around and over it, leaving a shadow of more slowly moving air behind the obstacle as well as a smaller zone of quieter air just in front of the obstacle. Some of the sand grains moving with the wind come to rest in these "wind shadows." As the accumulation of sand continues, it forms an increasingly efficient wind barrier to trap even more sand. If there is a sufficient supply of sand and the wind blows steadily long enough, the mound of sand grows into a dune.

Many dunes have an asymmetrical profile, with the leeward (sheltered) slope being steep and the windward slope more gently inclined. Sand is rolled up the gentle slope on the windward side by the force of the wind. Just beyond the crest of the dune, the wind velocity is reduced and the sand accumulates. As more sand collects, the slope steepens and eventually some of it slides under the pull of gravity. In this way, the leeward slope of the dune, called the **slip face**, maintains a relatively steep angle. Continued sand accumulation, coupled with periodic slides down the slip face, results in the slow migration of the dune in the direction of air movement.

As sand is deposited on the slip face, it forms layers inclined in the direction the wind is blowing. These sloping layers are called **cross beds**. When the dunes are eventually buried under layers of sediment and become part of the sedimentary rock record, their asymmetrical shape is destroyed, but the cross beds remain as a testimony to their origin. Nowhere is cross-bedding more prominent than in the sandstone walls of Zion Canyon in southern Utah (Figure 4.28).

The Chapter in Review

1. A *glacier* is a thick mass of ice originating on the land from the compaction and recrystallization of snow. It shows evidence of past or present movement. Today, *valley*, or *alpine glaciers* are found in mountain areas where they usually follow valleys originally occupied by streams. *Ice sheets* exist on a much larger scale, covering most of Greenland and Antarctica.

2. Glaciers move in part by flowing. On the surface of a glacier, ice is brittle. However, below about 50 meters, pressure is great and ice behaves like a *plastic material* and *flows*. A second important mechanism of glacial movement consists of the whole ice mass *slipping* along the ground.

3. Glaciers erode land by *plucking* (lifting pieces of bedrock out of place) and abrasion (grinding and scraping of a rock surface). Erosional features produced by valley glaciers include: *glacial troughs, hanging valleys, cirques, arêtes, horns,* and *fiords.*

4. Any sediment of glacial origin is called *drift.* The two distinct types of glacial drift are: (a) *till,* which is material deposited directly by the ice, and (b) *stratified drift,* which is sediment laid down by meltwater from a glacier.

5. The most widespread features created by glacial deposition are layers or ridges of till, called *moraines.* Associated with valley glaciers are *lateral moraines,* formed along the sides of the valley, and *medial moraines,* formed between two valley glaciers that have joined. *End moraines,* which mark the former position of the front of a glacier, and *ground moraine,* an undulating layer of till deposited as the ice front retreats, are common to both valley glaciers and ice sheets.

6. Perhaps the most convincing evidence for several glacial advances during the *Ice Age* is the widespread existence of *multiple layers of drift* on land and an uninterrupted record of climate cycles preserved in *seafloor sediments.* In addition to massive erosional and depositional work, other effects of Ice Age glaciers included the *forced migration of* animals, *changes in river courses, adjustment of the crust* by rebounding after the removal of the immense load of ice, and *climate changes* caused by the existence of the glaciers themselves. In the sea, the most far-reaching effect of the Ice Age was the *worldwide change in sea level* that accompanied each advance and retreat of the ice sheets.

7. Deserts in the lower latitudes coincide with zones of high air pressure known as *subtropical highs.* Middle latitude deserts exist because of their positions in the deep interiors of large continents, far removed from oceans. Mountains also act to shield these regions from marine air masses.

8. Practically all desert streams are dry most of the time, and are said to be *ephemeral.* Nevertheless, *running water is responsible for most of the erosional work in a desert.* Although wind erosion is more significant in dry areas than elsewhere, the main role of wind in a desert is in the transportation and deposition of sediment.

9. Many of the landscapes of the Basin and Range region of the western United States are the result of streams eroding uplifted mountain blocks and depositing the sediment in interior basins. *Alluvial fans, playas,* and *playa lakes,* are features often associated with these landscapes.

10. In order for wind erosion to be effective, dryness and scant vegetation are essential. *Deflation,* the lifting and removal of loose material, often produces shallow depressions called *blowouts* and can also lower the surface by removing sand and silt, leaving behind a stony veneer, called *desert pavement. Abrasion,* the "sandblasting" effect of wind, is often given too much credit for producing desert features. However, abrasion does cut and polish rock near the surface.

11. Wind deposits are of two distinct types: (a) extensive *blankets of silt,* called *loess,* that was carried by wind in *suspension,* and (b) *mounds and ridges of sand,* called *dunes,* formed from sediment that was carried as part of the wind's *bed load.*

Key Terms

alluvial fan (p. 106)	crevasse (p. 91)	drift (p. 97)	esker (p. 99)
alpine glacier (p. 90)	cross beds (p. 110)	drumlin (p. 99)	fiord (p. 95)
arête (p. 94)	deflation (p. 108)	dune (p. 110)	glacial erratic (p. 98)
blowout (p. 108)	desert (p. 103)	end moraine (p. 98)	glacial striations (p. 93)
cirque (p. 94)	desert pavement (p. 108)	ephemeral stream (p. 103)	glacial trough (p. 94)

glacier (p. 90)

ground moraine (p. 99)

hanging valley (p. 94)

horn (p. 94)

ice sheet (p. 91)

interior drainage (p. 105)

kame (p. 100)

kettle (p. 99)

lateral moraine (p. 98)

loess (p. 109)

medial moraine (p. 98)

outwash plain (p. 99)

playa lake (p. 106)

Pleistocene epoch (p. 101)

plucking (p. 93)

pluvial lake (p. 102)

rock flour (p. 93)

slip face (p. 110)

steppe (p. 103)

stratified drift (p. 98)

till (p. 97)

valley glacier (p. 90)

valley train (p. 99)

zone of accumulation (p. 91)

zone of wastage (p. 91)

Questions for Review

1. What is a glacier? What percentage of Earth's land area do glaciers cover?

2. Contrast valley glaciers and ice sheets.

3. Describe the two components of glacial flow. At what rates do glaciers move? In a valley glacier, does all of the ice move at the same rate? Explain.

4. Why do crevasses form in the upper portion of a glacier but not below a depth of about 50 meters?

5. Under what circumstances will the terminus of a glacier advance? Retreat? Remain stationary?

6. Describe two basic processes by which glaciers erode.

7. How does a glaciated mountain valley differ from an unglaciated mountain valley?

8. List and describe the erosional features you might expect to see in an area where alpine glaciers exist or have recently existed.

9. What is glacial drift? What is the difference between till and stratified drift? What general effect do glacial deposits have on the landscape?

10. List the four basic moraine types. What do all moraines have in common? Distinguish between terminal and recessional moraines.

11. List and briefly describe four depositional features other than moraines.

12. How does a kettle form?

13. About what percentage of Earth's land surface was covered at some time by Pleistocene glaciers? How does this compare to the area presently covered by ice sheets and glaciers? (Check your answer with Question 1.)

14. List three indirect effects of Ice Age glaciers.

15. How extensive are the desert and steppe regions of Earth?

16. Most desert streams are said to be ephemeral. What does that mean?

17. What is the most important erosional agent in deserts?

18. Describe the features and characteristics associated with each of the stages in the evolution of a mountainous desert.

19. Why is wind erosion relatively more important in arid regions than in humid areas?

20. List two types of wind erosion and the features which may result from each.

21. Although sand dunes are the best-known wind deposits, accumulations of loess are very significant in some parts of the world. What is loess? Where are such deposits found? What are the origins of this sediment?

22. How do sand dunes migrate?

Testing What You Have Learned

Multiple-Choice Questions

1. The combined areas of present-day continental ice sheets cover about what percentage of Earth's land surface?
   a. 5%   **c.** 15%   **e.** 25%
   b. 10%   **d.** 20%

2. The uppermost 50 meters of a glacier is referred to as the zone of _____
   a. flow.   **c.** fracture.   **e.** slipping.
   b. abrasion.   **d.** drift.

3. When ice accumulation equals ice wastage, the front of the glacier will _____
   a. reverse.   **c.** flow downhill.   **e.** remain stationary.
   b. disappear. **d.** flow uphill.

4. Which one of the following is NOT a feature of valley glaciation?
   a. glacial trough   **c.** horn   **e.** fiord
   b. cirque   **d.** levee

5. Which one of the following is deposited directly by a glacier?
   a. till   **c.** loess   **e.** stratified drift
   b. sandstone   **d.** granite

6. Layers or ridges of glacial till are _____
   a. kettles.   **c.** moraines.   **e.** valley trains.
   b. kames.   **d.** drumlins.

7. Most of the major glacial episodes of the recent Ice Age occurred during a division of geologic time called the _____
   a. Mesozoic era.   **c.** Eocene epoch.   **e.** Hadean eon.
   b. Pliocene epoch. **d.** Pleistocene epoch.

8. About what percentage of Earth's land surface do dry regions encompass?
   a. 10%   **c.** 30%   **e.** 50%
   b. 20%   **d.** 40%

9. Most of the erosional work in deserts is accomplished by _____
   a. running water.   **c.** wind.   **e.** ice.
   b. heat.   **d.** groundwater.

10. Many arid areas, including the Basin and Range region of the United States, are characterized by this type of drainage.
    a. slow   **c.** humid   **e.** swampy
    b. exterior   **d.** interior

Fill-In Questions

11. Two basic types of glaciers are _____ glaciers and _____.

12. The movement of glacial ice is generally referred to as _____.

13. Glacial sediments, no matter how, where, or in what form they were deposited, are called _____.

14. _____ moraines form at the terminus of a glacier.

15. The two climatic types commonly recognized within the dry regions of the world are _____, or arid, and _____, or semiarid.

True/False Questions

16. Under a pressure equivalent to the weight of about 50 meters of ice, glacial ice behaves as a plastic material. _____

17. Erosion by valley glaciers in mountainous areas tends to produce spectacular features and rugged scenery. _____

18. During the Ice Age, sea level was as much as 100 meters higher than today. _____

19. Middle latitude deserts and steppes exist principally because of their positions beneath zones of high air pressure known as subtropical highs. _____

20. Most desert streams carry water only in response to specific episodes of rainfall. _____

Answers:

1. b; 2. c; 3. e; 4. 5. a; 6. c; 7. d; 8. c; 9. a; 10. d; 11. valley (alpine); ice sheets; 12. flow; 13. drift; 14. End; 15. desert, steppe; 16.T; 17. T; 18. F; 19. F; 20. T.

Web Work

The following are informative and interesting World Wide Web sites that address topics related to those presented in this chapter:

- Glacial Internet Links
  http://boris.qub.ac.uk/ggg/resources/internet/index.html

- World's Dryland Exhibit
  http://drylands.nasm.edu:1995/

For direct access to these sites and other relevant Web locations, contact the *Foundations of Earth Science* WWW home page at http://www.prenhall.com/lutgens.

UNIT THREE
Earth's Internal Processes

El Capitan, Yosemite National Park, California.
Photo by Jeff Gnass Photography

CHAPTER 5

Plate Tectonics

A Unifying Theory

FOCUS ON LEARNING

To assist you in learning the important concepts
in this chapter, you will find it helpful to focus
on the following questions:

1. What lines of evidence were used to support the continental drift hypothesis?

2. What was one of the main objections to the continental drift hypothesis?

3. What is the theory of plate tectonics?

4. In what major way does the plate tectonics theory depart from the continental drift hypothesis?

5. What are the three types of plate boundaries?

6. What is the evidence used to support the plate tectonics theory?

7. What models have been proposed to explain the driving mechanism for plate motion?

Composite satellite image of part of North Africa and the Arabian Peninsula. *Image © Worldsat International, Inc./Jim Knighton, 1995. Mississauga, Ontario, Canada. All Rights Reserved.*

Early in this century, most geologists believed that the geographic positions of the ocean basins and continents were fixed. During the last few decades, however, vast amounts of new data have dramatically changed our understanding of the nature and workings of our planet. Earth scientists now realize that the continents gradually migrate across the globe. Where landmasses split apart, new ocean basins are created between the diverging blocks. Meanwhile, older portions of the seafloor are carried back into the mantle in regions where trenches occur in the deep ocean floor. Because of these movements, blocks of continental material eventually collide and form Earth's great mountain ranges. In short, a revolutionary new model of Earth's tectonic* processes has emerged.

This profound reversal of scientific understanding has been appropriately described as a scientific revolution. Like other scientific revolutions, considerable time elapsed between the idea's inception and its general acceptance. The revolution began early in the twentieth century as a relatively straightforward proposal that the continents drift about the face of Earth. After many years of heated debate, the idea of drifting continents was rejected by the vast majority of Earth scientists. However, during the 1950s and 1960s, new evidence rekindled interest in this proposal. By 1968, these new developments led to the unfolding of a far more encompassing theory than continental drift—a theory known as *plate tectonics*.

Continental Drift: An Idea Before Its Time

The idea that continents, particularly South America and Africa, fit together like pieces of a jigsaw puzzle originated with improved world maps. However, little significance was given this idea until 1915, when Alfred Wegener, a German meteorologist and geophysicist, published *The Origin of Continents and Oceans*. In this book, Wegener set forth his radical hypothesis of **continental drift**.†

Wegener suggested that a supercontinent he called **Pangaea** (meaning "all land") once existed (Figure 5.1). He further hypothesized that, about 200 million years ago, this supercontinent began breaking into smaller continents, which then "drifted" to their present positions.

Wegener and others collected substantial evidence to support these claims. The fit of South America and Africa, and the geographic distribution of fossils, rock

---

*Tectonics refers to the deformation of Earth's crust and results in the formation of structural features such as mountains.

†Wegeners's ideas were actually preceded by those of an American geologist, F. B. Taylor, who in 1910 published a paper on continental drift. Taylor's paper provided little supporting evidence for continental drift, which may have been the reason that it had a relatively small impact on the scientific community.

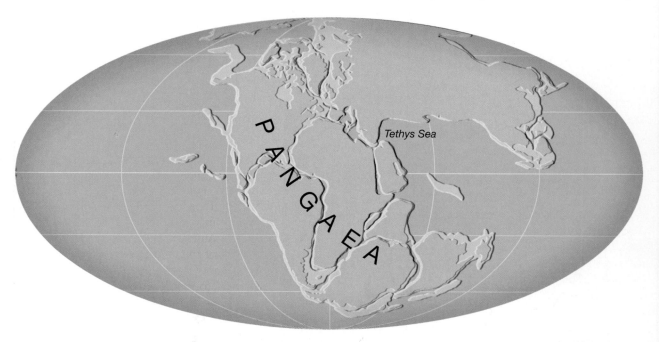

Figure 5.1 Reconstruction of Pangaea as it is thought to have appeared 200 million years ago. *(After R. S. Deitz and J. C. Holden.* Journal of Geophysical Research *75: 4943. Copyright by American Geophysical Union)*

structures, and ancient climates all seemed to support the idea that these now-separate landmasses were once joined. Let us examine their evidence.

Evidence: The Continental Jigsaw Puzzle

Like a few others before him, Wegener first suspected that the continents might have been joined when he noticed the remarkable similarity between the coastlines on opposite sides of the South Atlantic. However, his use of present-day shorelines to make a fit of the continents was challenged immediately by other Earth scientists. These opponents correctly argued that shorelines are continually modified by erosional processes, and even if continental displacement had taken place, a good fit today would be unlikely. Wegener appeared to be aware of this problem, and, in fact, his original jigsaw fit of the continents was only very crude.

A much better approximation of the true outer boundary of the continents is the continental shelf. Today, the seaward edge of the continental shelf lies submerged, several hundred meters below sea level. In the early 1960s, scientists produced a map that attempted to fit the edges of the continental shelves at a depth of 900 meters. The remarkable fit that was obtained is shown in Figure 5.2. Although the continents overlap in a few places, these are regions where streams have deposited large quantities of sediment, thus enlarging the continental shelves. The overall fit was even better than the supporters of continental drift suspected it would be.

Evidence: Fossils Match Across the Seas

Although Wegener was intrigued by the jigsaw fit of the continental margins that lie on opposite sides of the Atlantic, he at first thought the idea of a mobile Earth improbable. Not until he came across an article citing fossil evidence for the existence of a land bridge connecting South America and Africa did he begin to take his own idea seriously. Through a search of the literature, Wegener learned that most paleontologists were in agreement that some type of land connection was needed to explain the existence of identical fossils on the widely separated landmasses.

To add credibility to his argument for the existence of the supercontinent of Pangaea, Wegener cited documented cases of several fossil organisms that had been found on different landmasses but which could not have crossed the vast oceans presently separating the continents. The classic example is *Mesosaurus*, a presumably aquatic, snaggle-toothed reptile whose fossil remains are limited

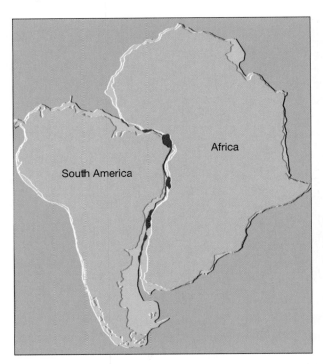

Figure 5.2 This shows the best fit of South America and Africa along the continental slope at a depth of 500 fathoms (about 900 meters). The areas where continental blocks overlap appear in brown. *(After A. G. Smith. "Continental Drift," in Understanding the Earth, edited by I.G. Gass. Courtesy of Artemis Press.)*

to eastern South America and southern Africa (Figure 5.3). If *Mesosaurus* had been able to swim well enough to cross the vast South Atlantic Ocean, its remains should be more widely distributed. As this is not the case, Wegener argued that South America and Africa must have been joined—somehow.

How did scientists explain the discovery of identical fossil organisms separated by thousands of kilometers of open ocean? The idea of land bridges was the most widely accepted solution to the problem of migration. We know, for example, that during the most recent glacial period, the lowering of sea level allowed animals to cross the narrow Bering Strait between Asia and North America. Was it possible, then, that one or more land bridges once connected Africa and South America? We are now quite certain that land bridges of this magnitude did not exist, for their remnants should still lie below sea level. But they are nowhere to be found.

Evidence: Rock Types and Structures Match

Anyone who has worked a picture puzzle knows that, in addition to the pieces fitting together, the picture

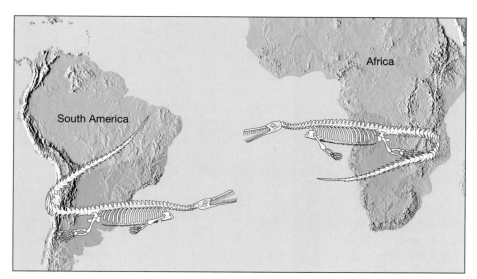

Figure 5.3 Fossils of *Mesosaurus* have been found on both sides of the South Atlantic and nowhere else in the world. Fossil remains of this and other organisms on the continents of Africa and South America appear to link these landmasses during the late Paleozoic and early Mesozoic eras.

must be continuous as well. The "picture" that must match in the "Continental Drift Puzzle" is one of rock types and mountain belts. If the continents were once together to form Pangaea, the rocks found in a particular region on one continent should closely match in age and type those in adjacent positions on the adjoining continent.

Such evidence exists in the form of several mountain belts that terminate at one coastline, only to reappear on a landmass across the ocean. For instance, the mountain belt that includes the Appalachians trends northeastward through the eastern United States and disappears off the coast of Newfoundland (Figure 5.4A). Mountains of comparable age and structure are found in the British Isles and Scandinavia. When these landmasses are reassembled, as in Figure 5.4B, the mountain chains form a nearly continuous belt. Numerous other rock structures exist that appear to have formed at the same time and were subsequently split apart.

Wegener was very satisfied that the similarities in rock structure on both sides of the Atlantic linked these landmasses. In his own words, "It is just as if we were to refit the torn pieces of a newspaper by matching their edges and then check whether the lines of print run smoothly across. If they do, there is nothing left but to conclude that the pieces were in fact joined in this way."*

---

*Alfred Wegener, *The Origin of Continents and Oceans.* Translated from the 4th revised German edition of 1929 by J. Birman (London: Methuen, 1966).

Evidence: Ancient Climates

Because Alfred Wegener was a meteorologist by training, he was keenly interested in obtaining paleoclimatic (ancient climatic) data in support of continental drift. His efforts were rewarded when he found evidence for dramatic global climatic changes. For instance, glacial deposits indicate that, near the end of the Paleozoic era (between 220 million and 300 million years ago), ice sheets covered extensive areas of the Southern Hemisphere. Layers of glacial till were found in southern Africa and South America, as well as in India and Australia. Below these beds of glacial debris lay striated and grooved bedrock. In some locations, the striations and grooves indicated that the ice had moved from what is now the sea onto land. Much of the land area containing evidence of this late Paleozoic glaciation presently lies within 30 degrees of the equator in a subtropical or tropical climate.

Could Earth have gone through a period sufficiently cold to have generated extensive continental glaciers in what is presently a tropical region? Wegener rejected this explanation because, during the late Paleozoic, large tropical swamps existed in the Northern Hemisphere. The lush vegetation of these swamps eventually became the major coal fields of the eastern United States, Europe, and Siberia.

Fossils from these coal fields indicate that the tree ferns which produced the coal deposits had large fronds. This indicates a tropical setting. Furthermore, unlike trees in colder climates, the tree trunks lacked growth rings. Growth rings do not

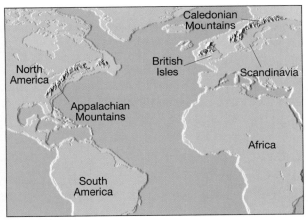

A.

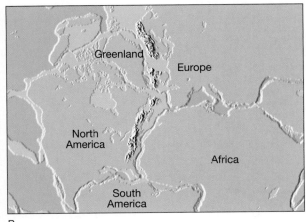

B.

Figure 5.4 Matching mountain ranges across the North Atlantic. **A.** The Appalachian Mountains trend along the eastern flank of North America and disappear off the coast of Newfoundland. Mountains of comparable age and structure are found in the British Isles and Scandinavia. **B.** When these landmasses are placed in their predrift locations, these ancient mountain chains form a nearly continuous belt. These folded mountain belts formed roughly 300 million years ago as the landmasses collided during the formation of the supercontinent of Pangaea.

form in tropical plants because there are minimal seasonal fluctuations in temperature.

Wegener believed that a better explanation for the paleoclimatic regimes he observed is provided by fitting together the landmasses as a supercontinent, with South Africa centered over the South Pole (Figure 5.5). This would account for the conditions necessary to generate extensive expanses of glacial ice over much of the Southern Hemisphere. At the same time, this geography would place the northern landmasses nearer the tropics and account for their vast coal deposits.

Wegener was so convinced that his explanation was correct that he wrote, "This evidence is so compelling that by comparison all other criteria must take a back seat."

How does a glacier develop in hot, arid Australia? How do land animals migrate across wide expanses of open water? As compelling as this evidence may have been, 50 years passed before most of the scientific community would accept it and the logical conclusions to which it led.

The Great Debate

Wegener's proposal did not attract much open criticism until 1924 when his book was translated into English. From this time on, until his death in 1930, his drift hypothesis encountered a great deal of hostile criticism. To quote the respected American geologist T.C. Chamberlin, "Wegener's hypothesis...takes considerable liberty with our globe, and is less bound by restrictions or tied down by awkward, ugly facts than most of its rival theories. Its appeal seems to lie in the fact that it plays a game in which there are few restrictive rules and no sharply drawn code of conduct."

One of the main objections to Wegener's hypothesis stemmed from his inability to provide a mechanism that was capable of moving the continents across the globe. Wegener proposed that the tidal influence of the Moon was strong enough to give the continents a westward motion. However, the prominent physicist Harold Jeffreys quickly countered with the argument that tidal friction of the magnitude needed to displace the continents would bring Earth's rotation to a halt in a matter of a few years.

Wegener also proposed that the larger and sturdier continents broke through the oceanic crust, much like ice breakers cut through ice. However, no evidence existed to suggest that the ocean floor was weak enough to permit passage of the continents without themselves being appreciably deformed in the process.

Although most of Wegener's contemporaries opposed his views, even to the point of open ridicule, a few considered his ideas plausible. For these few geologists who continued the search for additional evidence, the exciting concept of continents in motion held their interest. Others viewed continental drift as a solution to previously unexplainable observations.

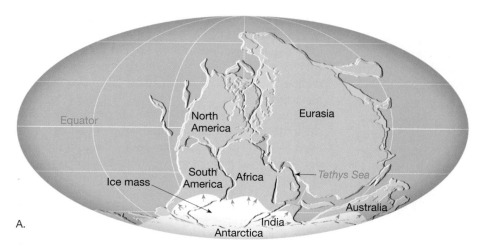

Figure 5.5 A. The supercontinent Pangaea showing the area covered by glacial ice 300 million years ago. **B.** The continents as they are today. The shading outlines areas where evidence of the old ice sheets exists.

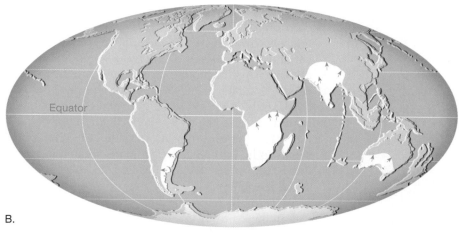

Plate Tectonics:
A Modern Version of an Old Idea

During the years that followed Wegener's proposal, major strides in technology permitted mapping of the ocean floor. Moreover, extensive data on seismic activity and Earth's magnetic field became available. By 1968, these developments led to the unfolding of a far more encompassing theory than continental drift, known as **plate tectonics**. The implications of plate tectonics are so far-reaching that this theory is today the framework within which to view most geologic processes.

According to the plate tectonics model, the uppermost mantle, along with the overlying crust, behave as a strong, rigid layer, known as the *lithosphere*. This outermost shell overlies a weaker region in the mantle known as the *asthenosphere*. Further, the lithosphere is broken into numerous segments called **plates**, which

are in motion and are continually changing in shape and size. As shown in Figure 5.6, seven major plates are recognized. They are the North American, South American, Pacific, African, Eurasian, Australian, and Antarctic plates. The largest is the Pacific plate, which is located mostly within the ocean. Notice from Figure 5.6 that several of the large plates include an entire continent plus a large area of seafloor (for example, the South American plate). This is a major departure from Wegener's continental drift hypothesis, which proposed that the continents moved through the ocean floor, not with it. Note also that none of the plates are defined entirely by the margins of a continent.

Intermediate-sized plates include the Caribbean, Nazca, Philippine, Arabian, Cocos, and Scotia plates. In addition, there are over a dozen smaller plates that have been identified but are not shown in Figure 5.6.

We know that lithospheric plates move at very slow, but continuous, rates of a few centimeters a year. This movement is ultimately driven by the unequal distribution of heat within Earth. The titanic grinding movements of Earth's lithospheric plates generate earthquakes, create volcanoes, and deform large masses of rock into mountains.

Plate Boundaries

Plates move as coherent units relative to all other plates. Although the interiors of plates may deform, all major interactions among individual plates (and therefore most deformation) occur along their *boundaries*. In fact, the first attempts to outline plate boundaries were made using locations of earthquakes. Later work showed that plates are bounded by three distinct types of boundaries, which are differentiated by the type of movement they exhibit. These boundaries are depicted in Figure 5.7 and are briefly described here:

1. **Divergent boundaries**—where plates move apart, resulting in upwelling of material from the mantle to create new seafloor (Figure 5.7A).
2. **Convergent boundaries**—where plates move together, resulting in the subduction (consumption) of oceanic lithosphere into the mantle (Figure 5.7B).
3. **Transform fault boundaries**—where plates grind past each other without the production or destruction of lithosphere (Figure 5.7C).

Each plate is bounded by a combination of these zones, as you can see in Figure 5.6. For example, the Nazca plate has a divergent boundary on the west, a convergent boundary on the east, and numerous small transform faults that offset segments of divergent boundaries on the north and south.

Divergent Boundaries (Spreading)

Most divergent boundaries, where plate spreading occurs, are situated along the crests of oceanic ridges (Figure 5.8). Here, as the plates move away from the ridge axis, the fractures created are immediately filled with molten rock that oozes up from the hot asthenosphere. This hot material cools to hard rock, producing new slivers of seafloor. In a continuous manner, successive plate spreading and upwelling of magma add new oceanic crust (lithosphere) between the diverging plates.

This mechanism is called **seafloor spreading**. It has produced the floor of the Atlantic Ocean during the past 165 million years. The rate of spreading at these ridges ranges between 2 and 20 centimeters per year, and averages about 5 centimeters (2 inches) per year. This extremely slow rate of lithosphere production is nevertheless rapid enough so that all of Earth's ocean basins could have been generated within the last 200 million years. In fact, none of the ocean floor that has been dated exceeds 180 million years in age.

Along divergent boundaries where molten rock emerges, the ocean floor is elevated. Worldwide, this ridge extends for over 70,000 kilometers through all major ocean basins. You can see parts of the mid-ocean ridge system in Figure 5.6. As new lithosphere is formed along the oceanic ridge it is slowly, yet continually, displaced away from the zone of upwelling found along the ridge axis. Thus, it begins to cool and contract, thereby increasing in density. This partially accounts for the greater depth of the older and cooler oceanic crust found in the deep ocean basins. In addition, the mantle rocks located below the oceanic crust cool and strengthen with increased distance from the ridge axis, thereby adding to the plate thickness. Stated another way, both the thickness and the density of oceanic lithosphere are age dependent. The older (cooler) it is, the greater its thickness and density.

Not all spreading centers are found in the middle of large oceans. The Red Sea is the site of a recently formed divergent boundary. Here the Arabian Peninsula separated from Africa and began to move toward the northeast. Consequently, the Red Sea is providing oceanographers with a view of how the Atlantic Ocean may have looked in its infancy. Another narrow, linear sea produced by seafloor spreading in the recent geologic past is the Gulf of California, which separates the Baja Peninsula from the rest of Mexico.

When a spreading center develops within a continent, the landmass may split into two or more smaller segments as Alfred Wegener had proposed for the breakup of Pangaea. The fragmentation of a continent is thought to be associated with the upward movement of hot rock from the mantle. The effect of this activity is doming the crust directly above the hot rising plume. This uplifting produces extensional forces that stretch and thin the crust as shown in Figure 5.9A. Extension of the crust is accompanied by alternate episodes of faulting and

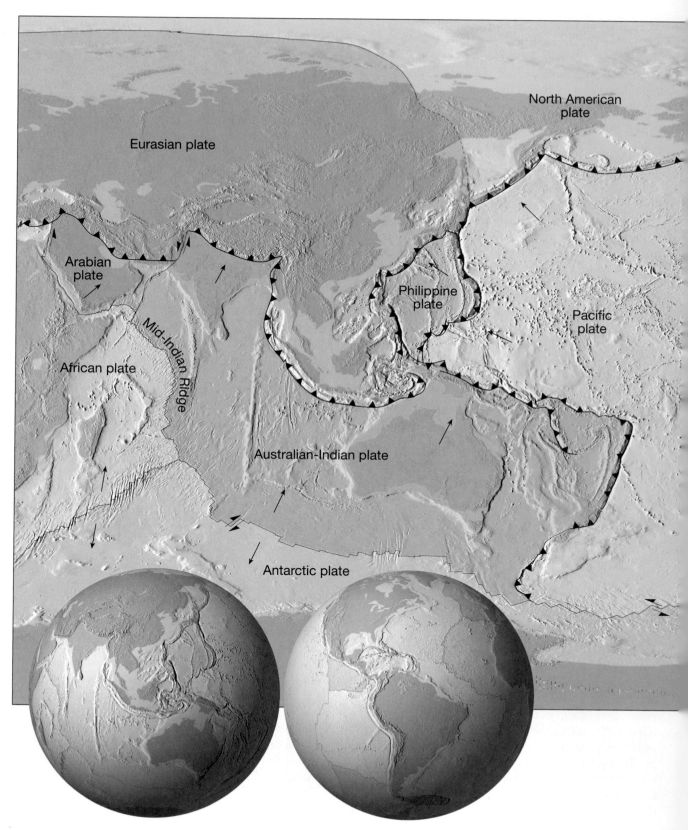

Figure 5.6 A mosaic of rigid plates constitutes Earth's outer shell. *(After W. B. Hamilton and others.)*

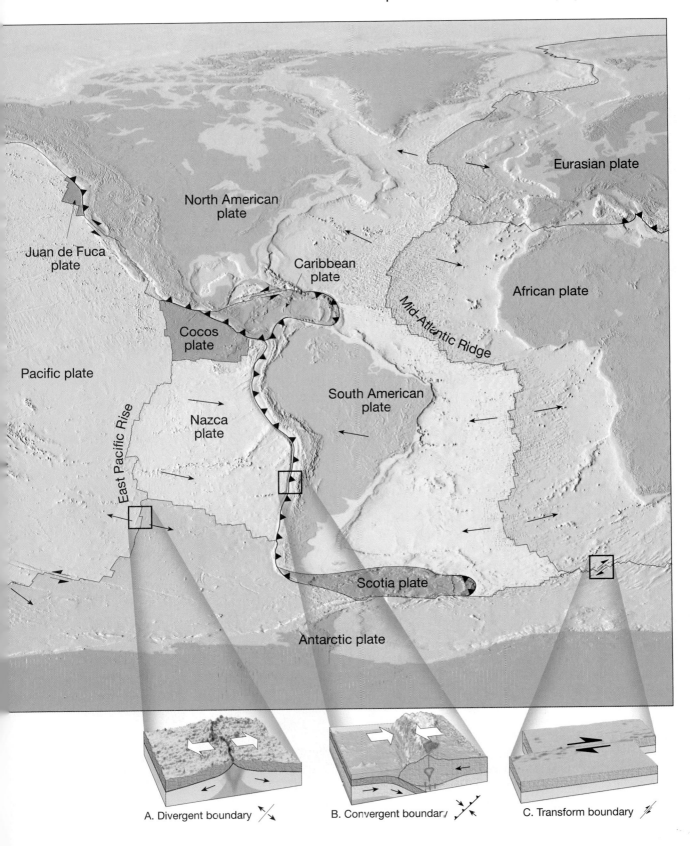

Juan de Fuca plate

North American plate

Eurasian plate

Caribbean plate

African plate

Mid-Atlantic Ridge

Cocos plate

Pacific plate

East Pacific Rise

Nazca plate

South American plate

Scotia plate

Antarctic plate

A. Divergent boundary

B. Convergent boundary

C. Transform boundary

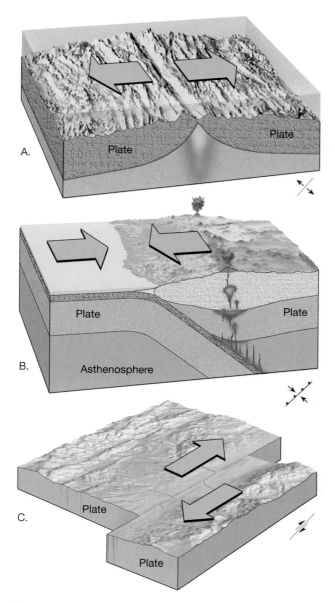

Figure 5.7 The three types of plate boundaries.
A. Divergent boundary. **B.** Convergent boundary.
C. Transform boundary.

volcanism. Adjacent to the spreading axis, faulted crustal blocks create downfaulted valleys called **rifts** or **rift valleys** (Figure 5.9B). As the spreading continues, the rift valley will lengthen and deepen, eventually extending out into the ocean. At this point the valley will become a narrow linear sea with an outlet to the ocean, similar to the Red Sea today (Figure 5.9C). The zone of rifting will remain the site of igneous activity, continually generating new seafloor in an ever-expanding ocean basin (Figure 5.9D).

The East African rift valleys represent the initial stage in the breakup of a continent as just described (Figure 5.10). The extensive volcanic activity that accompanies continental rifting is exemplified by large volcanic mountains such as Kilimanjaro and Mount Kenya. If the rift valleys in Africa remain active, East Africa will eventually separate from the mainland in much the same way the Arabian Peninsula split just a few million years ago.

Not all rift valleys develop into full-fledged spreading centers. Running through the central United States is an aborted rift zone extending from

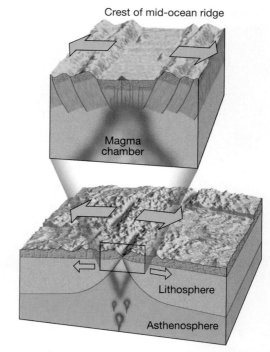

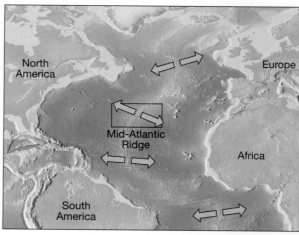

Figure 5.8 Most divergent plate boundaries are situated along the crests of oceanic ridges.

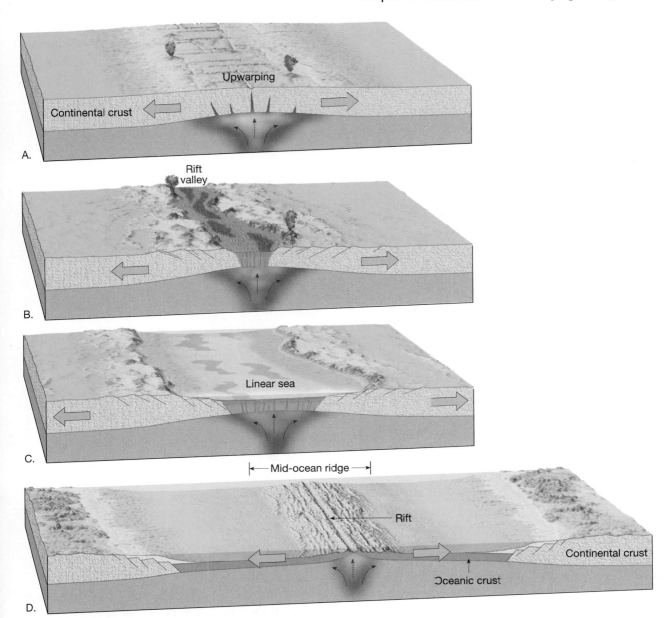

A.

B.

C.

D.

Figure 5.9 A. Rising magma forces the crust upward, causing numerous cracks in the rigid lithosphere. **B.** As the crust is pulled apart, large slabs of rock sink, generating a rift zone. **C.** Further spreading generates a narrow sea. **D.** Eventually, an expansive ocean basin and ridge system are created.

Lake Superior to Kansas. This once-active rift valley is filled with rock that was extruded onto the crust more than a billion years ago. Why one rift valley develops into a full-fledged oceanic spreading center while others are abandoned is not yet known.

Convergent Boundaries

At spreading centers, new oceanic lithosphere is continually being generated. However, because the total surface area of Earth remains constant, lithosphere must also be consumed. Zones of plate convergence are the sites where oceanic lithosphere is subducted and absorbed into the mantle. When two plates converge, the leading edge of one is bent downward, allowing it to descend beneath the other (Figure 5.11A). The region where an oceanic plate descends into the asthenosphere is called a **subduction zone**. As the oceanic plate slides beneath the overriding

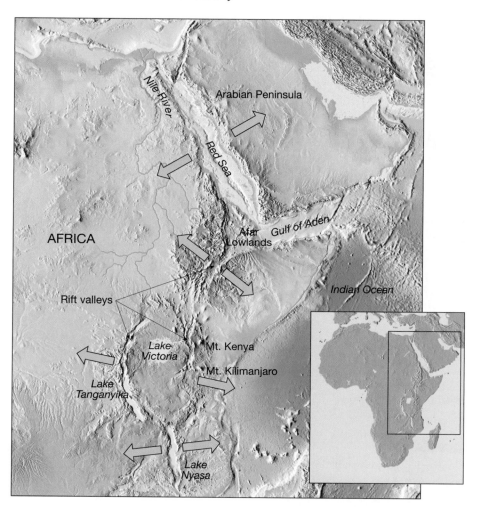

Figure 5.10 East African rift valleys and associated features.

plate, the oceanic plate bends, thereby producing a **deep-ocean trench**, like the Peru–Chile trench (Figure 5.12). Trenches formed in this manner may be thousands of kilometers long, 8 to 12 kilometers deep, and about 100 kilometers wide.

The average angle at which oceanic lithosphere descends into the asthenosphere is about 45 degrees. However, depending on its buoyancy, a plate may descend at an angle as small as a few degrees or it may plunge vertically (90 degrees) into the mantle. When a spreading axis is located near a subduction zone, the lithosphere is young and, therefore, warm and buoyant. Consequently, the angle of descent is small. This is the situation along parts of the Peru–Chile trench. Low dip angles are usually associated with a strong coupling between the descending slab and the overriding plate. As a result, these regions experience great earthquakes. By contrast, some subduction zones, such as

the Mariana trench, have steep angles of descent and few strong earthquakes.

Although all convergent zones have the same basic characteristics, they are highly variable features. Each is controlled by the type of crustal material involved and the tectonic setting. Convergent boundaries can form between two oceanic plates, one oceanic and one continental plate, or two continental plates. All three situations are illustrated in Figure 5.11.

Oceanic-Continental Convergence Whenever the leading edge of a plate capped with continental crust converges with a plate capped with oceanic crust, the plate with the less dense continental material remains "floating," while the denser oceanic slab sinks into the asthenosphere (Figure 5.11A). When a descending plate reaches a depth of about 100 to 150 kilometers, heat drives water and other

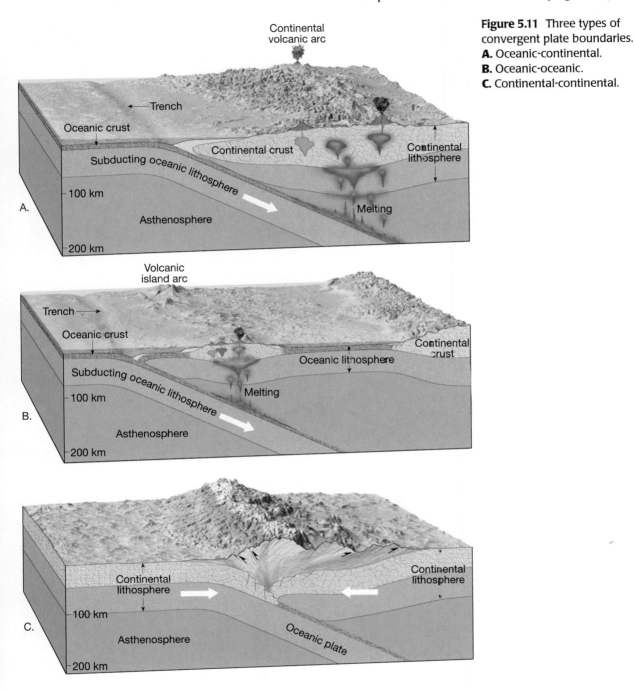

Figure 5.11 Three types of convergent plate boundaries.
A. Oceanic-continental.
B. Oceanic-oceanic.
C. Continental-continental.

volatile components from the subducted sediments into the overlying mantle. These substances act as a flux does at a foundry, inducing partial melting of mantle rocks at reduced temperatures. The partial melting of mantle rock generates magmas having a basaltic or possibly andesitic composition. The newly formed magma, being less dense than the rocks of the mantle, will buoyantly rise. Often the magma will pond (accumulate) beneath the overlying continental crust where it may melt some of the silica-enriched crustal rocks. Eventually some of this silica-rich magma may migrate to the surface, where it can give rise to volcanic eruptions, some of which are explosive.

The Andean arc that runs along the western flank of South America is the product of magma generated as the Nazca plate descends beneath the continent (see Figure 5.6). In the central section of the

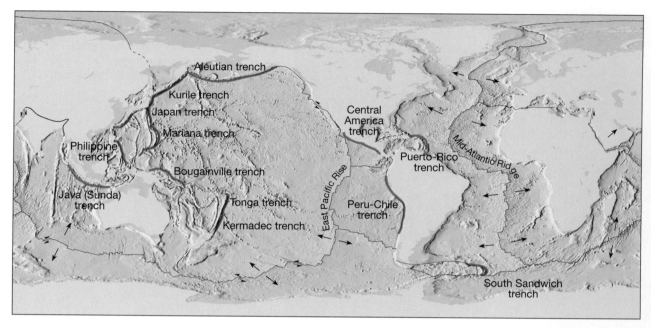

Figure 5.12 Distribution of the world's oceanic trenches, ridge system, and transform faults. Where transform faults offset ridge segments, they permit the ridge to change direction (curve) as can be seen in the Atlantic Ocean.

Southern Andes the subduction angle is very shallow, which probably accounts for the lack of volcanism in that area. As the South American plate moves westward, it overruns the Nazca plate. The result is a seaward migration of the Peru–Chile trench and a reduction in the size of the Nazca plate.

Mountains such as the Andes, which are produced in part by volcanic activity associated with the subduction of oceanic lithosphere, are called **continental volcanic arcs**. Another active continental volcanic arc is located in the western United States. The Cascade Range of Washington, Oregon, and California consists of several well-known volcanic mountains, including Mount Rainier, Mount Shasta, and Mount St. Helens. As the continuing activity of Mount St. Helens testifies, the Cascade Range is still active. The magma here arises from the melting of a small remaining segment of the Farallon plate, of which the Juan de Fuca plate is the largest northern segment.

A remnant of a formerly extensive continental volcanic arc is California's Sierra Nevada, in which Yosemite National Park is located. The Sierra Nevada is much older than the Cascade Range, and has been inactive for several million years as evidenced by the absence of volcanic cones. Here erosion has stripped away most of the obvious traces of volcanic activity and left exposed the large, crystallized magma chambers that once fed lofty volcanoes.

Oceanic-Oceanic Convergence When two oceanic slabs converge, one descends beneath the other, initiating volcanic activity in a manner similar to that which occurs at an oceanic-continental convergent boundary. In this case, however, the volcanoes form on the ocean floor rather than on a continent (Figure 5.11B). If this activity is sustained, it will eventually build volcanic structures that emerge as islands. The volcanic islands are spaced about 80 kilometers apart and are built upon submerged ridges a few hundred kilometers wide. This newly formed land consisting of an arc-shaped chain of small volcanic islands is called a **volcanic island arc**. The Aleutian, Mariana, and Tonga islands are examples of volcanic island arcs. Island arcs such as these are generally located 200 to 300 kilometers from an trench axis. Located adjacent to the island arcs just mentioned are the Aleutian trench, Mariana trench, and the Tonga trench (Figure 5.12).

Only two volcanic island arcs are located in the Atlantic, the Lesser Antilles arc adjacent to the Caribbean Sea, and the Sandwich Islands in the South Atlantic. The Lesser Antilles are a product of the subduction of the Atlantic plate beneath the Caribbean plate. Located within this arc is the island of Martinique where Mount Pelée erupted in 1902 destroying the town of St. Pierre and killing an estimated 28,000 people, and the island of Monstserrat, where volcanic activity has occurred very recently.

In a few places, volcanic arcs are built upon both oceanic and continental crust. For example, the western section of the Aleutian arc consists of numerous islands built on oceanic crust, whereas the volcanoes at the eastern end of the chain are located on the Alaska Peninsula. Further, some volcanic island arcs are built on fragments of continental crust that have been rifted from the mainland. This type of volcanic island arc is exemplified by the Philippines and Japan.

Continental-Continental Convergence   When two plates carrying continental crust converge, neither plate will subduct beneath the other because of the low density and thus the buoyant nature of continental

rocks. The result is a collision between the two continental blocks (Figure 5.11C). Such a collision occurred when the subcontinent of India "rammed" into Asia and produced the Himalayas—the most spectacular mountain range on Earth (Figure 5.13). During this collision, the continental crust buckled, fractured, and was generally shortened and thickened. In addition to the Himalayas, several other major mountain systems, including the Alps, Appalachians, and Urals, formed during continental collisions.

Prior to a continental collision, the landmasses involved are separated by an ocean basin. As the continental blocks converge, the intervening seafloor is subducted beneath one of the plates. Subduction initiates partial melting in the overlying mantle rocks, which, in turn, results in the growth of a volcanic arc.

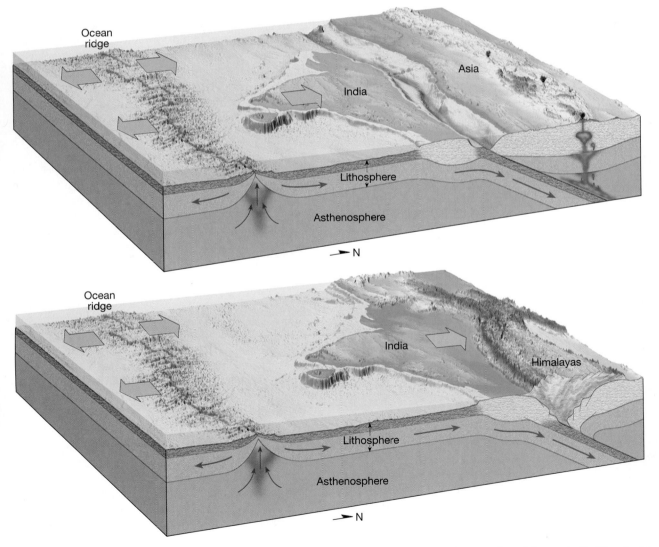

Figure 5.13 The ongoing collision of India and Asia, starting about 45 million years ago, produced the majestic Himalayas.

Depending on the location of the subduction zone, the volcanic arc could develop on either of the converging landmasses, or if the subduction zone developed several hundred kilometers seaward from the coast, a volcanic island arc would form. Eventually, as the intervening seafloor is consumed, these continental masses collide. This folds and deforms the accumulation of sediments along the continental margin as if they were placed in a gigantic vise. The result is the formation of a new mountain range, composed of deformed and metamorphosed sedimentary rocks, fragments of the volcanic arc, and possibly slivers of oceanic crust.

After continents collide, the subducted oceanic plate may separate from the continental block and continue its downward movement. However, because of its buoyancy, continental lithosphere cannot be carried very far into the mantle. In the case of the Himalayas, the leading edge of the Indian plate was forced partially under Asia, generating an unusually great thickness of continental lithosphere. This accumulation accounts, in part, for the high elevation of the Himalayas and helps explain the elevated Tibetan Plateau to the north.

Transform Fault Boundaries

The third type of plate boundary is the transform fault, where plates grind past one another without the production or destruction of lithosphere. Transform faults were first identified where they join offset segments of an ocean ridge. At first it was erroneously assumed that the ridge system originally formed a long continuous chain that was later offset by horizontal displacement along these large faults. However, the displacement along these faults was found to be in the exact opposite direction required to produce the offset ridge segments.

The true nature of transform faults was discovered in 1965 by a Canadian researcher who suggested that these large faults connect the global active belts (convergent boundaries, divergent boundaries, and other transform faults) into a continuous network that divides Earth's outer shell into several rigid plates. Thus, he became the first to suggest that Earth was made of individual plates, while at the same time identifying the faults along which relative motion between the plates is made possible.

Most transform faults join two segments of a mid-ocean ridge (Figure 5.14). Here, they are part of prominent linear breaks in the oceanic crust known as *fracture zones,* which include both the transform fault and their inactive extension into the plate interior. These fracture zones are present approximately every 100 kilometers along the trend of a ridge axis. As shown in Figure 5.14, active transform faults lie only between the two offset ridge segments. Here seafloor produced at one ridge axis moves in the opposite direction as seafloor produced at an opposing ridge segment. Thus, between the ridge segments these adjacent slabs of oceanic crust are grinding past each other along a transform fault. Beyond the ridge crests are the inactive zones, which are preserved as linear topographic scars. These fracture zones tend to curve such that small segments roughly parallel the direction of plate motion at the time of their formation.

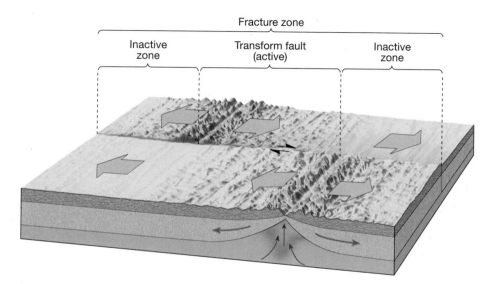

Fracture zone

Inactive zone — Transform fault (active) — Inactive zone

Figure 5.14 Illustration of a transform fault joining segments of an oceanic ridge.

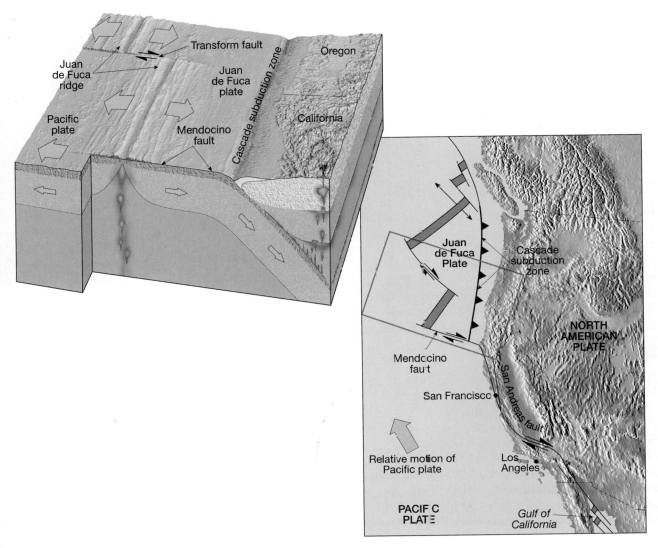

Figure 5.15 The role of transform faults. The Mendocino transform fault permits sea floor generated at the Juan de Fuca ridge to move southeastward past the Pacific plate and beneath the North American plate. Thus, this transform fault connects a divergent boundary to a subduction zone. Furthermore, the San Andreas fault, also a transform fault, connects two spreading centers; the Juan de Fuca ridge and a divergent zone located in the Gulf of California.

In another role, transform faults provide the means by which the oceanic crust created at ridge crests can be transported to a site of destruction, the deep-ocean trenches. Figure 5.15 illustrates this situation. Notice that the Juan de Fuca plate moves in a southeasterly direction, eventually being subducted under the western coast of the United States. The southern end of this relatively small plate is bounded by the Mendocino transform fault. This transform fault boundary connects the Juan de Fuca ridge to the Cascade subduction zone (Figure 5.15). Therefore, it facilitates the movement of the crustal

material created at the ridge crest to its destination beneath the North American continent (Figure 5.15). Another example of a *ridge-trench* transform fault is found southeast of the tip of South America. Here transform faults on the north and south margins of the South Sandwich plate connect the trench to a short spreading axis (see Figure 5.6).

Although most transform faults are located within the ocean basins, a few, including California's famous San Andreas fault, cut through continental crust. Notice in Figure 5.15 that the San Andreas fault connects a spreading center located in the Gulf of California to the

Cascade subduction zone and the Mendocino transform fault located along the northwest coast of the United States. Along the San Andreas fault, the Pacific plate is moving toward the northwest, past the North American plate. If this movement continues, that part of California west of the fault zone, including the Baja Peninsula, will become an island off the western coast of the United States and Canada. It could eventually reach Alaska. However, a more immediate concern is the earthquake activity triggered by movements along this fault system.

Testing the Plate Tectonics Model

With the birth of the plate tectonics model, researchers from all of the Earth sciences began testing it. Some of the evidence supporting continental drift and seafloor spreading has already been presented. Some of the evidence that was instrumental in solidifying the support for this new concept follows. Note that some of the data was not new; rather, it was a new interpretation of old data that swayed the tide of opinion.

Evidence: Paleomagnetism

Probably the most persuasive evidence to the geologic community for the acceptance of the plate tectonics theory comes from the study of Earth's magnetic field. Anyone who has used a compass to find direction knows that the magnetic field has a north pole and a south pole. These magnetic poles align closely, but not exactly, with the geographic poles. (The geographic poles are simply the top and bottom of the spinning sphere we live on, the points through which passes the imaginary axis of rotation.)

In many respects the magnetic field is very much like that produced by a simple bar magnet. Invisible lines of force pass through Earth and extend from one pole to the other. A compass needle, itself a small magnet free to move about, becomes aligned with these lines of force and thus points toward the magnetic poles.

The technique used to study ancient magnetic fields relies on the fact that certain rocks contain minerals that serve as fossil compasses. These iron-rich minerals, such as magnetite, are abundant in lava flows of basaltic composition. When heated above a certain temperature called the *Curie point*, these magnetic minerals lose their magnetism. However, when these iron-rich grains cool below their Curie point (about 580°C) they become magnetized in the direction parallel to the existing magnetic field. Once the minerals solidify, the magnetism they possess will remain "frozen" in this position. In this regard, they behave much like a compass needle

inasmuch as they "point" toward the existing magnetic poles. Then, if the rock is moved, or if the magnetic pole changes position, the rock magnetism will, in most instances, retain its original alignment. Rocks formed thousands of millions of years ago thus "remember" the location of the magnetic poles at the time of their formation and are said to possess fossil magnetism, or **paleomagnetism**.

Polar Wandering A study of lava flows conducted in Europe in the 1950s led to an amazing discovery. The magnetic alignment in the iron-rich minerals in lava flows of different ages was found to vary widely. A plot of the apparent positions of the magnetic north pole revealed that, during the past 500 million years, the location of the pole had gradually wandered from a spot near Hawaii northward through eastern Siberia and finally to its present site (Figure 5.16A). This was clear evidence that either the magnetic poles had migrated through time, an idea known as **polar wandering**, or that the lava flows had moved—in other words, the continents had drifted.

Although the magnetic poles are known to move, studies of the magnetic field indicated that the average positions of the magnetic poles correspond closely to the positions of the geographic poles. This is consistent with our knowledge of Earth's magnetic field, which is generated in part by the rotation of Earth about its axis. If the geographic poles do not wander appreciably, which we believe is true, neither can the magnetic poles. Therefore, a more acceptable explanation for the apparent polar wandering is provided by the plate tectonics theory. *If the magnetic poles remain stationary, their apparent movement was produced by the drifting of the continents.*

Further evidence for plate tectonics came a few years later when polar wandering curves were constructed for North America and Europe (Figure 5.16A). To nearly everyone's surprise, the curves for North America and Europe had similar paths, except that they were separated by about 24 degrees of longitude. When these rocks solidified, could there have been two magnetic north poles that migrated parallel to each other? This is very unlikely. The differences in these migration paths, however, can be reconciled if the two presently separated continents are placed next to one another, as we now believe they were prior to the opening of the Atlantic Ocean (Figure 5.16B).

Magnetic Reversals and Seafloor Spreading Another discovery came when geophysicists learned that Earth's magnetic field periodically reverses polarity; that is, the north magnetic pole becomes the south

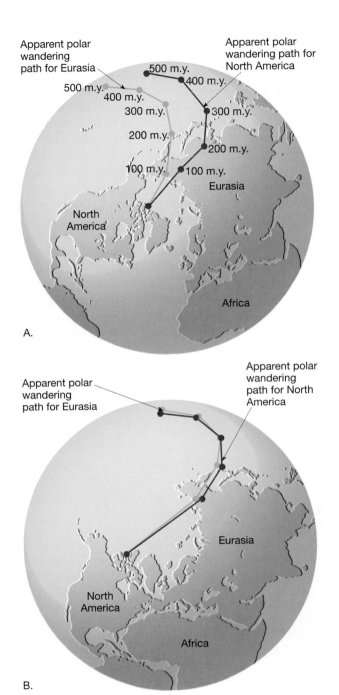

Figure 5.16 Simplified apparent polar wandering paths as established from North American and Eurasian paleomagnetic data. **A.** The more westerly path determined from North American data is thought to have been caused by the westward drift of North America by about 24 degrees from Eurasia. **B.** The positions of the wandering paths when the landmasses are reassembled in their predrift locations.

magnetic pole, and vice versa. A rock solidifying during one of the periods of reverse polarity will be magnetized with the polarity opposite that of rocks being formed today. When rocks exhibit the same magnetism as the present magnetic field, they are said to possess **normal polarity**, while those rocks exhibiting the opposite magnetism are said to have **reverse polarity**. Evidence for magnetic reversals was obtained from lavas and sediments from around the world.

Once the concept of magnetic reversals was confirmed, researchers set out to establish a time scale for polarity reversals. There are many areas where volcanic activity has occurred sporadically for periods of millions of years. The task was to measure the directions of paleomagnetism in numerous lava flows of various ages. These data were collected from several places and were used to determine the dates when the polarity of Earth's magnetic field changed. Figure 5.17 shows the time scale of the polarity reversals established for the last few million years.

A significant relationship was uncovered between the magnetic reversals and the seafloor spreading hypothesis. Very sensitive instruments

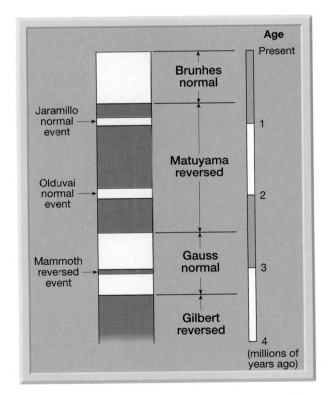

Figure 5.17 Time scale of Earth's magnetic field in the recent past. This time scale was developed by establishing the magnetic polarity for lava flows of known age. *(Data from Allen Cox and G. B. Dalrymple)*

called *magnetometers* were towed by research vessels across a segment of the ocean floor located off the western coast of the United States. Here workers from the Scripps Institute of Oceanography discovered alternating strips of high- and low-intensity magnetism that trended in roughly a north–south direction. This relatively simple pattern of magnetic variation defied explanation until 1963, when it was tied to the concept of seafloor spreading. The strips of high-intensity magnetism are regions where the paleomagnetism of the ocean crust is of the normal type (Figure 5.18). Consequently, these positively magnetized rocks *enhance* the existing magnetic field. Conversely, the low-intensity strips represent regions where the ocean crust is polarized in the reverse direction and, therefore, *weaken* the existing magnetic field (Figure 5.18). But how do parallel strips of normally and reversely magnetized rock become distributed across the ocean floor?

As new basalt is added to the ocean floor at the oceanic ridges, it becomes magnetized according to the existing magnetic field. Because new rock is added in approximately equal amounts to the trailing edges of both plates, we should expect strips of equal size and polarity to parallel both sides of the ocean ridges, as shown in Figure 5.18. This explanation of the alternating strips of normal and reverse polarity, which lay as mirror images across the ocean ridges, was the strongest evidence so far presented in support of the concept of seafloor spreading.

Now that the dates of the most recent magnetic reversals have been established, the rate at which spreading occurs at the various ridges can be determined accurately. In the Pacific Ocean, for example, the magnetic strips are much wider for corresponding time intervals than those of the Atlantic Ocean. Hence, we conclude that a faster spreading rate exists for the spreading center of the Pacific as compared to the Atlantic. When we apply absolute dates to these magnetic events, we find that the spreading rate for the North Atlantic Ridge is only 2 centimeters per year. The rate is somewhat faster for the South Atlantic. The spreading rates for the East Pacific Rise generally range between 6 and 12 centimeters per year, with a maximum rate of about 20 centimeters a year. Thus, we have a magnetic tape recorder that records changes in Earth's magnetic field. This recorder also permits us to determine the rate of seafloor spreading.

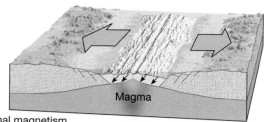

A. Period of normal magnetism

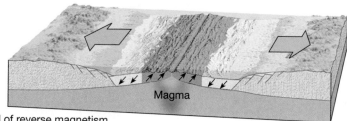

B. Period of reverse magnetism

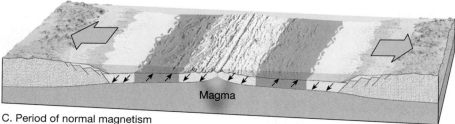

C. Period of normal magnetism

Figure 5.18 As new basalt is added to the ocean floor at the mid-oceanic ridges, it is magnetized according to Earth's existing magnetic field. Hence, it behaves much like a tape recorder as it records each reversal of the planet's magnetic field.

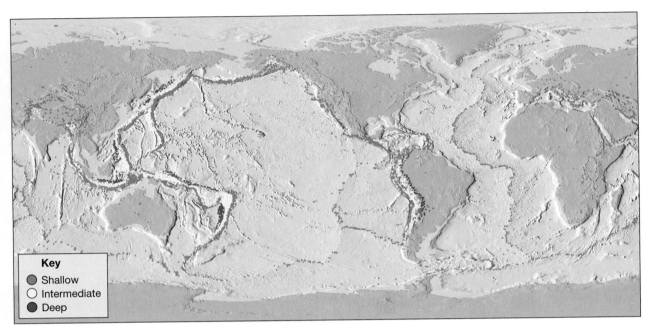

Figure 5.19 Distribution of shallow-, intermediate-, and deep-focus earthquakes. Note that deep focus earthquakes only occur in association with subduction zones. *(Data from NOAA)*

Evidence: Earthquake Patterns

By 1968, the basic outline of global tectonics was firmly established. In that same year, three seismologists published papers demonstrating how successfully the new plate tectonics model accounted for the global distribution of earthquakes (Figure 5.19). In particular, these scientists were able to account for the close association between deep-focus earthquakes and ocean trenches. Furthermore, the absence of deep-focus earthquakes along the oceanic ridge system was also shown to be consistent with the new theory.

The close association between plate boundaries and earthquakes can be seen by comparing the distribution of earthquakes shown in Figure 5.19 with the map of plate boundaries in Figure 5.6 (pp. 124–25) and trenches in Figure 5.12 (p. 130). In trench regions where dense slabs of lithosphere plunge into the mantle, this association is especially striking. When the depths of earthquake foci and their locations within the trench systems are plotted, an interesting pattern emerges. Figure 5.20, which shows the distribution of earthquakes in the vicinity of the Japan trench, is an example. Here most shallow-focus earthquakes occur within, or adjacent to, the trench, whereas intermediate- and deep-focus earthquakes occur toward the mainland.

In the plate tectonics model, deep-ocean trenches are produced where cool and dense slabs of oceanic lithosphere plunge into the mantle. Shallow-focus earthquakes are produced as the descending plate interacts with the overriding lithosphere. As the slab descends further into the asthenosphere, deeper-focus earthquakes are generated. Because

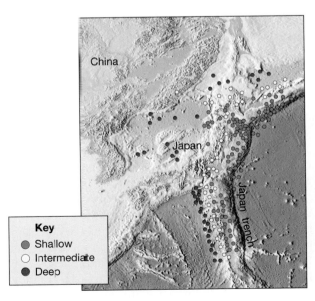

Figure 5.20 Distribution of earthquake foci in the vicinity of the Japan trench. *(Data from NOAA)*

the earthquakes occur within the rigid subducting plate rather than in the "plastic" mantle, they provide a method for tracking the plate's descent. Very few earthquakes have been recorded below 700 kilometers (435 miles), possibly because the slab has been heated sufficiently to lose its rigidity.

Evidence: Ocean Drilling

Some of the most convincing evidence confirming the plate tectonics theory has come from drilling directly into ocean-floor sediment. From 1968 until 1983, the source of these important data was the Deep Sea Drilling Project, an international program sponsored by several major oceanographic institutions and the National Science Foundation. A new drilling ship was built. The *Glomar Challenger* represented a significant technological breakthrough, because this ship could lower drill pipe thousands of meters to the ocean floor and then drill hundreds of meters into the sediments and underlying basaltic crust.

Operations began in August 1968, in the South Atlantic. At several sites, holes were drilled through the entire thickness of sediments to the basaltic rock below. An important objective was to gather samples of sediment from just above the igneous crust as a means of dating the seafloor at each site. (Radiometric dates of the ocean crust itself are unreliable because seawater alters basalt.)

When the oldest sediment from each drill site was plotted against its distance from the ridge crest, it was revealed that the age of the sediment increased with increasing distance from the ridge. This finding agreed with the seafloor spreading hypothesis, which predicted that the youngest oceanic crust would be found at the ridge crest, and that the oldest oceanic crust would be at the continental margins.

The data from the Deep Sea Drilling Project also reinforced the idea that the ocean basins are geologically youthful, because no sediment with an age in excess of 180 million years was found. By comparison, some continental crust has been dated at 3.9 billion years.

During its 15 years of operation, the *Glomar Challenger* drilled 1092 holes and obtained more than 96 kilometers (60 miles) of invaluable core samples. The Ocean Drilling Program has succeeded the Deep Sea Drilling Project and, like its predecessor, it is a major international program. A more technologically advanced drilling ship, the *JOIDES Resolution*, now continues the work of the *Glomar Challenger*.

Evidence: Hot Spots

Mapping of seafloor volcanoes called *seamounts* in the Pacific revealed a chain of volcanic structures extending from the Hawaiian Islands to Midway Island and then continuing northward toward the Aleutian trench (Figure 5.21). Radiometric dates of volcanoes in this chain showed that the volcanoes increase in age with increasing distance from Hawaii. Suiko Seamount, which is located near the Aleutian trench, is 65 million years old, Midway Island is 27 million years old, and the island of Hawaii built up from the seafloor less than 1 million years ago (Figure 5.21).

Researchers have proposed that a rising plume of mantle material is located below the island of Hawaii. Partial melting of this hot rock as it enters the low-pressure environment near the surface generates a volcanic area or **hot spot**. Presumably, as the Pacific plate moved over the hot spot, successive volcanic mountains have been built. The age of each volcano indicates the time when it was situated over the relatively stationary mantle plume.

This pattern is shown in Figure 5.21. Kauai is the oldest of the large islands in the Hawaiian chain. Five million years ago, when it was positioned over the hot spot, Kauai was the only Hawaiian Island in existence (Figure 5.21). Visible evidence of the age of Kauai can be seen by examining its extinct volcanoes, which have been eroded into jagged peaks and vast canyons. By contrast, the south slopes of the relatively youthful island of Hawaii consist of fresh lava flows, and two of Hawaii's volcanoes, Mauna Loa and Kilauea, remain active.

This evidence supports the fact that the plates do indeed move relative to Earth's interior. The hot spot "tracks" also trace the direction of plate motion. Notice, for example, in Figure 5.21 that the Hawaiian Island–Emperor Seamount chain bends. This particular bend in the trace occurred about 40 million years ago when the motion of the Pacific plate changed from nearly due north to its present northwesterly path.

The Driving Mechanism: Still Under Investigation

The plate tectonics theory *describes* plate motion and the *effects* of this motion. Because we accept the theory does not mean that we *understand the forces* that move the plates. In fact, none of the driving mecha-

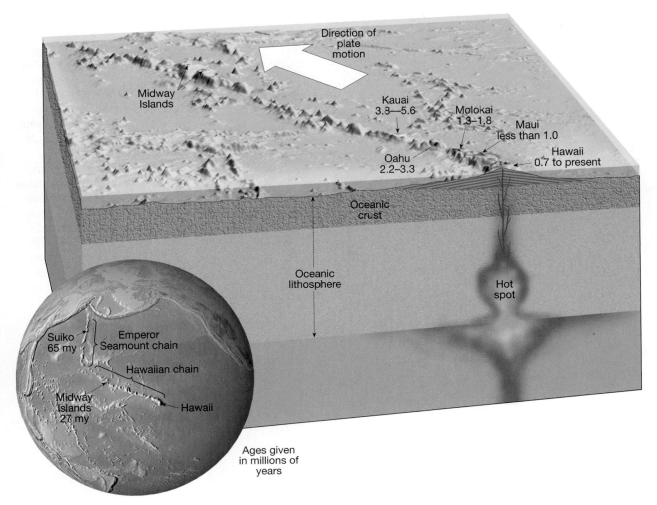

Figure 5.21 The chain of islands and seamounts that extends from Hawaii to the Aleutian trench results from the movement of the Pacific plate over an apparently stationary hot spot. Radiometric dating of the Hawaiian Islands shows that the volcanic activity decreases in age toward the island of Hawaii.

nisms yet proposed can account for *all* of the major facets of plate motion. Nevertheless, it is clear that the unequal distribution of heat within Earth is the underlying driving force for plate movement.

Convection Current Hypothesis One of the first models to explain the movements of plates was originally proposed as a possible driving mechanism for continental drift. Adapted to plate tectonics, this hypothesis suggests that large convection currents within the mantle drive plate motion (Figure 5.22A). The warm, less dense material of the mantle rises very slowly in the regions of oceanic ridges. As the material spreads laterally, it might drag the lithosphere along. Eventually, the material cools and begins to sink back into the mantle, where it is reheated. Partly because of

its simplicity, this proposal had wide appeal. However, modern research techniques reveal that the flow of material in the mantle is far more complex than such simple convection cells. Further, it is now known that the lithospheric plates are not passengers carried along by convective flow, but rather are part of the circulation.

Slab-Push and Slab-Pull Hypotheses Many other mechanisms that may contribute to plate motion have been suggested. One relies on the fact that, as a newly formed slab of oceanic crust moves away from the ridge crest, it gradually cools and becomes denser. Eventually, the cold oceanic slab becomes denser than the asthenosphere and begins to sink. When this occurs, the dense sinking slab pulls the trailing lithosphere along. This so-called *slab-pull hypothesis*

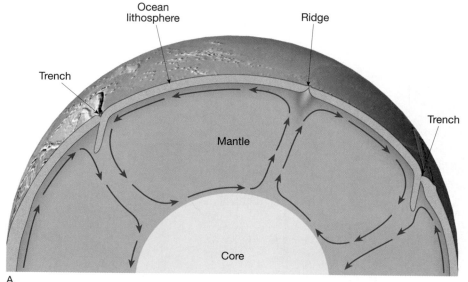

A.

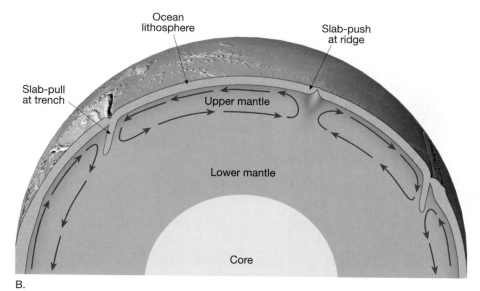

B.

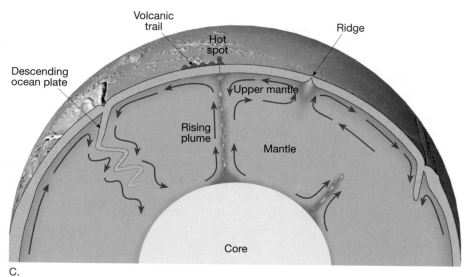

C.

Figure 5.22 Proposed models of the driving force for plate tectonics. **A.** Large convection cells in the mantle may carry the lithosphere in a conveyor-belt fashion. **B.** Slab-pull results because a subducting slab is more dense than the underlying material. Slab-push is a form of gravity sliding caused by the elevated position of lithosphere at a ridge crest. **C.** The hot plume model suggests that all upward convection is confined to a few plumes, while the downward limbs of these convection cells are the cold, dense subducting oceanic plates.

is thought to be an important mechanism for transporting cold material back into the mantle.

Another mechanism proposes that the elevated position of an oceanic ridge could cause the lithosphere to slide under the influence of gravity. However, some ridge systems are subdued, which would reduce the effectiveness of the *slab-push* model. Nevertheless, the slab-push phenomenon appears to be active in some ridge systems but is less effective than slab-pull (Figure 5.22B).

Hot Plumes Hypothesis One version of the thermal convection model suggests that hot, buoyant plumes of rock contribute to plate motion (Figure 5.22C). These hot plumes are presumed to extend upward from the vicinity of the mantle-core boundary. Upon reaching the lithosphere, they spread laterally and facilitate plate motion away from the zone of upwelling. These mantle plumes reveal themselves as long-lived volcanic areas (hot spots) in such places as Iceland. A dozen or so hot spots have been identified along ridge systems where they may contribute to plate divergence. Recall, however, that many hot spots, including the one which generated the Hawaiian Islands, are not located in ridge areas.

The downward limbs of these convection cells are the cold, dense subducting lithospheric plates (Figure 5.22C). Advocates of this view suggest that subducting slabs may descend all the way to the core–mantle boundary.

Although there is still much to be learned about the mechanisms that cause plates to move, some facts are clear. The unequal distribution of heat in Earth generates some type of thermal convection in the mantle, which ultimately drives plate motion. Furthermore, the descending lithospheric plates are active components of downwelling, and they serve to transport cold material into the mantle.

The Chapter in Review

1. In the early 1900s, *Alfred Wegener* set forth his *continental drift* hypothesis. One of its major tenets was that a supercontinent called *Pangaea* began breaking apart into smaller continents about 200 million years ago. The smaller continental fragments then "drifted" to their present positions. To support the claim that the now-separate continents were once joined, Wegener and others used the *fit of South America and Africa, the distribution of ancient climates, fossil evidence*, and *rock structures*.

2. One of the main objections to the continental drift hypothesis was the inability of its supporters to provide an acceptable mechanism for the movement of continents.

3. The theory of *plate tectonics*, a far more encompassing theory than continental drift, holds that Earth's rigid outer shell, called the *lithosphere*, consists of seven large and numerous smaller segments called *plates* that are in motion relative to each other. Most of Earth's *seismic activity, volcanism*, and *mountain building* occur along the dynamic margins of these plates.

4. A major departure of the plate tectonics theory from the continental drift hypothesis is that large plates contain both continental and ocean crust and the entire plate moves. By contrast, in continental drift, Wegener proposed that the sturdier continents "drifted" by breaking through the oceanic crust, much like ice breakers cut through ice.

5. The three distinct types of plate boundaries are (1) *divergent boundaries*—where plates move apart, (2) *convergent boundaries*—where plates move together as in *oceanic-continental convergence, oceanic-oceanic convergence*, or *continental-continental convergence*; and (3) *transform boundaries*—where plates slide past each other.

6. The theory of plate tectonics is supported by: (1) *paleomagnetism*, the direction and intensity of Earth's magnetism in the geologic past; (2) the global distribution of *earthquakes* and their close association with plate boundaries; (3) the ages of *sediments* from the floors of the deep-ocean basins; and (4) the existence of island groups that formed over *hot spots* and provide a frame of reference for tracing the direction of plate motion.

7. Several models for the driving mechanism of plates have been proposed. One model suggests that large *convection cells* within the mantle carry the overlying plates. Another proposes that dense *oceanic material descends and pulls* the lithosphere along. No single driving mechanism can account for all of the major facets of plate motion.

Key Terms

continental drift (p. 118)
continental volcanic arc
   (p. 130)
convergent boundary
   (p. 123)
deep-ocean trench (p. 128)

divergent boundary
   (p. 123)
hot spot (p. 138)
normal polarity (p. 135)
paleomagnetism (p. 134)
Pangaea (p. 118)

plate (p. 122)
plate tectonics (p. 122)
polar wandering (p. 134)
reverse polarity (p. 135)
rift (rift valley) (p. 126)
seafloor spreading (p. 123)

subduction zone (p. 127)
transform fault boundary
   (p. 123)
volcanic island arc (p. 130)

Questions for Review

1. Who is credited with developing the continental drift hypothesis?
2. What was probably the first evidence that led some to suspect the continents were once connected?
3. What was Pangaea?
4. List the evidence that Wegener and his supporters gathered to substantiate the continental drift hypothesis.
5. Explain why the discovery of the fossil remains of *Mesosaurus* in both South America and Africa, but nowhere else, supports the continental drift hypothesis.
6. Early in this century, what was the prevailing view of how land animals migrated across vast expanses of ocean?
7. How did Wegener account for the existence of glaciers in the southern landmasses, while at the same time areas in North America, Europe, and Siberia supported lush tropical swamps?
8. On what basis were plate boundaries first established?
9. What are the three major types of plate boundaries? Describe the relative plate motion at each of these boundaries.
10. What is seafloor spreading? Where is active seafloor spreading occurring today?
11. What is a subduction zone? With what type of plate boundary is it associated?
12. Where is lithosphere being consumed? Why must the production and destruction of lithosphere be going on at approximately the same rate?
13. Briefly describe how the Himalaya Mountains formed.
14. Differentiate between transform faults and the other two types of plate boundaries.
15. Some predict that California will sink into the ocean. Is this idea consistent with the theory of plate tectonics?
16. Define the term *paleomagnetism*.
17. How does the continental drift hypothesis account for the apparent wandering of Earth's magnetic poles?
18. Describe the distribution of earthquake epicenters and foci depths as they relate to oceanic trench systems.
19. What is the age of the oldest sediments recovered by deep-ocean drilling? How do the ages of these sediments compare to the ages of the oldest continental rocks?
20. How do hot spots and the plate tectonics theory account for the fact that the Hawaiian Islands vary in age?
21. With what type of plate boundary are the following places or features associated (be as specific as possible): Himalayas, Aleutian Islands, Red Sea, Andes Mountains, San Andreas fault, Iceland, Japan, Mount St. Helens?

Testing What You Have Learned

Multiple-Choice Questions

1. Which of the following refers to the deformation of Earth's crust that results in the formation of structural features such as mountains?
   a. tectonics   **c.** transformation   **e.** lithification
   b. destruction   **d.** catastrophics

2. One of the major tenets of the continental drift hypothesis suggested the existence of a supercontinent. What is the name of this continent?
   a. Antaustria   **c.** Soafric   **e.** Pangaea
   b. Euroam   **d.** Paneuro

3. Which one of the following was NOT used to support the continental drift hypothesis?
   a. fossils   **c.** paleoclimates   **e.** rock types and structures
   b. paleomagnetism   **d.** fit of continents

4. Today we know that Earth's rigid outer shell consists of about 20 segments called _____

   a. continents.   **c.** plates.   **e.** crust segments.

   b. basins.   **d.** layers.

5. Which of the following is NOT a distinct type of plate boundary?

   a. transform   **c.** divergent   **e.** both a. and d.

   b. convergent   **d.** zonal

6. As plates move apart along ridges, broken slabs are displaced downward, creating downfaulted valleys called _____ valleys.

   a. trend   **c.** hanging   **e.** rift

   b. displaced   **d.** matching

7. The theory of plate tectonics does NOT help to explain the origin and location of which one of the following?

   a. earthquakes   **c.** volcanoes   **e.** major seafloor features

   b. mountains   **d.** ocean currents

8. The basic outline of plate tectonics was firmly established by the late _____

   a. 1800s.   **c.** 1940s.   **e.** 1980s.

   b. 1920s.   **d.** 1960s.

9. Which one of the following provides a method for tracking a plate's descent into the "plastic" mantle along a deep-ocean trench?

   a. echo soundings   **c.** surveying   **e.** light reflections

   b. earthquakes   **d.** drilling

10. Researchers have proposed that the Hawaiian Islands are forming over a plume of rising mantle material called a _____ spot.

    a. hot   **c.** mantle   **e.** island

    b. core   **d.** convergence

Fill-In Questions

11. The basic idea of _____ is that Earth's rigid outer shell is made of several large segments that are slowly moving.
12. Earth's rigid outer shell, called the _____, lies over a hotter, weaker zone known as the _____.
13. _____ is the mechanism responsible for producing ocean floor material at the crest of oceanic ridges.
14. The region where an oceanic slab sinks into the asthenosphere because of convergence is called a _____.
15. When rocks exhibit the same magnetism as the present magnetic field, they are said to possess _____ polarity.

True/False Questions

16. According to the plate tectonics theory, continents are moving *through* the ocean basins. _____
17. The continental drift hypothesis can be used to explain the existence of identical fossils on widely separated landmasses. _____
18. The average rate of spreading along ocean ridges is about 60 centimeters per year. _____
19. Mountain ranges such as the Himalayas, Alps, and Appalachians formed because of continental-continental convergence. _____
20. One result of plate tectonics is that ocean basins are all older than continents. _____

Answers:

1. a; 2. e; 3. b; 4. c; 5. d; 6. e; 7. d; 8. d; 9. b; 10. a; 11. plate tectonics; 12. lithosphere, asthenosphere; 13. Seafloor spreading; 14. subduction zone; 15. normal; 16. F; 17. T; 18. F; 19. T; 20. F.

Web   Work

The following are informative and interesting World Wide Web sites that address topics related to those presented in this chapter:

- Dynamic Earth Tutorial (United States Geological Survey)

  http://pubs.usgs.gov/publications/text/dynamic.html

- Plate Tectonics Description (United States Geological Survey)

  http://vulcan.wr.usgs.gov/Glossary/Plate Tectonics/framework.html

For direct access to these sites and other relevant Web locations, contact the *Foundations of Earth Science* WWW home page at http://www.prenhall.com/lutgens.

CHAPTER 6

Restless Earth:

Earthquakes, Geologic Structures, and Mountain Building

FOCUS ON LEARNING

To assist you in learning the important concepts
in this chapter, you will find it helpful to focus
on the following questions:

1. What is an earthquake?
2. What are the types of earthquake waves?
3. How is the epicenter of an earthquake determined?
4. Where are Earth's principal earthquake zones?
5. How is earthquake magnitude expressed by the Richter scale?
6. What key factors determine the destructiveness of an earthquake?

7. What are the four major zones of Earth's interior?
8. How do continental crust and oceanic crust differ?
9. What are the two basic types of rock deformation?
10. How is mountain building associated with the various kinds of convergent plate boundaries?

Mount McKinley, Denali National Park, Alaska.
Photo by Carr Clifton Photography

On October 17, 1989, at 4:04 P.M. Pacific Daylight Time, millions of television viewers around the world were settling in to watch the third game of the World Series. Instead, they saw their TV sets go black as tremors hit San Francisco's Candlestick Park. Although the earthquake was centered in a remote section of the Santa Cruz Mountains, 100 kilometers to the south, major damage occurred in the Marina District of San Francisco.

The most tragic result of the violent shaking was the collapse of some double-decked sections of Interstate 880, also known as the Nimitz Freeway. The ground motions caused the upper deck to sway, shattering the concrete support columns along a mile-long section of the freeway. The upper deck then collapsed onto the lower roadway, flattening cars as if they were aluminum cans. This earthquake, named the Loma Prieta quake for its point of origin, claimed 67 lives.

In mid-January 1994, less than 5 years after the Loma Prieta earthquake devastated portions of the San Francisco Bay area, a major earthquake struck the Northridge area of Los Angeles (Figure 6.1). Although it was not the fabled "Big One," the moderate 6.6- to 6.9-magnitude earthquake left 51 dead, over 5000 injured, and tens of thousands of households without water and electricity. In total, damage in excess of $15 billion was attributed to an apparently unknown fault that ruptured at a depth of 14 kilometers (9 miles) beneath Northridge.

The Northridge earthquake, which occurred northwest of downtown Los Angeles, began at 4:31 A.M. and lasted roughly 40 seconds. During this brief period, the quake terrorized the entire Los Angeles area. The three-story Northridge Meadows apartment complex where 16 people died was devastated. Here, most of the deaths resulted when sections of the upper floors collapsed onto the first-floor units. In addition, nearly 300 schools were seriously damaged and a dozen major roadways buckled. Among these were two of California's major arteries—the Golden State Freeway (Interstate 5), where an overpass collapsed, completely blocking the roadway, and the Santa Monica Freeway. Fortunately, these roadways had practically no traffic at this early morning hour.

In nearby Granada Hills, broken gas lines were set ablaze while the streets were flooded from broken water mains. In Los Angeles County alone, over 100 calls were made to the fire department. Seventy homes burned in the Sylmar area. A 64-car freight train derailed, including some cars carrying hazardous cargo.

Despite the huge economic losses, it is remarkable that the destruction was not greater. A year later, an earthquake of equal magnitude caused over 5000 deaths and left 300,000 homeless in Kobe, Japan. Unquestionably, the upgrading of structures to meet the requirements of building codes developed for this earthquake-prone area helped minimize what could have been a much greater human tragedy.

Figure 6.1 Damage to a home in Pacific Palisades, California, caused by the January 17, 1994, Northridge earthquake. *(Photo by ChromoSohm/Sohm/The Stock Market)*

Over 30,000 earthquakes that are strong enough to be felt occur worldwide annually. Fortunately, most are minor tremors and do very little damage. Generally, only about 75 significant earthquakes take place each year, and many of these occur in remote regions. However, occasionally a large earthquake occurs near a major population center. Under these conditions, an earthquake is among the most destructive natural forces on Earth.

The shaking of the ground, coupled with the liquefaction of some soils, wreaks havoc on buildings and other structures. In addition, when a quake occurs in a populated area, power and gas lines are often ruptured, causing numerous fires. In the 1906 San Francisco earthquake, much of the damage was caused by fires, which quickly became uncontrollable when broken water mains left firefighters with only trickles of water.

What Is an Earthquake?

An **earthquake** is the vibration of Earth produced by a rapid release of energy. This energy radiates in all directions from its source, the **focus**, in the form of waves. The waves are like those produced when a stone is dropped into a calm pond (Figure 6.2). Just as the impact of the stone sets water waves in motion, an earthquake generates seismic waves that radiate throughout Earth. Even though the energy dissipates rapidly with increasing distance from the focus, sensitive instruments worldwide record the event.

Earthquakes and Faults

The tremendous energy released by atomic explosions or by volcanic eruptions can produce an earthquake, but these events are comparatively weak and infrequent. What mechanism produces a destructive earthquake? Ample evidence exists that Earth is not a static planet. Worldwide, scientists have found places where forces have elevated sections of the crust. Other regions exhibit evidence of extensive subsidence. In addition to these vertical displacements, offsets in fence lines, roads, and other structures indicate that horizontal movement is common (Figure 6.3). These movements are usually associated with large fractures in the earth called **faults**.

Most of the motion along faults can be satisfactorily explained by the plate tectonics theory. Recall that this theory describes large slabs of Earth's crust in continual motion. These mobile plates interact with neighboring plates, straining and deforming the rocks at their edges. It is along faults associated with plate boundaries that most earthquakes occur.

Elastic Rebound

The actual mechanism of earthquake generation eluded geologists until H.F. Reid of Johns Hopkins University conducted a study following the great 1906 San Francisco earthquake. The earthquake was accompanied by very noticeable displacements of

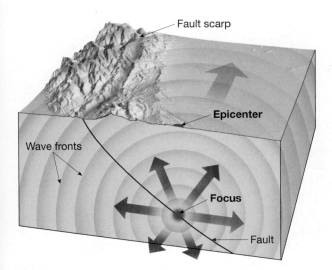

Figure 6.2 The focus of an earthquake is located at depth. The surface location directly above it is called the epicenter.

Figure 6.3 Slippage along a fault produced an offset in this orange grove located east of Calexico, California. *(Photo by John S. Shelton)*

several meters along the northern portion of the San Andreas fault. This 1300-kilometer (780-mile) fracture runs north–south through southern California. The San Andreas fault zone separates two great sections of Earth's crust, the North American plate and the Pacific plate. Field investigations determined that during this single earthquake, the Pacific plate slid as much as 4.7 meters (15 feet) in a northward direction past the adjacent North American plate.

Using land surveys conducted several years apart, Reid discovered that, during the 50 years prior to the 1906 earthquake, the land at distant points on both sides of the San Andreas fault showed a relative displacement of slightly more than 3 meters (10 feet). The mechanism for earthquake formation, which Reid deduced from this information, is illustrated in Figure 6.4.

In Figure 6.4A, you can see an existing fault, or break in the rocks. In Figure 6.4B, tectonic forces ever so slowly deform the crustal rocks on both sides of the fault, as demonstrated by the bent features. Under these conditions, rocks are bending and storing elastic energy, much like a wooden stick would if bent. Eventually, the frictional resistance holding the rocks together is overcome. As slippage occurs at the weakest point (the focus), displacement there will exert stress farther along the fault where additional slippage will occur until most of the built-up strain is released (Figure 6.4C). This slippage allows the deformed rock to "snap back." The vibrations we know as an earthquake occur as the rock elastically returns to its original shape. The "springing back" of the rock was termed **elastic rebound** by Reid because the rock behaves elastically, much like a stretched rubber band does when released.

We have just described slippage along an existing fault. New faults also are created when the strain in rocks exceeds their strength.

The San Andreas is undoubtedly the most studied fault system in the world. Over the years, investigators have shown that displacement occurs along discrete segments that are 100 to 200 kilometers long. Further, each fault segment behaves somewhat differently from the others. Some portions of the San Andreas exhibit a slow, gradual displacement known as *fault creep*, which occurs with little noticeable seismic activity. Other segments regularly slip, producing small earthquakes. Still other segments remain locked and store elastic energy for hundreds of years before rupturing in great earthquakes. The latter process is described as *stick-slip* motion,

Deformation of rocks

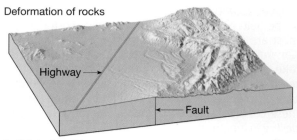

A. Original position

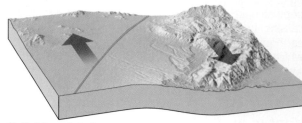

B. Buildup of strain

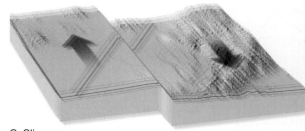

C. Slippage

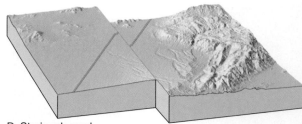

D. Strain released

Figure 6.4 Elastic rebound. As rock is deformed it bends, storing elastic energy. Once the rock is strained beyond its breaking point it ruptures, releasing the stored-up energy in the form of earthquake waves.

because the fault exhibits alternating periods of locked behavior followed by sudden slippage.

It is estimated that great earthquakes should occur about every 50 to 200 years along those sections of the San Andreas fault that exhibit stick-slip motion. This knowledge is useful when assigning a potential earthquake risk to a given segment of the fault zone.

Foreshocks and Aftershocks

The intense vibrations of the 1994 Northridge earthquake lasted about 40 seconds. Although most of the displacement along the fault occurred in this rather short period, additional movements along this and other nearby faults lasted for several days following the main quake. The adjustments that follow a major earthquake often generate smaller earthquakes called **aftershocks**. Although aftershocks are usually much weaker than the main earthquake, they can cause further damage to badly weakened structures. This occurred, for example, during the 1988 Armenian earthquake, when a large aftershock of magnitude 5.8 collapsed many structures that had been weakened by the main tremor. That disaster killed about 25,000 people.

In addition, small earthquakes called **foreshocks** often precede a major earthquake by days or, in some cases, by as much as several years. Monitoring of these foreshocks has been used as a means of predicting forthcoming major earthquakes, with mixed success.

In summary, most earthquakes are produced by the rapid release of elastic energy stored in rock that has been subjected to great differential stress. Once the strength of the rock is exceeded, it suddenly ruptures, which results in the vibrations of an earthquake. Earthquakes also occur along existing faults when the frictional forces on the fault surfaces are overcome.

Earthquake Waves

The study of earthquake waves, **seismology**, dates back almost 2000 years. Modern **seismographs** are instruments that record earthquake waves. Their principle is simple. A weight is freely suspended from a support that is attached to bedrock (Figure 6.5). When waves from a distant earthquake reach the instrument, the inertia of the weight keeps it stationary, while Earth and the support vibrate. The movement of Earth in relation to the stationary weight is recorded on a rotating drum. (Inertia is the tendency of a stationary object to hold still, or a moving object to stay in motion.)

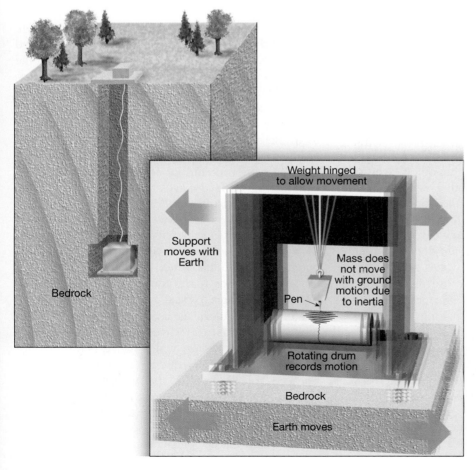

Figure 6.5 Principle of the seismograph. The inertia of the suspended mass tends to keep it motionless, while the recording drum, which is anchored to bedrock, vibrates in response to seismic waves. Thus, the stationary mass provides a reference point from which to measure the amount of displacement occurring as the seismic wave passes through the ground below.

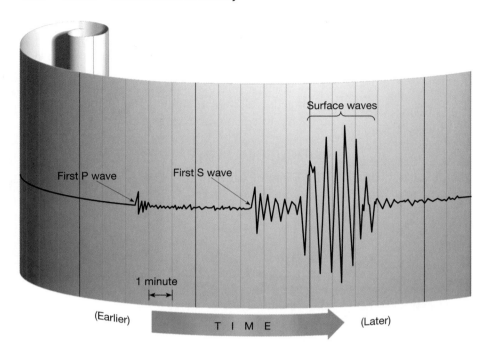

Figure 6.6 Typical seismic record. Note the time interval between the arrival of the first P wave and the arrival of the first S wave.

Surface waves

First P wave

First S wave

1 minute

(Earlier)

T I M E

(Later)

Modern seismographs amplify and record ground motion, producing a trace as shown in Figure 6.6. This record, called a **seismogram**, provides a great deal of information about the behavior of seismic waves. Simply stated, seismic waves result from the abrupt release of stored energy. The waves radiate outward in all directions from the focus, as you saw in Figure 6.2. The transmission of this energy can be compared to the shaking of gelatin in a bowl. Seismograms reveal that two main types of seismic waves are generated by the slippage of a rock mass. One wave type travels around the outer layer of Earth, called **surface waves**. Other waves travel through Earth's interior and are called **body waves**. Body waves are further divided into two types called **primary (P) waves** and **secondary (S) waves**.

The two types of body waves are divided on the basis of their mode of travel. P waves *push* (compress) and *pull* (expand) rocks in the direction the wave is traveling (Figure 6.7A). Imagine holding someone by the shoulders and shaking them. This push–pull movement is how P waves move through Earth materials. The wave energy doesn't move up-and-down or side-to-side, but through the rocks. This wave motion is analogous to that generated by human vocal cords as they move air to create sound. Solids, liquids, and gases resist a change in volume when compressed and will elastically spring back once the force is removed. Therefore, P waves,

which are compressional waves, can travel through all these materials.

Conversely, S waves, "shake" the particles at right angles to their direction of travel. This can be illustrated by tying one end of a rope to a post and shaking the other end, as shown in Figure 6.7B. Unlike P waves, which change the volume of the intervening material by alternately compressing and expanding it, S waves change only the shape of the material that transmits them. Because fluids (gases and liquids) do not resist changes in shape, they cannot transmit S waves.

The motion of surface waves is somewhat more complex. As surface waves travel along the ground, they cause the ground and anything resting on it to move, much like ocean swells toss a ship. In addition to their up-and-down motion, surface waves have a side-to-side motion similar to an S wave oriented in a horizontal plane. This latter motion is particularly damaging to the foundations of structures.

By observing a "typical" seismic record, as shown in Figure 6.6, you can see a major difference between these seismic waves: P waves arrive at the recording station *before* S waves, which arrive before surface waves. This is a consequence of their speeds. For purposes of illustration, the velocity of P waves through granite within the crust is about 6 kilometers per second. Under the same conditions, S waves travel only 3.5 kilometers per second. Differences in density and elastic properties of the transmitting

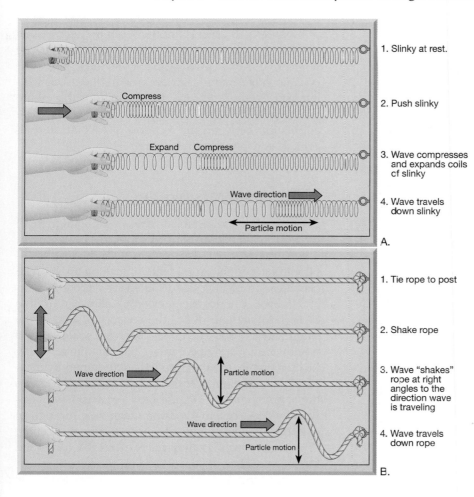

1. Slinky at rest.

Compress

2. Push slinky

Expand   Compress

3. Wave compresses and expands coils cf slinky

Wave direction

4. Wave travels down slinky

Particle motion

A.

1. Tie rope to post

2. Shake rope

Wave direction   Particle motion

3. Wave "shakes" rope at right angles to the direction wave is traveling

Wave direction

4. Wave travels down rope

Particle motion

B.

Figure 6.7 Types of seismic waves and their characteristic motion. **A.** P waves compress the particles in the material, causing it to vibrate back and forth in the same direction as the waves move. **B.** S waves cause particles to oscillate at right angles to the direction of wave motion.

material greatly influence the velocities of these waves. Nevertheless, in any solid material, P waves travel about 1.7 times faster than S waves, and surface waves are about 10 percent slower than S waves.

As you shall see, seismic waves allow us to determine the location and magnitude of earthquakes. In addition, seismic waves provide us with a tool for probing Earth's interior.

Finding Where Earthquakes Occur

Recall that the *focus* is the place within Earth where the earthquake waves originate. The **epicenter** is the location *on the surface directly above the focus* (see Figure 6.2, p. 147).

The difference in velocities of P and S waves provides a method for determining the epicenter. The principle used is analogous to a race between two autos, one faster than the other. The P wave always wins the race, arriving ahead of the S wave.

But, the greater the distance of the race, the greater will be the *difference* in the arrival times at the finish line (the seismic station). Therefore, the greater the interval measured on a seismogram between the arrival of the first P wave and the first S wave, the greater is the distance to the earthquake source.

The system for locating earthquake epicenters was developed by using seismograms from earthquakes whose epicenters could be easily pinpointed from physical evidence. From the seismograms, travel-time graphs were constructed (Figure 6.8). The first travel-time graphs were greatly improved when seismograms became available from nuclear explosions, because the precise location and time of detonation were known.

Using the simple seismogram in Figure 6.6 and the travel-time curves in Figure 6.8, we can determine the distance separating the recording station from the earthquake in two steps: (1) determine the time interval between the arrival of the first P wave and the first S wave, and (2) find on the travel-time graph the

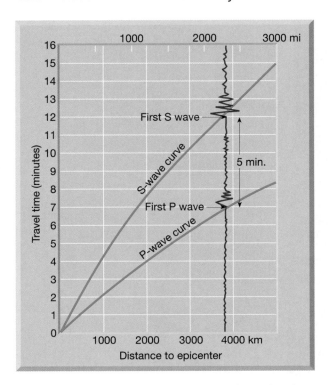

Figure 6.8 A travel-time graph is used to determine the distance to the epicenter. The difference in arrival times of the first P wave and the first S wave in the example is 5 minutes. Thus, the epicenter is about 3800 kilometers (2350 miles) away.

equivalent time spread between the P- and S-wave curves. This is shown in Figure 6.8. From this information, we can determine that this earthquake occurred 3800 kilometers (2350 miles) from the recording instrument.

Now we know the distance, but which way? Its location could be in any direction from the seismic station. As shown in Figure 6.9, the precise location can be found when the distance is known from three or more different seismic stations. On a globe, we draw a circle around each seismic station. Each circle represents the epicenter distance for each station. The point where the three circles intersect is the epicenter of the quake.

About 95 percent of the energy released by earthquakes originates in a few relatively narrow zones (Figure 6.10). The greatest energy is released along a path around the outer edge of the Pacific Ocean known as the *circum-Pacific belt*. Included in this zone are regions of great seismic activity, such as Japan, the Philippines, Chile, and numerous volcanic island chains, as exemplified by Alaska's Aleutian Islands.

Figure 6.10 reveals another continuous belt that extends for thousands of kilometers through the world's oceans. This zone coincides with the oceanic ridge system, an area of frequent but low-intensity seismic activity. By comparing this figure with Figure 5.6 (pp. 124-25), you can see a close correlation between the location of earthquake epicenters and plate boundaries.

Earthquake Magnitude

Early attempts to establish the size or strength of earthquakes relied heavily on subjective descriptions. There was an obvious problem with this method—people's accounts varied widely, making an accurate classification of the quake's intensity difficult.

In 1935, Charles Richter of the California Institute of Technology introduced the concept of earthquake **magnitude**. Today a refined **Richter scale** is used worldwide to describe earthquake magnitude. Using Richter's scale, the magnitude is determined by measuring the amplitude of the largest wave recorded on the seismogram (see Figure 6.6). For seismic stations worldwide to obtain the same magnitude for a given earthquake, adjustments are made for the weakening of seismic waves as they move from the focus and for the sensitivity of the recording instrument.

The largest earthquakes ever recorded have Richter magnitudes near 8.6. These great shocks released energy roughly equivalent to the detonation of a billion tons of TNT. Conversely, earthquakes with a Richter magnitude of less than 2.0 are usually not felt by humans. Table 6.1 shows how earthquake magnitudes and their effects are related.

As we have seen, earthquakes vary enormously in strength. To accommodate this wide variation, Richter did not use a linear scale but a logarithmic scale to express magnitude. On this scale, a tenfold increase in recorded wave amplitude corresponds to an increase of one number on the Richter scale (for example, 5.0 to 6.0). More importantly, each unit of magnitude increase on the Richter scale equates to roughly a 30-fold increase in the energy released. Thus, an earthquake with a magnitude of 6.5 releases 30 times more energy than one with a magnitude of 5.5, and roughly 900 times (30 × 30) more energy than a 4.5-magnitude quake.

A major earthquake with a magnitude of 8.5 releases millions of times more energy than the smallest earthquake felt by humans. This dispels the

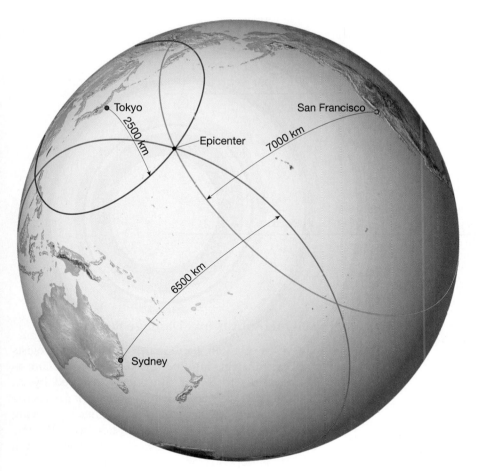

Figure 6.9 Earthquake epicenter is located using the distances obtained from three seismic stations.

Figure 6.10 Distribution of the 14,229 earthquakes with magnitudes equal to or greater than 5.0 for the period 1980–1990. *(Data from National Geophysical Data Center/NOAA)*

Table 6.1 ■ Earthquake magnitude and expected world incidence

Richter Magnitudes	Effects Near Epicenter	Estimated Number per Year
<2.0	Generally not felt, but recorded	600,000
2.0–2.9	Potentially perceptible	300,000
3.0–3.9	Felt by some	49,000
4.0–4.9	Felt by most	6200
5.0–5.9	Damaging shocks	800
6.0–6.9	Destructive in populous regions	266
7.0–7.9	Major earthquakes. Inflict serious damage	18
≥8.0	Great earthquakes. Destroy communities near epicenter	1–2

Source: Earthquake Information Bulletin and others.

notion that a moderate earthquake acts as a "pressure relief valve," decreasing the chances for the occurrence of a major quake in the same region. A moderate quake certainly relieves some strain, but thousands of moderate tremors would be needed to release the vast energy unleashed by one "great" earthquake. Some of the world's great historic earthquakes and their corresponding Richter magnitudes are listed in Table 6.2.

Destruction from Earthquakes

The most violent earthquake to jar North America this century—the Good Friday Alaskan earthquake—occurred in 1964. Felt throughout that state, the earthquake had a magnitude of 8.3–8.4 on the Richter scale and reportedly lasted 3 to 4 minutes. This brief event left 131 people dead, thousands homeless, and the economy of the state badly disrupted. Had the schools and business districts been open on this holiday, the toll surely would have been higher. Within 24 hours of the initial shock, 28 aftershocks were recorded, 10 of which exceeded a magnitude of 6 on the Richter scale.

Many factors determine the degree of destruction that accompanies an earthquake. The most obvious is the magnitude of the tremor and its proximity to a populated area. Fortunately, most quakes are small and occur in remote regions. However, about 20 major earthquakes are reported annually, one or two of which are catastrophic.

Seismic Vibrations Damage Structures

The 1964 Alaskan earthquake provided geologists with new insights into the role of *ground shaking* as a destructive force. As the energy released by an earthquake travels along Earth's surface, it causes the ground to vibrate in a complex manner by moving up and down as well as from side to side. The amount of structural damage attributable to the vibrations depends on several factors, including (1) the intensity and (2) duration of the vibrations, (3) the nature of the material upon which the structure rests, and (4) the design of the structure.

All of the multistory structures in Anchorage were damaged by the vibrations. The more flexible wood-frame residential buildings fared best. A striking example of how construction variations affect earthquake damage is shown in Figure 6.11. You can see that the steel-frame building on the left withstood the vibrations, whereas the relatively rigid concrete structure was badly damaged.

Most large structures in Anchorage were damaged, even though they were built to the earthquake provisions of the Uniform Building Code of California. Perhaps some of that destruction can be attributed to the unusually long duration of this earthquake. Most earthquakes consist of tremors that last less than one minute. For example, the 1994 Northridge earthquake was felt for about 40 seconds, and the strong vibrations of the 1989 Loma Prieta earthquake lasted less than 15 seconds. But the Alaska quake reverberated for 3 or 4 minutes.

Table 6.2 ■ Some notable earthquakes

Year	Location	Deaths (est.)	Magnitude	Comments
1556	Shensi, China	830,000	Unknown	Possibly the greatest natural disaster.
1755	Lisbon, Portugal	70,000	Unknown	Tsunami damage extensive.
*1811–1812	New Madrid, Missouri	Few	Unknown	Three major earthquakes.
*1886	Charleston, South Carolina	60	Unknown	Greatest historical earthquake in the eastern United States.
*1906	San Francisco, California	1500	8.1–8.2	Fires caused extensive damage.
1908	Messina, Italy	120,000	Unknown	
1923	Tokyo, Japan	143,000	7.9	Fire caused extensive destruction.
1960	Southern Chile	5700	8.5–8.6	Possibly the largest-magnitude earthquake ever recorded.
*1964	Alaska	131	8.3–8.4	Greatest North American earthquake.
1970	Peru	66,000	7.8	Great rockslide.
*1971	San Fernando, California	65	6.5	Damage exceeded $1 billion.
1975	Liaoning Province, China	1328	7.5	First major earthquake to be predicted.
1976	Tangshan, China	240,000	7.6	Not predicted.
1985	Mexico City	9500	8.1	Major damage occurred 400 km from epicenter.
1988	Armenia	25,000	6.9	Poor construction practices.
*1989	San Francisco Bay area	62	7.1	Damages exceeded $6 billion.
1990	Iran	50,000	7.3	Landslides and poor construction practices caused great damage.
1993	Latur, India	10,000	6.4	Located in stable continental interior.
*1994	Northridge, California	51	6.6–6.9	Damages in excess of $15 billion.
1995	Kobe, Japan	5472	6.9	Damages estimated to exceed $100 billion.

*U.S. earthquakes.
Source: U.S. National Oceanic and Atmospheric Administration.

Figure 6.11 Damage to the five-story J. C. Penney Co. building, Anchorage, Alaska. Very little structural damage was incurred by the adjacent building. *(Courtesy of NOAA)*

Although the region within 20 to 50 kilometers (12 to 30 miles) of an epicenter will experience about the same degree of ground shaking, the destruction will vary considerably within this area. This difference is mainly attributable to the nature of the ground on which the structures are built. Soft sediments, for example, generally amplify the vibration more than does solid bedrock. Thus, the buildings in Anchorage, which were situated on unconsolidated sediments, experienced heavy structural damage. By contrast, most of the town of Whittier, although much nearer the epicenter, rests on a firm foundation of granite and hence suffered much less damage. However, Whittier was damaged by a seismic sea wave, which we describe in the following section.

In areas where unconsolidated materials are saturated with water, earthquakes can generate a phenomenon known as **liquefaction**. Under these conditions, what had been a stable soil turns into a fluid that is not capable of supporting buildings or other structures (Figure 6.12). As a result, underground objects such as storage tanks and sewer lines may literally float toward the surface, while buildings and other structures may settle and collapse. During the 1989 Loma Prieta earthquake, in San Francisco's Marina District, foundations failed and geysers of sand and water shot from the ground, indicating that liquefaction had occurred.

Tsunamis

Most deaths associated with the 1964 Alaskan quake were caused by **seismic sea waves**, or **tsunamis**.* These destructive waves often are called "tidal waves" by the media. However, this name is incorrect, for these waves are generated by earthquakes, not the tidal effect of the Moon or Sun.

Most tsunamis result from vertical displacement of the ocean floor during an earthquake (Figure 6.13). Once formed, a tsunami resembles the ripples formed when a pebble is dropped into a pond. In contrast to ripples, tsunamis advance across the ocean at amazing speeds between 500 and 950 kilometers (300 and 600 miles) per hour. Despite this, a tsunami in the open ocean can pass undetected because its height is usually less than a meter and the distance between wave crests is great, ranging from 100 to 700 kilometers. However, upon entering shallower coastal water, these destructive waves are slowed and the water begins to pile up to heights that occasionally exceed 30 meters (100 feet), as shown in Figure 6.13. As the crest of a tsunami approaches shore, it appears as a rapid rise in sea level with a turbulent and chaotic surface. Tsunamis are quite destructive.

---

*Seismic sea waves were given the name *tsunami* by the Japanese, who have suffered a great deal from them. The term tsunami is now used worldwide.

Figure 6.12 Effects of liquefaction. This tilted building rests on unconsolidated sediment that imitated quicksand during a major 1985 Mexican earthquake. *(Photo by James L. Beck)*

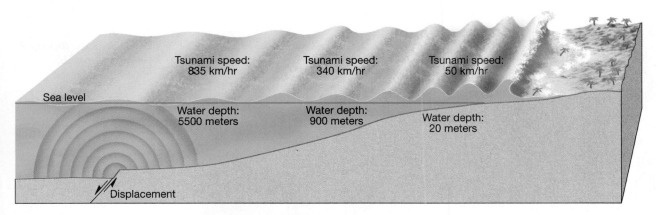

Figure 6.13 Schematic drawing of a tsunami generated by displacement of the ocean floor. The speed of a wave column correlates with ocean depth. As shown, waves moving in deep water advance at speeds in excess of 800 kilometers per hour. Speed gradually slows to 50 kilometers per hour at depths of 20 meters. Decreasing depth slows the movement of the wave column. As waves slow in shallow water, they grow in height until they topple and rush onto shore with tremendous force. The size and spacing of these swells are not to scale.

The tsunami generated in the 1964 Alaskan earthquake inflicted heavy damage to communities around the Gulf of Alaska. The deaths of 107 persons have been attributed to this tsunami. By contrast, only 9 persons died in Anchorage as a direct result of the vibrations.

Fire

The 1906 earthquake in San Francisco reminds us of the formidable threat of fire. The central city contained mostly large, older wooden structures and brick buildings. Although many of the unreinforced brick buildings were extensively damaged by vibrations, the greatest destruction was caused by fires that started when gas and electrical lines were severed. The fires raged out of control for 3 days and devastated over 500 blocks of the city. The problem was compounded by the initial ground shaking, which broke the city's water lines into hundreds of unconnected pieces.

The fire was finally contained when buildings were dynamited along a wide boulevard to provide a *fire break*. Only a few deaths were attributed to the San Francisco fire, but this is not always the case. An earthquake that rocked Japan in 1923 triggered an estimated 250 fires, which devastated the city of Yokohama and destroyed more than half the homes in Tokyo. Over 100,000 deaths were attributed to the fires, which were driven by unusually high winds.

Landslides and Ground Subsidence

In the 1964 Alaskan earthquake, the greatest damage to structures was from landslides and ground subsidence triggered by the vibrations. At Valdez

and Seward, the violent shaking caused river-delta materials to experience liquefaction; the subsequent slumping carried both waterfronts away. Because it could happen again in another quake, the entire town of Valdez was relocated about 7 kilometers away on more stable ground. In Valdez, 31 people on a dock died when it slid into the sea.

Most of the damage in Anchorage was attributed to landslides caused by the shaking and lurching ground. Many homes were destroyed in Turnagain Heights when a layer of clay lost its strength and over 200 acres of land slid toward the ocean (Figure 6.14). A portion of this landslide was left in its natural condition as a reminder of this destructive event. The site was named "Earthquake Park." Downtown Anchorage was also disrupted as sections of the main business district dropped by as much as 3 meters (10 feet).

Earthquakes East of the Rockies

When we think of earthquake destruction, places near plate boundaries usually come to mind—California, Mexico, Japan. If you don't live near earthquake-prone plate boundaries, are you immune from these destructive events? Not necessarily.

At least six major earthquakes have occurred in the central and eastern United States since colonial times. Three of these had estimated Richter magnitudes of 7.5, 7.3, and 7.8, and they were centered near the Mississippi River Valley in southeastern Missouri. Occurring in December 1811 and January and February 1812, these earthquakes, plus numerous smaller tremors, destroyed the town of New

Figure 6.14 Photo of a small portion of the Turnagain Heights slide. *(Photo courtesy of U.S. Geological Survey, U.S. Department of the Interior)*

Madrid, Missouri, triggered massive landslides, and caused damage over a six-state area. The course of the Mississippi River was altered, and Tennessee's Reelfoot Lake was enlarged. The distance over which these earthquakes were felt is truly remarkable. Reportedly, chimneys were toppled in Cincinnati, Ohio, and Richmond, Virginia, while Boston residents, located 1770 kilometers (1100 miles) to the northeast, felt the tremor.

The greatest historical earthquake in the eastern states occurred in 1886 in Charleston, South Carolina (Figure 6.15). This 1-minute event caused 60 deaths, numerous injuries, and great economic loss within a radius of 200 kilometers (120 miles) of Charleston. Within 8 minutes, effects were felt as far away as Chicago and St. Louis, where strong vibrations shook the upper floors of buildings, causing people to rush outdoors. In Charleston alone, over

Figure 6.15 Damage to Charleston, South Carolina, caused by the August 31, 1886, earthquake. Damage ranged from toppled chimneys and broken plaster to total collapse. *(Photo courtesy of U.S. Geological Survey, U.S. Department of the Interior)*

a hundred buildings were destroyed, and 90 percent of the remaining structures were damaged. It was difficult to find a chimney still standing.

Numerous other strong earthquakes have been recorded in the central and eastern United States. New England and adjacent areas have experienced sizable shocks since colonial times. The first reported earthquake in the Northeast took place in Plymouth, Massachusetts, in 1683, and was followed in 1755 by the destructive Cambridge, Massachusetts, quake. Moreover, ever since records have been kept, New York State alone has experienced over 300 earthquakes large enough to be felt.

Earthquakes in the central and eastern United States occur far less frequently than in California and other plate-boundary areas. Yet history indicates that the East is vulnerable. Further, the shocks that have occurred east of the Rockies have generally produced structural damage over a larger area than their counterparts of similar magnitude in California. The reason is that the underlying bedrock in the central and eastern United States is older and more rigid. As a result, seismic waves travel greater distances with less attenuation than in the western United States. It is estimated that for similar earthquakes, the region of maximum ground motion in the East may be up to ten times larger than in the West. Consequently, the higher rate of earthquake occurrence in the western United States is somewhat balanced by the capability of eastern earthquakes to create damage over a larger area.

Can Earthquakes Be Predicted?

The vibrations that shook Northridge, California, in 1994 inflicted 61 deaths and about $15 billion in damage (Figure 6.16). This was from a brief earthquake (about 40 seconds) of moderate rating (6.6–6.9 on the Richter scale). Seismologists warn that earthquakes of comparable or greater strength will occur along the San Andreas fault, which cuts a 1300-kilometer (800-mile) path through the state. Following the 1995 earthquake in Kobe, Japan, a U.S. Geological Survey physicist cautioned: "Kobe is almost a dress rehearsal for an earthquake on the Hayward fault [a fault parallel to the San Andreas, near San Francisco]." The obvious question is, can earthquakes be predicted?

Short-Range Predictions

The goal of short-range earthquake prediction is to provide a warning of the location and magnitude of a

Figure 6.16 Damage to Interstate 5 during the January 17, 1994, Northridge earthquake. *(Photo by Ted Soqui/Sygma)*

large earthquake within a narrow time frame. Substantial efforts to achieve this objective are being put forth in Japan, the United States, China, and Russia—countries where earthquake risks are high. This research has concentrated on monitoring possible *precursors*—phenomena that precede, and thus provide a warning of, a forthcoming earthquake. In California, for example, some seismologists are measuring uplift, subsidence, and strain in the rocks near active faults, and in Japan scientists are studying peculiar anomalous behavior that may precede a quake.

One claim of a successful short-range prediction was made by Chinese seismologists after the February 4, 1975, earthquake in Liaoning Province. According to reports, very few people were killed, although more than a million lived near the epicenter, because the earthquake was predicted and the population was evacuated. Recently, some Western seismologists have questioned this claim and suggest instead that an intense swarm of foreshocks that began 24 hours before the main earthquake may have caused many people to evacuate spontaneously. Further, an official Chinese government report issued 10 years later stated that 1328 people died and 16,980 injuries resulted from this earthquake.

One year after the Liaoning earthquake, at least 240,000 people were killed in the Tangshan, China, earthquake, which was not predicted. The Chinese have also issued false alarms. In a province near Hong Kong, people reportedly left their dwellings for over a month, but no earthquake followed. Clearly, whatever method the Chinese employ for short-range predictions, it is *not* reliable.

In order for a prediction scheme to warrant general acceptance, it must be both accurate and reliable. Thus, *it must have a small range of uncertainty as regards to location and timing, and it must produce few failures, or false alarms.* Can you imagine the debate that would precede an order to evacuate a large city in the United States, such as Los Angeles or San Francisco? The cost of evacuating millions of people, arranging for living accommodations, and providing for their lost work time and wages would be staggering.

Long-Range Forecasts

In contrast to short-range predictions, which aim to predict earthquakes within a time frame of hours or at most days, long-range forecasts give the probability of a certain magnitude earthquake occurring on a time scale of 30 to 100 years, or more. Stated another way, these forecasts give statistical estimates of the expected intensity of ground motion for a given area over a specified time frame. Although long-range forecasts may not be as informative as we might like, this data is important for updating the Uniform Building Code, which contains nationwide standards for designing earthquake-resistant structures.

Long-range forecasts are based on the premise that earthquakes are repetitive or cyclical, like the weather. In other words, as soon as one earthquake is over, the continuing motions of Earth's plates begin to build strain in the rocks again, until they fail once more. This has led seismologists to study historical records of earthquakes, for patterns, so their probability of recurrence might be established.

One study conducted by the U.S. Geological Survey gives the probability of a rupture occurring along various segments of the San Andreas fault for the 30 years between 1988 and 2018 (Figure 6.17). From this investigation, the Santa Cruz Mountains area was given a 30 percent probability of producing a 6.5-magnitude earthquake during this time period. In fact, it produced the Loma Prieta quake in 1989, of 7.1 magnitude.

The region along the San Andreas fault given the highest probability (90 percent) of generating a quake is the Parkfield section. This area has been called the "Old Faithful" of earthquake zones because activity here has been very regular since record keeping began in 1857. Another section between Parkfield and the Santa Cruz Mountains is given a very low probability of generating an earthquake. This area has experienced very little seismic activity in historical times; rather, it exhibits a slow, continual movement known as *fault creep*. Such movement is beneficial because it prevents strain from building to high levels in the rocks.

In summary, it appears that the best prospects for making useful earthquake predictions involve forecasting magnitudes and locations on time scales of years, or perhaps even decades. These forecasts are important because they provide information used to develop the Uniform Building Code and assist in land-use planning.

Earthquakes and Earth's Interior

Earth's interior lies just below us, yet its accessibility to direct observation is very limited. Much of the knowledge of our planet's interior comes from the study of P and S waves that travel through Earth and

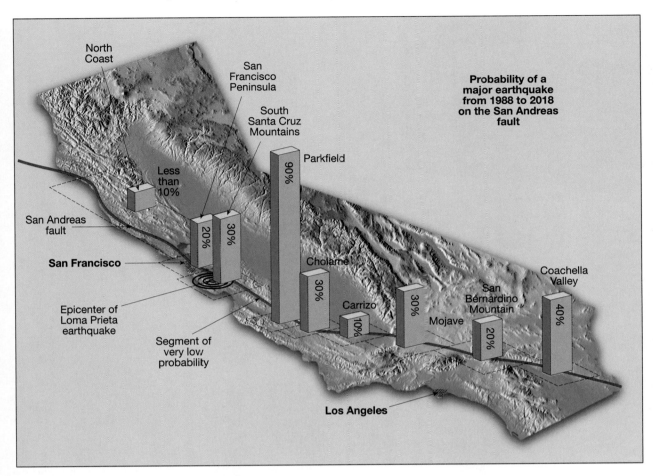

Figure 6.17 Probability of a major earthquake from 1988 to 2018 along the San Andreas fault.

vibrate the surface at some distant point. Simply stated, the technique involves accurately *measuring the time* required for seismic waves to travel from the focus of an earthquake or nuclear explosion to a seismographic station. Because the time required for P and S waves to travel through Earth depends on the properties of the materials encountered, seismologists search for variations in travel times that cannot be accounted for simply by differences in the distances traveled. These variations correspond to changes in the properties of the materials encountered.

Based on seismological data, geologists divide Earth into four major layers: (1) the **crust**, a very thin outer layer; (2) the **mantle**, a rocky layer located below the crust and having a thickness of 2885 kilometers (1789 miles); (3) the **outer core**, a layer about 2270 kilometers (1407 miles) thick, which exhibits the characteristics of a mobile liquid; and (4) the **inner core**, a solid metallic sphere about 1216 kilometers (754 miles) in radius (Figure 6.18).

Discovering Earth's Major Layers

In 1909, a pioneering Yugoslavian seismologist, Andrija Mohorovičić, presented the first convincing evidence for layering within Earth. By studying seismic records, he found that the velocity of seismic waves increases abruptly below about a 50-kilometer depth. This boundary separates the crust from the under-lying mantle and is known as the **Mohorovičić discontinuity** in his honor. For reasons that are obvious, the name for this boundary was quickly shortened to **Moho.**

A few years later another major boundary was discovered by the German seismologist Beno Gutenberg. This discovery was based primarily on the observation that P waves diminish and eventually die out completely about 105 degrees from an earthquake. Then, about 140 degrees away, the P waves reappear, but about 2 minutes later than would be expected based on the distance traveled. This belt

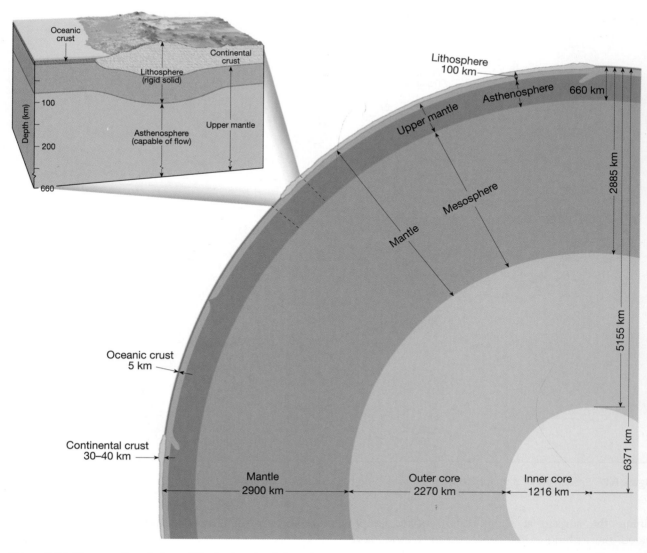

Figure 6.18 Cross-sectional view of Earth showing internal structure.

where direct seismic waves are absent is about 35 degrees wide and has been named the **shadow zone** (Figure 6.19).

Gutenberg realized that the shadow zone could be explained if Earth contained a core composed of material unlike the overlying mantle. The core must somehow hinder the transmission of P waves in a manner similar to the light rays being blocked by an opaque object that casts a shadow. However, rather than actually stopping the P waves, the shadow zone is produced by the bending of P waves that enter the core, as shown in Figure 6.19.

It was further learned that S waves could not propagate through the core. Therefore, geologists concluded that at least a portion of this region is liquid.

In 1936, the last major subdivision of Earth's interior was predicted by Inge Lehmann, a Danish seismologist. Lehmann discovered that seismic waves were reflected from yet another boundary within the core. Hence, a core-within-a-core was discovered. The actual size of the inner core was not accurately calculated until the early 1960s when underground nuclear tests were conducted in Nevada. Because the precise locations and times of the explosions were known, echoes from seismic waves that bounced off the inner core provided an accurate means of determining its size.

Over the past few decades, advances in seismology have allowed for much refinement of the gross view of Earth's interior presented so far. Of major

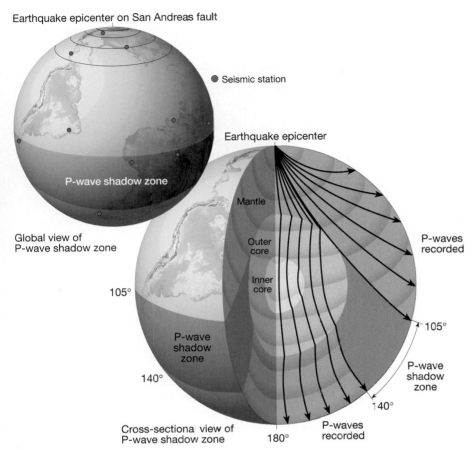

Earthquake epicenter on San Andreas fault

● Seismic station

Global view of
P-wave shadow zone

P-wave shadow zone

Earthquake epicenter

Mantle

Outer core

Inner core

P-waves recorded

105°

105°

140°

P-wave shadow zone

140°

Cross-sectiona view of
P-wave shadow zone

P-waves recorded

180°

Figure 6.19 The abrupt change in physical properties at the mantle–core boundary causes the wave paths to bend sharply. This abrupt change in wave direction results in a shadow zone for P waves between about 105 and 140 degrees.

importance was the discovery cf the lithosphere and asthenosphere.

The Lithosphere and Asthenosphere

Earth's outer layer, consisting of the uppermost mantle and crust, forms a relatively cool, rigid shell. Although this layer consists of materials with markedly different chemical compositions, it tends to act as a unit that behaves similarly to mechanical deformation. This outermost rigid unit of Earth is called the **lithosphere** ("sphere of rock"). Averaging about 100 kilometers in thickness, the lithosphere may be 250 kilometers or more in thickness below the older portions of the continents (Figure 6.20). Within the ocean basins the lithosphere is only a few kilometers thick along the oceanic ridges and increases to perhaps 100 kilometers in regions of older and cooler crustal rocks.

Beneath the lithosphere (to a depth of about 660 kilometers) lies a soft, relatively weak layer located in the upper mantle known as the **asthenosphere** ("weak sphere"). The upper 150 kilometers or so of

the asthenosphere has a temperature/pressure regime in which a small amount of melting takes place (perhaps 1 to 5 percent). Within this very weak zone, the lithosphere is effectively detached from the asthenosphere located below. The result is that the lithosphere is able to move independently of the asthenosphere.

Thus, the discovery of the asthenosphere was an important contribution to the theory of plate tectonics. It is the weak nature of the asthenosphere that facilitates the motion of Earth's rigid, broken outer shell of lithospheric plates.

Earth's Composition

Earth's crust varies in thickness, being greater than 70 kilometers in some mountainous regions and less than 5 kilometers in some oceanic regions (see Figure 6.18). Early seismic data indicated that the continental crust, which is mostly made of lighter, granitic rocks, is quite different in composition from the oceanic crust. Until recently, however, scientists had only seismic evidence from which to determine the composition of oceanic

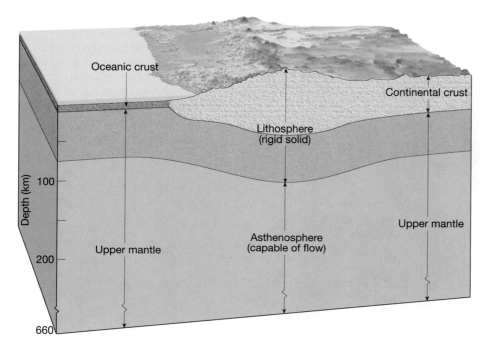

Figure 6.20 Relative positions of the asthenosphere and lithosphere.

crust, because it lies beneath 3 kilometers of water as well as hundreds of meters of sediment. The deep-sea drilling ship *Glomar Challenger* made possible recovery of ocean floor samples. These samples were of basaltic composition—very different from the rocks that make up the continents.

Our knowledge of rocks in the mantle and core is much more speculative. However, we do have some clues. Recall that some of the lava that reaches Earth's surface originates in the partially melted asthenosphere within the mantle. In the laboratory, experiments have shown that partial melting of a rock called peridotite results in magma that has a basaltic composition similar to that found during volcanic activity on oceanic islands. Denser rocks like peridotite are thought to make up the mantle and provide the lava for oceanic eruptions.

Surprisingly, meteorites or "shooting stars" that fall to Earth from space provide evidence of Earth's inner composition. Because meteorites are part of the solar system, they are assumed to be representative samples. Their composition ranges from metallic meteorites made of iron and nickel to stony meteorites composed of dense rock similar to peridotite.

Because Earth's crust contains a much smaller percentage of iron than do meteorites, geologists believe that the heavy minerals sank during Earth's early history. By the same token, the lighter minerals may have floated to the top, creating the crust. Thus,

Earth's core is thought to be mainly iron and nickel, similar to metallic meteorites, while the surrounding mantle is believed to be composed of dense rocks, similar to stony meteorites.

The concept of a molten iron outer core is further supported by Earth's magnetic field. Our planet acts as a large bar magnet. The most widely accepted mechanism explaining why Earth has a magnetic field requires that the core be made of a material that conducts electricity, such as iron, and that is mobile enough that circulation can occur. Both of these conditions are met by the model of Earth's core that was established on the basis of seismic data.

Not only does an iron core explain Earth's magnetic field but it also explains the high density of the interior, about 14 times that of water at Earth's center. Even under the extreme pressure at those depths, average crustal rocks with densities 2.8 times that of water would not have the density calculated for the core. But iron, which is three times more dense than crustal rocks, has the required density.

Geologic Structures

Earth is a dynamic planet. Evidence for the enormous forces that operate within it are seen in the Northern Rockies, where massive rock units have been thrust great distances over others. On a smaller scale, crustal movements of a few meters occur along faults during large earthquakes. The result of these tectonic activities is Earth's major mountain belts.

Before examining mountain building, we first must look at the nature of rock deformation and the structures that result. The most basic geologic structures associated with rock deformation are *folds* and *faults*.

Folds

During mountain building, flat-lying sedimentary and volcanic rocks often become slowly bent into a series of wavelike **folds** (Figure 6.21). Folds in sedimentary strata are much like those that would form if you were to pick up a sheet of paper, hold it by the edges, and then push the edges together. In nature, folds come in a wide variety of sizes and shapes.

The two most common types of folds are anticlines and synclines (Figure 6.21). An **anticline** is most commonly formed by the upfolding, or arching, of rock layers. Usually found in association with anticlines are downfolds, or troughs, called **synclines**. Depending on their orientation, anticlines and synclines are said to be *symmetrical, asymmetrical,* or, if one limb has been tilted beyond the vertical, *overturned*. These folds are sometimes spectacularly displayed where highways have been cut through deformed strata.

Most folds result from compressional stresses that shorten and thicken the crust. Occasionally, folds are found singly, but most often they occur as a series of undulations. Folds do not continue forever; they die out over distances much like the wrinkles in cloth.

Although most folds are caused by compressional stresses that squeeze and crumble strata, some folds are a consequence of vertical displacement. When upwarping produces a circular or somewhat elongated structure, the feature is called a **dome** (Figure 6.22). The Black Hills of western South Dakota are one such domal structure in which erosion has stripped away the upwarped sedimentary beds, exposing older igneous and metamorphic rocks in the center.

Downwarped structures with a similar shape are termed **basins**. Several basins also exist in the United States. The basins of Michigan and Illinois have very gently sloping beds, similar to saucers. Because large basins contain layers of sedimentary rock that slope at very low angles, they are usually identified by the age of the rocks that comprise them. The youngest rocks are found near the center, and the oldest rocks are at the flanks. This is just the opposite order of a domal structure such as the Black Hills, where the oldest rocks form the core.

Faults

Faults are fractures in the crust along which displacement has occurred, from millimeters to kilometers. Occasionally, small faults can be recognized in road cuts where sedimentary beds have been offset a few meters, as shown in Figure 6.23. Large-scale faults, like the San Andreas fault in California, have displacements of hundreds of kilometers and consist of many interconnecting fault surfaces. These *fault zones* can be several kilometers wide and are often easier to identify from high-altitude images than at ground level.

Faults in which the movement is primarily vertical are called **dip-slip faults** because the displacement is along the tilt (dip) of the fault plane (Figure 6.24A). Because movement along dip-slip faults can

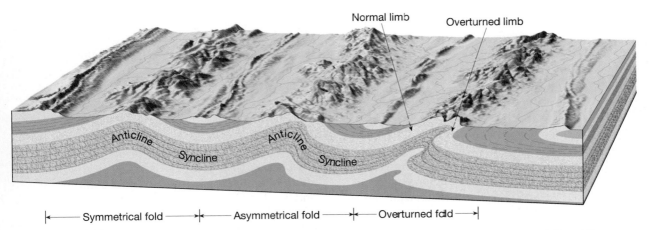

Normal limb   Overturned limb

Anticline   Syncline   Anticline   Syncline

|← Symmetrical fold →|← Asymmetrical fold →|← Overturned fold →|

Figure 6.21 Block diagram of principal types of folded strata. The upfolded or arched structures are anticlines. The downfolds or troughs are synclines. Notice that the limb of an anticline is also the limb of the adjacent syncline.

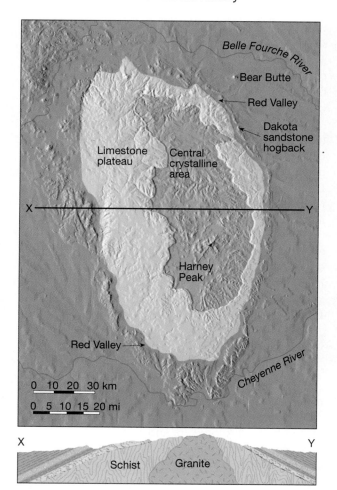

Figure 6.22 The Black Hills of South Dakota, an example of a domal structure in which the resistant igneous and metamorphic central core has been exposed by erosion. **A.** Map view. **B.** Cross-sectional view from point X to point Y.

Figure 6.23 Faulting caused vertical displacement of these sedimentary beds in southern Nevada. *(Photo by E. J. Tarbuck)*

be either up or down the fault plane, two types of dip-slip faults are recognized. To distinguish between the two, it has become practice to call the rock immediately above the fault surface the *hanging wall* and to call the rock below the *footwall*. This nomenclature arose from prospectors and miners who excavated shafts along fault zones because such zones are frequently sites of mineral deposition. During these operations, the miners would walk on the rocks below the fault trace (the footwall), and hang their lanterns on the rocks above (the hanging wall).

Dip-slip faults are classified as **normal faults** when the hanging wall moves down relative to the footwall (Figure 6.24A). Conversely, **reverse faults** occur when the hanging wall moves up relative to

the footwall (Figure 6.24B). In mountainous regions such as the Alps and the Appalachians, reverse faults with dips less than 45 degrees are called **thrust faults**; they have displaced rock as far as 100 kilometers over adjacent strata. Thrust faults of this type result from strong compressional stresses.

Faults in which the dominant displacement is along the trend, or strike, of the fault are called **strike-slip faults** (Figure 6.24D). Many large strike-slip faults are associated with plate boundaries and are called *transform faults*. Transform faults have nearly vertical dips and serve to connect large structures, such as segments of an oceanic ridge. The San Andreas fault in California is a well-known transform fault in which the total displacement exceeds 560 kilometers.

Fault motion provides the geologist with a method of determining the nature of the forces at work. Normal faults indicate *tensional stresses* that pull the crust apart. This "pulling apart" can be accomplished either by uplifting, which causes the surface to stretch and break, or by horizontal forces that actually pull the crust apart. Normal faulting is known to occur at spreading centers, because plate divergence (spreading apart) is prevalent. Here a central block called a

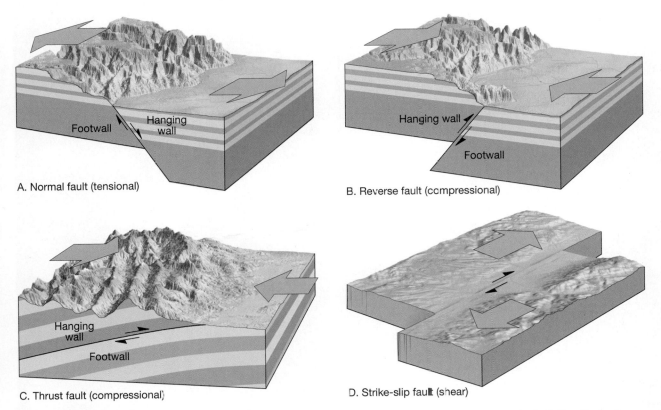

A. Normal fault (tensional)

B. Reverse fault (ccmpressional)

C. Thrust fault (compressional)

D. Strike-slip fault (shear)

Figure 6.24 Block diagrams of four types of faults. **A.** Normal fault. **B.** Reverse fault. **C.** Thrust fault. **D.** Strike-slip fault. (Note: FW = Footwall and HW = Hanging wall)

graben is bounded by normal faults and drops as the plates separate (Figure 6.25). These grabens produce an elongated valley flanked by upfaulted **horsts**. The Great Rift Valley of East Africa consists of several large grabens, above which tilted horsts produce a linear mountainous topography (see Figure 5.9, p. 127). The Great Rift Valley, nearly 6000 kilcmeters (3700 miles) long, contains the excavation sites of some of the ear-

liest human fcssils. Other rift valleys include the Rhine Valley in Germany and the valley of the Dead Sea in the Middle East.

Because tne blocks involved in reverse and thrust faulting are displaced *toward* one another, geologists conclude that *compressional forces* are at work. The primary regions of this activity are convergent zones, where plates are colliding. Compressional forces generally produce folds as well as faults, and they result in a general tłickening and shortening of the crust.

Mountain Building

Mountains are spectacular features that often rise several hundred meters or more above the surrounding terrain (Figure 6.26). All mountain systems show evidence of enormous forces that have folded, faulted, and generally ceformed large sections of Earth's crust. Although this folding and faulting have contributed to the majestic structure of mountains, much of the credit for their beauty must be given to erosion by running water and glacial ice, which sculpture these uplifted masses in an unending effort to lower them

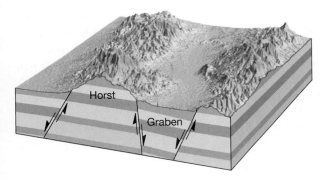

Figure 6.25 Diagrammatic sketch of cownfaulted block (graben) and upfaulted block (horst).

Figure 6.26 Teton Range, Grand Teton National Park, Wyoming. *(Photo by Carr Clifton Photography)*

to sea level. **Mountain building**, however, refers to the processes that deform and uplift the mountains.

Mountain building has operated during the recent geologic past in several locations around the world. These relatively young mountainous belts include the American Cordillera, which runs along the western margin of the Americas from Cape Horn to Alaska (Rockies, Andes); the Alpine-Himalayan chain, which extends from the Mediterranean through Iran to northern India and into Indochina; and the mountainous terrains of the western Pacific, which include volcanic island arcs such as Japan, the Philippines, and Sumatra. Most of these young mountain belts have come into existence within the last 100 million years. Some, including the Himalayas, began their growth as recently as 45 million years ago.

In addition to these recently formed mountains, several chains of much older mountains exist. Although these structures are deeply eroded and topographically less prominent, they clearly possess the same structural features found in younger mountains. Typical of this older group are the Appalachians in the eastern United States and the Urals in Russia.

The first encompassing explanation of mountain building came a little over three decades ago as part of plate tectonic theory. It is now known that Earth's major mountain systems formed along convergent plate boundaries. Recall that convergence can occur between one oceanic and one continental plate, between two oceanic plates, or between two continental plates. We will consider these sites of mountain building in the following sections.

Mountain Building at Convergent Boundaries

To unravel the events that produce mountains, many studies have been conducted along active subduction zones where plates are converging. At most modern-day subduction zones, volcanic arcs are forming. This situation is typified by Alaska's Aleutian Islands and by the Andes Mountains of western South America. Although all volcanic arcs are similar, *Aleutian-type subduction zones* occur where *two oceanic plates* converge (Figure 6.27). *Andean-type subduction zones* are situated where *oceanic crust is being thrust beneath a continental mass*. Consequently, the events generating these volcanic arcs followed different evolutionary paths. Further, although the development of a volcanic arc does result in the formation of mountainous topog-

raphy, this activity is viewed as just one of the phases in the development of a major mountain belt.

Mountain Building Where Oceanic and Continental Crust Converge   Mountain building along continental margins involves the convergence of an oceanic plate and a plate whose leading edge contains continental crust. Exemplified by the Andes Mountains, this *Andean type* of convergence generates structures resembling those of a developing volcanic island arc.

The first stage in the development of an Andean-type mountain belt occurs prior to the formation of the subduction zone. During this period the continental margin is *passive*. It is not a plate boundary but a part of the same plate as the adjoining oceanic crust. The East Coast of the United States is a present-day example of a passive continental margin. Here, as at other passive continental margins surrounding the Atlantic, deposition of sediment along the continental margin is producing a thick wedge of sandstones, limestones, and shales (Figure 6.28A).

At some point the continental margin becomes active. A subduction zone forms and the deformation process begins (Figure 6.28B). A good place to examine such an active continental margin is the western coast of South America. Here, an oceanic plate is being subducted eastward beneath the South American plate along the Peru–Chile trench (see Figure 5.6, pp. 124-125).

In an idealized Andean-type subduction, convergence of the continental block and the subducting oceanic plate leads to deformation and metamorphism of the continental margin. Once the oceanic plate descends to about 100 kilometers, partial melting of mantle rock located above the subducting slab generates magma that slowly migrates upward, intruding and further deforming these strata (Figure 6.28B). During the development of the volcanic arc, sediment derived from the land as well as that scraped from the subducting plate becomes plastered against the landward side of the trench. This chaotic accumulation of sedimentary and metamorphic rocks, including occasional scraps of ocean crust, is called an **accretionary wedge** (Figure 6.28B). Prolonged subduction can build an accretionary wedge that is large enough to stand above sea level (Figure 6.28C).

Andean-type mountain belts, like island arcs, are made up of two roughly parallel zones. The landward segment is the volcanic arc, consisting of volcanoes and large bodies of intrusive igneous rock, intermixed with high-temperature metamorphic

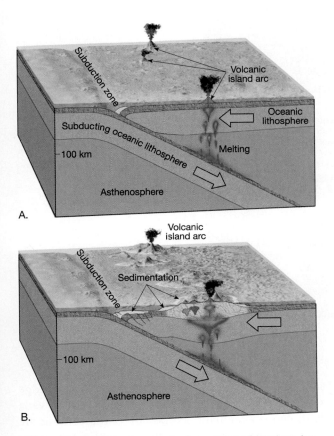

Figure 6.27 The development of a volcanic island arc by the convergence of two oceanic plates. These Aleutian-type subduction zones produce a mountainous topography that is similar to the volcanic arc that comprises the Aleutian islands of Alaska.

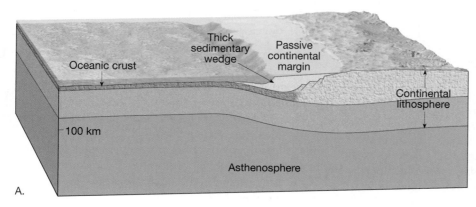

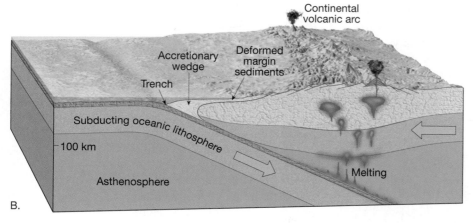

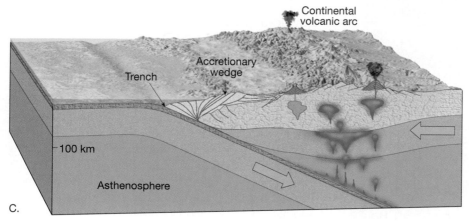

Figure 6.28 Mountain building along an Andean-type subduction zone. **A.** Passive continental margin with extensive wedge of sediments. **B.** Plate convergence generates a subduction zone, and partial melting produces a developing continental volcanic arc. **C.** Continued convergence and igneous activity further deform and thicken the crust, elevating the mountain belt, while the accretionary wedge grows.

rocks. The seaward segment is the accretionary wedge. It consists of folded, faulted, and metamorphosed sediments and volcanic debris (Figure 6.28C).

One of the best examples of an inactive Andean-type mountain belt is found in the western United States. It includes the Sierra Nevada and the Coast Ranges in California (Figure 6.29). These parallel mountain belts were produced by the subduction of

a portion of the Pacific basin under the western edge of the North American plate. The Sierra Nevada batholith is a remnant of a portion of the volcanic arc that was produced by several surges of magma over tens of millions of years. Subsequent uplifting and erosion have removed most evidence of past volcanic activity and exposed a core of crystalline metamorphic and igneous rocks.

Figure 6.29 Map of the California Coast Ranges and the Sierra Nevada.

In the trench region, sediments scraped from the subducting plate, plus those provided by the eroding volcanic arc, were intensely folded and faulted into an accretionary wedge. This chaotic mixture of rock presently constitutes part of California's Coast Ranges. Uplifting of the Coast Ranges took place only recently, as evidenced by the young unconsolidated sediments that still mantle portions of these highlands.

Mountain Building Where Continents Converge So far, we have discussed the formation of mountain belts where the leading edge of just one of the two converging lithospheric plates was capped by continental crust. However, it is possible for two converging plates to be carrying continental crust. Because continental crust is too buoyant to undergo any appreciable subduction, a collision between the continental fragments results (Figure 6.30).

An example of such a collision began about 45 million years ago when India collided with the Eurasian plate. India was once part of Antarctica but split from that continent and moved a few thousand kilometers due north. The result of this collision was the formation of the spectacular Himalaya Mountains and the Tibetan Plateau.

Although most of the oceanic crust that separated these landmasses prior to the collision was subducted, some was caught up in the squeeze, along with sediment that lay offshore (Figure 6.30). Today these sedimentary rocks and slivers of oceanic crust are elevated high above sea level. Geologists think that, following such a collision, the subducted oceanic plate decouples from the rigid continental plate and continues its descent into the mantle.

A similar but much older collision is believed to have taken place when the European continent collided with the Asian continent to produce the Ural Mountains, which extend north–south through Russia. Prior to the discovery of plate tectonics, geologists had difficulty explaining the existence of mountain ranges like the Urals, which are located deep within continental interiors. How could thousands of meters of marine sediment be deposited and then become highly deformed while situated in the middle of a large landmass?

Other mountain ranges showing evidence of continental collisions are the Alps and the Appalachians The Appalachians resulted from collisions among North America, Europe, and northern Africa. Although they have since separated, these landmasses were juxtaposed as part of the supercontinent Pangaea less than 200 million years ago. Detailed studies in the southern Appalachians indicate that the formation of this mountain belt was more complex than once thought. Rather than forming during a single continental collision, the Appalachians resulted from several distinct episodes of mountain building occurring over a period of nearly 300 million years.

Mountain Building and Continental Accretion
Plate tectonics theory originally suggested two mechanisms for mountain building: (1) continental collisions were proposed to explain the formation of such mountainous terrains as the Appalachians, Himalayas, and Urals; and (2) subduction of oceanic lithosphere was thought to be the underlying tectonic process for many circum-Pacific mountain chains, as typified by the Andes. Recent investigations, however, indicate a third mechanism of mountain building: *Smaller crustal fragments collide and accrete to*

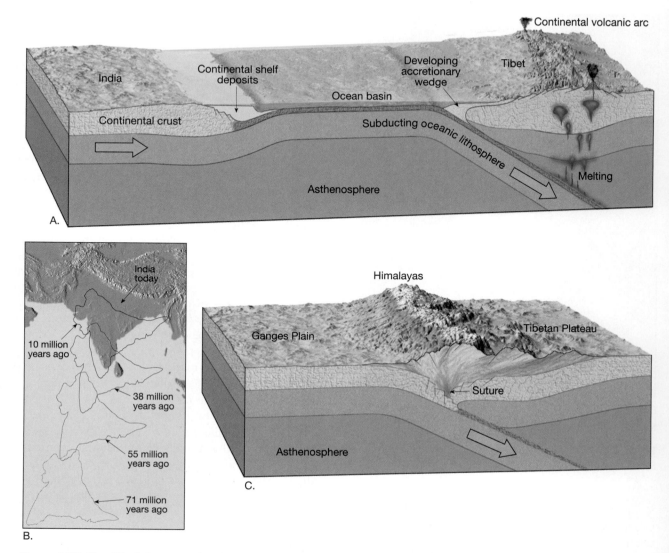

Figure 6.30 Simplified diagrams showing the northward migration and collision of India with the Eurasian plate. **A.** Converging plates generated a subduction zone, while partial melting of the subducting oceanic slab produced a continental volcanic arc. Sediments scraped from the subducting plate were added to the accretionary wedge. **B.** Position of India in relation to Eurasia at various times. (Modified after Peter Molnar) **C.** Eventually the two landmasses collided, deforming and elevating the accretionary wedge and continental shelf deposits. In addition, slices of the Indian crust were thrust up onto the Indian plate.

continental margins. Through this process of collision and accretion, many of the mountainous regions rimming the Pacific have been generated.

What is the nature of these small crustal fragments, and where do they come from? Researchers suggest that some of the fragments, prior to their accretion to a continental block, may have been microcontinents similar to the present-day island of Madagascar located just east of Africa. Others were island arcs like Japan, the Philippines, and the Aleutian Islands. Still others may have been submerged crustal fragments that extended high above the ocean floor. It is believed that these fragments originated as submerged continental fragments, extinct volcanic island arcs, or submerged volcanic plateaus associated with hotspot activity.

Accretion of Foreign Terranes The widely accepted view today is that, as oceanic plates move, they carry the embedded oceanic plateaus or

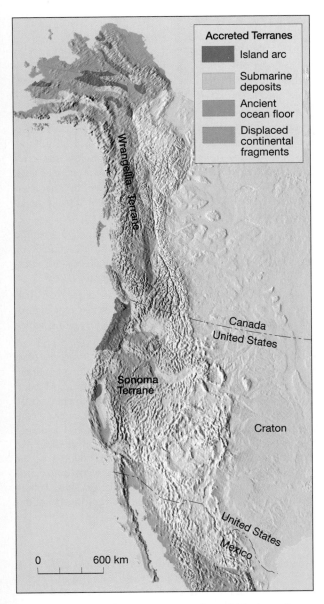

Figure 6.31 Map showing terranes thought to have been added to western North America during the past 200 million years. *(Redrawn after D. R. Hutchinson and others)*

microcontinents to a subduction zone. Here the upper portions of these thickened zones are peeled from the descending plate and thrust in relatively thin sheets onto the adjacent continental block. This newly added material increases the width of the continent. The material may later be overridden and displaced further inland by the addition of other fragments.

Geologists refer to these accreted crustal blocks as terranes. Simply, the term **terrane** designates any crustal fragment whose geologic history is distinct from that of the adjoining terranes. (Do not confuse the term *terrane* with the word *terrain*, which indicates the topography or lay of the land.) Terranes come in varied shapes and sizes; some are no larger than volcanic islands. Others, such as the one making up the entire Indian subcontinent, are very large.

Accretion and Mountain Building The idea that mountain building occurs in association with the accretion of small crustal fragments to a continental mass arose principally from studies conducted in the North American Cordillera (Figure 6.31). Some mountainous areas, principally those of Alaska and British Columbia, contain fossil and magnetic evidence that these strata once lay much nearer the equator.

It is now assumed that many other terranes found in the North American Cordillera were once scattered throughout the eastern Pacific, much as we find island arcs and oceanic plateaus distributed in the western Pacific today. Since the breakup of Pangaea 200 million years ago, North America has been migrating westward overriding the Pacific basin (Figure 6.31). Apparently, this activity resulted in the piecemeal addition of fragments to the entire Pacific Coast, from the Baja Peninsula to northern Alaska. In a like manner, many modern microcontinents will eventually be accreted to active continental margins, thus resulting in the formation of new mountainous belts.

The Chapter in Review

1. *Earthquakes* are vibrations produced by the rapid release of energy from rocks that rupture when subjected to stresses beyond their limit. The movements that produce earthquakes occur along large fractures, called *faults*, which are associated with plate boundaries.

2. Two main groups of *seismic waves* are generated during an earthquake: (1) *surface waves*, which travel along the outer layer of Earth, and (2) *body waves*, which travel through Earth's interior. Body waves are further divided into *primary waves* (P waves), which push (compress) and pull (expand) rocks in the direction the wave is traveling, and *secondary waves* (S waves), which "shake" the particles in rock at right angles to their direction of travel. The P waves can travel through solids, liquids, and gases; the S waves travel only through solids. In any solid material, P waves travel about 1.7 times faster than S waves.

3. The location on Earth's surface directly above the focus of an earthquake is the *epicenter*. An epicenter is determined using the difference in velocities of P and S waves.

4. *There is a close correlation between earthquake epicenters and plate boundaries.* The principal earthquake epicenter zones are along the margin of the Pacific Ocean, known as the *circum-Pacific belt*, and through the world's oceans along the *oceanic ridge system*.

5. The *Richter scale* indicates the *magnitude* (a measure of the total amount of energy released) of an earthquake. Magnitude is determined by measuring the *amplitude* (maximum displacement) of the largest seismic wave recorded, with adjustment for the weakening of seismic waves over distance and for the sensitivity of the recording instrument. Magnitude is expressed on a logarithmic scale, where each unit on the Richter scale corresponds to a tenfold increase in wave amplitude.

6. The most obvious factors that determine the amount of earthquake destruction are the *magnitude* of the earthquake and its *proximity* to a populated area.

7. As indicated by the behavior of P and S waves as they travel through Earth, the four major zones of Earth's interior are: (1) *crust* (the very thin outer layer, 5–40 kilometers); (2) *mantle* (a rocky layer located below the crust with a thickness of 2885 kilometers); (3) *outer core* (a layer about 2270 kilometers thick, which exhibits the characteristics of a mobile liquid); and (4) *inner core* (a solid metallic sphere with a radius of about 1216 kilometers).

8. The *continental crust* is primarily made of *granitic* rocks, while the *oceanic crust* is of *basaltic* composition. *Rocks* such as *peridotite* are thought to make up the *mantle*. The *core* is composed mainly of *iron* and *nickel*.

9. The most basic types of rock deformation are *folds* (flatlying rocks that have been bent into a series of wavelike undulations) and *faults* (fractures in the crust along which appreciable displacement has occurred).

10. *Major mountain systems form along convergent plate boundaries. Andean-type mountain building* along continental margins involves the convergence of an oceanic plate and a plate whose leading edge contains continental crust. *Continental collisions*, in which both plates are carrying continental crust, have most recently formed the Himalayan Mountains and Tibetan Highlands. Investigations indicate that *accretion*, a third mechanism of mountain-building occurs when *smaller crustal fragments collide and merge with continental margins* along some plate boundaries. Many mountainous regions rimming the Pacific have formed in this manner.

Key Terms

accretionary wedge (p. 169)

aftershock (p. 149)

anticline (p. 165)

asthenosphere (p. 162)

basin (p. 165)

body wave (p. 150)

crust (p. 161)

dip-slip fault (p. 165)

dome (p. 165)

earthquake (p. 147)

elastic rebound (p. 148)

epicenter (p. 151)

fault (pp. 147, 165)

focus (p. 147)

fold (p. 165)

foreshock (p. 149)

graben (p. 167)

horst (p. 167)

inner core (p. 161)

liquefaction (p. 156)

lithosphere (p. 162)

magnitude (p. 152)

mantle (p. 161)

Mohorovičić discontinuity (Moho) (p. 161)

mountain building (p. 168)

normal fault (p. 166)

outer core (p. 161)

primary (P) wave (p. 150)

reverse fault (p. 166)

Richter scale (p. 152)

secondary (S) wave
   (p. 150)
seismic sea wave (tsunami)
   (p. 156)

seismogram (p. 150)
seismograph (p. 149)
seismology (p. 149)

shadow zone (p. 162)
strike-slip fault (p. 165)
surface wave (p. 150)

syncline (p. 165)
terrane (p. 173)
thrust fault (p. 166)

Questions for Review

1. What is an earthquake? Under what circumstances do earthquakes occur?
2. How are faults, foci, and epicenters related?
3. Who was first to explain the actual mechanism by which earthquakes are generated?
4. Explain what is meant by elastic rebound.
5. Faults that are experiencing no active creep may be considered "safe." Rebut or defend this statement.
6. Describe the principle of a seismograph.
7. List the major differences between P and S waves.
8. Most strong earthquakes occur in a zone on the globe known as the _____.
9. An earthquake measuring 7 on the Richter scale releases about _____ times more energy than an earthquake with a magnitude of 6.
10. List four factors that affect the amount of destruction caused by seismic vibrations.
11. In addition to the destruction created directly by seismic vibrations, list three other types of destruction associated with earthquakes.
12. What is a tsunami? How is one generated?
13. What evidence do we have that the outer core is molten?
14. Contrast the physical makeup of the asthenosphere and the lithosphere.
15. Why are meteorites considered important clues to the composition of Earth's interior?
16. Describe the composition (mineral makeup) of the following:
    a. Continental crust   **c.** Mantle
    b. Oceanic crust   **d.** Core
17. What conditions favor rock deformation by folding? By faulting?
18. The San Andreas fault is an excellent example of a _____ fault.
19. Compare the movement of normal and reverse faults. What type of force produces each?
20. Compare and contrast anticlines and synclines; domes and basins; anticlines and domes.
21. What do we call the site where sediments are deposited along the margin of the continent? (Where they have a good chance of being squeezed into a mountain range.)
22. Which type of plate boundary is most directly associated with mountain building?
23. Suppose a sliver of oceanic crust was discovered 1000 kilometers into the interior of a continent. Would this support the theory of plate tectonics? Why?

Testing What You Have Learned

Multiple-Choice Questions

1. The movements that cause earthquake vibrations are usually associated with large fractures called
   - **a.** domes.
   - **b.** anticlines.
   - **c.** faults.
   - **d.** tsunamis.
   - **e.** seismograms.

2. Which of the following is true of P waves?
   - **a.** stopped by liquid
   - **b.** greatest velocity
   - **c.** compression wave
   - **d.** a. and c.
   - **e.** b. and c.

3. Which one of the following lists the order in which earthquake waves will be recorded by a seismograph located several hundred kilometers from an earthquake epicenter?
   - **a.** P, S, surface
   - **b.** S, P, surface
   - **c.** surface, S, P
   - **d.** surface, P, S
   - **e.** P, surface, S

4. Each unit of magnitude increase on the Richter scale equates to roughly how much of an increase in the energy released?
   - **a.** 2-fold
   - **b.** 10-fold
   - **c.** 15-fold
   - **d.** 20-fold
   - **e.** 30-fold

5. Seismic sea waves are also called _____
   - **a.** swells.
   - **b.** tsunami.
   - **c.** breakers.
   - **d.** surf.
   - **e.** currents.

6. The zone in the upper mantle consisting of hot, weak rock that is easily deformed is the _____
   - **a.** lithosphere.
   - **b.** hydrosphere.
   - **c.** shadow zone.
   - **d.** crust.
   - **e.** asthenosphere.

7. The rigid layer of Earth, which includes the crust as well as the uppermost mantle, is the _____
   - **a.** lithosphere.
   - **b.** hydrosphere.
   - **c.** shadow zone.
   - **d.** asthenosphere.
   - **e.** outer core.

8. Dip-slip faults in which the hanging wall moves down relative to the footwall are called _____ faults.
   - **a.** strike-slip
   - **b.** normal
   - **c.** reverse
   - **d.** complex
   - **e.** oblique-slip

9. Faults in which the dominant displacement is along the trend of the fault are called _____ faults.
   - **a.** strike-slip
   - **b.** normal
   - **c.** reverse
   - **d.** complex
   - **e.** oblique-slip

10. A chaotic accumulation of sedimentary and metamorphic rocks with occasional scraps of ocean crust scraped from a subducting plate and found on the landward side of an ocean trench is called a(n) _____
    - **a.** fault.
    - **b.** dome.
    - **c.** plateau.
    - **d.** hot spot.
    - **e.** accretionary wedge.

Fill-In Questions

11. The actual place of origin of an earthquake within Earth is called the _____, while the location on Earth's surface directly above is called the _____.

12. Small earthquakes that precede a major earthquake are called _____, while adjustments that follow a major earthquake often generate smaller quakes called _____.

13. The four major zones of Earth's interior, from the surface to the center, are the _____.

14. _____ are upfolded rock layers and _____ are downfolded rock layers that form from the bending of previously flat-lying sedimentary and volcanic rocks.

15. Most major mountain ranges form along _____ plate boundaries.

True/False Questions

16. Most of the motions that produce earthquakes can be satisfactorily explained by the plate tectonics theory. _____

17. The slow, gradual displacement, with little noticeable seismic activity, along a fault is known as fault creep. _____

18. Where unconsolidated materials are saturated with water, earthquakes may produce a phenomenon known as liquefaction. _____

19. At least a portion of Earth's outer core is liquid because S waves can not propagate through it. _____

20. Tensional stresses shorten and thicken the crust. _____

Answers:

1. c; 2. e; 3. a; 4. e; 5. b; 6. e; 7. a; 8. b; 9. a; 10. e; 11. focus, epicenter; 12. foreshocks, aftershocks; 13. crust, mantle, outer core, inner core; 14. Anticlines, synclines; 15. convergent; 16. T; 17. T; 18. T; 19. T; 20. F.

Web Work

The following are informative and interesting World Wide Web sites that address topics related to those presented in this chapter:

- National Earthquake Information Center
  http://wwwneic.cr.usgs.gov/
- National Geophysical Data Center
  http://www.ngdc.noaa.gov/seg/segd.html

For direct access to these sites and other relevant Web locations, contact the *Foundations of Earth Science* WWW home page at http://www.prenhall.com/lutgens.

CHAPTER 7

Fires Within

Igneous Activity

FOCUS ON LEARNING

To assist you in learning the important concepts
in this chapter, you will find it helpful to focus
on the following questions:

1. What primary factors determine the nature of volcanic eruptions? How do these factors affect a magma's viscosity?

2. What materials are associated with a volcanic eruption?

3. What are the eruptive patterns and characteristic shapes of the three groups of volcanoes generally recognized by volcanologists?

4. What are some other Earth features formed by volcanic activity?

5. What criteria are used to classify intrusive igneous bodies? What are some of these features?

6. What is the relation between volcanic activity and plate tectonics?

Mount Shasta, California, with Shastina in the foreground. *Photo © by Carr Clifton Photography*

Figure 7.1 In May 1980, Washington State's Mount St. Helens erupted explosively, sending huge quantities of volcanic ash and debris into the atmosphere. *(Photo by R. Hoblitt, U.S. Geological Survey, U.S. Department of the Interior)*

On Sunday, May 18, 1980, the largest volcanic eruption to occur in North America in historic times transformed a picturesque volcano into a decapitated remnant. On this date in southwestern Washington State, Mount St. Helens erupted with tremendous force (Figure 7.1). The blast blew out the entire north flank of the volcano, leaving a gaping hole. In one brief moment, a prominent volcano whose summit had been more than 2900 meters (9500 feet) above sea level was lowered by more than 400 meters (1350 feet).

The event devastated a wide swath of timber-rich land on the north side of the mountain. Trees within a 400-square-kilometer area lay intertwined and flattened, stripped of their branches and appearing from the air like toothpicks strewn about (Figure 7.2). The accompanying mudflows carried ash, trees, and water-saturated rock debris 29 kilometers down the Toutle River. The eruption claimed 59 lives, some dying from the intense heat and the suffocating cloud of ash and gases.

Not all volcanic eruptions are as violent as the 1980 Mount St. Helens event. Some volcanoes, such as Hawaii's Kilauea volcano, generate relatively quiet outpourings of fluid lavas (Figure 7.3). These "gentle" eruptions are not without some fiery displays; occasionally, fountains of incandescent lava spray hundreds of meters into the air. Such events, however, are typically short-lived and harmless, and the lava generally falls back into a lava pool.

Testimony to the quiet nature of Kilauea's eruptions is the fact that the Hawaiian Volcanoes Observatory has operated on its summit since 1912—this despite the fact that Kilauea has had more than 50 eruptive phases since record keeping began in 1823. Further, the longest and largest of Kilauea's rift eruptions began in 1983 and remains active at the time of this writing, although it has received only modest media attention.

Why do volcanoes like Mount St. Helens erupt explosively, whereas others like Kilauea are relatively quiet? Why do volcanoes occur in chains like the Aleutian Islands or the Cascade Range? Why do some volcanoes form on the ocean floor, while others occur on the continents? This chapter will deal with these and other questions as we explore the nature and movement of magma and lava.

The Nature of Volcanic Eruptions

Volcanic activity is commonly perceived as a process that produces a picturesque, cone-shaped structure that periodically erupts in a violent manner. Although some eruptions are very explosive, many others are not. What determines whether a volcano extrudes magma violently or "gently"? The primary factors include the magma's *composition*, its *temperature*, and the amount of *dissolved gases* it contains. To varying degrees these factors affect the magma's **viscosity**. The more viscous ("thicker") the material, the greater its resistance to flow. For example, syrup is more viscous than water. The viscosity of magma associated with an explosive eruption may be five times greater than magma that is extruded in a quiescent manner.

Figure 7.2 Forest lands and battered van destroyed by the lateral blast of Mount St. Helens on May 18, 1980. *(Photo courtesy of USDA)*

Factors Affecting Viscosity

The effect of temperature on viscosity is easily seen. Just as heating syrup makes it more fluid (less viscous), the mobility of lava is strongly influenced by temperature. As a lava flow cools and begins to congeal, its mobility decreases and eventually the flow halts.

A more significant influencing volcanic behavior is the chemical composition of magmas. This was discussed in Chapter 2 along with the classification of igneous rocks. Recall that a major difference among various igneous rocks is their silica (SiO_2) content. The same is true of the magmas from which rocks form. Magmas that produce basaltic rocks contain about 50 percent silica, whereas magmas that produce granitic rocks contain over 70 percent silica (Table 7.1).

Figure 7.3 Fluid basaltic lava from the flank of Kilauea Volcano, Hawaii. *(Photo by G. Brad Lewis/Liaison Agency)*

Table 7.1 ■ Variations in properties among magmas of differing compositions

Property	Basaltic magma	Andesitic magma	Granitic magma
Silica content	Least (about 50%)	Intermediate (about 60%)	Most (about 70%)
Viscosity	Least ("thinnest")	Intermediate	Greatest ("thickest")
Tendency to form lavas	Highest	Intermediate	Least
Tendency to form pyroclastics	Least	Intermediate	Greatest
Melting temperature	Highest	Intermediate	Lowest

A magma's viscosity is directly related to its silica content. In general, the more silica in magma, the greater is its viscosity. The flow of magma is impeded because silicate structures link together into long chains, even before crystallization begins. Consequently, because of high silica content, rhyolitic lavas are very viscous and tend to form comparatively short, thick flows. By contrast, basaltic lavas, which contain less silica, tend to be more fluid and have been known to travel distances of 150 kilometers (90 miles) or more before congealing (Figure 7.3).

In Hawaiian eruptions, the magmas are hot and basaltic, so they are extruded with ease. By contrast, highly viscous magmas are more difficult to force through a vent. On occasion, the vent may become plugged with viscous magma, which results in a buildup of gases and a great pressure increase, so a potentially explosive eruption may result. However, a viscous magma is not explosive by itself. It is the gas content that puts the "bang" into a violent eruption.

Importance of Dissolved Gases in Magma

Dissolved gases in magma provide the force that extrudes molten rock from the vent. These gases are mostly water vapor and carbon dioxide. As magma moves into a near-surface environment, such as within a volcano, the confining pressure in the uppermost portion of the magma body is greatly reduced. This reduction in confining pressure allows dissolved gases to be released suddenly, just as opening a soda bottle allows dissolved carbon dioxide gas bubbles to escape.

At high temperatures and low, near-surface pressures, these gases will expand to occupy hundreds of times their original volume. Very fluid basaltic magmas allow the expanding gases to bubble upward and

escape from the vent with relative ease. As they escape, the gases will often propel incandescent lava hundreds of meters into the air, producing lava fountains. Although spectacular, such fountains are mostly harmless and not generally associated with major explosive events that cause great loss of life and property. Rather, eruptions of fluid basaltic lavas, such as those that occur in Hawaii, are relatively quiescent.

At the other extreme, highly viscous magmas impede the upward migration of expanding gases. The gases collect in bubbles and pockets that increase in size and pressure until they explosively eject the semimolten rock from the volcano. The result is a Mount St. Helens or Mt. Pinatubo in the Philippines.

To summarize, the viscosity of magma, plus the quantity of dissolved gases and the ease with which they can escape, determines the nature of a volcanic eruption. We can now understand the "gentle" volcanic eruptions of hot, fluid lavas in Hawaii and the explosive, violent, dangerous eruptions of viscous lavas in some volcanoes.

What Is Extruded During Eruptions?

Lava may appear to be the primary material extruded from a volcano, but this is not always the case. Just as often, explosive eruptions eject huge quantities of broken rock, lava "bombs," fine ash, and dust. Moreover, all volcanic eruptions emit large amounts of gas. In this section we will examine each of these materials associated with a volcanic eruption.

Lava Flows

Because of their low silica content, basaltic lavas are usually very fluid. They flow in thin, broad sheets or streamlike ribbons. On the island of Hawaii, such lavas have been clocked at speeds of 30 kilometers (20 miles) per hour down steep slopes. These velocities are rare, however, and flow rates of 10 to 300 meters per hour are more common. In contrast, the movement of silica-rich lava is on occasion too slow to be perceptible.

When fluid basaltic lavas of the Hawaiian type congeal, they commonly form a relatively smooth skin that wrinkles as the still-molten subsurface lava continues to advance (Figure 7.4A). These are known as **pahoehoe flows** (pronounced *pah-hoy-hoy*) and resemble the twisted braids in ropes.

Another common type of basaltic lava has a surface of rough, jagged blocks with dangerously sharp edges and spiny projections (Figure 7.4B). The name **aa** (pronounced *ah-ah*) is given to these flows. Active aa flows are relatively cool and thick and, depending on the slope, advance at rates from 5 to 50 meters per hour. Further, escaping gases fragment the cool surface and produce numerous voids and sharp spines in the congealing lava. As the molten interior advances, the outer crust is broken further, giving the flow the appearance of an advancing mass of lava rubble.

Gases

Magmas contain varied amounts of dissolved gases held in the molten rock by confining pressure, just as carbon dioxide is held in soft drinks. As with soft drinks, as soon as the pressure is reduced, the gases begin to escape. Obtaining gas samples from an erupting volcano is often difficult and dangerous, so geologists can often only estimate the amount of gas originally contained within the magma.

The gaseous portion of most magmas is believed to make up 1 to 5 percent of the total weight, with most of this being water vapor. Although the percentage may be small, the actual quantity of emitted gas can exceed thousands of tons daily.

The composition of volcanic gases is important to scientists because these gases are thought to be the original source of the water for the oceans. Further, volcanic eruptions have contributed significantly to gases that make up the atmosphere. Analysis of samples taken during Hawaiian eruptions indicates that the gases are about 70 percent water vapor, 15 percent carbon dioxide, 5 percent nitrogen, 5 percent sulfur, and lesser amounts of chlorine, hydrogen, and argon. Sulfur compounds are easily recognized by their pungent odor and because they readily form sulfuric acid. Volcanoes are a natural source of air pollution, and they can cause serious problems.

Pyroclastic Materials

When basaltic lava is extruded, the dissolved gases escape quite freely and continually. As stated earlier,

A.

B.

Figure 7.4 A. Typical pahoehoe (ropy) lava flow, Kilauea, Hawaii. **B.** Typical slow-moving aa lava flow. *(Photos by J. D. Griggs, U.S. Geological Survey, U.S. Department of the Interior)*

these gases propel incandescent blobs of lava to great heights, thereby producing spectacular lava fountains. Some ejected material may land near the vent and build a cone structure, whereas smaller particles will be carried great distances by the wind.

The gases in highly viscous magmas, on the other hand, become superheated, and upon release they expand a thousandfold as they blow pulverized rock, lava, and glass fragments from the vent. The particles produced by these processes are called **pyroclasts** (meaning "fire fragments"). These ejected lava fragments range in size from very fine dust and sand-sized volcanic ash, to large pieces (Figure 7.5).

The *ash* particles are produced when the extruded lava contains so many gas bubbles that it resembles the froth flowing from a newly opened bottle of champagne. As the hot gases expand explosively, the lava is disseminated into very fine glassy fragments. When the hot ash falls, the glassy shards often fuse to form *welded tuff*. Sheets of this material, as well as ash deposits that later consolidated, cover vast portions of the western United States. Sometimes the frothlike lava is ejected in larger pieces called *pumice*. This material has so many voids (air spaces) that it is often light enough to float in water.

Pyroclasts the size of walnuts, called *lapilli* ("little stones"), and pea-sized particles called *cinders* are also very common. Cinders contain numerous voids, and they form when ejected lava blocks are pulverized by escaping gases. Particles larger than lapilli are called *blocks* when they are made of hardened lava and *bombs* when they are ejected as incandescent lava. Because volcanic bombs are semimolten upon ejection, they often take on a streamlined shape, as shown in Figure 7.6.

Volcano Types

Successive eruptions from a central vent result in a mountainous accumulation of material known as a **volcano**. Located at the summit of many volcanoes is a steep-walled depression called a **crater**. The crater is connected to a magma chamber via a pipelike conduit, or **vent**. Some volcanoes have unusually large summit depressions that exceed 1 kilometer in diameter and are known as **calderas**.

When fluid lava leaves a conduit, it is often stored in the crater or caldera until it overflows. Conversely, viscous lava forms a plug in the pipe. It rises slowly or is blown out, often enlarging the

Figure 7.5 Pyroclasts and lava ejected from a flank eruption at Kilauea volcano, Hawaii. *(Photo by Douglas Peebles)*

Figure 7.6 Volcanic bomb. Ejected lava fragments take on a steamlined shape as they sail through the air.

crater. However, lava does not always issue from a central crater. Sometimes it is easier for the magma or escaping gases to push through fissures on the volcano's flanks. Mount Etna in Italy, for example, has more than 200 secondary vents. Some of these emit only gases and are appropriately called *fumaroles*.

The eruptive history of each volcano is unique. Consequently, all volcanoes are somewhat different in form and size. Nevertheless, volcanologists recognize three general eruptive patterns and characteristic forms: shield volcanoes, cinder cones, and composite cones.

Shield Volcanoes

When fluid lava is extruded, the volcano takes the shape of a broad, slightly domed structure called a **shield volcano** (Figure 7.7). They are so called because they roughly resemble the shape of a warrior's shield. Shield volcanoes are built primarily of basaltic lava flows and contain only a small percentage of pyroclastic material.

Mauna Loa, probably the largest volcano on Earth, is one of the five shield volcanoes that together make up the island of Hawaii. Its base rests on the ocean floor 5000 meters below sea level, and its summit is 4170 meters (13,680 feet) above the water. Nearly a million years and numerous eruptive cycles built this truly gigantic pile of volcanic rock. Many other volcanic structures, including Midway Island and the Galapagos Islands, have been built in a similar manner from the ocean's depths.

Perhaps the most active and intensively studied shield volcano is Kilauea, located on the island of Hawaii on the southeastern flank of the larger volcano

Mauna Loa. Kilauea has erupted more than 50 times in recorded history and is still active today. Several months before an eruptive phase, Kilauea's summit inflates as magma rises from its source 60 kilometers or more below the surface. This molten rock gradually works its way upward and accumulates in smaller reservoirs 3 to 5 kilometers below the summit. For up to 24 hours in advance of each eruption, swarms of small earthquakes warn of the impending activity.

The longest and largest rift eruption ever recorded at Kilauea began in 1983 and continues as this is written in 1998. The eruption began along a 6.5-kilometer (4-mile) fissure in an inaccessible forested area east of the summit caldera (Figure 7.8). Between 1983 and 1986, nearly 50 eruptive phases were observed. Each consisted of a period of high lava fountaining that lasted a few hours to a few days, followed by an inactive period that averaged one month. After a few months, the eruptions localized in one spot along the fissure, building a major spatter cone 250 meters (820 feet) high.

Lava falling back from these fountains formed ash flows that migrated a few hundred meters per hour. Although inhabitants in a nearby settlement had time to be evacuated, many of their homes were burned and buried in the aa flow.

By the summer of 1986, the eruptions shifted 3 kilometers downslope from the cone. Here, smooth-surfaced pahoehoe lava formed a lava lake. Eventually, it overflowed, and the fast-moving pahoehoe destroyed nearly a hundred rural homes, covered a major roadway, and eventually flowed into the sea.

Eruptions of the type just described are typical of shield volcanoes and have been occurring sporadically on the Island of Hawaii for nearly a million

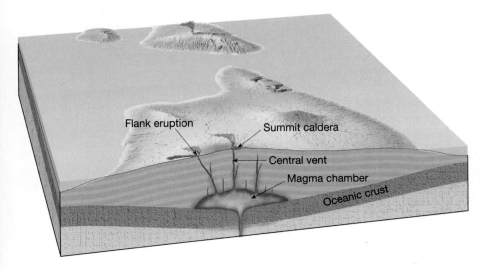

Figure 7.7 Shield volcanoes are built primarily of fluid basaltic lava flows and contain only a small percentage of pyroclastic materials. These broad, slightly domed structures, exemplified by the Hawaiian Islands, are the largest volcanoes on Earth.

Figure 7.8 Fissure eruption ("Curtain of Fire") on Kilauea, Hawaii, in 1986. *(Photo by Greg Vaughn Photography)*

years. The result is the formation of mountains with summits over 9000 meters (30,000 feet) above the seafloor, the tallest mountains on Earth.

Cinder Cones

As the name suggests, **cinder cones** are built from ejected lava fragments. Loose pyroclastic material has a high angle of repose (between 30 and 40 degrees), the steepest angle at which the material remains stable. Thus, volcanoes of this type have very steep slopes (Figure 7.9). Cinder cones are rather small, usually less than 300 meters (1000 feet) high, and they often form as parasitic cones on or near larger volcanoes. In addition, they frequently occur in groups.

One of the very few volcanoes whose formation has been observed by geologists from beginning to end is a cinder cone called Parícutin. In 1943, about 200 miles west of Mexico City, the volcano Parícutin was born (Figure 7.10). It erupted in a cornfield owned by Dionisio Pulido, who witnessed the event as he prepared the field for planting. For 2 weeks prior to the first eruption, numerous earth tremors caused apprehension in the village of Parícutin about 3.5 kilometers away. Then sulfurous smoke began billowing from a small hole that had been in the cornfield for as long as Señor Pulido could remember.

During the night, hot, glowing rock fragments thrown into the air from the hole produced a spectacular fireworks display. In one day the cone grew to 40 meters (130 feet) and by the fifth day it was over 100 meters (330 feet) high. Explosive eruptions threw hot fragments 1000 meters (3300 feet) above the crater rim. Larger fragments fell near the crater, some remaining incandescent as they rolled down the slope. These built an aesthetically pleasing cone, while finer ash fell over a much larger area, burning and eventually covering the village of Parícutin. Within 2 years the cone attained its final height of about 400 meters (1300 feet).

The first lava flow came from a fissure that opened just north of the cone, but after a few months flows began to emerge from the base of the cone itself. In June 1944, a clinkery aa flow 10 meters thick moved over much of the village of San Juan Parangaricutiro, leaving the church steeple exposed (Figure 7.10). After 9 years, the activity ceased almost as quickly as it had begun. Today, Parícutin is just another one of the numerous quiet cinder cones dotting the landscape in this region of Mexico. Like the others, it will probably not erupt again.

Composite Cones

Earth's most picturesque volcanoes are composite cones. Most active composite cones are in a narrow zone that encircles the Pacific Ocean, appropriately named the *Ring of Fire*. Found in this region are Fujiyama (Mt. Fuji) in Japan, Mount Mayon in the Philippines, and the picturesque volcanoes of the Cascade Range in the northwestern United States, including Mount St. Helens, Mount Rainier, and Mount Shasta (Figure 7.11).

A **composite cone** or **stratovolcano** is a large, nearly symmetrical structure composed of interbedded lava flows and pyroclastic deposits, emitted mainly from a central vent. Just as shield volcanoes owe their shape to the highly fluid nature of the extruded lavas, so too do composite cones reflect the nature of the erupted material.

Composite cones are produced when relatively viscous lavas of andesitic composition are extruded. A composite cone may extrude viscous lava for long periods. Then, suddenly, the eruptive style changes and the volcano violently ejects pyroclastic material. Occasionally, both activities occur simultaneously, and the resulting structure consists of alternating layers of lava and pyroclasts. Two of the most perfect cones, Mount Mayon in the Philippines and Fujiyama in Japan, exhibit the classic form of the stratovolcano with its steep summit area and more gently sloping flanks.

Composite cones represent the most violent type of volcanic activity. This eruption can be unexpected and devastating, as was the A.D. 79 eruption of the Italian volcano we now call Vesuvius. Prior to this eruption, Vesuvius had been dormant for centuries. Although minor earthquakes probably warned of the events to follow, Vesuvius was covered with dense

Pyroclastic material

Crater

Central vent filled with rock fragments

Figure 7.9 SP Crater, a cinder cone north of Flagstaff, Arizona. Cinder cones are built from ejected lava fragments and are usually less than 300 meters (1000 feet) high. *(Photo by Michael Collier)*

Figure 7.13 St. Pierre as it appeared shortly after the eruption of Mount Pelée, 1902. *(Reproduced from the collection of the Library of Congress)*

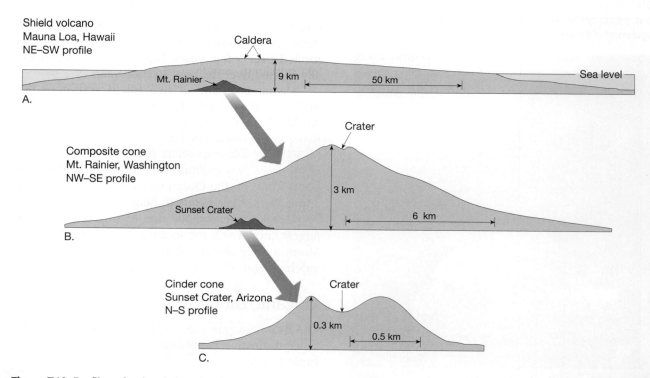

Figure 7.14 Profiles of volcanic landforms. **A.** Profile of Mauna Loa, Hawaii, the largest shield volcano in the Hawaiian chain. Note size comparison with Mt. Rainier, Washington, a large composite cone. **B.** Profile of Mt. Rainier, Washington. Note how it dwarfs a typical cinder cone. **C.** Profile of Sunset Crater, Arizona, a typical steep-sided cinder cone.

easily eroded because they are made of unconsolidated materials. As erosion progresses, the rock occupying the vent is often more resistant and may remain standing above the surrounding terrain long after most of the cone has vanished. Ship Rock, New Mexico, is such a feature, called a **volcanic neck** (Figure 7.15). This structure, higher than many skyscrapers, is but one of many such landforms that protrude conspicuously from the red desert landscapes of the Southwest. Ultimately, even these resistant necks will succumb to relentless erosion over geologic time.

Craters and Calderas

As noted, most volcanoes have a steep-walled *crater*. A crater is called a *caldera* when it exceeds 1 kilometer in diameter. Calderas are usually circular, with rather flat floors and steep walls. Several large calderas exist; the largest in the United States, LaGanita in Colorado, is about 32 kilometers (19 miles) across and 80 kilometers (50 miles) long.

Most calderas probably form when the summit of a volcanic structure collapses into the partially emptied magma chamber below. Crater Lake in Oregon, 8 to 10 kilometers (5 to 6 miles) wide and 1175 meters (over 3800 feet) deep, is located in such a depression (Figure 7.16). The formation of Crater Lake began about 7000 years ago when the volcano, later to be named Mount Mazama, put forth a violent ash eruption that extruded an estimated 50 to 70 cubic kilometers of volcanic material. With the loss of support, 1500 meters (4900 feet) of this once-prominent 3600-meter cone collapsed. After the collapse, rainwater filled the caldera. Later volcanic activity built a small cinder cone called Wizard Island, which today provides a mute reminder of past activity.

Not all calderas are produced following explosive eruptions. For example, the summits of Hawaii's active shield volcanoes, Mauna Loa and Kilauea, have large calderas that formed as a result of relatively quiet activity. These calderas, which measure 3 to 5 kilometers (2 to 3 miles) across and nearly 200 meters (650 feet) deep, formed by collapse as magma slowly drained from the summit magma chambers during flank eruptions.

Fissure Eruptions and Lava Plateaus

We think of volcanic eruptions as occurring from a central vent and building a cone or mountain, but by far the greatest volume of volcanic material is extruded from fractures in the crust called **fissures**. Rather than building a cone, these long, narrow cracks pour forth lava, blanketing a wide area. The extensive Columbia Plateau in the northwestern United States was formed this way (Figure 7.17). Here, numerous fissure eruptions extruded very fluid basaltic lava. Successive flows, some 50 meters thick, buried the existing landscape as they built a lava plateau, nearly a mile thick in places (Figure 7.18). The fluidity is evident, because some lava remained molten long enough to flow 150

Figure 7.15 Ship Rock, New Mexico, a volcanic neck. This structure, which stands over 420 meters (1380 feet) high, consists of igneous rock that crystallized in the vent of a volcano that has long since been eroded away. *(Photo by David Muench Photography)*

Figure 7.16 Crater Lake in Oregon occupies a caldera about 10 kilometers (6 miles) in diameter. *(Photo by Greg Vaughn/Tom Stack & Associates)*

kilometers (90 miles) from its source. The term **flood basalts** appropriately describes these flows.

Pyroclastic Flows

When silica-rich magma is extruded, **pyroclastic flows** of ash and pumice fragments usually result. They are propelled away from the vent at high speeds and may blanket extensive areas before coming to rest. Once deposited, the pyroclastic materials may closely resemble lava flows.

Extensive pyroclastic flow deposits exist in many parts of the world, most associated with large calderas. Perhaps best known is the Yellowstone Plateau in northwestern Wyoming. Here a large magma body, rich in silica, still exists a few kilometers below the surface. Several times over the past 2 million years, fracturing of the rocks overlying the magma chamber has resulted in huge outpourings, accompanied by the formation of calderas. In Yellowstone National Park, numerous buried fossil forests have been discovered. During volcanic activity, a forest would develop on a newly formed volcanic surface, only to be covered by ash from the next eruptive phase. Fortunately, no eruption of this type has occurred in historic times.

Intrusive Igneous Activity

Volcanic eruptions can be among the most violent and spectacular events in nature and thus receive much scientific attention. Yet most magma is emplaced at depth, intruding into existing rocks. An understanding of this intrusive igneous activity is therefore as important to geologists as the study of volcanic events. In mythology, Pluto was lord of the underworld. Thus, underground igneous rock bodies are called *plutons*.

Intrusive Igneous Bodies

Figure 7.19 shows intrusive igneous bodies (plutons) that form when magma crystallizes within Earth's crust. Notice that some of these structures have a tabular or sheetlike shape, while others are bulky masses. Some cut across existing structures, such as

192

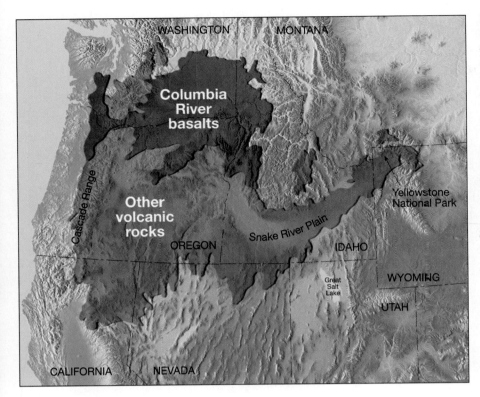

Figure 7.17 Volcanic areas in the northwestern United States. The Columbia River basalts cover an area of nearly 200,000 square kilometers (80,000 square miles). Activity here began about 17 million years ago as lava began to pour out of large fissures, eventually producing a basalt plateau with an average thickness of more than 1 kilometer. *(After U.S. Geological Survey)*

layers of sedimentary rocks; others form when magma is injected between sedimentary layers. Because of these differences, plutons are generally classified according to their shape as either *tabular* (sheetlike) or *massive*, and by their orientation with respect to the host rock. Intrusive igneous bodies are said to be *discordant* if they cut across existing sedimentary beds, and *concordant* if they form parallel to the existing sedimentary beds.

Figure 7.18 Basalt flows of the Columbia Plateau. *(Photo by E. J. Tarbuck)*

Dikes are sheetlike bodies produced when magma is injected into fractures that cut across rock layers. They range in thickness from less than 1 centimeter to more than 1 kilometer. The largest are hundreds of kilometers long. Dikes are often oriented vertically, forming in pathways followed by molten rock that fed ancient lava flows. Frequently, dikes are more resistant to weathering than is the surrounding rock. When exposed, such dikes have the appearance of a wall, as shown in Figures 7.15 and 7.19.

Sills are tabular plutons formed when magma is injected along sedimentary bedding surfaces (Figures 7.19 and 7.20). Horizontal sills are the most common, although all orientations, even vertical, are known to exist where the strata have been tilted. Because of their relatively uniform thickness and large extent, sills must form from very fluid basaltic magma. The emplacement of a sill requires that the overlying sedimentary rock be lifted to a height equal to the thickness of the sill. Consequently, sills form only at relatively shallow depths where the pressure exerted by the weight of overlying strata is relatively low.

One of the largest and best-known sills in the United States is the Palisades Sill, which is exposed along the Hudson River near New York City. This resistant, 300-meter-thick sill has formed an imposing cliff easily seen from across the river.

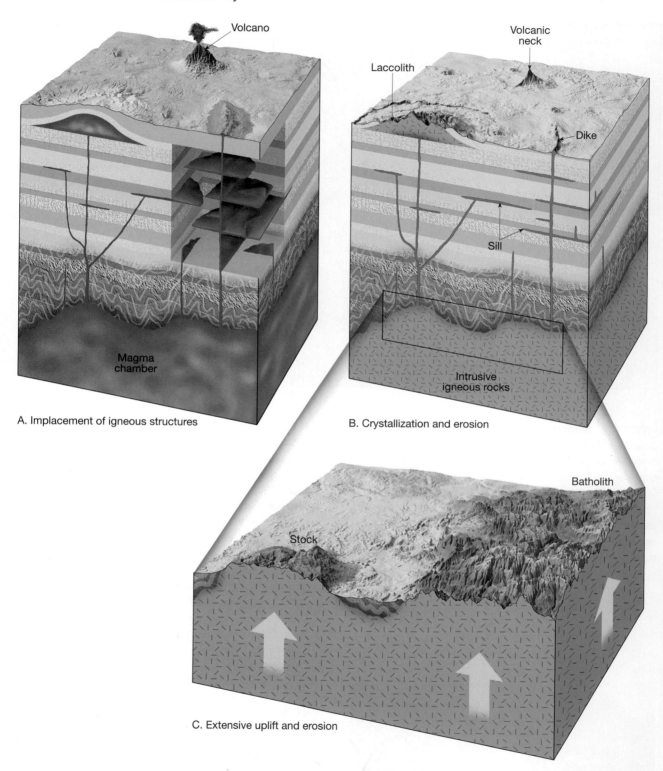

Figure 7.19 Illustrations showing basic igneous structures. **A.** This cross-sectional view shows the relationship between volcanism and intrusive igneous activity. **B.** This view illustrates the basic intrusive igneous structures, some of which have been exposed by erosion long after their formation. **C.** After millions of years of uplifting and erosion a batholith is exposed at the surface.

Sill

Figure 7.20 Salt River Canyon, Arizona. The dark, essentially horizontal band is a sill of basaltic composition that intruded into horizontal layers of sedimentary rock. *(Photo by E. J. Tarbuck)*

Laccoliths form the same way as sills, but from more viscous magma, which collects as a lens-shaped mass that arches the overlying strata upward. Consequently, a laccolith can occasionally be detected because of the dome it creates at the surface (Figure 7.19).

Batholiths are by far the largest intrusive igneous bodies. The Idaho batholith, for example, encompasses more than 40,000 square kilometers. Indirect evidence from gravitational studies indicates that batholiths are also very thick, possibly extending tens of kilometers into the crust. Geologists define a batholith as an intrusive body with a surface exposure of more than 100 square kilometers (40 square miles).

Batholiths frequently form the cores of mountain systems. Here, uplift and erosion have removed the surrounding rock to expose the resistant igneous body. Some of the highest mountain peaks, such as Mount Whitney in the Sierra Nevada, are carved from such a granitic mass. Large batholiths form over millions of years. The intrusive activity that created the Sierra Nevada batholith, for example, occurred nearly continuously over a 130-million-year span (Figure 7.21).

Large expanses of granitic rock are also exposed in the stable interiors of the continents, such as the Canadian Shield of North America. These relatively flat outcrops are believed to be the remnants of ancient mountains that erosion has long since leveled.

Igneous Activity and Plate Tectonics

For many years geologists have realized that the global distribution of igneous activity is not random, but rather exhibits a definite pattern. In particular, igneous rocks having a granitic or andesitic composition (those rich in silica) are confined largely to the continents and continental margins. By contrast, volcanoes located within the deep-ocean basins, such as those in Hawaii and Iceland, extrude lavas that are basaltic in composition (low in silica). Moreover, basaltic rocks are also commonly found in continental settings. This pattern puzzled geologists until the development of the plate tectonics theory, which greatly clarified the picture.

Most of the more than 600 known active volcanoes are located along those continental margins

Figure 7.21 Half Dome in Yosemite National Park, California. This feature is just a tiny portion of the Sierra Nevada batholith, a huge structure that extends for approximately 400 kilometers. (Photo by Lewis Kemper/DRK Photo)

adjacent to oceanic trenches. Further, volcanic activity occurs along spreading centers of the oceanic ridge system. The later activity, although extensive, is hidden from view by the world's oceans.

In this section we examine three zones of igneous activity and relate them to global tectonics. These active areas are found along the oceanic ridges (spreading centers), adjacent to ocean trenches (subduction zones), and within the plates themselves (intraplate volcanism). These three settings are shown in Figure 7.22, which highlights two examples of each.

Igneous Activity at Spreading Centers

The greatest volume of volcanic rock is produced along the oceanic ridge system, where seafloor spreading is active (Figure 7.22). As plates of rigid lithosphere pull apart, the pressure on the underlying rocks is lessened. This reduced pressure, in turn, lowers the melting temperature of the mantle rocks. Partial melting of these mantle materials (primarily peridotite), produces large quantities of basaltic magma that move upward to fill the newly formed cracks between the diverging plates. In this way new slivers of oceanic crust are generated.

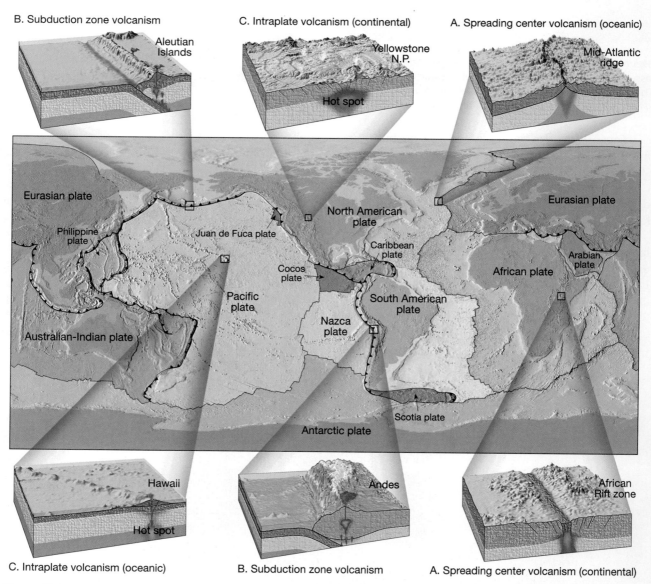

Figure 7.22 Three zones of volcanism. Two of these zones are plate boundaries, and the third includes areas within the plates.

Figure 7.23 Eruption near the fishing port of Heimaey, off the south coast of Iceland, 1973. (Photo courtesy of Vulcain-Explorer/Photo Researchers, Inc.)

Some of the molten basalt reaches the ocean floor, where it produces extensive lava flows or occasionally grows into a volcanic cone. Sometimes this activity produces a volcanic structure large enough to rise above sea level such as the islands of Vestmann off the southern coast of Iceland (Figure 7.23). Numerous submerged volcanoes also dot the flanks of the ridge system and the adjacent deep-ocean floor. Many of these formed along the ridge crest and gradually moved away as new oceanic crust was created by the seemingly unending process that generates new seafloor.

Igneous Activity at Subduction Zones

Recall that deep-ocean trenches are sites where slabs of water-rich oceanic crust are bent and descend into the mantle. As a slab sinks, volatiles (mainly water and carbon dioxide) are driven from the oceanic crust and migrate upward into the wedge-shaped piece of mantle located directly above. At a depth of 100 to 150 kilometers, these water-rich fluids reduce the melting point of mantle rocks sufficiently to promote partial melting. In some environments, silica and other components of the subducted sediments may become incorporated into the magma body. This process generates basaltic magma, and may produce some andesitic magma.

After a sufficient quantity of magma has accumulated, it slowly migrates upward because it is less dense than the surrounding rock. When igneous activity occurs along a subduction zone in the ocean, a chain of volcanoes called a *volcanic island arc* is produced. These structures, which usually form 200 to 300 kilometers from an oceanic trench, include such island chains as the Aleutians, the Tongas, and the Marianas.

When subduction of oceanic crust occurs beneath a continent, the magma that forms may become contaminated with silica-rich rocks as it moves through the crust. The assimilation of crustal fragments into the ascending basaltic magma will change it into one exhibiting an andesitic to granitic composition. The *continental volcanic arc* consisting of South America's Andes Mountains is one place where andesitic magma having such an origin is extruded.

Many subduction-zone volcanoes border the Pacific basin. Because of this pattern, the region has come to be called the *Ring of Fire*. Here volcanism is associated with subduction of the Pacific seafloor. As oceanic plates sink, they carry sediments and oceanic crust containing abundant water to great depths. The presence of water contributes to the high gas content and explosive nature of volcanoes that make up the Ring of Fire. The volcanoes of the Cascade Range in the northwestern United States, including Mount St. Helens, Mount Rainier, and Mount Shasta, are of this type.

Intraplate Igneous Activity

We know why igneous activity occurs along plate boundaries. But why do eruptions occur in the middle of plates?

Because basalts having relatively similar compositions are found in the ocean basins as well as on the continents, partial melting of mantle rocks is the most probable source for these rocks. The source of some intraplate basaltic magma are hot **mantle plumes** that originate at great depth. Upon reaching the crust, these structures begin to spread laterally. The result is localized volcanic regions a few hundred kilometers across called **hot spots**. More than a hundred hot spots have been identified, and most appear to have persisted for a few tens of millions of years. One hot spot is situated beneath the island of Hawaii. Another is responsible for the large outpourings of lava that make up Iceland.

With few exceptions, lavas and ash of granitic composition are restricted to vents located landward of the continental margins. This suggests that remelting of the continental crust may be the mechanism responsible for the formation of these silica-rich magmas. But what mechanism causes large quantities of continental material to be melted? One explanation is that a thick unit of continental crust occasionally becomes situated over a rising plume of hot mantle material. Rather than producing vast outpourings of basaltic lava as occurs at oceanic sites such as Hawaii, the magma from the rising plume is emplaced at depth. Here melting and assimilation of the surrounding host rock result in the formation of a highly altered *secondary magma*. As this buoyant silica-rich magma slowly migrates upward, continued hot-spot activity supplies heat to the rising mass, thereby aiding its ascent. Volcanism in the Yellowstone region may have resulted from this process.

Although the plate tectonics theory has answered may questions regarding the distribution of igneous activity, many new questions have arisen: Why does seafloor spreading occur in some areas but not others? How do mantle plumes and associated hot spots originate? These and other questions are the subject of continuing geologic research.

The Chapter in Review

1. The primary factors that determine the nature of volcanic eruptions include the magma's *temperature*, its *composition*, and the *amount of dissolved gases* it contains. As lava cools, it begins to congeal. As *viscosity* (thickness) increases, its mobility decreases. *The viscosity of magma is directly related to its silica content.* *Granitic* lava, high in silica, is very viscous and forms short, thick flows. *Basaltic* lava, lower in silica, is more fluid and may travel a long distance before congealing. Dissolved gases provide the force that propels molten rock from the vent of a volcano.

2. The materials associated with a volcanic eruption include *lava flows* (*pahoehoe* and *aa* flows for basaltic lavas), *gases* (primarily *water vapor*), and *pyroclastic material* (pulverized rock, glass, and lava fragments blown from the volcano's vent, which include *ash, pumice, cinders, blocks,* and *bombs*).

3. *Shield volcanoes* are broad, slightly domed volcanoes built primarily of fluid, basaltic lava. *Cinder cones* have very steep slopes composed of pyroclastic material. *Composite cones*, or *stratovolcanoes*, are large, nearly symmetrical structures built of interbedded lavas and pyroclastic deposits. Composite cones have the most violent eruptions.

4. Other than volcanoes, regions of volcanic activity may contain *volcanic necks* (rocks that once occupied the vents of volcanoes but are now exposed by erosion), *craters* (steep-walled depressions at the summit of most volcanoes), *calderas* (craters that exceed 1 kilometer in diameter), *fissure eruptions* (volcanic material extruded from fractures in the crust), and *pyroclastic flows*.

5. Igneous intrusive bodies are classified according to their *shape* and by their *orientation with respect to the host rock*, which generally is sedimentary rock. The two general shapes are *tabular* (sheetlike) and *massive*. Intrusive igneous bodies that cut across existing sedimentary beds are said to be *discordant*, whereas those that form parallel to existing sedimentary beds are *concordant*.

6. *Dikes* are tabular, discordant igneous bodies produced when magma is injected into fractures that cut across rock layers. Tabular, concordant bodies, called *sills*, form when magma is injected along the bedding surfaces of sedimentary rocks. *Laccoliths* are similar to sills but form from less-fluid magma that collects as a lens-shaped mass that pushes the overlying strata upward. *Batholiths*, the largest intrusive igneous bodies with surface exposures more than 100 square kilometers (40 square miles), frequently compose the cores of mountains.

7. Active *volcanism* occurs *along oceanic ridges, adjacent to ocean trenches*, and *within the plates* themselves. Most active volcanoes are associated with plate boundaries.

Key Terms

aa flow (p. 183)

batholith (p. 195)

caldera (p. 184)

cinder cone (p. 186)

composite cone (stratovolcano) (p. 187)

crater (p. 184)

dike (p. 193)

fissure eruption (p. 191)

flood basalt (p. 192)

hot spot (p. 198)

laccolith (p. 195)

mantle plumes (p. 198)

nuée ardente (p. 188)

pahoehoe flow (p. 183)

pyroclastic flow (p. 192)

pyroclasts (p. 184)

shield volcano (p. 185)

sill (p. 193)

vent (p. 184)

viscosity (p. 180)

volcanic neck (p. 191)

volcano (p. 184)

Questions for Review

1. What is the difference between magma and lava?
2. What three factors determine the nature of a volcanic eruption? What role does each play?
3. Why is a volcano fed by highly viscous magma likely to be a greater threat than a volcano supplied with very fluid magma?
4. Distinguish pahoehoe and aa lava.
5. List the main gases released during a volcanic eruption.
6. Analysis of samples taken during Hawaiian eruptions on the island of Hawaii indicate that _____ was the most abundant gas released.
7. Describe three examples of pyroclastic material.
8. Compare and contrast the main types of volcanoes (size, shape, eruptive style, and so forth).
9. Name one example of each of the three types of volcanoes.
10. Compare the formation of Mauna Loa with that of Parícutin volcano.
11. How is a caldera different from a crater?
12. Describe the formation of Crater Lake. Compare it to the caldera formed during the eruption of Kilauea.
13. What is Ship Rock, New Mexico, and how did it form?
14. Describe each of the four intrusive features discussed in the text.
15. What is the largest of all intrusive igneous bodies? Is it tabular or massive? Concordant or discordant (see Figure 7.19, p. 194)?
16. Spreading center volcanism is associated with which rock type? What causes rocks to melt in regions of spreading center volcanism?
17. What is the *Ring of Fire?*
18. Are volcanic eruptions in the *Ring of Fire* generally quiet or violent? Name a volcano that would support your answer.
19. The Hawaiian Islands and Yellowstone are associated with which of the three zones of volcanism?
20. Volcanic islands in the deep ocean are made primarily of what igneous rock type?

Testing What You Have Learned

Multiple-Choice Questions

1. Which of the following is NOT a primary factor that affects the viscosity of magma?
   - **a.** composition
   - **b.** age
   - **c.** temperature
   - **d.** dissolved gases
   - **e.** a. and b.
2. Which one of the following provides the force that propels molten rock from the vent of a volcano?
   - **a.** basalt
   - **b.** lapilli
   - **c.** escaping gases
   - **d.** viscosity
   - **e.** silica
3. What is the most abundant gas found in magma?
   - **a.** oxygen
   - **b.** hydrogen
   - **c.** nitrogen
   - **d.** methane
   - **e.** water vapor
4. The Hawaiian Islands consist primarily of what type of volcanic cone?
   - **a.** shield
   - **b.** composite
   - **c.** pyroclastic
   - **d.** cinder
   - **e.** laccolith

5. Intrusive igneous bodies that cut across existing sedimentary beds are said to be _____
   - **a.** concordant.
   - **b.** parallel.
   - **c.** homogenous.
   - **d.** disinclined.
   - **e.** discordant.

6. Tabular igneous intrusive bodies formed when magma is injected along sedimentary bedding surfaces are _____
   - **a.** dikes.
   - **b.** sills.
   - **c.** batholiths.
   - **d.** laccoliths.
   - **e.** stocks.

7. Igneous intrusive bodies with a surface exposure of more than 100 square kilometers (40 square miles) that frequently make up the cores of mountain systems are _____
   - **a.** dikes.
   - **b.** sills.
   - **c.** batholiths.
   - **d.** laccoliths.
   - **e.** stocks.

8. The area commonly called the *Ring of Fire* surrounds the _____
   - **a.** United States.
   - **b.** Atlantic Ocean.
   - **c.** Mediterranean Sea.
   - **d.** Indian Ocean.
   - **e.** Pacific Ocean.

9. Volcanoes located within the ocean extrude lavas that are primarily of _____ composition.
   - **a.** rhyolitic
   - **b.** granitic
   - **c.** andesitic
   - **d.** basaltic
   - **e.** gabbroic

10. Subduction zone volcanism in the ocean produces a chain of volcanoes called an island _____.
    - **a.** reef
    - **b.** arc
    - **c.** group
    - **d.** string
    - **e.** system

Fill-In Questions

11. Basaltic lava with a surface of rough, jagged blocks is called _____.
12. Although volcanic eruptions from a central vent are most familiar, by far the largest amounts of volcanic material are extruded from fractures in the crust called _____.
13. _____ are sheetlike intrusive igneous bodies that form when magma is injected into fractures that cut across rock layers.
14. Most of Earth's more than 600 active volcanoes are near _____ plate margins.
15. Rising plumes of hot mantle material produce intraplate volcanic regions a few hundred kilometers across called _____.

True/False Questions

16. The greatest volume of volcanic rock is produced along the oceanic ridge system. _____
17. Shield volcanoes represent the most violent type of volcanic activity. _____
18. Pyroclastic flows consist largely of ash and pumice fragments produced when silica-rich magma is extruded. _____
19. Volcanoes that extrude lava that has a high silica content are generally located landward of the continental margins. _____
20. Most of the 600 known active volcanoes are located along the mid-oceanic ridge. _____

Answers:

1. b; 2. c; 3. e; 4. a; 5. e; 6. b; 7. c; 8. e; 9. d; 10. b; 11. aa; 12. fissures; 13. Dikes; 14. convergent; 15. hot spots; 16. T; 17. F; 18. T; 19. T; 20. F.

Web Work

The following are informative and interesting World Wide Web sites that address topics related to those presented in this chapter:

- Global Volcanism Program
  http://nmnhwww.si.edu/gvp/

- Volcano World
  http://volcano.und.edu/

For direct access to these sites and other relevant Web locations, contact the *Foundations of Earth Science* WWW home page at http://www.prenhall.com/lutgens.

UNIT FOUR
Deciphering Earth's History

Grand Canyon National Park. Vishnu Temple in clouds at sunset from Mather Point on the South Rim.
Photo by Jack W. Dykinga & Associates

CHAPTER 8
Geologic Time

FOCUS ON LEARNING

To assist you in learning the important concepts
in this chapter, you will find it helpful to focus
on the following questions:

1. What is the doctrine of uniformitarianism? How does it differ from
catastrophism?

2. What are the two types of dates
used by geologists to interpret
Earth history?

3. What are the laws, principles, and
techniques used to establish relative dates?

4. What are fossils? What conditions
favor the preservation of organisms as fossils?

5. How are fossils used to correlate
rocks of similar ages that are in
different places?

6. What is radioactivity and how are
radioactive isotopes used in
radiometric dating?

7. What is the geologic time scale
and what are its principal subdivisions?

8. Why is it difficult to assign reliable absolute dates to samples of
sedimentary rock?

The strata exposed in the Grand Canyon contain clues to millions of years of Earth history.
Photo by Tom Bean

Figure 8.1 The Grand Canyon viewed from West Rim Drive. *(Photo © by Carr Clifton Photography)*

In 1869, John Wesley Powell, who was later to head the U.S. Geological Survey, led a pioneering expedition down the Colorado River and through the Grand Canyon. Writing about the strata that were exposed by the downcutting of the river, Powell observed that "the canyons of this region would be a Book of Revelations in the rock-leaved Bible of geology." He was undoubtedly impressed with the millions of years of Earth history exposed along the walls of the Grand Canyon (Figure 8.1).

Powell realized that the evidence for an ancient Earth is concealed in its rocks. Like the pages in a long and complicated history book, rocks record the geological events and changing life-forms of the past. The book, however, is not complete. Many pages, especially in the early chapters, are missing. Others are tattered, torn, or smudged. Yet enough of the book remains to allow much of the story to be deciphered.

Interpreting Earth history is an important goal of geology. Like a modern-day sleuth, the geologist must interpret clues found preserved in the rocks. By studying rocks, especially sedimentary rocks, and the features they contain, geologists can unravel the complexities of the past.

Geological events by themselves, however, have little meaning until they are put into a time perspective. Studying history, whether it be the Civil War or the Age of Dinosaurs, requires a calendar. Among geology's major contributions is a calendar called the *geologic time scale* and the discovery that Earth history is exceedingly long.

Some Historical Notes About Geology

The nature of our Earth—its materials and processes—has been a focus of study for centuries. However, the late 1700s is generally regarded as the beginning of modern geology. It was during this time that James Hutton published his important work, *Theory of the Earth*. Prior to the time of Hutton, most explanations about Earth's history relied on supernatural explanations.

Catastrophism

During the seventeenth and eighteenth centuries the doctrine of **catastrophism** strongly influenced people's thinking about Earth. Briefly stated, catastrophists believed that Earth's varied landscapes had been fashioned primarily by great catastrophes. Features such as mountains and canyons, which today we know take great periods of time to form, were explained as having been produced by sudden and often worldwide disasters produced by unknowable causes that no longer operate. This philosophy was an attempt to fit the rate of Earth processes to the prevailing ideas on the age of Earth.

In the mid-1600s, James Ussher, Anglican Archbishop of Armagh, Primate of all Ireland, published a work that had immediate and profound influence. A respected scholar of the Bible, Ussher constructed a chronology of human and Earth history in which he determined that Earth was only a few thousands of years old, having been created in 4004 B.C. Ussher's treatise earned widespread acceptance among scientific and religious leaders alike, and his chronology was soon printed in the margins of the Bible itself.

The Birth of Modern Geology

The late 1700s is generally regarded as the beginning of modern geology, for it was during this time that James Hutton, a Scottish physician and gentleman farmer, published his *Theory of the Earth* (Figure 8.2). In this work, Hutton put forth a principle that came to be known as the doctrine of **uniformitarianism**. Uniformitarianism is a fundamental principle of modern geology. It simply states that *the physical, chemical, and biological laws that operate today have also operated in the geologic past*. This means that the forces and processes that we observe presently shaping our planet have been at work for a very long time. Thus, to understand ancient rocks, we must first understand present-day processes and their results. This idea is commonly stated by saying "the present is the key to the past."

Prior to Hutton's *Theory of the Earth*, no one had effectively demonstrated that geological processes occur over extremely long periods of time. However, Hutton persuasively argued that processes that appear weak and slow-acting could, over long spans of time, produce effects that were just as great as those resulting from sudden catastrophic events. Unlike his predecessors, Hutton cited verifiable observations to support his ideas.

Figure 8.2 James Hutton, the eighteenth-century Scottish geologist who is often called the "founder of modern geology." *(Photo courtesy of The Natural History Museum, London)*

Hutton's literary style was cumbersome and difficult, so his work was not widely read nor easily understood. It is the English geologist Sir Charles Lyell who is given the most credit for advancing the basic principles of modern geology. Between 1830 and 1872, Lyell produced 11 editions of his great work, *Principles of Geology*. As was customary in those days Lyell's book had a lengthy subtitle that outlined the main theme of the work: *Being an Attempt to Explain the Former Changes of the Earth's Surface, by Reference to Causes Now in Operation*.

Lyell painstakingly illustrated the uniformity of nature through time. He was able to show more convincingly than his predecessors that geologic processes observed today can be assumed to have operated in the past. Although the doctrine of uniformitarianism did not originate with Lyell, he was most successful in interpreting and publicizing it for society at large.

Geology Today

Today uniformitarianism is just as viable as in Lyell's day. We realize more strongly than ever that the present gives us insight into the past and that the physical, chemical, and biological laws that govern geological processes remain unchanging through time. However, we also understand that the doctrine should not be taken too literally. To say that geological processes in the past were the same as those occurring today is not to suggest that they always had the same relative importance or that they operated at precisely the same rate. Although the same processes have prevailed through time, their rates have undoubtedly varied.

The acceptance of the concept of uniformitarianism, however, meant the acceptance of a very long history for Earth. Although processes vary in their intensity, they still take a very long time to create or destroy major landscape features.

For example, geologists have established that mountains once existed in portions of present-day Minnesota, Wisconsin, and Michigan. Today the region consists of low hills and plains. Erosion (processes that wear land away) gradually destroyed these peaks. Estimates indicate that the North American continent is being lowered at a rate of about 3 centimeters per 1000 years. At this rate, it would take 100 million years for water, wind, and ice to lower mountains that were 3000 meters (10,000 feet) high.

But even these time spans are relatively short on the time scale of Earth history, for the rock record contains evidence that shows Earth has experienced many cycles of mountain building and erosion. Concerning the ever-changing nature of Earth through great expanses of geologic time, Hutton stated: "We find no vestige of a beginning, no prospect of an end."

It is important to remember that, although many features of our physical landscape may seem to be unchanging over the decades we might observe them, they are nevertheless changing, but on time scales of hundreds, thousands, or even many millions of years.

Relative Dating—Key Principles

During the late 1800s and early 1900s, various attempts were made to determine Earth's age. Although some of the methods appeared promising at the time, none proved to be reliable. What these scientists were seeking was an **absolute date**. Such dates pinpoint the time in history when something took place—for example, the extinction of the dinosaurs about 65 million years ago. Today our understanding of radioactivity allows us to accurately determine absolute dates for many rocks that represent important events in Earth's distant past. We will study radioactivity later in this chapter. Prior to the

discovery of radioactivity, geologists had no accurate and dependable method of absolute dating and had to rely solely on relative dating.

Relative dating means placing rocks in their proper *sequence of formation*—which formed first, second, third, and so on. Relative dating cannot tell us how long ago something took place, only that it followed one event and preceded another. The relative dating techniques that were developed are valuable and still widely used. Absolute dating methods did not replace these techniques; they simply supplemented them. To establish a relative time scale, a few simple principles or rules had to be discovered and applied. Although they may seem obvious to us today, they were major breakthroughs in thinking at the time, and their discovery and acceptance was an important scientific achievement.

Law of Superposition

Nicolaus Steno, a Danish anatomist, geologist, and priest (1636–1686), is credited with being the first to recognize a sequence of historical events in an outcrop of sedimentary rock layers. Working in the mountains of western Italy, Steno applied a very simple rule that became the most basic principle of relative dating—the **law of superposition**. The law simply states that in an undeformed sequence of sedimentary rocks, each bed is older than the one above it and younger than the one below. Although it may seem obvious that a rock layer could not be deposited unless it had something older beneath it for support, it was not until 1669 that Steno clearly stated the principle.

This rule also applies to other surface-deposited materials such as lava flows and beds of ash from volcanic eruptions. Applying the law of superposition to the beds exposed in the upper portion of the Grand Canyon (Figure 8.3), you can easily place the layers in their proper order. Among those that are shown, the sedimentary rocks in the Supai Group must be the oldest, followed in order by the Hermit Shale, Coconino Sandstone, Toroweap Formation, and Kaibab Limestone.

Principle of Original Horizontality

Steno is also credited with recognizing the **principle of original horizontality**. This principle simply states that most layers of sediment are deposited in a horizontal position. Thus, if we observe rock layers that are flat, it means they have not been disturbed, and still have their *original* horizontality

A.

B.

Figure 8.3 Applying the law of superposition to these layers exposed in the upper portion of the Grand Canyon, the Supai Group is oldest and the Kaibab Limestone is youngest. *(Photo by E. J. Tarbuck)*

(Figure 8.4A). But if they are folded or inclined at a steep angle, they must have been moved into that position by crustal disturbances sometime *after* their deposition (Figure 8.4B).

Principle of Cross-Cutting Relationships

When a fault cuts through other rocks, or when magma intrudes and crystallizes, we can assume that the fault or intrusion is younger than the rocks

A.

B.

Figure 8.4 A. The principle of original horizontality states that most layers of sediment are deposited in a nearly horizontal position. **B.** When we see folded rock layers such as these, we can assume they must have been moved into that position by crustal disturbances *after* their deposition. *(Photos by E. J. Tarbuck)*

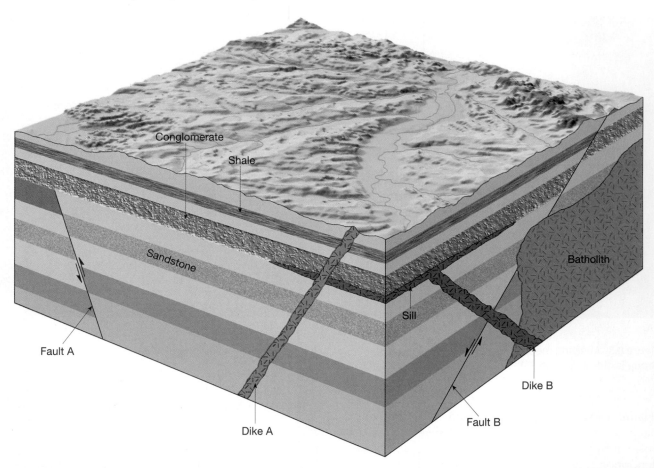

Figure 8.5 Cross-cutting relationships are an importnat principle used in relative dating. An intrusive rock body is younger than the rocks it intrudes. A fault is younger than the rock layers it cuts.

affected. For example, in Figure 8.5, the faults and dikes clearly must have occurred *after* the sedimentary layers were deposited.

This is the **principle of cross-cutting relationships**. By applying the cross-cutting principle you can see that fault A occurred *after* the sandstone layer was deposited, because it "broke" the layer. However, fault A occurred *before* the conglomerate was laid down, because that layer is unbroken.

We can also state that dike B and its associated sill are older than dike A, because dike A cuts the sill. In the same manner, we know that the batholith was emplaced after movement occurred along fault B, but before dike B was formed. This is true because the batholith cuts across fault B and dike B cuts across the batholith.

Inclusions

Sometimes inclusions can aid the relative dating process. **Inclusions** are pieces of one rock unit that are contained within another. The basic principle is logical and straightforward. The rock mass adjacent to the one containing the inclusions must have been there first in order to provide the rock fragments. Therefore, the rock mass containing inclusions is the younger of the two. Figure 8.6 provides an example. Here the inclusions of granite in the adjacent sedimentary layer indicate the sedimentary layer was deposited on top of an eroded mass of granite rather than being intruded from below by a mass of magma that later crystallized.

Unconformities

When we observe layers of rock that have been deposited essentially without interruption, we call them **conformable**. Many areas exhibit conformable beds representing certain spans of geologic time. However, no place on Earth has a complete set of conformable strata.

Throughout Earth's history, the deposition of sediment has been interrupted again and again.

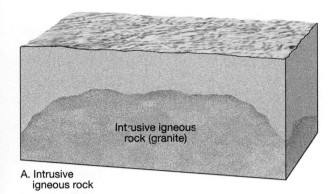

A. Intrusive
igneous rock

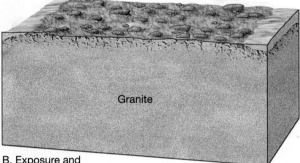

B. Exposure and
weathering of granite

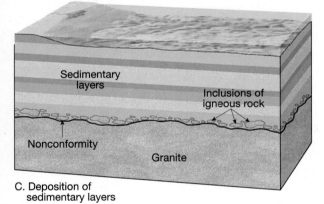

C. Deposition of
sedimentary layers

Figure 8.6 These three diagrams illustrate the formation of inclusions and a type of unconformity termed a *nonconformity*. In the third diagram we know the granite must be older because pieces of granite are included in the overlying sedimentary bed. When older intrusive igneous rocks are overlain by younger sedimentary layers, a nonconformity is said to exist.

All such breaks in the rock record are termed *unconformities*.

An **unconformity** represents a long period during which deposition ceased, erosion removed previously formed rocks, and then deposition resumed. In each case Earth's crust has undergone uplift and erosion, followed by subsidence and renewed sedimentation. Unconformities are important features because they represent significant geologic events in Earth's history. Moreover, their recognition helps us identify what intervals of time are not represented by strata and thus are missing from the geologic record.

Three distinct types of unconformities are recognized: angular unconformity, disconformity, and nonconformity. The most easily recognized unconformity is an **angular unconformity**. It consists of tilted or folded sedimentary rocks that are overlain by younger, more flat-lying strata. An angular unconformity indicates that during the pause in deposition, a period of deformation (folding or tilting) as well as erosion occurred. Study of the steps in Figure 8.7 will illustrate this process.

When contrasted with angular unconformities, **disconformities** are more common, but usually far less conspicuous because the strata on either side are essentially parallel. Many disconformities are difficult to identify because the rocks above and below are similar and there is little evidence of erosion. Such a break often resembles an ordinary bedding plane. Other disconformities are easier to identify because the ancient erosion surface is cut deeply into the older rocks below.

The third basic type of unconformity is a **nonconformity**. Here the break separates older metamorphic or intrusive igneous rocks from younger sedimentary strata (see Figure 8.6). Intrusive igneous masses and metamorphic rocks originate far below the surface. Thus, for a nonconformity to develop, there must be a period of uplift and the erosion of overlying rocks. Once exposed at the surface, the igneous or metamorphic rocks are subjected to weathering and erosion prior to subsidence and the renewal of sedimentation.

Using Relative Dating Principles

By applying the principles of relative dating to the hypothetical geologic cross section shown in Figure 8.8, the rocks and the events in Earth's history they represent can be placed into their proper sequence. The statements within the figure summarize the logic used to interpret the cross section. In this example, we establish a relative time scale for the rocks and events in the area of the cross section. Remember, we have no idea how many years of Earth history are represented, nor do we know how this area compares to any other.

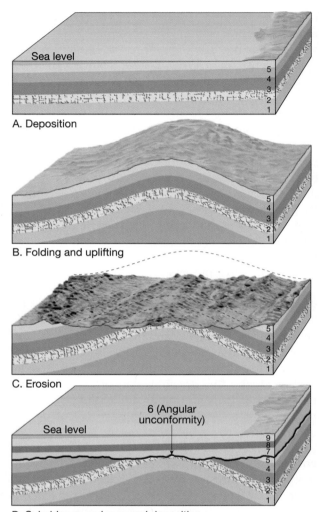

A. Deposition

B. Folding and uplifting

C. Erosion

6 (Angular unconformity)

D. Subsidence and renewed deposition

Figure 8.7 Formation of an angular unconformity. An angular unconformity represents a period during which deformation and erosion occurred.

Correlation of Rock Layers

To develop a geologic time scale that applies to the entire Earth, rocks of similar age in different regions must be matched up. Such a task is referred to as **correlation**. Within a limited area, correlating the rocks of one locality with those of another may be done simply by walking along the outcropping edges. However, this may not be possible when a bed is not continuously exposed. Correlation over short distances is often achieved by noting the position of a rock layer in a sequence of strata. Or, a layer may be identified in another location if it consists of very distinctive or uncommon minerals.

By correlating the rocks from one place to another, a more comprehensive view of the geologic history of a region is possible. Figure 8.9, for example, shows the correlation of strata at three sites on the Colorado Plateau. No single locale exhibits the entire sequence, but correlation reveals a more complete picture of the sedimentary rock record.

Many geologic studies involve relatively small areas. Such studies are important in their own right, but their full value is realized only when the rocks are correlated with those of other regions. Although the methods just described are sufficient to trace a rock formation over relatively short distances, they are not adequate for matching rocks that are separated by great distances. When correlation between widely separated areas or between continents is the objective, the geologist must rely on fossils.

Fossils: Evidence of Past Life

Fossils, the remains or traces of prehistoric life, are important inclusions in sediment and sedimentary rocks. Examples of different types of fossils are shown in Figure 8.10. Fossils are important tools for interpreting the geologic past. Knowing the nature of the life-forms that existed at a particular time helps researchers understand past environmental conditions. Further, fossils are important time indicators and play a key role in correlating rocks of similar ages that are from different places.

Conditions Favoring Preservation

Only a tiny fraction of the organisms that have lived during the geologic past have been preserved as fossils. Normally the remains of an animal or plant are destroyed. Under what circumstances are they preserved? Two special conditions appear to be necessary: rapid burial and the possession of hard parts.

When an organism perishes, its soft parts usually are quickly eaten by scavengers or decomposed by bacteria. Occasionally, however, the remains are buried by sediment. When this occurs, the remains are protected from the environment where destructive processes operate. Rapid burial therefore is an important condition favoring preservation.

In addition, animals and plants have a much better chance of being preserved as part of the fossil record if they have hard parts. Although traces and imprints of soft-bodied animals such as jellyfish, worms, and insects exist, they are rare. Flesh usually decays so rapidly that preservation is exceedingly

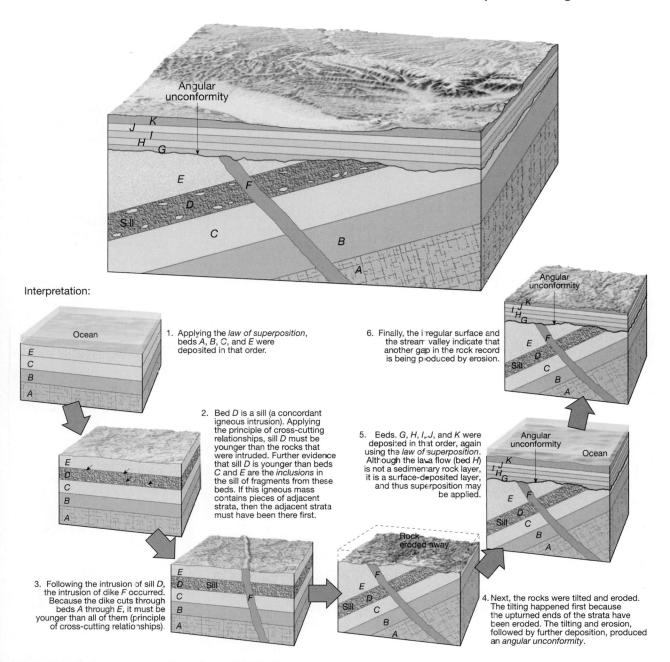

Figure 8.8 Geologic cross section of a hypothetical region.

unlikely. Hard parts such as shells, bones, and teeth predominate in the record of past life.

Because preservation is contingent on special conditions, the record of life in the geologic past is biased. The fossil record of those organisms with hard parts that lived in areas of sedimentation is quite abundant. However, we get only an occasional glimpse of the vast array of other life-forms that did not meet the special conditions favoring preservation.

Fossils and Correlation

The existence of fossils had been known for centuries, yet it was not until the late 1700s and early 1800s that their significance as geologic tools was made evident. During this period an English engineer and canal builder, William Smith, discovered that each rock formation in the canals he worked on contained fossils unlike those in the beds either

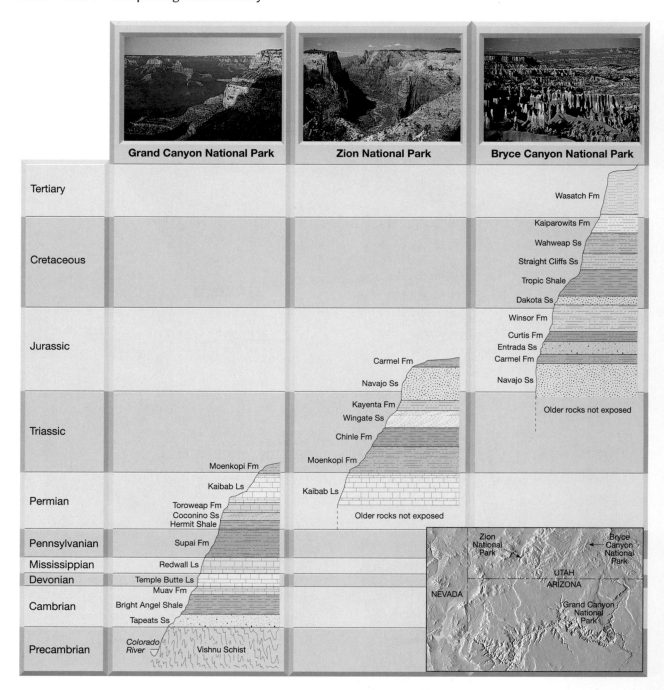

Figure 8.9 Correlation of strata at three locations on the Colorado Plateau reveals the total extent of sedimentary rocks in the region. *(After U.S. Geological Survey)*

above or below. Further, he noted that sedimentary strata in widely separated areas could be identified—and correlated—by their distinctive fossil content.

Based on Smith's classic observations and the findings of many geologists who followed, one of the most important and basic principles in historical geology was formulated: *Fossil organisms succeed*

one another in a definite and determinable order, and therefore any time period can be recognized by its fossil content. This has come to be known as the **principle of fossil succession**. In other words, when fossils are arranged according to their age by applying the law of superposition to the rocks in which they are found, they do not present a random

Figure 8.10 There are many types of fossilization. Six examples are shown here. **A.** Petrified wood in Petrified Forest National Park, Arizona. **B.** Natural casts of shelled invertebrates. **C.** A fossil bee preserved as a thin carbon film. **D.** Impressions are common fossils and often show considerable detail. **E.** Insect in amber. **F.** Dinosaur footprint in fine-grained limestone near Tuba City, Arizona. *(Photo A by David Muench Photography; Photos B, D, and F by E. J. Tarbuck; Photo C courtesy of Florissant Fossil Beds National Monument; Photo E by Breck P. Kent)*

or haphazard picture. To the contrary, fossils document the evolution of life through time.

For example, an Age of Trilobites is recognized quite early in the fossil record. Then, in succession, paleontologists recognize an Age of Fishes, an Age of Coal Swamps, an Age of Reptiles, and an Age of Mammals. These "ages" pertain to groups that were especially plentiful and characteristic during particular time periods. Within each of the "ages," there are many subdivisions based, for example, on certain species of trilobites, and certain types of fish, reptiles, and so on. This same succession of dominant organisms, never out of order, is found on every major landmass.

Once fossils were recognized as time indicators, they became the most useful means of correlating rocks of similar age in different regions. Geologists pay particular attention to certain fossils called **index fossils**. These fossils are widespread geographically and are limited to a short span of geologic time, so their presence provides an important method of matching rocks of the same age. Rock formations, however, do not always contain a specific index fossil. In such situations, groups of fossils, called *fossil assemblages*, are used to establish the age of the bed. Figure 8.11 illustrates how a group of just two fossils can be used to date rocks more precisely than could be accomplished by the use of only one of the fossils.

In addition to being important and often essential tools for correlation, fossils are important environmental indicators. Although much can be deduced about past environments by studying the

nature and characteristics of sedimentary rocks, a close examination of the fossils present can usually provide a great deal more information.

For example, when the remains of certain clam shells are found in limestone, the geologist can assume that the region was once covered by a shallow sea, because that is where clams live today. Also, by using what we know of living organisms, we can conclude that fossil animals with thick shells capable of withstanding pounding and surging waves must have inhabited shorelines. On the other hand, animals with thin, delicate shells probably indicate deep, calm offshore waters. Hence, by looking closely at the types of fossils, the approximate position of an ancient shoreline may be identified.

Further, fossils can indicate the former temperature of the water. Certain present-day corals require warm and shallow tropical seas like those around Florida and the Bahamas. When similar corals are found in ancient limestones, they indicate that a Florida-like marine environment must have existed when they were alive. These are just a few examples of how fossils can help unravel the complex story of Earth's history.

Radioactivity and Radiometric Dating

In addition to establishing relative dates by using the principles described in the preceding sections, it is also possible to obtain reliable absolute dates for events in the geologic past. For example, we know that Earth is about 4.6 billion years old and that the dinosaurs became extinct about 66 million years ago. Dates that are expressed in millions and billions of years truly stretch our imagination because our personal calendars involve time measured in hours, weeks, and years. Nevertheless, the vast expanse of geologic time is a reality, and it is radiometric dating that allows us to measure it accurately. In this section you will learn about radioactivity and its application in radiometric dating.

Recall from Chapter 1 that each atom has a *nucleus* containing protons and neutrons, and that the nucleus is orbited by electrons. *Electrons* have a negative electrical charge, and *protons* have a positive charge. A *neutron* is actually a proton and an electron combined, so it has no charge (it is neutral).

The *atomic number* (the element's identifying number) is the number of protons in the nucleus. Every element has a different number of protons in the nucleus, and thus a different atomic number (hydrogen = 1; oxygen = 8; uranium = 92, etc.). Atoms of the

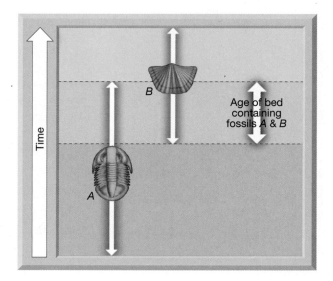

Figure 8.11 Overlapping ranges of fossils help date rocks more exactly than using a single fossil.

same element always have the same number of protons, so the atomic number is constant (uranium = 92).

Practically all (99.9 percent) of an atom's mass is found in the nucleus, indicating electrons have practically no mass at all. By adding together the number of protons and neutrons in the nucleus, the *mass number* of the atom is determined. The number of neutrons in the nucleus can vary. These variants, called *isotopes*, have different mass numbers. For example, uranium has three isotopes—U-234, U-235, and U-238.

To summarize with an example, uranium's nucleus always has 92 protons, so its atomic number always is 92. But its neutron population varies, so uranium has three isotopes: uranium-234 (mass of protons + neutrons = 234), uranium-235, and uranium-238. All three isotopes are mixed in nature. They look the same and behave the same in chemical reactions.

Radioactivity

The forces that bind protons and neutrons together in the nucleus usually are strong. However, in some isotopes, the nuclei are unstable because the forces binding protons and neutrons together are not strong enough. As a result, the nuclei spontaneously break apart (decay), a process called **radioactivity**. What happens when unstable nuclei break apart? Three common types of radioactive decay are illustrated in Figure 8.12 and are summarized as follows:

1. *Alpha particles* (α particles) may be emitted from the nucleus. An alpha particle consists of 2 protons and 2 neutrons. Consequently, the emission of an alpha particle means (a) the mass number of the isotope is reduced by 4, and (b) the atomic number is decreased by 2.
2. When a *beta particle* (β particle), or electron, is given off from a nucleus, the mass number remains unchanged, because electrons have practically no mass. However, because the electron has come from a neutron (remember, a neutron is a combination of a proton and an electron), the nucleus contains one more proton than before. Therefore, the atomic number increases by 1.
3. Sometimes an electron is captured by the nucleus. The electron combines with a proton and forms an additional neutron. As in the last example, the mass number remains unchanged. However, because the nucleus now contains one less proton, the atomic number decreases by 1.

An unstable (radioactive) isotope is referred to as the *parent*. The isotopes resulting from the decay of the parent are the *daughter products*. Figure 8.13 provides an example of radioactive decay. Here it can be seen that when the radioactive parent, uranium-238 (atomic number 92, mass number 238), decays, it follows a number of steps, emitting 8 alpha particles and 6 beta particles before finally becoming the stable daughter product lead-206 (atomic number 82, mass number 206).

Certainly among the most important results of the discovery of radioactivity is that it provided a reliable method of calculating the ages of rocks and minerals that contain particular radioactive isotopes. The procedure is called **radiometric dating**. Why is radiometric dating reliable? Because the rates of decay for many isotopes have been precisely measured and do not vary under the physical conditions that exist in Earth's outer layers. Therefore, each radioactive isotope used for dating has been decaying at a fixed rate ever since the formation of the rocks in which it occurs, and the products of decay have been accumulating at a corresponding rate. For example, when uranium is incorporated into a mineral that crystallizes from magma, there is no lead (the stable daughter product) from previous decay. The radiometric "clock" starts at this point. As the uranium in this newly formed mineral disintegrates, atoms of the daughter product are trapped and measurable amounts of lead eventually accumulate.

Half-Life

The time required for one-half of the nuclei in a sample to decay is called the **half-life** of the isotope. Half-life is a common way of expressing the rate of radioactive disintegration. Figure 8.14 illustrates what occurs when a radioactive parent decays directly into its stable daughter product. When the quantities of parent and daughter are equal (ratio 1:1), we know that one half-life has transpired. When one-quarter of the original parent atoms remain and three-quarters have decayed to the daughter product, the parent/daughter ratio is 1:3 and we know that two half-lives have passed. After three half-lives, the ratio of parent atoms to daughter atoms is 1:7 (one parent for every seven daughter atoms).

If the half-life of a radioactive isotope is known and the parent/daughter ratio can be measured, the age of the sample can be calculated. For example, assume that the half-life of a hypothetical unstable isotope is 1 million years and the parent/daughter

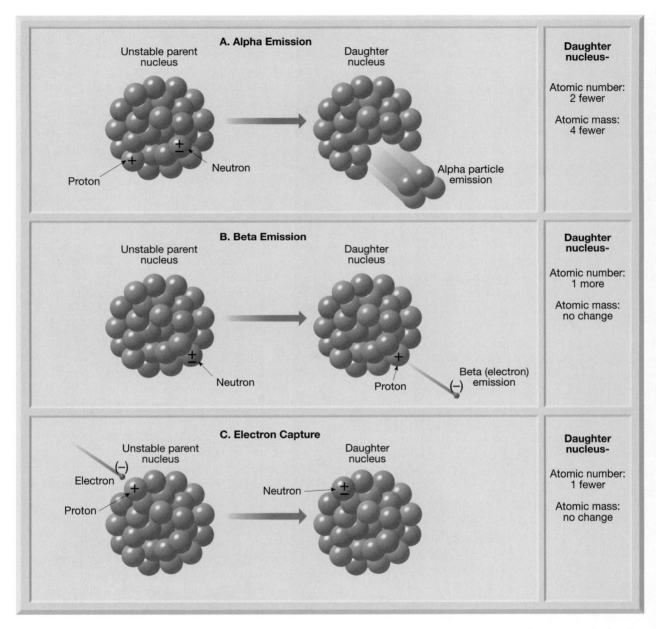

Figure 8.12 Common types of radioactive decay. Notice that in each case the number of protons (atomic number) in the nucleus changes, thus producing a different element.

ratio in a sample is 1:15. Such a ratio indicates that four half-lives have passed and that the sample must be 4 million years old.

Radiometric Dating

Notice that the *percentage* of radioactive atoms that decay during one half-life is always the same: 50 percent. However, the *actual number* of atoms that decay with the passing of each half-life continually decreases. Thus, as the percentage of radioactive parent atoms declines, the proportion of stable daughter atoms rises, with the increase in daughter atoms just matching the drop in parent atoms. This fact is the key to radiometric dating.

Of the many radioactive isotopes that exist in nature, five have proven particularly important in providing radiometric ages for ancient rocks (Table 8.1).

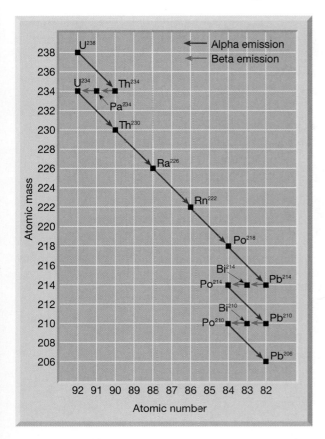

Figure 8.13 The most common isotope of uranium (U-238) is an example of a radioactive decay series. Before the stable end product (Pb-206) is reached, many different isotopes are produced as intermediate steps.

Rubidium-87, uranium-238, and uranium-235 are used for dating rocks that are millions of years old, but potassium-40 is more versatile. Although the half-life of potassium-40 is 1.3 billion years, recent analytical techniques have made possible the detection of tiny amounts of its stable daughter product, argon-40, in some rocks that are younger than 100,000 years. Another important reason for its frequent use is that potassium is abundant in many common minerals, particularly micas and feldspars.

Sources of Error It is important to realize that an accurate radiometric date can be obtained only if the mineral remained a closed system during the entire period since its formation. A correct date is not possible unless there was neither the addition nor loss of parent or daughter isotopes. This is not always the case. In fact, an important limitation of the potassium–argon method arises from the fact that argon is a gas, and it may leak from minerals, throwing off measurements. Cross-checking of samples, using two different radiometric methods, is done where possible to ensure accurate age determinations.

Bear in mind that, although the basic principle of radiometric dating is simple, the actual procedure is quite complex. The analysis that determines the quantities of parent and daughter must be painstakingly precise. In addition, some radioactive materials do not decay directly into the stable daughter product. As you saw in Figure 8.13, uranium-238 produces 13 intermediate unstable daughter products before the fourteenth and final daughter product, the stable isotope lead-206, is produced.

Value of Radiometric Dating Radiometric dating methods have produced literally thousands of dates for events in Earth's history. Rocks from several localities have been dated at more than 3 billion years, and geologists realize that still-older rocks exist. For example, a granite from South Africa has been dated at 3.2 billion years—and it contains inclusions of quartzite that must be older. Quartzite itself is a metamorphic rock that originally was the sedimentary rock sandstone. Sandstone, in turn, is the product of the lithification of sediments produced by the weathering of existing rocks. Thus, we have a positive indication that much older rocks existed.

Radiometric dating has vindicated the ideas of Hutton, Darwin, and others who inferred that geologic time must be immense. Indeed, modern dating methods have proven that there has been enough time for the processes we observe to have accomplished tremendous tasks.

The Geologic Time Scale

Geologists have divided the whole of geologic history into units of varying magnitude. Together they comprise the **geologic time scale** of Earth's history (Figure 8.15). The major units of the time scale were delineated during the 1800s, principally by workers in Western Europe and Great Britain. Because absolute dating was unavailable at that time, the entire time scale was created using methods of relative dating. It was only in the twentieth century that radiometric dating permitted absolute dates to be added.

The geologic time scale subdivides the 4.6-billion-year history of Earth into many different units and provides a meaningful time frame within which the events of the geologic past are arranged. As shown in Figure 8.15, **eons** represent the greatest

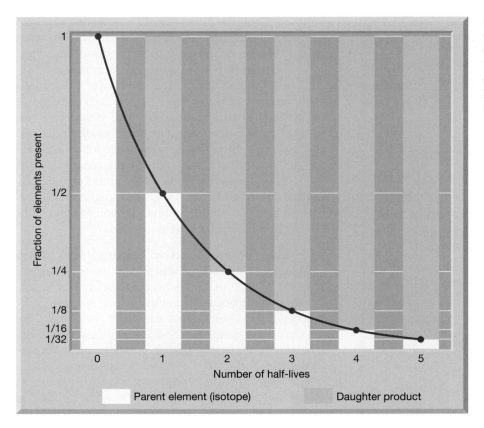

expanses of time. The eon that began about 570 million years ago is the **Phanerozoic**, meaning *visible life*. It is an appropriate description because the rocks and deposits of the Phanerozoic eon contain abundant fossils that document major evolutionary trends (Figure 8.16).

Another glance at the time scale reveals that the Phanerozoic eon is divided into **eras**. The three eras within the Phanerozoic are the **Paleozoic** ("ancient life"), the **Mesozoic** ("middle life"), and the **Cenozoic** ("recent life"). As the names imply, the eras are bounded by profound worldwide changes in life-forms. Each era is subdivided into **periods**. The Paleozoic has seven, the Mesozoic three, and the Cenozoic two. Each of these dozen periods is characterized by a somewhat less profound change in life-forms as compared with the eras.

Finally, periods are divided into still smaller units called **epochs**. As you can see in Figure 8.15, seven epochs have been named for the periods of the Cenozoic era. The epochs of other periods, however, are not usually referred to by specific names. Instead, the terms *early, middle*, and *late* are generally applied to the epochs of these earlier periods.

Table 8.1 ■ Radioactive isotopes frequently used in radiometric dating

Radioactive Parent	Stable Daughter Product	Currently Accepted Half-life Values
Uranium-238	Lead-206	4.5 billion years
Uranium-235	Lead-207	713 million years
Thorium-232	Lead-208	14.1 billion years
Rubidium-87	Strontium-87	47.0 billion years
Potassium-40	Argon-40	1.3 billion years

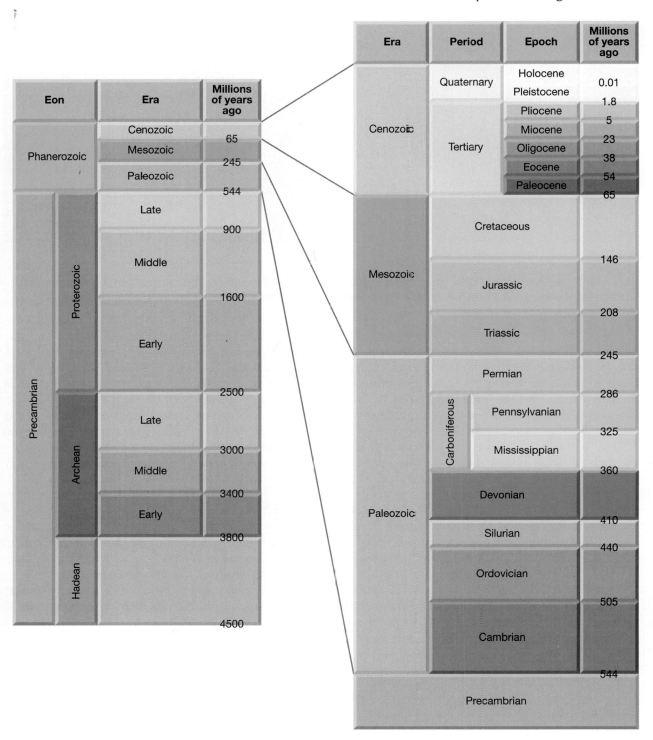

Figure 8.15 The geologic time scale. Numbers on the time scale represent time in millions of years before the present. These dates were added long after the time scale had been established using relative dating techniques. The Precambrian accounts for more than 85 percent of geologic time. *(Data from Geological Society of America)*

Notice that the detail of the geologic time scale does not begin until about 570 million years ago, the

date for the beginning of the Cambrian period. The more than 4 billion years prior to the Cambrian is

A.

B.

Figure 8.16 Unlike Precambrian time, which has a relatively meager fossil record, the Phanerozoic eon is characterized by abundant fossil evidence. **A.** Fossil fish, Eocene epoch, Kemmerer, Wyoming. **B.** Fossil ferns, Pennsylvanian period, St. Clair, Pennsylvania. *(Photos by Breck P. Kent)*

divided into three eons, the *Hadean*, the *Archean*, and the *Proterozoic*. It is also common for this vast expanse of time to simply be referred to as the **Precambrian**. Although it represents more than 85 percent of Earth history, the Precambrian is not divided into nearly as many smaller time units as is the Phanerozoic eon.

Why is the huge expanse of Precambrian time not divided into numerous eras, periods, and epochs? The reason is that Precambrian history is not yet known in great enough detail. The quantity of information geologists have deciphered about Earth's past is somewhat analogous to the detail of human history. The further back we go, the less that we know. Certainly more data and information exist about the past ten years than for the first decade of the twentieth century; the events of the nineteenth century have been documented much better than the events of the first century A.D., and so on. So it is with Earth history. The more recent past has the freshest, least disturbed, and most observable record. The further back in time the geologist goes, the more fragmented the record and clues become.

Difficulties in Dating the Geologic Time Scale

Although reasonably accurate absolute dates have been worked out for the periods of the geologic time scale (see Figure 8.15), the task is not without difficulty. The primary problem in assigning absolute

dates is the fact that not all rocks can be dated radiometrically. Recall that for a radiometric date to be useful, all minerals in the rock must have formed at approximately the same time. For this reason, radioactive isotopes can be used to determine when minerals in an igneous rock crystallized and when pressure and heat created new minerals in a metamorphic rock.

However, samples of sedimentary rock can only rarely be dated directly by radiometric means. A sedimentary rock may include particles that contain radioactive isotopes, but the rock's age cannot be accurately determined because the grains that make up the rock are not the same age as the rock in which they occur. Rather, the sediments have been weathered from rocks of diverse ages.

Radiometric dates obtained from metamorphic rocks may also be difficult to interpret, because the age of a particular mineral in a metamorphic rock does not necessarily represent the time when the rock initially formed. Instead, the date may indicate any one of a number of subsequent metamorphic phases.

If samples of sedimentary rocks rarely yield reliable radiometric ages, how can absolute dates be assigned to sedimentary layers? Usually the geologist must relate them to datable igneous masses, as in Figure 8.17. In this example, radiometric dating has determined the ages of the volcanic ash bed within the Morrison Formation and the dike cutting the Mancos Shale and Mesaverde Formation. The sedimentary beds below the ash are obviously older than the ash, and all the layers above the ash are younger (principle of

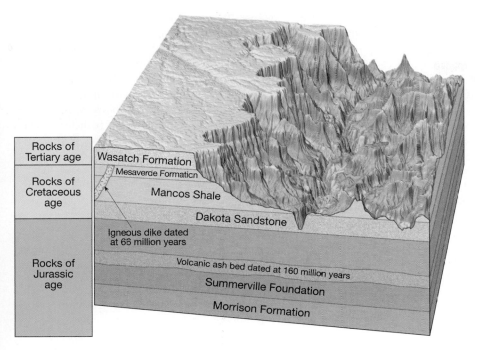

Figure 8.17 Absolute dates for sedimentary layers are usually determined by examining their relationship to igneous rocks. *(After U.S. Geological Survey)*

Rocks of Tertiary age

Rocks of Cretaceous age

Rocks of Jurassic age

Wasatch Formation

Mesaverde Formation

Mancos Shale

Dakota Sandstone

Igneous dike dated at 66 million years

Volcanic ash bed dated at 160 million years

Summerville Foundation

Morrison Formation

superposition). The dike is younger than the Mancos Shale and the Mesaverde Formation but older than the Wasatch Formation because the dike does not intrude the Tertiary rocks (cross-cutting relationships).

From this kind of evidence, geologists estimate that a part of the Morrison Formation was deposited about 160 million years ago, as indicated by the ash bed. Further, they conclude that the Tertiary period began after the intrusion of the dike, 66 million years ago. This is one example of literally thousands that illustrates how datable materials are used to *bracket* the various episodes in Earth history within specific time periods. It shows the necessity of combining laboratory dating methods with field observations of rocks.

The Chapter in Review

1. The *doctrine of uniformitarianism*, a fundamental principle of modern geology put forth by James Hutton in the late 1700s, states that the physical, chemical, and biological laws that operate today have also operated in the geologic past. The idea is often summarized as "the present is the key to the past." Hutton argued that processes that appear to be slow-acting could, over long spans of time, produce effects that were just as great as those resulting from sudden catastrophic events. *Catastrophism,* on the other hand, states that Earth's landscapes have been developed primarily by great catastrophes.

2. The two types of dates used by geologists to interpret Earth history are: (a) *relative dates*, which put events in their *proper sequence of formation*, and (b) *absolute dates*, which pinpoint the *time in years* when an event took place.

3. Relative dates can be established using the *law of superposition, principle of original horizontality, principle of* *cross-cutting relationships, inclusions*, and *unconformities*.

4. *Correlation,* matching up two or more geologic phenomena in different areas, is used to develop a geologic time scale that applies to the entire Earth.

5. *Fossils* are the remains or traces of prehistoric life. The special conditions that favor preservation are *rapid burial* and the possession of *hard parts* such as shells, bones, or teeth.

6. Fossils are used to correlate sedimentary rocks that are widely separated by using the rocks' distinctive fossil content and applying the *principle of fossil succession*. The principle of fossil succession states that fossil organisms succeed one another in a definite and determinable order, and therefore any time period can be recognized by its fossil content.

7. *Radioactivity* is the spontaneous breaking apart (decay) of certain unstable atomic nuclei. Three common forms of radioactive decay are: (a) emission of an alpha particle from the nucleus, (b) emission of a beta particle (or electron) from the nucleus, and (c) capture of an electron by the nucleus.

8. An unstable *radioactive isotope*, called the *parent*, will decay and form *daughter products*. The length of time for one-half of the nuclei of a radioactive isotope to decay is called the *half-life* of the isotope. If the half-life of the isotope is known, and the parent/daughter ratio can be measured, the age of a sample can be calculated.

9. The *geologic time scale* divides Earth history into units of varying magnitude. It is commonly presented in chart form, with the oldest time and event at the bottom and the youngest at the top. The principal subdivisions of the geologic time scale, called *eons*, include the *Hadean, Archean, Proterozoic* (together, these three eons are commonly referred to as the *Precambrian*), and, beginning about 570 million years ago, the *Phanerozoic*. The Phanerozoic (meaning "visible life") eon is divided into the following *eras: Paleozoic* ("ancient life"), *Mesozoic* ("middle life"), and *Cenozoic* ("recent life").

10. A significant problem in assigning absolute dates to units of time is that *not all rocks can be dated radiometrically*. A sedimentary rock may contain particles of many ages that have been weathered from different rocks that formed at various times. One way geologists assign absolute dates to sedimentary rocks is to relate them to datable igneous masses, such as volcanic ash beds.

Key Terms

absolute date (p. 206)

angular unconformity (p. 209)

catastrophism (p. 205)

Cenozoic era (p. 218)

conformable (p. 208)

correlation (p. 210)

cross-cutting relationships, principle of (p. 208)

disconformity (p. 209)

eon (p. 217)

epoch (p. 218)

era (p. 218)

fossil (p. 210)

fossil succession, principle of (p. 212)

geologic time scale (p. 217)

half-life (p. 215)

inclusions (p. 208)

index fossil (p. 214)

Mesozoic era (p. 218)

nonconformity (p. 209)

original horizontality, principle of (p. 206)

Paleozoic era (p. 218)

period (p. 218)

Phanerozoic eon (p. 218)

Precambrian (p. 220)

radioactivity (p. 215)

radiometric dating (p. 215)

relative dating (p. 206)

superposition, law of (p. 208)

unconformity (p. 209)

uniformitarianism (p. 205)

Questions for Review

1. Contrast the philosophies of catastrophism and uniformitarianism. How did the proponents of each perceive the age of Earth?

2. Distinguish between absolute and relative dating.

3. What is the law of superposition? How are cross-cutting relationships used in relative dating?

4. When you observe an outcrop of steeply inclined sedimentary layers, what principle allows you to assume that the beds became tilted after they were deposited?

5. Refer to Figure 8.5 (p. 208) and answer the following questions:

   a. Is fault A older or younger than the sandstone layer?

   b. Is dike A older or younger than the sandstone layer?

   c. Was the conglomerate deposited before or after fault A?

   d. Was the conglomerate deposited before or after fault B?

   e. Which fault is older, A or B?

   f. Is dike A older or younger than the batholith?

6. A mass of granite is in contact with a layer of sandstone. Using a principle described in this chapter, explain how you might determine whether the sandstone was deposited on top of the granite or the granite was intruded from below after the sandstone was deposited.

7. Distinguish among angular unconformity, disconformity, and nonconformity.

8. What is meant by the term correlation?

9. List two conditions that improve an organism's chances of being preserved as a fossil.

10. Why are fossils such useful tools in correlation?

11. In addition to being important aids in dating and correlating rocks, how else are fossils helpful in geologic investigations?

12. If a radioactive isotope of thorium (atomic number 90, mass number 232) emits 6 alpha particles and 4 beta particles during the course of radioactive decay, what are the atomic number and mass number of the stable daughter product?

13. Why is radiometric dating the most reliable method of dating the geologic past?

14. Assume that a hypothetical radioactive isotope has a half-life of 10,000 years. If the ratio of radioactive parent to stable daughter product is 1:3, how old is the rock containing the radioactive material?

15. To make calculations easier, let us round the age of Earth to 5 billion years.

   a. What fraction of geologic time is represented by recorded history (assume 5000 years for the length of recorded history)?

   b. The first abundant fossil evidence does not appear until the beginning of the Cambrian period (approximately 600 million years ago). What percentage of geologic time is represented by abundant fossil evidence?

16. What subdivisions make up the geologic time scale? What is the primary basis for differentiating the eras?

17. Briefly describe the difficulties in assigning absolute dates to layers of sedimentary rock.

Testing What You Have Learned

Multiple-Choice Questions

1. The statement "The physical, chemical, and biological laws that operate today have also operated in the geological past" is known as the doctrine of _____

   a. catastrophism.   c. understanding.   e. uniformitarianism.

   b. history.   d. constancy.

2. Within a sequence of undisturbed sedimentary rocks, the bed at the bottom can be assumed to be the oldest because of the law of _____

   a. superposition.   c. inclusions.   e. uniformitarianism.

   b. cross-cutting.   d. unconformities.

3. Which one of the following is NOT used to determine relative dates?

   a. superposition   c. inclusions   e. radioactive isotopes

   b. cross-cutting   d. fossils

4. What type of unconformity consists of folded sedimentary rocks overlain by younger, more flat-lying strata?

   a. nonconformity   c. reversed   e. angular unconformity

   b. conformity   d. disconformity

5. The task of matching rocks of similar age in different regions is called _____

   a. inclusions.   c. dating.   e. connecting.

   b. conforming.   d. correlation.

6. Fossils that are widespread geographically and are limited to a short span of geologic time are called _____ fossils.

   a. base   c. term   e. key

   b. index   d. matching

7. The number of protons plus neutrons in an atom's nucleus is the atom's _____
   - **a.** isotope.
   - **b.** electrons.
   - **c.** form.
   - **d.** mass number.
   - **e.** atomic number.

8. Isotopes of the same atom will have a different number of _____
   - **a.** protons.
   - **b.** ions.
   - **c.** neutrons.
   - **d.** molecules.
   - **e.** electrons.

9. The time required for one-half of the nuclei in a radioactive isotope to decay is called the _____
   - **a.** period.
   - **b.** half-life.
   - **c.** era.
   - **d.** decay time.
   - **e.** radiometric time.

10. The entire geologic time scale was originally created using only methods of _____ dating.
    - **a.** absolute
    - **b.** relative
    - **c.** reverse
    - **d.** inclusive
    - **e.** precise

Fill-In Questions

11. The English geologist Sir Charles _____ is given the most credit for advancing the basic principles of modern geology.

12. _____ dates pinpoint the time in history when something took place, while _____ dates put events in their proper sequence.

13. The principle of _____ _____ states that layers of sediment are deposited in a horizontal position.

14. An _____ represents a long period during which deposition ceased, erosion removed previously formed rocks, and then deposition resumed.

15. The fact that fossil organisms succeed one another in a definite and determinable order is known as the principle of _____.

True/False Questions

16. Inclusions are pieces of one rock unit that are contained within another. _____
17. Rapid burial is the only condition necessary for an organism to become a fossil. _____
18. The spontaneous breaking apart of unstable atomic nuclei is called radioactivity. _____
19. Radiometric dates are more reliable for samples of igneous rocks than for samples of sedimentary rock. _____
20. Together, the Hadean, Archean, and Proterozoic eons are often referred to simply as the Precambrian. _____

Answers:

1. e; 2. a; 3. e; 4. e; 5. d; 6. b; 7. d; 8. c; 9. b; 10. b; 11. Lyell; 12. Absolute, relative; 13. original horizontality; 14. unconformity; 15. fossil succession; 16. T; 17. F; 18. T; 19. T; 20. T.

Web   Work

The following are informative and interesting World Wide Web sites that address topics related to those presented in this chapter:

- Geological Time Machine
  http://www.ucmp.berkeley.edu/help/ timeform.html

- Museum of Paleontology, University of California
  http://www.ucmp.berkeley.edu/

For direct access to these sites and other relevant Web locations, contact the *Foundations of Earth Science* WWW home page at http://www.prenhall.com/lutgens.

UNIT FIVE
The Global Ocean

Sea stacks and surf-rounded stones, Garrapata State Beach, California.
Photo by Carr Clifton Photography

CHAPTER 9

Oceans

The Last Frontier

FOCUS ON LEARNING

To assist you in learning the important concepts
in this chapter, you will find it helpful to focus
on the following questions:

1. What is oceanography?
2. What is the extent and distribution of the world's oceans?
3. What are the principal elements that contribute to the ocean's salinity? What are the sources for these elements?
4. How do temperature and salinity change with depth in the open ocean?
5. What are the zones and features that collectively make up the continental margin?

6. What are the features of the ocean basin floor? How are mid-ocean ridges related to seafloor spreading?
7. What are coral reefs and atolls?
8. What are the various types of seafloor sediments? How can these sediments be used to study worldwide climatic changes?

A coral reef in the tropical Pacific Ocean.
Photo by Nancy Sefton/Photo Researchers, Inc.

Calling Earth the "water planet" is certainly appropriate, because nearly 71 percent of its surface is covered by the global ocean. Although the ocean comprises a much greater percentage of Earth's surface than the continents, it was only in the relatively recent past that the ocean became an important focus of study. Recently there has been a virtual explosion of data about the oceans, and with it, oceanography has grown dramatically. **Oceanography** is a composite science that draws on the methods and knowledge of biology, chemistry, physics, and geology to study all aspects of the world ocean.

The Vast World Ocean

A glance at a globe or a view of Earth from space reveals a planet dominated by the world ocean (Figure 9.1). It is for this reason that Earth is often referred to as the blue planet. The area of Earth is about 510 million square kilometers (197 million square miles). Of this total, approximately 360 million square kilometers (140 million square miles), or 71 percent, is represented by oceans and marginal seas (meaning seas around the ocean's margin, like the Mediterranean Sea and Caribbean Sea). Continents and islands comprise the remaining 29 percent, or 150 million square kilometers (58 million square miles).

By studying a globe or world map, it is readily apparent that the continents and oceans are not evenly divided between the Northern and Southern hemispheres (Figure 9.1). When we compute the percentages of land and water in the Northern Hemisphere, we find that nearly 61 percent of the surface is water, and about 39 percent is land. In the Southern Hemisphere, on the other hand, almost 81 percent of the surface is water, and only 19 percent is land. It is no wonder then that the Northern Hemisphere is called the *land hemisphere*, and the Southern Hemisphere the *water hemisphere*.

Figure 9.2 shows the distribution of land and water in the Northern and Southern hemispheres by way of a graph. Between latitudes 45 degrees north and 70 degrees north, there is actually more land than water, whereas between 40 degrees south and 65 degrees south there is almost no land to interrupt the oceanic and atmospheric circulation.

The volume of the ocean basins is many times greater than the volume of the continents above sea level. In fact, the volume of all land above sea level is only 1/18 that of the ocean.

An obvious difference between continents and the ocean basins is their relative levels. The average

Figure 9.1 These views of Earth show the uneven distribution of land and water between the Northern and Southern hemispheres. Almost 81 percent of the Southern Hemisphere is covered by the oceans—20 percent more than the Northern Hemisphere.

elevation of the continents above sea level is about 840 meters (2750 feet), whereas the average depth of the oceans is more than 4.5 times this figure—3800 meters (12,500 feet). If Earth's solid mass were perfectly smooth (level) and spherical, the oceans would cover it to a uniform depth of more than 2000 meters (1.2 miles).

A comparison of the three major oceans reveals that the Pacific is by far the largest; it is nearly as large as the Atlantic and Indian oceans combined (Figure 9.2). The Pacific contains slightly more than

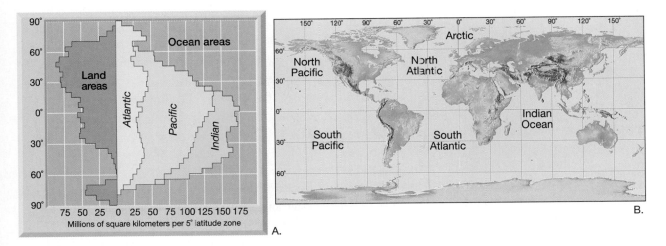

Figure 9.2 Distribution of land and water. **A.** The graph shows the amount of and and water in each 5-degree latitude belt. **B.** The world map provides a more familiar view.

half of the water in the world ocean, and because it includes few shallow seas along its margins, it has the greatest average depth—3940 meters (12,900 feet).

The Atlantic, bounded by almost parallel continental margins, is a relatively narrow ocean when compared to the Pacific. When the Arctic Ocean is included, the Atlantic has the greatest north–south extent and connects the two polar regions. Because the Atlantic has many shallow adjacent seas, including the Caribbean, Gulf of Mexico, Baltic, and Mediterranean, as well as wide continental shelves along its borders, it is the shallowest of the three oceans, with an average depth of 3310 meters (10,860 feet).

As Figure 9.2 shows, the Indian Ocean covers less area than the Pacific or the Atlantic. Unlike the others, it is largely a Southern Hemisphere water body.

Composition of Seawater

Seawater is a complex solution of salts, consisting of about 3.5 percent (by weight) dissolved mineral substances. Although the percentage of salts may seem small, the actual quantity is huge. If all of the water were evaporated from the oceans, a layer of salt approaching 60 meters (200 feet) thick would cover the entire ocean floor.

Salinity and Its Variations

Salinity is the proportion of dissolved salts to pure water. In many things, we express proportion in parts-per-hundred, or percent, %. But because the proportion

of salts in seawater is a small number, scientists express salinity in parts-per-thousand, noted ‰. Thus, the average salinity of the ocean is about 35‰.

The principal elements that contribute to the ocean's salinity are shown in Figure 9.3. If we made our own seawater, we could come reasonably close by following the recipe shown in Table 9.1. From this table it is evident that most of the salt is sodium chloride (common table salt). Sodium chloride together with the next four most abundant salts comprise 99 percent of the salt in the sea. Although only eight elements make up these five most abundant salts, seawater contains more than 70 of Earth's other naturally occurring elements. Despite their presence in minute quantities, many of these elements are very important in maintaining the necessary chemical environment for life in the sea.

The relative abundances of the major components in sea salt are essentially constant, no matter where the ocean is sampled. Variations in salinity, therefore, are primarily a consequence of changes in the water content of the solution. As a result, high salinities are found where evaporation is high, as is the case in the dry subtropics. Conversely, where heavy precipitation dilutes ocean waters, as in the mid-latitudes and near the equator, lower salinities prevail (Figure 9.4).

Whereas salinity variations in the open ocean normally range from 33‰ to 37‰, some marginal seas demonstrate extraordinary extremes. For example, in the restricted waters of the Persian Gulf and the Red Sea, where evaporation is far greater than precipitation, salinity may exceed 42‰. Conversely, very low

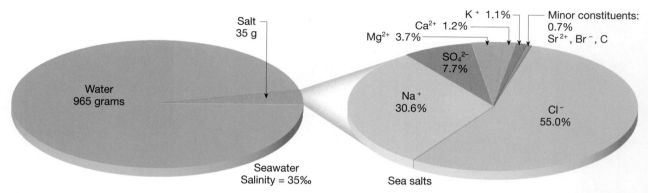

Figure 9.3 Relative proportions of water and dissolved salts in seawater. Ions shown by chemical symbol are chlorine (Cl), sodium (Na), sulfate (SO_4), magnesium (Mg), calcium (Ca), potassium (K), strontium (Sr), bromine (Br), and carbon (C).

salinities occur where large quantities of freshwater are supplied by rivers and precipitation. Such is the case for the Baltic Sea, where salinity is often below 10‰.

Sources of Sea Salts

What are the primary sources for the vast quantities of salts in the ocean? Chemical weathering of rocks on the continents is one source. These soluble materials are delivered to the oceans by streams at an estimated rate of more than 2.5 billion tons annually. The second major source of elements found in ocean water is Earth's interior. Through volcanic eruptions, large quantities of water and dissolved gases have been emitted during much of geologic time. This process, called **outgassing**, is the principal source of water in the oceans and in the atmosphere. Certain elements, notably chlorine, bromine, sulfur, and boron, are much

more abundant in the ocean than in Earth's crust. Because they result from outgassing, their abundance in the sea tends to confirm the hypothesis that volcanic action is largely responsible for the present oceans.

Although rivers and volcanic activity continually contribute salts to the oceans, the salinity of seawater is not increasing. In fact, many oceanographers believe that the composition of seawater has been relatively stable for a large span of geologic time. Why doesn't the sea get saltier? Obviously, material is being removed just as rapidly as it is added. Some elements are withdrawn from seawater by plants and

Table 9.1 ■ Recipe for Artificial Seawater

MIX:

Sodium chloride (NaCl)	23.48 grams
Magnesium chloride ($MgCl_2$)	4.98
Sodium sulfate (Na_2SO_4)	3.92
Calcium chloride ($CaCl_2$)	1.10
Potassium chloride (KCl)	0.66
Sodium bicarbonate ($NaHCO_3$)	0.192
Potassium bromide (KBr)	0.096
Hydrogen borate (H_3BO_3)	0.026
Strontium chloride ($SrCl_2$)	0.024
Sodium fluoride (NaF)	0.003

ADD:

Water (H_2O) to form 1000 grams of solution.

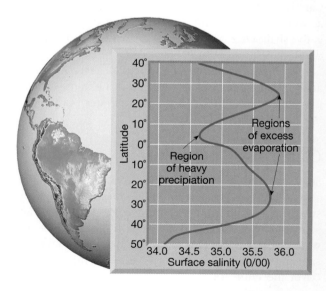

Figure 9.4 The subtropics are dry. On the continents, deserts such as the Sahara occur in this latitude belt. Over the oceans, high evaporation rates remove more water than is replaced by the meager rainfall. Thus, surface salinities are high here. By contrast, near the equator rainfall is heavy, therefore surface salinities are reduced here.

animals as they build hard parts. Others are removed when they chemically precipitate out of the water as sediment. The net effect is that the overall makeup of seawater remains relatively constant.

The Ocean's Layered Structure

By sampling ocean waters, oceanographers have found that temperature and salinity change with depth. Generally, they recognize a three-layered structure in the open ocean: a shallow surface mixed zone, a transition zone, and a deep zone (Figure 9.5).

Because solar energy is received at the ocean surface, it is here that water temperatures are warmest. The mixing of these waters by waves as well as the turbulence from currents creates a mixed surface zone that has nearly uniform temperatures. The thickness and temperature of this layer vary, depending on latitude and season.

Below the Sun-warmed zone of mixing, the temperature falls abruptly with depth. This layer of rapid temperature change, known as the **thermocline**, marks the transition between the warm surface layer and the deep zone of cold water below. Below the thermocline temperatures fall only a few more degrees. At depths greater than about 1500 meters (5000 feet), ocean-water temperatures are consistently below 4°C (39°F). Because the average ocean depth is approximately 3800 meters (12,500 feet), you can see that the temperature of most seawater is not much above freezing. In the high polar latitudes, surface waters are cold and temperature changes with depth are slight. Consequently, the three-layered structure is not present (Figure 9.5).

Salinity variations with depth correspond to the general three-layered system described for temperatures. Generally, in the low and middle latitudes, a surface zone of higher salinity is created when freshwater is removed by evaporation. Below the surface zone, salinity decreases rapidly. This layer of rapid change, the **halocline**, corresponds closely to the thermocline. Below the halocline salinity variations are small.

Earth Beneath the Sea

If all water were removed from the ocean basins, a great diversity of features would be seen, including linear chains of volcanoes, deep canyons, rift valleys, and large submarine plateaus. The scenery would be nearly as varied as that on the continents (Figure 9.6).

Our knowledge of the diverse topography of the ocean floor is relatively recent. Until the 1920s, ocean depths were determined laboriously, by lowering a weighted rope until it touched bottom. As a result, our knowledge of the seafloor remained scant. In the 1920s, a technological breakthrough occurred with the invention of electronic depth-sounding equipment—the echo sounder (also referred to as *sonar*).

The **echo sounder** works by transmitting sound waves toward the ocean bottom (Figure 9.7A). A sensitive receiver intercepts the echo reflected from the bottom, and a clock precisely measures how long it took for the sound waves to make the round trip to the bottom and back up to the ship. We know how fast sound moves in water (about 1500 meters or 5000 feet per second). So if sound takes 1 second to make the round trip from the ship to the bottom and back, you know the distance traveled was about 1500 meters for the round trip, so the distance to the bottom is half of that, or 750 meters.

The depths determined from continuous monitoring of these echoes are plotted to produce a profile of the ocean floor. By laboriously combining profiles from several adjacent traverses, a chart of the seafloor is produced. Although much more complete and detailed than anything available before, these charts only show the largest topographic features of the ocean floor (see Figure 9.6).

In the last few decades, researchers have designed even more sophisticated echo sounders to map the ocean floor. In contrast to simple echo sounders, *multibeam sonar* employs an array of

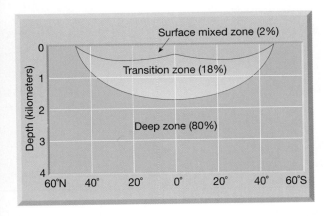

Figure 9.5 Oceanographers recognize three "layers" of the ocean based on temperature and salinity. The thickness of each varies with latitude, as shown here, and with the season of the year. Clearly, most of the ocean's water is in the deep zone.

Figure 9.6 The topography of Earth's solid surface is shown on these two pages.

sound sources and listening devices. Thus, rather than obtaining the depth of a single point every few seconds, this technique obtains a profile of a narrow strip of seafloor (Figure 9.7B).

Despite their greater efficiency and enhanced detail, research vessels equipped with multibeam sonar travel at a mere 10 to 20 kilometers per hour. It would take at least 100 vessels outfitted with this

equipment hundreds of years to map the entire seafloor. By contrast, in 1991 and 1992, the *Magellan* spacecraft, traveling at 19,500 kilometers per hour, used radar to map more than 90 percent of the surface of Venus. Someday we hope to be able to view the ocean floor with the same detail we presently have for the Moon and some of the planets.

Oceanographers studying the topography of the ocean basins have delineated three major units: *continental margins*, the *ocean basin floor*, and *mid-ocean*

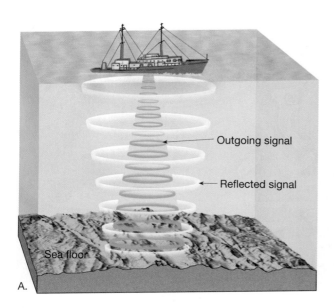

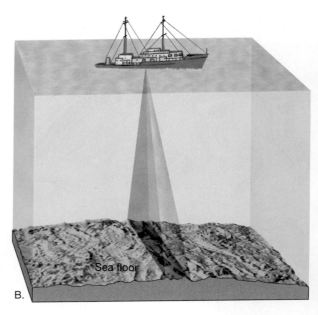

Figure 9.7 Echo sounders. **A.** An echo sounder determines water depth by measuring the time interval required for an acoustic wave to travel from a ship to the seafloor and back. The speed of sound in water is 1500 m/sec. Therefore, water depth = 1/2 (1500 m/sec × echo travel time). **B.** Modern multibeam sonar obtains a profile of a narrow swath of seafloor every few seconds.

ridges. The map in Figure 9.8 shows these provinces for the North Atlantic, and the profile at the bottom of the illustration shows the varied topography. Such profiles usually have their vertical dimension exaggerated many times—40 times in this case—to make topographic features more conspicuous. Because of this, the slopes in the profile of the seafloor in Figure 9.8 appear to be much steeper than they actually are.

Continental Margins

We will now look at the three major topographic regions of the seafloor (continental margin, ocean basin floor, and mid-ocean ridge). To keep your perspective during this discussion, refer to Figures 9.6, 9.8, and 9.9.

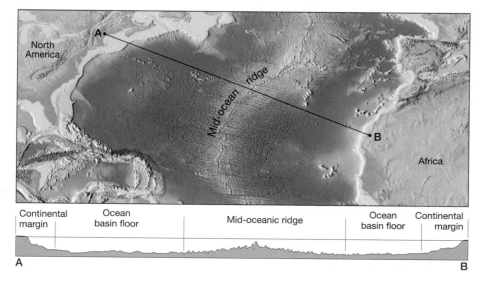

Figure 9.8 This map and profile show the major topographic divisions of the North Atlantic.

Continental Shelf

The zones that collectively make up the **continental margin** include the continental shelf, continental slope, and continental rise (Figure 9.9). The first of these parts, the **continental shelf**, is a gently sloping submerged surface extending from the shoreline toward the deep-ocean basin. It is a flooded extension of the continent.

The continental shelf varies greatly in width, as you can see in Figures 9.6 and 9.8. Almost nonexistent along some continents, the shelf may extend seaward as far as 1500 kilometers (900 miles) along others. On the average, the continental shelf is about 80 kilometers (50 miles) wide and 130 meters (423 feet) deep at its seaward edge, the shelf break. The average inclination of the continental shelf is less than one-tenth of one degree, a drop of only about 2 meters per kilometer (10 feet per mile). The slope is so slight that it appears horizontal.

The continental shelves represent 7.5 percent of the total ocean area, which is equivalent to about 18 percent of Earth's total land area. These areas have taken on increased economic and political significance because they contain important mineral deposits, including large reservoirs of petroleum and natural gas, as well as huge sand and gravel deposits. The waters of the continental shelf also contain many important fishing grounds.

When compared with many parts of the deep-ocean floor, the surface of the continental shelf is relatively featureless. This is not to say that the shelves are completely smooth. In many regions, long valleys run from the coastline into deeper waters. Many of these shelf valleys are seaward extensions of river valleys on the adjacent landmass. Such valleys were carved by streams during the Pleistocene epoch (Ice Age).

During this time, great quantities of water were stored in vast ice sheets on the continent. The result was that sea level dropped by 100 meters (330 feet) or more, exposing large portions of the continental shelves (see Figure 4.17, p. 102). Because of the lower sea level, rivers extended their courses, and land-dwelling plants and animals moved onto the newly exposed portions of the continents. Today these areas are again covered by the sea and inhabited by marine organisms. Scientists dredging along the eastern coast of North America have discovered the remains of numerous land dwellers, including mammoths, mastodons, and horses. Bottom sampling has also revealed that freshwater peat bogs existed, adding to the evidence that the continental shelves were once above sea level.

Continental Slope and Rise

Marking the seaward edge of the continental shelf is the **continental slope** (Figure 9.9). It leads into deep water and has a steep gradient compared to the continental shelf. The gradient varies from place to place, but it has an average drop of about 70 meters per kilometer (370 feet per mile). The continental slope is the true edge of the continent.

In regions where trenches do not exist, the steep continental slope merges into a more gradual incline known as the **continental rise** (Figure 9.9). Here the gradient lessens to between 4 and 8 meters per kilometer (20 to 40 feet per mile). Whereas the width of

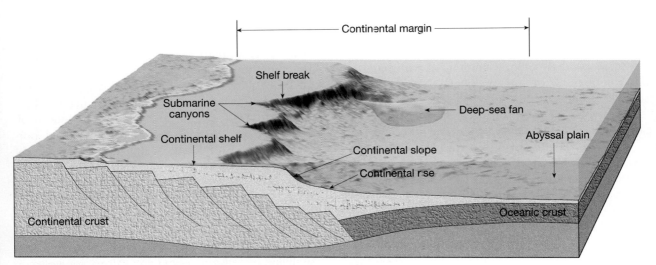

Figure 9.9 Schematic view showing the parts of the continental margin.

the continental slope averages about 20 kilometers (12 miles), the continental rise may extend for hundreds of kilometers. This feature consists of a thick accumulation of sediment that moved downslope from the continental shelf to the deep-ocean floor. The sediments are delivered to the base of the continental slope by *turbidity currents* that follow submarine canyons (Figure 9.9). (We will discuss these shortly.) When these muddy currents emerge from the mouth of a canyon onto the relatively flat ocean floor, they deposit sediment that forms a **deep-sea fan**. Deep-sea fans have the same basic shape as alluvial fans, which form at the foot of steep mountain slopes on land. As fans from adjacent submarine canyons grow, they coalesce to produce the continuous apron of sediment at the base of the continental slope.

Along some mountainous coasts, the continental slope does not grade gently into the continental rise, but descends abruptly into a *deep-ocean trench* that parallels the landmass. In such cases, the shelf is very narrow or does not exist at all. The side of the trench and the continental slope are essentially the same feature and grade into the adjacent mountains that tower thousands of meters above sea level. These narrow continental margins are primarily located around the Pacific Ocean where plates are converging. The stress between colliding plates results in deformation of the continental margin. An example of this activity is found along the western coast of South America (see Figure 9.6, pp. 232–33). Here the vertical distance from the peaks of the Andes Mountains to the floor of the deep Peru–Chile trench bordering the continent exceeds 12,000 meters (7.5 miles!).

Submarine Canyons and Turbidity Currents

Deep, steep-sided valleys known as **submarine canyons** originate on the continental slope and may extend to depths of 3 kilometers (see Figure 9.9). Although some of these canyons appear to be the seaward extensions of river valleys, many others do not line up in this manner. Furthermore, these canyons extend to depths far below the maximum lowering of sea level during the Ice Age, so we cannot attribute their formation to stream erosion.

These submarine canyons probably have been excavated by turbidity currents. **Turbidity currents** are downslope movements of dense, sediment-laden water. They are created when sand and mud on the continental shelf and slope are dislodged and thrown into suspension. Because the mud-choked water is denser than normal seawater, it flows downslope as a mass, eroding and accumulating more sediment as it goes. The erosional work repeatedly carried on by these muddy torrents is thought to be the major force in the excavation of most submarine canyons.

Turbidity currents usually originate along the continental slope and continue across the continental rise, still cutting channels. Eventually they lose momentum and come to rest along the ocean basin floor. As these currents slow, suspended sediments begin to settle out. First, the coarser sand is dropped, followed by successively finer accumulations of silt and then clay. Consequently, these deposits display a decrease in sediment grain size from bottom to top.

Turbidity currents are a very important mechanism of sediment transport in the ocean. By the action of turbidity currents, submarine canyons are excavated and sediments are carried to the deep-ocean floor.

The Ocean Basin Floor

Between the continental margin and the mid-ocean ridge system lies the ocean basin floor (see Figure 9.8). The size of this region—almost 30 percent of Earth's surface—is roughly comparable to the percentage of the surface that presently projects above the sea as land. On the ocean basin floor, we find dramatically deep grooves, which are *deep ocean trenches*; remarkably flat regions, known as *abyssal plains*; and steep-sided volcanic peaks, called *seamounts*. We will now examine each of these features.

Deep-Ocean Trenches

Deep-ocean trenches are long, narrow troughs that are the deepest parts of the ocean. Most trenches are located in the Pacific Ocean where some approach or exceed 10,000 meters (33,000 feet) in depth. A portion of one, the Challenger Deep in the Mariana trench, is more than 11,000 meters (36,000 feet) below sea level.

Deep-ocean trenches represent only a small portion of the ocean floor area, but they are significant geological features. Trenches are the sites where moving lithospheric plates plunge back into the mantle (see Figure 5.11, p. 129). In addition to the earthquakes created as one plate descends beneath another, volcanic activity is also associated with these regions. Thus, trenches in the open ocean are often paralleled by volcanic island arcs. Moreover, volcanic mountains, such as those making up portions of the

Andes, are located parallel to trenches that lie adjacent to continental margins.

Abyssal Plains

Abyssal plains are incredibly flat features; in fact, these regions are likely the most level places on Earth. The abyssal plain found off the coast of Argentina, for example, has less than 3 meters (10 feet) of relief over a distance exceeding 1300 kilometers (800 miles). The monotonous topography of abyssal plains is occasionally interrupted by the protruding summit of a buried volcanic structure.

By employing seismic profilers, instruments whose signals penetrate far below the ocean floor, researchers have determined that abyssal plains consist of thick accumulations of sediment that have buried an otherwise rugged ocean floor. The nature of the sediment indicates that these plains consist primarily of sediments transported far out to sea by turbidity currents.

Abyssal plains occur in all of the oceans. However, they are more widespread where there are no deep-ocean trenches adjacent to the continents. Because the Atlantic Ocean has fewer trenches to act as traps for the sediments carried down the continental slope, it has more extensive abyssal plains than does the Pacific.

Seamounts

Dotting the ocean floors are isolated volcanic peaks called **seamounts**, which may rise hundreds of meters above the seafloor. Although these steep-sided conical peaks are found on the floors of all the oceans, the greatest number have been identified in the Pacific (see Figure 9.6).

Some, like the Emperor Seamount chain that stretches from the Hawaiian Islands to the Aleutian Trench, form in association with volcanic hot spots. Others are born near mid-ocean ridges, divergent plate boundaries where the plates of the lithosphere move apart (see Chapter 5). If a volcano grows rapidly, it may emerge as an island. Examples in the Atlantic include the Azores, Ascension, Tristan da Cunha, and St. Helena.

When they exist as islands, some of these volcanoes are eroded to near sea level by running water and wave action. Over a span of millions of years the islands gradually sink as the moving plate slowly carries them away from the elevated oceanic ridge or hot spot where they originated. These submerged, flat-topped seamounts are called **guyots**.

Mid-Ocean Ridges

Our knowledge of **mid-ocean ridges**, the sites of seafloor spreading, comes from soundings taken of the ocean floor, core samples obtained from deep-sea drilling, and visual inspection using deep-diving submersibles (Figure 9.10). We also have firsthand inspection of slices of ocean floor that have been shoved up onto dry land by plate movements. Ocean ridge systems are characterized by an elevated position, extensive faulting, and numerous volcanic structures that have developed on the newly formed crust. This is quite evident as you follow the path of mid-ocean ridges in Figure 9.6, pp. 232-33.

The interconnected ocean ridge system is the longest topographic feature on Earth, exceeding

Figure 9.10 The deep-diving submersible *Alvin* is 7.6 meters long, weighs 16 tons, has a cruising speed of 1 knot, and can reach depths as great as 4000 meters. A pilot and two scientific observers are along during a normal 6- to 10-hour dive. *(Courtesy of Rod Catanach/©Woods Hole Oceanographic Institution)*

70,000 kilometers (43,000 miles) in length. Representing 20 percent of Earth's surface, the ocean ridge system winds through all major oceans in a manner similar to the seam on a baseball (Figure 9.6). The term *ridge* may be misleading as these features are not narrow but have widths from 3000 to 4000 kilometers and, in places, may occupy as much as one-half the total area of the ocean floor.

Although ocean ridges stand above the adjacent deep-ocean basins, they are much different from the mountains found on the continents. Rather than containing thick sequences of folded and faulted sedimentary rocks, oceanic ridges consist of layer upon layer of basaltic rocks that have been faulted and uplifted.

Ridges are broad features, yet much of the geologic activity occurs along a relatively narrow region on the ridge crest, called the **rift zone**. As plates move apart, magma wells up into the newly created fractures and generates new slivers of oceanic crust. Active rift zones have frequent but generally weak earthquakes and exhibit a rate of heat flow that is greater than most other crustal segments. Here vertical displacement of large slabs of oceanic crust, caused by faulting and the growth of volcanic piles, contributes to the rugged topography of the mid-ocean ridge. Further, the rocks along the ridge axis appear very fresh and are nearly void of sediment. Away from the ridge axis the topography becomes more subdued, and the thickness of the sediments and the depth of the water increase. Gradually the ridge system grades into the less rugged sediment-laden regions of the deep-ocean floor.

During seafloor spreading, new material is added about equally to the two diverging plates. Hence, we would expect new ocean floor to grow symmetrically on both sides of a centrally located ridge. Indeed, the ridge systems of the Atlantic and Indian oceans are located near the middles of these water bodies and as a consequence are named mid-ocean ridges (Figure 9.6). However, the East Pacific Rise is situated far from the center of the Pacific Ocean. Despite uniform spreading along the East Pacific Rise, much of the Pacific basin that once lay east of this spreading center has been overridden by the westward migration of the American plate.

Partly because of its accessibility to both American and European scientists, the Mid-Atlantic Ridge has been studied more thoroughly than other ridge systems. The Mid-Atlantic Ridge is a broad submerged structure standing 2500 to 3000 meters (8200 to 9800 feet) above the adjacent floor of the deep-ocean basins. In a few places, such as Iceland, the ridge has actually grown above sea level (Figure 9.6, pp. 232-33). Throughout most of its length, however, this divergent plate boundary lies 2500 meters below sea level.

Another prominent feature of the Mid-Atlantic Ridge is a deep linear rift valley extending along the ridge axis. In places, this rift valley is deeper than the Grand Canyon of the Colorado River and two or three times as wide. The name *rift valley* has been applied to this feature because it is strikingly similar to continental rift valleys such as those in East Africa. An examination of Figure 9.6 reveals that this central rift is broken into sections that are offset by transform faults.

Coral Reefs and Atolls

Coral reefs are among the most picturesque features in the ocean. They are constructed primarily from the skeletal remains and secretions of corals and certain algae, built up over thousands of years. The term *coral reef* is somewhat misleading in that it makes no mention of the skeletons of many other small animals and plants found inside the branching framework built by the corals, or the fact that limy secretions of algae help bind the entire structure together.

Coral thrive in the warm water of the Pacific and Indian oceans, so most reefs are in these oceans, although a few occur elsewhere. Reef-building corals grow best in waters with an average annual temperature of about 24°C (75°F). They can survive neither sudden temperature changes nor prolonged exposure to temperatures below 18°C (65°F). In addition, these reef-builders require clear sunlit water. For this reason, the limiting depth of active reef growth is about 45 meters (150 feet).

In 1831, the naturalist Charles Darwin set out aboard the British ship H.M.S. Beagle on its famous 5-year expedition that circumnavigated the globe. One outcome of Darwin's studies during the voyage was his hypothesis on the formation of coral islands, called **atolls**. As Figure 9.11 illustrates, atolls consist of a continuous or broken ring of coral reef surrounding a central lagoon.

Atolls present a paradox: how can corals, which require warm, shallow, sunlit water no deeper than 45 meters, create structures that reach thousands of meters to the floor of the ocean? The essence of Darwin's hypothesis was that coral reefs form on the flanks of sinking volcanic islands. As an island slowly

A.

B.

Figure 9.11 A. An aerial view of Tetiaroa Atoll in the Pacific. The light blue waters of the relatively shallow lagoon contrast with the dark blue color of the deep ocean surrounding the atoll. **B.** View from space of a group of atolls in the Pacific Ocean. *(Photo A by Douglas Peebles Photography) (Photo B courtesy of NASA Headquarters)*

sinks, the corals continue to build the reef complex upward (Figure 9.12).

Thus atolls, like guyots, owe their existence to the gradual sinking of oceanic crust. Drilling on atolls has revealed that volcanic rock does indeed underlie the thick coral reef structure. This finding helped to confirm Darwin's explanation.

Seafloor Sediments

Except for a few areas, such as near the crests of mid-ocean ridges, the ocean floor is mantled with sediment. Part of this material has been deposited by turbidity currents, and the rest has slowly settled to the bottom from above. The thickness of this carpet of debris varies greatly. In some trenches, which act as traps for sediments originating on the continental margin, accumulations may approach 10 kilometers (6 miles). In general, however, sediment accumulations are considerably less. In the Pacific Ocean, uncompacted sediment measures about 600 meters (2000 feet) or less, whereas on the floor of the Atlantic, the thickness varies from 500 to 1000 meters (1500 to 3000 feet).

Although deposits of sand-sized particles are found on the deep-ocean floor, mud is the most common sediment covering this region. Muds also predominate on the continental shelf and slope, but the sediments in these areas are coarser overall because of greater quantities of sand.

Types of Seafloor Sediments

Seafloor sediments can be classified according to their origin into three broad categories: (1) terrigenous ("derived from land") sediment, (2) biogenous ("derived from organisms") sediment, and (3) hydrogenous ("derived from water") sediment. Although we discuss each category separately, remember that all seafloor sediments are mixtures. No body of sediment comes entirely from a single source.

Terrigenous sediment consists primarily of mineral grains that were weathered from continental rocks and transported to the ocean. The sand-sized particles settle near shore. Because the very smallest particles take years to settle to the ocean floor, they may be carried for thousands of kilometers by ocean currents. As a consequence, virtually every area of the ocean receives some terrigenous sediment. The rate at which this sediment accumulates on the deep-ocean floor is indeed very slow. From 5000 to 50,000 years are necessary for a 1-centimeter layer to form. Conversely, on the continental margins near the mouths of large rivers, terrigenous sediment accumulates rapidly. In the Gulf of Mexico, for example, the sediment has reached a depth of many kilometers.

Because fine particles remain suspended in the water for a very long time, there is ample opportunity for chemical reactions to occur. Because of this, the colors of deep-sea sediments are often red or

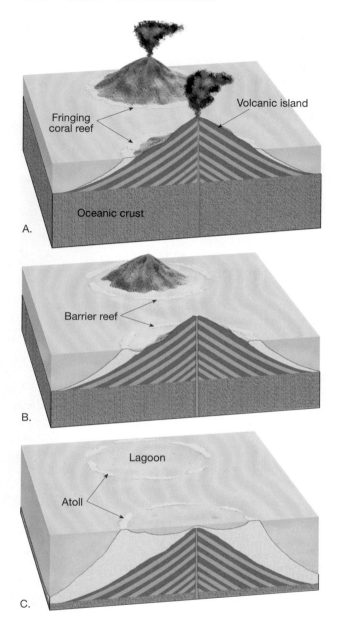

A.

B.

C.

Figure 9.12 Formation of a coral atoll due to the gradual sinking of oceanic crust.

brown. This results when iron in the particle or in the water reacts with dissolved oxygen in the water and produces a coating of iron oxide (rust).

Biogenous sediment consists of shells and skeletons of marine animals and plants (Figure 9.13). This debris is produced mostly by microscopic organisms living in the sunlit waters near the ocean surface. The remains continually "rain" down on the seafloor.

The most common biogenous sediments are known as *calcareous* ($CaCO_3$) *oozes*, and, as the name implies, they have the consistency of thick mud. These sediments are produced by organisms that inhabit warm surface waters. When calcareous hard parts slowly sink through a cool layer of water, they begin to dissolve. This results because cold seawater contains more carbon dioxide and is thus more acidic than warm water. In seawater deeper than about 4500 meters (15,000 feet), calcareous shells will completely dissolve before they reach bottom. Consequently, calcareous ooze does not accumulate where depths are great.

Other biogenous sediments include *siliceous* (SiO_2) *oozes*. They are composed primarily of skeletons of diatoms (single-celled algae) and radiolarians (single-celled animals). Other sediments are derived from the bones, teeth, and scales of fish and other marine organisms.

Hydrogenous sediment consists of minerals that crystallize directly from seawater through various chemical reactions. For example, some limestones are formed when calcium carbonate precipitates directly from the water; however, most limestone is composed of biogenous sediment.

One type of hydrogenous sediment is significant for its economic potential. *Manganese nodules* are rounded blackish lumps consisting of a complex mixture of minerals that form very slowly on the floor of the ocean basins. In fact, their formation rate is one of the slowest chemical reactions known.

Although manganese nodules may contain more than 20 percent manganese, the interest in them as a potential resource lies in the fact that other more valuable metals may be concentrated in them. In addition to manganese, nodules may contain significant iron, copper, nickel, and cobalt. Possible mining locations must have abundant nodules and contain the economically optimum mix of cobalt, copper, and nickel. Sites meeting these criteria are relatively limited. Furthermore, before such areas prove to be valuable commercial sources for these metals, the technical problems of extracting nodules from the floor of the deep-ocean basins must be solved. Also, the political and legal ramifications of mining the sea bed must be worked out.

Seafloor Sediments and Climate Change

Reliable climate records go back only a couple of hundred years, at best. How do scientists learn about

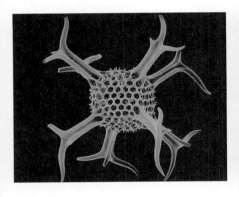

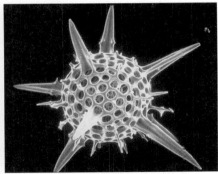

Figure 9.13 Microscopic radiolarian hard parts are examples of biogenous sediments. These photomicrographs have been enlarged *hundreds* of times. *(Photo by Manfred Kage/Peter Arnold, Inc.)*

climates and climatic changes prior to that time? The obvious answer is that they must reconstruct past climates from *indirect evidence*; that is, they must analyze phenomena that respond to and reflect changing atmospheric conditions. An interesting and important technique for analyzing Earth's climate history is the study of sediments from the ocean floor.

Seafloor sediments are of many types and most contain the remnants of organisms that once lived near the sea surface. When such near-surface organisms die, their shells slowly settle to the ocean floor where they become part of the sedimentary record. One reason that seafloor sediments are useful recorders of worldwide climate change is that the numbers and types of organisms living near the sea surface change as the climate changes.

Thus, in seeking to understand climate change as well as other environmental transformations, scientists are tapping the huge reservoir of data in seafloor sediments. The sediment cores gathered by drilling ships and other research vessels have significantly expanded our knowledge and understanding of past climates (Figure 9.14).

Figure 9.14 The *JOIDES Resolution* is the drill ship for the Ocean Drilling Program. It is capable of lowering drill pipe thousands of meters to the ocean floor and then drilling hundreds of meters into the sediments and underlying crust. It is not just a drilling rig but a floating laboratory with a complete array of scientific equipment and facilities. *(Photo courtesy of Ocean Drilling Program)*

The Chapter in Review

1. *Oceanography* is a composite science that draws on the methods and knowledge of biology, chemistry, physics, and geology to study all aspects of the world ocean.

2. *Earth is a planet dominated by oceans.* Seventy-one percent of Earth's surface area is oceans and marginal seas. In the Southern Hemisphere, often called the *water hemisphere*, about 81 percent of the surface is water. Of the three major oceans, Pacific, Atlantic, and Indian, the *Pacific Ocean* is the *largest*, contains slightly *more than half of the water* in the world ocean, and has the *greatest average depth*—3940 meters (12,900 feet).

3. *Salinity* is the proportion of dissolved salts to pure water, usually expressed in parts per thousand (‰). The average salinity in the open ocean ranges from 35‰ to 37‰. The principal elements that contribute to the ocean's salinity are *chlorine* (55‰) and *sodium* (31‰). The primary *sources for the salts* in the ocean are *chemical weathering* of rocks on the continents and *outgassing* through submarine volcanism. Outgassing is also considered to be the principal source of water in the oceans as well as the atmosphere.

4. In most regions, open oceans exhibit a *three-layered temperature and salinity structure*. Ocean water temperatures are warmest at the surface because solar energy is absorbed there. Mixing can distribute this heat to a depth of about 450 meters or more. Beneath the Sun-warmed zone of mixing, a layer of rapid temperature change, called the *thermocline*, occurs. Below the thermocline, in the deep zone, temperatures fall only a few more degrees. Salinity changes with depth correspond to the general three-layered temperature structure. In the low and middle latitudes, a surface zone of higher salinity is underlain by a layer of rapidly decreasing salinity, called the *halocline*. Below the halocline, salinity changes are small.

5. The zones that collectively make up the *continental margin* include the *continental shelf* (a gently sloping, submerged surface extending from the shoreline toward the deep-ocean basin), *continental slope* (the true edge of the continent, which has a steep slope that leads from the continental shelf into deep water), and in regions where trenches do not exist, the steep continental slope merges into a gradual incline known as the *continental rise*. The continental rise consists of sediments that have moved downslope from the continental shelf to the deep-ocean floor.

6. *Submarine canyons* are deep, steep-sided valleys that originate on the continental slope and may extend to depths of three kilometers. Many submarine canyons have been excavated by *turbidity currents* (downslope movements of dense, sediment-laden water).

7. The *ocean basin floor* lies between the continental margin and the mid-ocean ridge system. The features of the ocean basin floor include: *deep-ocean trenches* (the deepest parts of the ocean, where moving crustal plates descend into the mantle), *abyssal plains* (very level regions consisting of thick accumulations of sediments that were deposited atop the low, rough portions of the ocean floor by turbidity currents), and *seamounts* (isolated volcanic peaks on the ocean floor that originate near mid-ocean ridges or in association with volcanic hot spots).

8. *Mid-ocean ridges*, the sites of seafloor spreading, are found in all major oceans. These broad features are characterized by an elevated position, extensive faulting, and volcanic structures that have developed on newly formed oceanic crust. Most of the geologic activity associated with ridges occurs along a narrow region on the ridge crest, where magma moves upward to create new slivers of oceanic crust.

9. *Coral reefs*, which are confined largely to the warm, sunlit waters of the Pacific and Indian oceans, are constructed over thousands of years primarily from the skeletal remains and secretions of corals and certain algae. Coral islands, called *atolls*, consist of a continuous or broken ring of coral reef surrounding a central lagoon. Atolls form from corals that grow on the flanks of sinking volcanic islands, where the corals continue to build the reef complex upward as the island sinks.

10. *There are three broad categories of seafloor sediments.* *Terrigenous sediment* consists primarily of mineral grains that were weathered from continental rocks and transported to the ocean. *Biogenous sediment* consists of shells and skeletons of marine animals and plants. *Hydrogenous sediment* includes minerals that crystallize directly from seawater through various chemical reactions. Seafloor sediments are helpful when studying worldwide climate changes because they often contain the remains of organisms that once lived near the sea surface. The numbers and types of these organisms change as the climate changes, and their remains in the sediments record these changes.

Key Terms

abyssal plain (p. 237)

atoll (p. 238)

biogenous sediment
 (p. 240)

continental margin (p. 235)

continental rise (p. 235)

continental shelf (p. 235)

continental slope (p. 235)

coral reef (p. 238)

deep-ocean trench (p. 236)

deep-sea fan (p. 236)

echo sounder (p. 231)

guyot (p. 237)

halocline (p. 231)

hydrogenous sediment
 (p. 240)

mid-ocean ridge (p. 237)

oceanography (p. 228)

outgassing (p. 230)

rift zone (p. 238)

salinity (p. 229)

seamount (p. 237)

submarine canyon (p. 236)

terrigenous sediment
 (p. 239)

thermocline (p. 231)

turbidity current (p. 236)

Questions for Review

1. How does the area covered by the oceans compare with that of the continents? Describe the distribution of land and water on Earth.

2. Answer the following questions about the Atlantic, Pacific, and Indian oceans:
   a. Which is the largest in area? Which is smallest?
   b. Which has the greatest north–south extent?
   c. Which is shallowest? Explain why its average depth is least.

3. How does the average depth of the ocean compare to the average elevation of the continents?

4. What is meant by *salinity?* What is the average salinity of the ocean?

5. Why do variations in surface salinity occur? Give some specific examples to illustrate your answer.

6. What is the origin of the elements that make up sea salt?

7. Assuming that the average speed of sound waves in water is 1500 meters per second, determine the water depth if the signal sent out by an echo sounder requires 6 seconds to strike bottom and return to the recorder (see Figure 9.7, p. 234).

8. List the three subdivisions of the continental margin. Which subdivision is considered a flooded extension of the continent? Which has the steepest slope?

9. How does the continental margin along the west coast of South America differ from the continental margin along the east coast of North America?

10. Defend or rebut the statement "Most submarine canyons found on the continental slope and rise were formed during the Ice Age when rivers extended their valleys seaward."

11. Why are abyssal plains more extensive on the floor of the Atlantic than on the floor of the Pacific?

12. How are mid-ocean ridges and deep-ocean trenches related to plate tectonics?

13. What is an atoll? Describe Darwin's proposal on the origin of atolls. Was it ever confirmed?

14. Distinguish among the three basic types of seafloor sediment.

15. If you were to examine recently deposited biogenous sediment taken from a depth in excess of 4500 meters (15,000 feet), would it more likely be rich in calcareous materials or siliceous materials? Explain.

16. Why are seafloor sediments useful in studying climates of the past?

Testing What You Have Learned

Multiple-Choice Questions

1. What percentage of Earth's surface is covered by oceans and marginal seas?
   - **a.** 29%
   - **c.** 67%
   - **e.** 82%
   - **b.** 43%
   - **d.** 71% *(circled)*

2. The proportion of dissolved salts to pure water in the ocean is called _____
   - **a.** salinity. *(circled)*
   - **c.** chemistry.
   - **e.** fetch.
   - **b.** chlorinity.
   - **d.** halinity.

3. The layer of rapid temperature change below the ocean surface is the _____
   - **a.** halocline.
   - **c.** slope.
   - **e.** thermal.
   - **b.** thermocline. *(circled)*
   - **d.** thermohaline.

4. Using an echo sounder, scientists can determine the depth of water by measuring the time it takes for _____ to reflect off the ocean floor.
   - **a.** light
   - **c.** tsunamis
   - **e.** sound *(circled)*
   - **b.** radio waves
   - **d.** microwaves

5. Downslope movements of dense, sediment-laden ocean water are called _____ currents.
   - **a.** mud
   - **c.** turbidity *(circled)*
   - **e.** submarine
   - **b.** gravity
   - **d.** velocity

6. These ocean basin floor features are among the most level places on Earth.
   - **a.** seamounts
   - **c.** continental slopes
   - **e.** mid-ocean ridges
   - **b.** abyssal plains *(circled)*
   - **d.** deep-ocean trenches

7. The sites where moving plates plunge back into the mantle are _____
   - **a.** seamounts.
   - **c.** continental slopes.
   - **e.** mid-ocean ridges.
   - **b.** abyssal plains.
   - **d.** deep-ocean trenches. *(circled)*

8. Isolated volcanic peaks found on the ocean floor are called _____
   - **a.** seamounts. *(circled)*
   - **c.** continental slopes.
   - **e.** mid-ocean ridges.
   - **b.** abyssal plains.
   - **d.** deep-ocean trenches.

9. Which one of the following types of seafloor sediments includes calcareous and siliceous oozes?
   - **a.** fluvial
   - **c.** biogenous *(circled)*
   - **e.** lithogenous
   - **b.** terrigenous
   - **d.** hydrogenous

10. Sediments from the ocean floor are useful in analyzing Earth's _____ history.
    - **a.** population
    - **c.** industrial
    - **e.** recorded
    - **b.** climate *(circled)*
    - **d.** precambrian

Fill-In Questions

11. The ___North___ Hemisphere is often referred to as the "land hemisphere."

12. The zones that collectively make up the continental margin include the continental ___shelf___, continental ___slope___, and continental ___rise___.

13. Deep, long, and relatively narrow ocean basin floor features that are often paralleled by volcanic island arcs are deep-ocean ___trench___.

14. Charles Darwin developed a hypothesis explaining the formation of coral islands called ___atolls___.

15. The three broad categories of seafloor sediments are: ___terrigenous___ sediment, ___biogenous___ sediment, and ___hydrogenous___ sediment.

True/False Questions

16. The most abundant salt dissolved in seawater is sodium sulfate. _____F_____

17. High surface salinities are found in the subtropics because of the excessive precipitation that occurs in these regions. _____F_____

18. The continental shelf represents the true edge of the continent. _____F_____

19. Rift zones, located on the crests of mid-ocean ridges, are the active sites of seafloor spreading. _____T_____

20. Mud is the most common sediment found on the deep-ocean floor. _____T_____

Answers:

15. terrigenous, biogenous, hydrogenous; 16. F; 17. F; 18. F; 19. T; 20. T.
1. d; 2. a; 3. b; 4. e; 5. c; 6. b; 7. d; 8. a; 9. c; 10. b; 11. Northern; 12. shelf, slope, rise; 13. trenches; 14. atolls;

Web Work

The following are informative and interesting World Wide Web sites that address topics related to those presented in this chapter:

- Ocean Drilling Program
  http://www-odp.tamu.edu/

- Ocean Planet: Smithsonian
  http://seawifs.gsfc.nasa.gov/ocean_planet.html

For direct access to these sites and other relevant Web locations, contact the *Foundations of Earth Science* WWW home page at http://www.prenhall.com/lutgens.

CHAPTER 10
The Restless Ocean

FOCUS ON LEARNING

To assist you in learning the important concepts
in this chapter, you will find it helpful to focus
on the following questions:

1. What forces create and influence surface ocean currents?
2. What two factors are most significant in creating a dense mass of ocean water?
3. How are tides produced?
4. What factors determine the height, length, and period of a wave?

5. What are some typical shoreline features produced by wave erosion and from sediment deposited by beach drift and longshore currents?
6. What is the difference between a submergent and an emergent coast?

Drakes Bay, Point Reyes National Seashore, California. *Photo © by Carr Clifton Photography.*

The restless waters of the ocean are constantly in motion. Winds generate surface currents, the Moon and Sun produce tides, and density differences create deep-ocean circulation. Further, waves carry the energy from storms to distant shores, where their impact erodes the land. This chapter examines the movements of ocean waters and their importance (Figure 10.1).

Surface Currents

There is a river in the ocean. In the severest droughts it never fails, and in the mightiest floods…it never overflows. Its banks and its bottoms are of cold water, while its current is of warm. The Gulf of Mexico is its fountain, and its mouth is in the Arctic Sea. It is the Gulf Stream. There is in the world no other such majestic flow of waters.

This quote from Matthew Fontaine Maury's 1855 book, *The Physical Geography of the Sea,* describes perhaps the best-known and most-studied of the surface ocean currents, the Gulf Stream. This current is just a portion of a huge, slowly moving, circular whorl, or **gyre.** Figure 10.2 shows the entire gyre, which fills the North Atlantic, and other major gyres of the world ocean.

Ocean Circulation Patterns

Where the atmosphere and ocean are in contact, energy is passed from moving air to the water through friction. The drag exerted by winds blowing steadily across the ocean causes the surface layer of water to move. Thus, *winds are the primary driving force of surface currents.* This means there is a clear relationship between the oceanic circulation and the general atmospheric circulation. A comparison of Figures 10.2 and 13.13 (p. 338) illustrates this.

A further clue to the influence of winds on ocean circulation is provided by the currents in the northern Indian Ocean. Here, seasonal wind shifts known as the summer and winter monsoons occur. When the winds change direction, the surface currents also reverse direction.

Although wind is the force that generates surface ocean currents, other factors also influence these water movements. The most significant is the **Coriolis effect.** Because of Earth's rotation, currents are deflected to the right in the Northern Hemisphere and to the left in the Southern Hemisphere. (The Coriolis effect will be explained in Chapter 13, p. 330.) As a

Figure 10.1 Wind is not only responsible for creating waves, but it also provides the force that drives the ocean's surface circulation. *(Photo by David Muench Photography, Inc.)*

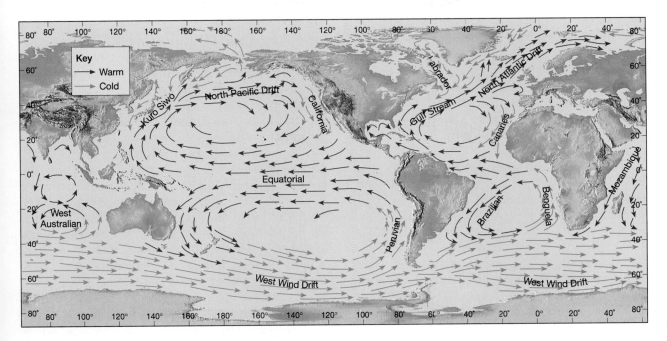

Figure 10.2 Average position and extent of the principal surface ocean currents.

consequence, the direction of surface currents does not coincide precisely with the wind direction. Rather, they are altered by Earth's rotation.

North and south of the equator are two westward-moving currents, the Equatorial Currents (Figure 10.2). They derive their energy principally from the trade winds that blow from the northeast and southeast, respectively, toward the equator. The equatorial currents can be thought of as the "backbone" of the system of ocean currents. Because of the Coriolis effect, these currents are deflected poleward to form clockwise gyres in the Northern Hemisphere and counterclockwise gyres in the Southern Hemisphere. These nearly circular patterns are found in each major ocean basin (Figure 10.2).

In the North Atlantic, the equatorial current is deflected northward through the Caribbean, where it becomes the Gulf Stream (Figure 10.2). As the Gulf Stream moves along the East Coast of the United States, it is strengthened by the prevailing westerly winds and is deflected to the east (to the right) between the Carolinas and New England. As it continues northeastward, it gradually widens and slows until it becomes a vast, slowly moving current known as the North Atlantic Drift (Figure 10.2).

As the North Atlantic Drift approaches Western Europe, it splits, with part moving northward past Great Britain, Norway, and Iceland, carrying heat to these otherwise chilly areas. The other part is deflected

southward as the cool Canaries Current. As the Canaries Current moves southward, it eventually merges into the North Equatorial Current, completing the gyre.

The clockwise circulation of the North Atlantic leaves a large central area that has no well-defined currents. This zone of calmer waters is known as the Sargasso Sea, named for the large quantities of *Sargassum,* a type of seaweed, encountered there.

In the South Atlantic, surface ocean circulation is very much the same as in the North Atlantic. And, circulation in the North and South Pacific generally parallels that of the North and South Atlantic (Figure 10.2).

Around the ice-covered continent of Antarctica, with no large landmasses in the way, cold surface waters circulate continuously. They move in response to prevailing westerly winds, giving this circulation the name West Wind Drift.

Ocean Currents and Upwelling

In addition to producing surface currents, winds may also cause vertical water movements. **Upwelling,** the rising of cold water from deeper layers to replace warmer surface water, is a common wind-induced vertical movement. It is most characteristic along the eastern shores of the oceans, most notably along the coasts of California, South America, and West Africa.

Upwelling occurs in these areas when winds blow toward the equator and parallel to the coast.

Owing to the Coriolis effect, the surface water movement is deflected away from the shore. As the surface layer moves away from the coast, it is replaced by water that rises (upwells) from below the surface. This slow upward flow from depths of 50 to 300 meters (165 to 1000 feet) brings water that is cooler than the original surface water and creates a characteristic zone of lower temperatures near the shore.

For swimmers who are accustomed to the warm waters along the mid-Atlantic shore of the United States, a swim in the Pacific off the coast of central California can be a chilling surprise. In August, when temperatures in the Atlantic are 21°C (70°F) or higher, central California's surf is only about 15°C (60°F). Coastal upwelling also brings to the ocean surface greater concentrations of dissolved nutrients, such as nitrates and phosphates. These nutrient-enriched waters from below promote the growth of plankton, which in turn supports extensive populations of fish.

The Importance of Ocean Currents

Anyone who navigates the oceans needs to be aware of currents. By understanding the direction and strength of ocean currents, sailors soon realize that their voyage time can be reduced if they travel with a current, or increased if they travel against a current.

Currents also have an important effect on climates. The moderating effect of poleward-moving warm ocean currents is well known. The North Atlantic Drift, an extension of the warm Gulf Stream, keeps Great Britain and much of northwestern Europe warmer during the winter than one would expect for their latitudes, which are similar to the latitudes of Alaska and Labrador. The prevailing westerly winds carry the moderating effects far inland. For example, Berlin (52 degrees north latitude) has an average January temperature similar to that experienced at New York City, which lies 12 degrees closer to the equator.

In contrast to warm ocean currents whose effects are felt most in the middle-to-high latitudes in winter, the influence of cold currents is most pronounced in the tropics or during summer months in the middle latitudes. Cold currents, such as the Benguela Current along Africa and the Peruvian Current (see Figure 10.2), moderate the tropical heat.

In addition to influencing temperatures of adjacent land areas, cold currents have other climatic influences. For example, where tropical deserts exist along the western coasts of continents, cold ocean currents have a dramatic impact. The principal west coast deserts are the Atacama in Peru and Chile, and the Namib in southern Africa (see Figure 4.19 p. 104). The aridity along these coasts is intensified because the lower air is chilled by cold offshore waters. When this occurs, the air becomes very stable and resists the upward movement necessary to create precipitation-producing clouds.

Moreover, the presence of cold currents causes surface air temperatures to approach and often reach the dew point (the temperature at which water vapor condenses). As a result, these areas are characterized by high relative humidities and much fog. Thus, not all tropical deserts are hot with low humidities and clear skies. Rather, the presence of cold currents transforms some tropical deserts into relatively cool, damp places that are often shrouded in fog.

Ocean currents also play a major role in maintaining Earth's heat balance. They accomplish this task by transferring heat from the tropics, where there is an excess of heat, to the polar regions, where a deficit exists (Figure 10.3). Ocean water movement accounts for about a quarter of this heat transport, and winds transport the remaining three-quarters.

Deep-Ocean Circulation

Ocean circulation is not confined to surface currents. Significant deep-water movements occur as well. Unlike the wind-induced movements of surface and near-surface waters, deep-ocean circulation is governed by gravity and driven by density differences.

Two factors—temperature and salinity—are most significant in creating a dense mass of water. Water that is cold and salty is denser than warmer, less salty water. Consequently, deep-ocean circulation is called **thermohaline circulation** (thermo—"heat," haline—"salt"). After sinking from the surface of the ocean, waters will not reappear at the surface for an average of 500 to 2000 years.

The basic idea of thermohaline circulation is easily understood. Water at the surface of the ocean is made colder (by heat loss to the atmosphere) and/or more salty (by removal of water by evaporation or freezing) and becomes more dense. This denser water then sinks toward the ocean bottom and displaces lighter (less dense) water, which moves back toward the zone where the denser water formed. In this manner, cold, dense waters flow away from their source near the poles and are replaced by warmer waters from lower latitudes.

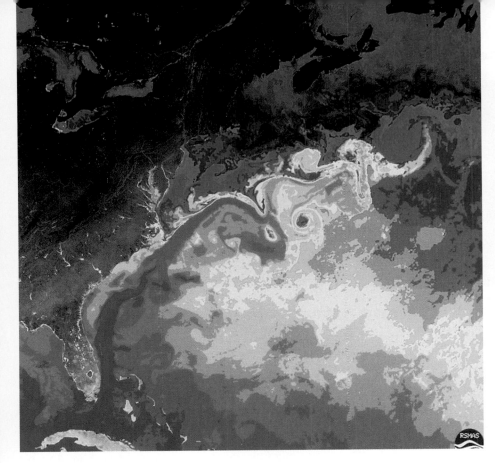

Figure 10.3 Satellite image of the Gulf Stream. Reds and yellows indicate warmer waters. The current transports heat from lower latitudes toward the North Pole. Meanders of the stream pinch off to form eddies that may move about the ocean for up to 2 years before dissipating. *(Courtesy of National Oceanic and Atmospheric Administration/ Seattle)*

Arctic and Antarctic waters represent the two major regions where dense water masses are created. Antarctic waters are chilled during the winter. The temperatures here are low enough to form sea ice, and because sea salts do not get frozen into the ice, the remaining water becomes saltier. The result is the densest water in all of the oceans (Figure 10.4). This cold saline brine slowly sinks to the sea floor, where it moves through the ocean basins in sluggish currents.

Centuries later, subsurface waters return to the sea surface. The return flow is accomplished by slow, upward movements of deep waters. Much of this return flow seems to be concentrated in upwelling along the equator and in coastal areas. Because water movements in the deep ocean are difficult to measure directly, relatively little is known about the details of deep-ocean circulation.

Tides

Tides are daily changes in the elevation of the ocean surface. Their rhythmic rise and fall along coastlines have been known since antiquity. Other than waves, they are the easiest ocean movements to observe. An exceptional example of extreme daily tides is shown in Figure 10.5.

Although known for centuries, tides were not explained satisfactorily until Sir Isaac Newton applied the law of gravitation to them. Newton showed that there is a mutual attractive force between two bodies, as between Earth and the

Figure 10.4 Sea ice in the Matha Strait, Antarctica. When seawater freezes, sea salts do not become part of the ice. Consequently, the salt content of the remaining seawater becomes more concentrated, which makes it denser and prone to sink. *(Photo by Kim Heacox/DRK Photo)*

Figure 10.5 High tide and low tide on Nova Scotia's Minas Basin in the Bay of Fundy. *(Courtesy of Nova Scotia Department of Tourism and Culture)*

Moon. Because the atmosphere and the ocean both are fluids, and free to move, both are deformed by this force. Hence, ocean tides result from the gravitational attraction exerted upon Earth by the Moon, and to a lesser extent by the Sun.

Causes of Tides

To illustrate how tides are produced, we will assume Earth is a rotating sphere covered to a uniform depth with water (Figure 10.6). It is easy to see how the Moon's gravitational force can cause the water to bulge on the side of Earth nearest the Moon. In addition, however, an equally large tidal bulge is produced on the side of Earth directly opposite the Moon.

Both tidal bulges are caused, as Newton discovered, by the pull of gravity. Gravity is inversely proportional to the square of the distance between two objects, meaning simply that it quickly weakens with distance. In this case, the two objects are the Moon and Earth. Because the force of gravity decreases with distance, the Moon's gravitational pull on Earth is slightly greater on the near side of Earth than on the far side. The result of this differential pulling is to stretch (elongate) the solid Earth, very slightly. In contrast, the world ocean, which is mobile, is deformed quite dramatically by this effect to produce the two opposing tidal bulges.

Because the position of the Moon changes only moderately in a single day, the tidal bulges remain in place while Earth rotates "through" them. For this reason, if you stand on the seashore for 24 hours, Earth will rotate you through alternating areas of deeper and shallower water. As you are

carried into regions of deeper water, the tide rises, and as you are carried away, the tide falls. Therefore, during one day you experience two high tides and two low tides.

Further, the tidal bulges migrate as the Moon revolves around Earth every 29 days. As a result, the tides, like the time of moonrise, shift about 50 minutes later each day. After 29 days, the cycle is complete and a new one begins.

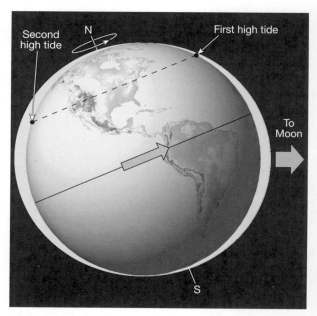

Figure 10.6 Tides on an Earth that is covered to a uniform depth with water. Depending on the Moon's position, tidal bulges may be inclined to the equator. In this situation an observer will experience two unequal high tides.

There may be an inequality between the high tides during a given day. Depending on the Moon's position, the tidal bulges may be inclined to the equator, as in Figure 10.6. This figure illustrates that the first high tide experienced by an observer in the Northern Hemisphere is considerably lower than the high tide half a day later. On the other hand, a Southern Hemisphere observer would experience the opposite effect.

Spring and Neap Tides

The Sun also influences tides. It is far larger than the Moon, but because the Sun is so far away, the effect is considerably less than that of the Moon. In fact, the tide-generating potential of the sun is slightly less than half that of the Moon.

Near the times of new and full moons, the Sun and Moon are aligned and their forces are added together (Figure 10.7A). Accordingly, the combined gravity of these two tide-producing bodies cause higher tidal bulges (high tides) and lower tidal troughs (low tides). These are called the *spring tides*. Spring tides create the largest daily tidal range—that is, the largest variation between high and low tides. Conversely, at about the time of the first and third quarters of the Moon, the gravitational forces of the Moon and Sun act on Earth at right angles, and each partially offsets the influence of the other (Figure 10.7B). As a result, the daily tidal range is less. These are called *neap tides*.

So far, we have explained the basic causes and patterns of tides. But keep in mind these theoretical considerations cannot be used to predict either the height or the time of actual tides at a particular place. The shape of coastlines and the configuration of ocean basins greatly influence the tides. Consequently, tides at various locations respond differently to the tide-producing forces. This being the case, the nature of the tide at any place can be determined most accurately by actual observation. The predictions in tidal tables and the tidal data on nautical charts are based on such observations.

Tidal Currents

Tidal current is the term used to describe the *horizontal* flow of water accompanying the rise and fall of the tide. These water movements induced by tidal forces can be important in some coastal areas. Tidal currents that advance into the coastal zone as the tide rises are called *flood currents*. As the tide falls, seaward-moving water generates *ebb currents*.

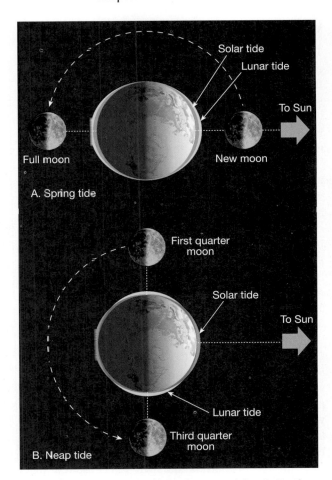

Figure 10.7 Relationship of the Moon and Sun to Earth during **A.** spring tides and **B.** neap tides.

Periods of little or no current, called *slack water,* separate flood and ebb. The areas affected by these alternating tidal currents are called **tidal flats.** Depending on the nature of the coastal zone, tidal flats vary from narrow strips seaward of the beach to zones that may extend for several kilometers.

Although tidal currents are not important in the open sea, they can be rapid in bays, river estuaries, straits, and other narrow places. Off the coast of Brittany in France, for example, tidal currents that accompany a high tide of 12 meters (40 feet) may attain a speed of 20 kilometers (12 miles) per hour. Tidal currents are not generally considered to be major agents of erosion and sediment transport, yet notable exceptions occur where tides move through narrow inlets. Here they scour the small entrances to many good harbors that would otherwise be blocked.

Waves Modify the Shoreline

The waters of the ocean are constantly in motion. The restless nature of the water is most noticeable along the shore—the dynamic interface between land and sea. Here we can observe the rhythmic rise and fall of tides and see waves constantly rolling in and breaking. Sometimes the waves are low and gentle. At other times, they pound the shore with an awesome fury.

Although it may not be obvious to occasional visitors, the shoreline is constantly being modified by waves. For example, south of Boston at Cape Cod, Massachusetts, wave activity is causing cliffs of poorly consolidated glacial sediment to retreat at nearly 1 meter per year. By contrast, north of San Francisco at Point Reyes, California, the durable bedrock cliffs are far less susceptible to wave attack and therefore are retreating much more slowly. Along both coasts, wave activity is moving sediment near the shore and building narrow sand bars that protrude across some bays.

The nature of present-day shorelines is not just the result of the relentless attack of the land by the sea. The shore is a complex zone whose unique character results from many geologic processes. For example, practically all coastal areas were affected by the worldwide rise in sea level that accompanied the melting of glaciers at the close of the Pleistocene epoch. As the sea edged landward, the shoreline became superimposed upon landscapes that resulted from such diverse processes as stream erosion, glaciation, volcanic activity, and the forces of mountain building.

Wind-generated waves provide most of the energy that shapes and modifies shorelines. Where the land and sea meet, waves that may have traveled unimpeded for hundreds of kilometers suddenly encounter a barrier that will not allow them to advance farther and which must absorb their energy. Stated another way, the shore is where a practically irresistible force confronts an almost immovable object. The conflict that results is never-ending and sometimes dramatic.

Characteristics of Waves

The undulations of the water surface, called waves, derive their energy and motion from the wind. If a breeze of less than 3 kilometers (2 miles) per hour starts to blow across still water, small wavelets appear almost instantly. When the breeze dies, the ripples disappear as suddenly as they formed. However, if the wind exceeds 3 kilometers per hour, more stable waves gradually form and progress with the wind.

All waves have the characteristics illustrated in Figure 10.8. The tops of the waves are the *crests,* which are separated by *troughs.* The vertical distance between trough and crest is called the **wave height,** and the horizontal distance separating successive crests is the **wavelength.** The **wave period** is the time interval between the passage of successive crests at a stationary point. The height, length, and period that are eventually achieved by a wave depend on three factors: (1) wind speed; (2) length of time the wind has blown; and (3) **fetch,** the distance that the

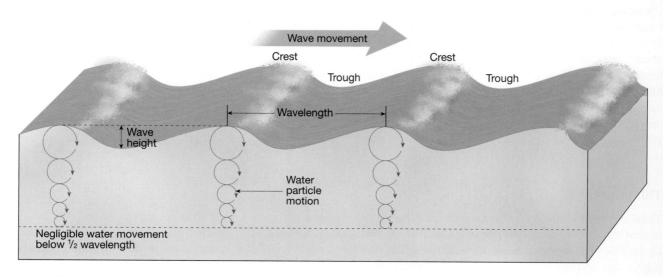

Figure 10.8 This diagram illustrates the basic parts of a wave as well as the movement of water particles with the passage of a wave. Negligible water movement occurs below a depth equal to one-half the wavelength (the level of the dashed line).

wind has traveled across open water. As the quantity of energy transferred from the wind to the water increases, the height and steepness of the waves increase as well. Eventually, a critical point is reached and ocean breakers called *whitecaps* form.

When the wind stops or changes direction, or the waves leave the stormy area where they were created, they continue on without relation to local winds. The waves also undergo a gradual change to *swells,* which are lower in height and longer in length and may carry a storm's energy to distant shores. Because many independent wave systems exist at the same time, the sea surface acquires a complex and irregular pattern. Hence, the sea waves we watch from the shore are usually a mixture of swells from faraway storms and waves created by local winds.

Types of Waves

It is important to realize that in the open sea the motion of the wave is different from the motion of the water particles within it. It is the wave form or shape that moves forward, not the water itself, despite the way it looks. Waves of *energy* simply move through the nearly stationary medium of the water. Each water particle moves in a nearly circular path during the passage of a wave (Figure 10.9). As a wave passes, a water particle returns almost to its original position. When water is part of the wave crest, it moves in the same direction as the advancing wave form. When it is in the trough, the water actually moves backward. This is demonstrated by observing the behavior of a floating cork as a wave passes. The cork merely seems to bob up and down and sway slightly back and forth without advancing appreciably from its original position. (The wind does drag the water forward slightly, causing the surface circulation of the oceans.) For this reason, waves in the open sea are called **waves of oscillation.**

The energy contributed by the wind to the water is transmitted not only along the surface of the sea but also downward. However, beneath the surface, the circular motion rapidly diminishes until, at a depth equal to about one-half the wavelength, the movement of water particles becomes negligible. This is shown by the rapidly diminishing diameters of water-particle orbits in Figure 10.8.

As long as a wave is in deep water it advances smoothly across the ocean (Figure 10.10, left). However, when a wave approaches the shore, the water becomes shallower and influences wave behavior. The wave begins to "feel bottom" at a water

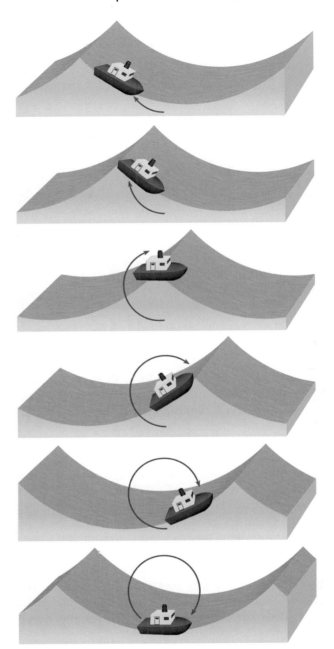

Figure 10.9 The movements of the toy boat show that the wave form advances, but the water does not advance appreciably from its original position. In this sequence, the wave moves from left to right as the boat (and the water in which it is floating) rotates in an imaginary circle. The boat moves slightly to the left up the front of the approaching wave, then after reaching the crest, slides to the right down the back of the wave.

depth equal to about one-half its wavelength. Such depths interfere with water movement at the base of the wave and slow its advance (Figure 10.10, center).

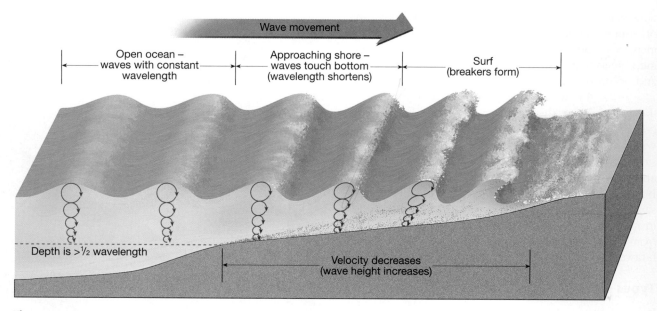

Figure 10.10 Changes that occur when a wave moves onto shore.

As a wave moves toward the shore, the slightly faster seaward waves catch up, decreasing the wavelength. As the speed and length of the wave diminish, the energy must go somewhere, and the only way to go is up, so the wave steadily grows higher. Finally a critical point is reached when the steep wave front collapses or *breaks* (Figure 10.10, right). What had been a wave of oscillation now becomes a **wave of translation** in which the water advances up the shore.

The turbulent water created by breaking waves is called **surf.** On the landward margin of the surf zone, the turbulent sheet of water from collapsing breakers, called *swash,* moves up the slope of the beach. When the energy of the swash has been expended, the water flows back down the beach toward the surf zone as *backwash*.

Wave Erosion

During periods of calm weather, wave action is at a minimum. However, just as streams do most of their work during floods, so do waves perform most of their work during storms. The impact of high, storm-induced waves against the shore can be awesome in its violence. Each breaking wave might hurl thousands of tons of water against the land, sometimes causing the ground literally to tremble.

It is no wonder that cracks and crevices are quickly opened in cliffs, seawalls, breakwaters, and anything else that is subjected to these enormous shocks. Water is forced into every opening, causing air in the cracks to become highly compressed by the thrust of crashing waves. When the wave subsides, the air expands rapidly, dislodging rock fragments and enlarging and extending preexisting fractures (Figure 10.11A).

In addition to the erosion caused by wave impact and pressure, **abrasion,** the sawing and grinding action of the water armed with rock fragments, is also important. In fact, abrasion is probably more intense in the surf zone than in any other environment. Smooth and rounded stones and pebbles along the shore are obvious reminders of the relentless grinding action of rock against rock in the surf zone. Further, such fragments are used as "tools" by the waves as they cut horizontally into the land (Figure 10.11B).

Wave Refraction

The bending of waves, called **wave refraction,** plays an important part in shoreline processes. It affects the distribution of energy along the shore, and thus strongly influences where and to what degree erosion, sediment transport, and deposition will take place.

A.   **B.**

Figure 10.11 A. When waves break against the shore, the force of the water can be powerful and the erosional work that is accomplished can be great. *(Photo by David Muench Photography, Inc.)* **B.** Cliff undercut by wave erosion along the Oregon coast. *(Photo by E. J. Tarbuck)*

Waves seldom approach the shore straight on. Rather, most waves move toward the shore at an angle. When they reach the shallow water of a smoothly sloping bottom, however, they are bent and tend to become parallel to the shore. Such bending occurs because the part of the wave nearest the shore touches bottom and slows first, whereas the end that is still in deeper water continues forward at its full speed. The net result is a wave front that may approach nearly parallel to the shore regardless of the original direction of the wave.

Because of refraction, wave impact is concentrated against the sides and ends of headlands that project into the water, whereas wave attack is weakened in bays. This differential wave attack along irregular coastlines is illustrated in Figure 10.12. Because the waves reach the shallow water in front of the headland sooner than they do in adjacent bays, they are bent more nearly parallel to the protruding land and strike it from all three sides. By contrast, refraction in the bays causes waves to diverge and expend less energy. In these zones of weakened wave activity, sediments can accumulate and form sheltered sandy beaches. Over a long period, erosion of the headlands and deposition in the bays will straighten an irregular shoreline.

Moving Sand Along the Beach

Although waves are refracted, most still reach the shore at an angle, however slight. Consequently, the uprush of water from each breaking wave (the swash) is not head-on, but oblique. However, the backwash is straight down the slope of the beach. The effect of this pattern of water movement is to transport particles of sediment in a zigzag pattern along the beach (Figure 10.13). This movement, called **beach drift,** can transport sand and pebbles hundreds or even thousands of meters each day.

Oblique waves also produce currents within the surf zone that flow parallel to the shore. Because the water here is turbulent, these **longshore currents** easily move the fine suspended sand, and roll larger sand and gravel along the bottom. When the sediment transported by longshore currents is added to the quantity moved by beach drift, the total amount can be very large. At Sandy Hook, New Jersey, for example, the quantity of sand transported along the shore over a 48-year period averaged almost 750,000 tons annually. For a 10-year period at Oxnard, California, more than 1.5 million tons of sediment moved along the shore each year.

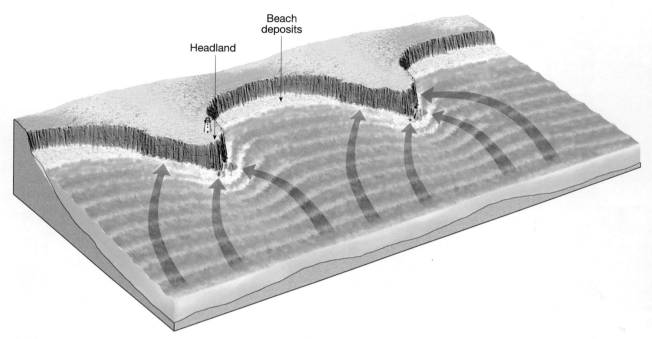

Figure 10.12 Because of wave refraction, the greatest erosional power is concentrated on the headlands. In the bays, the force of the waves is much weaker.

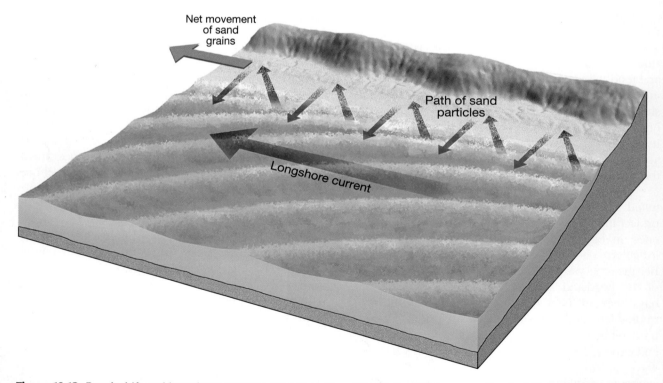

Figure 10.13 Beach drift and longshore currents are created by obliquely breaking waves. Beach drift occurs as incoming waves carry sand obliquely up the beach, while the water from spent waves carries it directly down the slope of the beach. Similar movements occur offshore in the surf zone to create the longshore current. These processes transport large quantities of material along the beach and in the surf zone.

There should be little wonder that beaches have been characterized as "rivers of sand." At any point along a beach there is likely to be more sediment that was derived elsewhere than material eroded from the land immediately behind it. It is also worth noting that much of the sediment making up beaches is not wave-eroded debris. Rather, in many areas sediment-laden rivers that discharge into the ocean are the major source of material. For that reason, if it were not for beach drift and longshore currents, many beaches would be nearly sandless.

Shoreline Features

Shoreline features vary, depending on the rocks of the shore, currents, wave intensity, and whether the coast is stable, sinking, or rising. This section summarizes many shoreline features, and several are shown in Figure 10.17 on p. 262.

Wave-Cut Cliffs and Platforms

Whether along the rugged and irregular New England coast or along the steep shorelines of the West Coast, the effects of wave erosion are often easily seen.

Wave-cut cliffs, as their name implies, originate by the cutting action of the surf against the base of coastal land. As erosion progresses, rocks overhanging the notch at the base of the cliff crumble into the surf and the cliff retreats. A relatively flat, benchlike surface, the **wave-cut platform,** is left behind by the receding cliff (Figure 10.14). The platform broadens as wave attack continues. Some debris produced by the breaking waves remains along the water's edge as part of the beach, while the remainder is transported farther seaward.

Arches, Stacks, Spits, and Bars

As explained, headlands that extend into the sea are vigorously attacked by waves because of refraction. The surf erodes the rock selectively, wearing away the softer or more highly fractured rock at the fastest rate. At first, sea caves may form. When two caves on opposite sides of a headland unite, a **sea arch** results (Figure 10.15). Finally the arch falls in, leaving an isolated remnant, or **sea stack,** on the wave-cut platform (Figure 10.15). Eventually it too will be consumed by the action of the waves.

Figure 10.14 Elevated wave-cut platform along the California coast north of San Francisco. A new platform is being created at the base of the cliff. *(Photo by John S. Shelton)*

Figure 10.15 Sea stack (left) and sea arches along the rugged coastline of Washington State. *(Photo by David Muench Photography, Inc.)*

Where beach drift and longshore currents are active, several features related to the movement of sediment along the shore may develop. **Spits** are elongated ridges of sand that project from the land into the mouth of an adjacent bay. Often the end in the water hooks landward in response to wave-generated currents. In Figure 10.16, Sandy Hook is a prominent example of such a feature. The term **baymouth bar** is applied to a sand bar that completely crosses a bay, sealing it off from the open ocean. Such a feature tends to form across bays where currents are weak, allowing a spit to extend to the other side. A **tombolo,** which is a ridge of sand that connects an island to the mainland or to another island, forms in much the same manner as does a spit.

Barrier Islands

The Atlantic and Gulf Coastal Plains are relatively flat and slope gently seaward. The shore zone is characterized by **barrier islands.** These low ridges of sand parallel the coast at distances from 3 to 30 kilometers offshore. From Cape Cod, Massachusetts, to Padre Island, Texas, nearly 300 barrier islands rim the coast (Figure 10.16).

Most barrier islands are from 1 to 5 kilometers wide and between 15 and 30 kilometers long. The highest features are sand dunes, which usually reach heights of 5 to 10 meters. The lagoons that separate these narrow islands from the shore are zones of relatively quiet water that allow small craft traveling between New York and northern Florida to avoid the rough waters of the North Atlantic.

Barrier islands probably form in several ways. Some originated as spits that were subsequently severed from the mainland by wave erosion or by the general rise in sea level following the last episode of glaciation. Others are created when turbulent waters in the line of breakers heap up sand that has been scoured from the bottom. Also, some barrier islands may be former sand dune ridges that originated along the shore during the last glacial period, when sea level was lower. As the ice sheets melted, sea level rose and flooded the area behind the beach-dune complex.

The Evolving Shore

A shoreline continually undergoes modification regardless of its initial configuration. At first most coastlines are irregular, although the degree of and

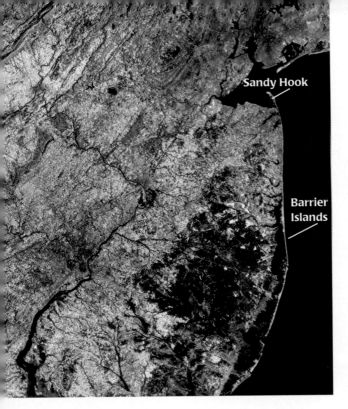

Figure 10.16 This satellite image of New Jersey shows a large spit, known as Sandy Hook, extending north from the coast into the bay. Across the bay is New York City. Farther south, barrier islands parallel the coast. Atlantic City is located on one of these islands. *(LANDSAT image courtesy of Phillips Petroleum Company, Exploration Projects Section)*

reason for the irregularity may differ considerably from place to place. Along a coastline of varied geology, the pounding surf may at first increase its irregularity because the waves will erode the weaker rocks more easily than the stronger ones. However, if a shoreline remains stable, marine erosion and deposition will eventually produce a straighter, more regular coast. Figure 10.17 illustrates the evolution of an initially irregular coast. As waves erode the headlands, creating cliffs and a wave-cut platform, sediment is carried along the shore. Some material is deposited in the bays, while other debris is formed into spits and baymouth bars. At the same time rivers fill the bays with sediment. Ultimately, a generally straight, smooth coast results.

Shoreline Erosion Problems

Compared with other natural hazards such as earthquakes, volcanic eruptions, and landslides, shoreline erosion appears to be a more continuous and predictable process that causes relatively modest damage to limited areas. In reality, the shoreline is a dynamic place that can change rapidly in response to natural forces. Exceptional storms are capable of eroding beaches and cliffs at rates that far exceed the long-term average. Such bursts of accelerated erosion not only have a significant impact on the natural evolution of a coast, but can also have a profound impact on people who reside in the coastal zone. Erosion along our coasts causes significant property damage. Large sums are spent annually not only to repair damage, but also to control erosion. Already a problem at many sites, shoreline erosion is certain to become increasingly serious as extensive coastal development continues.

Although the same processes cause change along every coast, not all coasts respond in the same way. Interactions among different processes and the relative importance of each process depend on local factors. These factors include: (1) the proximity of a coast to sediment-laden rivers, (2) the degree of tectonic activity, (3) the topography and composition of the land, (4) prevailing winds and weather patterns, and (5) the configuration of the coastline and nearshore areas.

During the past 100 years, growing affluence and increasing demands for recreation have brought unprecedented development to many coastal areas. Both the number and the value of buildings have increased; so too have efforts to protect property from storm waves. Also, controlling the natural migration of sand is an ongoing struggle in many coastal areas. Such interference can result in unwanted changes that are difficult and expensive to correct.

Groins

To maintain or widen beaches that are losing sand, **groins** are sometimes constructed. A groin is a barrier built at a right angle to the beach to trap sand that is moving parallel to the shore (Figure 10.18). These structures often do their job so effectively that the longshore current beyond the groin becomes sand-starved. As a result, the current erodes sand from the beach on the leeward side of the groin.

To offset this effect, property owners downcurrent from the structure may erect a groin on their property. In this manner, the number of groins multiplies. An example of such proliferation is the shoreline of New Jersey, where hundreds of these structures have been built (Figure 10.18). Because it has been shown that groins often do not provide a satisfactory solution, they are no longer the preferred method of keeping beach erosion in check.

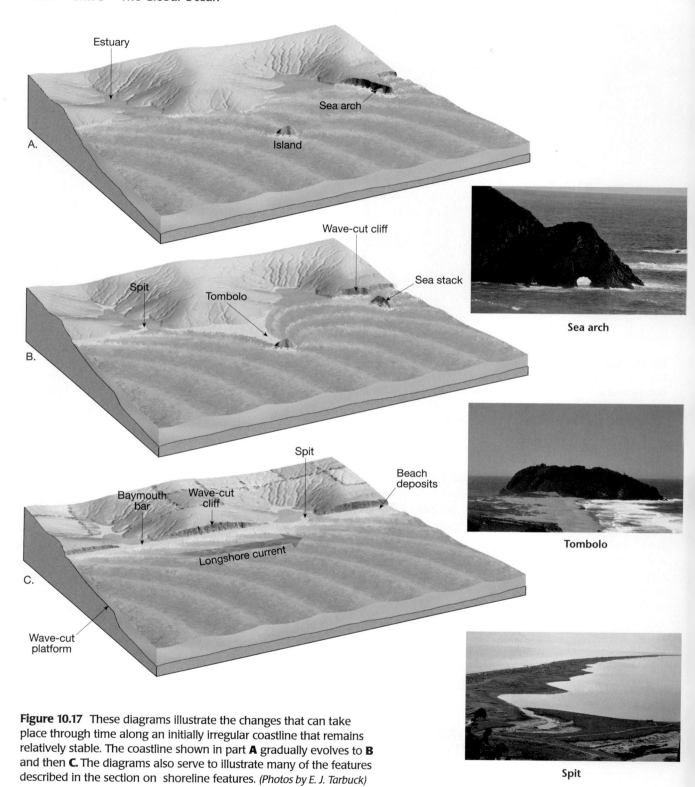

Figure 10.17 These diagrams illustrate the changes that can take place through time along an initially irregular coastline that remains relatively stable. The coastline shown in part **A** gradually evolves to **B** and then **C**. The diagrams also serve to illustrate many of the features described in the section on shoreline features. *(Photos by E. J. Tarbuck)*

Seawalls

As development has moved ever closer to the beach, seawalls are sometimes built to defend property from the force of breaking waves. **Seawalls** are simply massive barriers intended to prevent waves from reaching the areas behind the wall. Waves expend much of their energy as they move across an open beach. Seawalls cut this process short by reflecting the force of unspent waves seaward. As a consequence, the beach to the seaward side of the seawall experiences significant erosion and may, in some instances, be eliminated entirely. Once the width of the beach is reduced, the seawall is subjected to even greater pounding by the waves. Eventually this battering will cause the wall to fail and a larger, more expensive wall must be built to take its place.

The wisdom of building temporary protective structures along shorelines is increasingly questioned. The feeling of many coastal scientists is expressed in the following excerpt from a position paper that grew out of a conference on America's Eroding Shoreline:

> It is now clear that halting the receding shoreline with protective structures benefits only a few and seriously degrades or destroys the natural beach and the value it holds for the majority. Protective structures divert the ocean's energy temporarily from private properties, but usually refocus that energy on the adjacent natural beaches. Many interrupt the natural sand flow in coastal currents, robbing many beaches of vital sand replacement.*

Beach Nourishment

Beach nourishment represents another approach to stabilizing shoreline sands. As the term implies, this practice simply involves the addition of large quantities of sand to the beach system. By building the beaches seaward, beach quality and storm protection are both improved. Beach nourishment, however, is not a permanent solution to the problem of shrinking beaches. The same processes that removed the sand in the first place will eventually remove the replacement sand as well. In addition, beach nourishment is very expensive. When beach nourishment was used to renew 24 kilometers of Miami Beach, the cost was $64 million. Here the restoration must be redone every 10 to 12 years.

*"Strategy for Beach Preservation Proposed," *Geotimes* 30 (12) (December 1985):15.

Figure 10.18 A series of groins at Ship Bottom, New Jersey. Because groins trap sand on the upcurrent side, the movement of sand along this coast, caused by the longshore current, must be toward the bottom of the photograph. *(Photo by John S. Shelton)*

In some instances, beach nourishment can lead to unwanted environmental effects. For example, beach replenishment at Waikiki Beach, Hawaii, involved replacing coarse calcareous sand with softer, muddier calcareous sand. Destruction of the soft beach sand by breaking waves increased the water's turbidity and killed offshore coral reefs. At Miami Beach, increased turbidity also damaged local coral communities.

Beach nourishment appears to be an economically viable long-range solution to the beach preservation problem only in areas where there exists dense development, large supplies of sand, relatively low wave energy, and reconcilable environmental issues. Unfortunately, few areas possess all these attributes.

Abandonment and Relocation

So far, two basic responses to shoreline erosion problems have been considered: (1) the building of structures, such as groins and seawalls, to hold the shoreline in place; and (2) the addition of sand to replenish eroding beaches. However, a third option is also available. Many coastal scientists and planners are calling for a policy shift from defending and rebuilding beaches and coastal property in high hazard areas to removing storm-damaged buildings in those places and letting nature reclaim the beach. This approach is similar to that adopted by the federal government for river floodplains following the devastating 1993 Mississippi River floods in which vulnerable structures are abandoned and relocated on higher, safer ground. Such ideas will no doubt be the focus of much study and debate as states and communities evaluate and revise coastal land-use policies.

Emergent and Submergent Coasts

The great variety of present-day shorelines suggests that they are complex areas. To understand the nature of any particular coastal area, many factors must be considered, including rock types, size and direction of waves, number of storms, tidal range, and submarine profile. Moreover, recent tectonic events and changes in sea level must also be taken into account. These many variables make shoreline classification difficult.

Many geologists classify coasts based on changes that have occurred with respect to sea level. This commonly used classification divides coasts into two very general categories: emergent and submergent. **Emergent coasts** develop either because an area

experiences uplift or as a result of a drop in sea level. Conversely, **submergent coasts** are created when sea level rises or the land adjacent to the sea subsides.

In some areas the coast is clearly emergent because either rising land or a falling water level exposes wave-cut cliffs and platforms above sea level. Excellent examples include portions of coastal California where uplift has occurred in the recent geological past. The elevated wave-cut platform shown in Figure 10.14 (p. 259) illustrates this. In the case of the

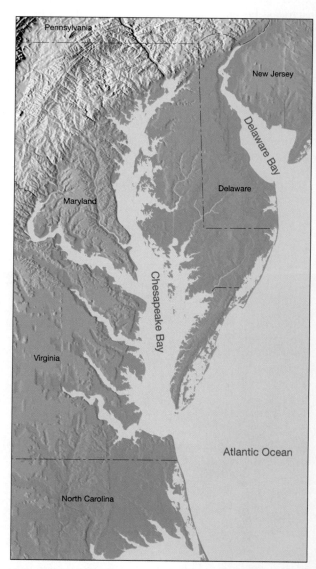

Figure 10.19 Estuaries along the East Coast of the United States. The lower portions of many river valleys were submerged by the rise in sea level that followed the end of the Ice Age. Chesapeake Bay and Delaware Bay are especially prominent examples.

Palos Verdes Hills, south of Los Angeles, seven different terrace levels exist, indicating seven episodes of uplift. The ever-persistent sea is now cutting a new platform at the base of the cliff. If uplift follows, it too will become an elevated marine terrace.

Other examples of emergent coasts include regions that were once buried beneath great ice sheets. When glaciers were present, their weight depressed the crust, and when the ice melted, the crust began to spring back gradually. As a result, prehistoric shoreline features today are found high above sea level. The Hudson Bay region of Canada is one such area, portions of which are still rising at a rate of more than 1 centimeter per year.

In contrast to the preceding examples, other coastal areas show definite signs of submergence. The shoreline of a coast that has been submerged in the relatively recent past is often highly irregular because the sea typically floods the lower reaches of river valleys flowing into the ocean. The ridges separating the valleys, however, remain above sea level and project into the sea as headlands. These drowned river mouths, which are called **estuaries,** characterize many coasts today. Along the Atlantic coast, the Chesapeake and Delaware bays are examples of estuaries created by submergence (Figure 10.19). The picturesque coast of Maine, particularly in the vicinity of Acadia National Park, is another excellent example of an area that was flooded by the postglacial rise in sea level and transformed into a highly irregular submerged coastline.

Keep in mind that most coasts have a complicated geologic history. With respect to sea level, many coasts have at various times emerged and then submerged again. Each time they retain some of the features created during the previous situation.

The Chapter in Review

1. *Surface ocean currents* are parts of huge, slowly moving, circular whirls, or *gyres,* that begin near the equator in each ocean. *Wind is the driving force* for the ocean's surface currents. Where wind is in contact with the ocean, it passes energy to the water through friction and causes the surface layer to move. The most significant factor other than wind that influences the movement of ocean waters is the *Coriolis effect,* the deflective effect of Earth's rotation that causes free-moving objects to be deflected to the right in the Northern Hemisphere and to the left in the Southern Hemisphere. Because of the Coriolis effect, surface currents form clockwise gyres in the Northern Hemisphere and counterclockwise gyres in the Southern Hemisphere.

2. Ocean currents are important in navigation and travel and for the effect that they have on climates. The moderating effect of poleward-moving warm ocean currents during the winter in middle latitudes is well known. Cold currents depress air temperatures and cause increased fog and lower rainfall totals.

3. In contrast to surface currents, *deep-ocean circulation* is governed by gravity and driven by *density* differences. The two factors that are most influential in creating a dense mass of water are colder *temperature* and higher *salinity.*

4. *Tides,* the daily rise and fall in the elevation of the ocean surface at a specific location, are caused by the *gravitational attraction* of the Moon and, to a lesser extent, by the Sun. Near the times of new and full moons, the Sun and Moon are aligned and their gravitational forces are added together to produce especially high tides. These are called the *spring tides.* Conversely, at about the times of the first and third quarters of the moon, when the gravitational forces of the Moon and Sun are at right angles, the daily tidal range is less. These are called *neap tides.*

5. The three factors that influence the *height, wavelength,* and *period* of a wave are (a) *wind speed,* (b) *length of time the wind has blown,* and (c) *fetch,* the distance that the wind has traveled across the open water.

6. The two types of wind-generated waves are (a) *waves of oscillation,* waves in the open sea, in which the wave form advances as the water particles move in circular orbits, and (b) *waves of translation,* the turbulent advance of water formed near the shore as waves of oscillation collapse, or *break,* and form *surf.*

7. Features produced by *shoreline erosion* include: *wave-cut cliffs* (which originate from the cutting action of the surf against the base of coastal land), *wave-cut platforms* (relatively flat, benchlike surfaces left behind by receding cliffs), *sea arches* (formed when a headland is eroded and two caves from opposite sides unite), and *sea stacks* (formed when the roof of a sea arch collapses).

8. Some of the features formed when sediment is moved by *beach drift* and *longshore currents* are: *spits* (narrow ridges of sand that project from the land into the mouth of an adjacent bay), *baymouth bars* (sand bars that completely cross a bay), and *tombolos* (ridges of sand that connect an island to the mainland or to another island).

9. Local factors that influence shoreline erosion are (a) the proximity of a coast to sediment-laden rivers, (b) the degree of tectonic activity, (c) the topography and composition of the land, (d) prevailing winds and weather patterns, and (e) the configuration of the coastline and nearshore areas.

10. Three basic responses by people to shoreline erosion are (a) building *structures* such as *groins* (short walls built at a right angle to the shore to trap moving sand) and *seawalls* (barriers constructed to prevent waves from reaching the area behind the wall) to hold the shoreline in place, (b) *beach nourishment,* which involves the addition of sand to replenish eroding beaches, and (c) *relocating* buildings away from the beach.

11. One frequently used classification of coasts is based on changes that have occurred with respect to sea level. *Emergent coasts,* often with wave-cut cliffs and wave-cut platforms above sea level, develop either because an area experiences uplift or as a result of a drop in sea level. Conversely, *submergent coasts,* with their drowned river mouths, called *estuaries,* are created when sea level rises or the land adjacent to the sea subsides.

Key Terms

abrasion (p. 256)
barrier island (p. 260)
baymouth bar (p. 260)
beach drift (p. 257)
beach nourishment (p. 263)
Coriolis effect (p. 248)
emergent coast (p. 264)
estuary (p. 265)

fetch (p. 254)
groin (p. 261)
gyre (p. 248)
longshore current (p. 257)
sea arch (p. 259)
sea stack (p. 259)
seawall (p. 263)
spit (p. 260)

submergent coast (p. 264)
surf (p. 256)
thermohaline circulation (p. 250)
tidal current (p. 253)
tidal flat (p. 253)
tide (p. 251)
tombolo (p. 260)
upwelling (p. 249)

wave-cut cliff (p. 259)
wave-cut platform (p. 259)
wave height (p. 254)
wavelength (p. 254)
wave of oscillation (p. 255)
wave of translation (p. 256)
wave period (p. 254)
wave refraction (p. 256)

Questions for Review

1. What is the primary driving force of surface ocean currents? How does the Coriolis effect influence these currents?
2. Describe the process of coastal upwelling.
3. How do ocean currents influence climate?
4. What is the driving force of deep-ocean circulation?
5. Discuss the cause of ocean tides.
6. Explain why an observer can experience two unequal high tides during a day (see Figure 10.6, p. 252).
7. How does the Sun influence tides?
8. Distinguish between flood current and ebb current.
9. List three factors that determine the height, length, and period of a wave.
10. Describe the motion of a water particle as a wave passes (see Figure 10.8, p. 254).
11. Explain what happens when a wave breaks.
12. List two ways that waves erode the land.
13. What is wave refraction? What is the effect of this process along irregular coastlines?
14. Why are beaches often called "rivers of sand"?
15. Describe the formation of the following features: wave-cut cliff, wave-cut platform, sea stack, spit, baymouth bar, tombolo.
16. List three ways that barrier islands may form.

17. For what purpose is a groin built? Why might the building of one groin lead to the building of others?
18. How might a seawall increase beach erosion?
19. What observable features would lead you to classify a coastal area as emergent?
20. Are estuaries associated with submergent or emergent coasts? Why?

Testing What You Have Learned

Multiple-Choice Questions

1. Surface ocean currents derive their energy from _____
   - **a.** salinity.
   - **c.** tectonics.
   - **e.** gravity.
   - **b.** wind.
   - **d.** tsunamis.

2. The vertical movement of cold water from deeper layers to replace warmer surface waters is called _____
   - **a.** turbidity.
   - **c.** upwelling.
   - **e.** halocline.
   - **b.** emergence.
   - **d.** rising.

3. Because temperature and salinity are most significant in creating a dense mass of seawater, deep-ocean circulation is called _____ circulation.
   - **a.** halocline
   - **c.** thermocline
   - **e.** thermohaline
   - **b.** tempsal
   - **d.** thermosal

4. Especially high and low tides that occur near the times of new and full moons are called _____ tides.
   - **a.** spring
   - **c.** neap
   - **e.** super
   - **b.** fall
   - **d.** diurnal

5. Waves derive their energy and motion from _____
   - **a.** salinity
   - **c.** tectonics.
   - **e.** gravity.
   - **b.** wind.
   - **d.** tsunamis.

6. The vertical distance from a wave's trough to its crest is termed.
   - **a.** fetch
   - **c.** wave period
   - **e.** wavelength
   - **b.** wavelength
   - **d.** gyre

7. A deep-water wave begins to "feel bottom" at a water depth equal to about _____ its wavelength.
   - **a.** one-fourth
   - **c.** one-half
   - **e.** twice
   - **b.** one-third
   - **d.** two-thirds

8. Which one of the following is NOT a feature produced by wave erosion?
   - **a.** sea arch
   - **c.** spit
   - **e.** wave-cut platform
   - **b.** sea stack
   - **d.** wave-cut cliff

9. The Atlantic and Gulf coasts are characterized by low ridges of sand that parallel the coast at distances of 3 to 30 kilometers offshore, called _____
   - **a.** tombolos.
   - **c.** sea stacks.
   - **e.** baymouth bars.
   - **b.** dune arcs.
   - **d.** barrier islands.

10. Drowned river mouths associated with a submergent coast are called _____
    - **a.** estuaries.
    - **c.** sea stacks.
    - **e.** tombolos.
    - **b.** spits.
    - **d.** sea zones.

Fill-In Questions

11. Because of Earth's rotation, the _____ _____ causes surface ocean currents to be deflected to the right in the Northern Hemisphere and to the left in the Southern Hemisphere.

12. At about the times of the first and third quarters of the Moon, _neap_ _____ tides produce lower daily tidal ranges.

13. The turbulent water created by breaking waves is called _surf_ _____.

14. The bending of waves, called wave _refraction_ _____, along a coast influences where and to what degree erosion, sediment transport, and deposition will take place.

15. Based on changes that have occurred with respect to sea level, geologists often classify coasts into two categories: _____ coasts and _____ coasts.

emergent + submergent

True/False Questions

16. When sea ice forms, salts are excluded from the ice and the remaining water becomes saltier. _T_ _____

17. Because the Sun is much larger than the Moon, its gravity influences the tides more than the moon's gravity. _F_ _____

18. _Run_ is the term used to describe the distance traveled by wind across the open water. _F_ _____

19. In the open sea, it is the wave form that moves forward, not the water itself. _T_ _____

20. If an irregular shoreline remains stable, marine erosion and deposition will eventually produce a straighter, more regular coast. _T_ _____

Answers:

1. b; 2. c; 3. e; 4. a; 5. b; 6. e; 7. c; 8. c; 9. d; 10. a; 11. Coriolis effect; 12. neap; 13. surf; 14. refraction; 15. emergent, submergent; 16. T; 17. F; 18. F; 19. T; 20. T.

Web Work

The following are informative and interesting World Wide Web sites that address topics related to those presented in this chapter:

- Coastal Briefs
  http://www.whoi.edu/coastal-briefs/table-of-contents.html

- Tsunami
  http://www.geophys.washington.edu/tsunami/welcome.html

For direct access to these sites and other relevant Web locations, contact the *Foundations of Earth Science* WWW home page at http://www.prenhall.com/lutgens.

UNIT SIX
The Atmosphere

Colorful clouds above the Owens River in the eastern Sierra Nevada.
Photo by Galen Rowell/Mountain Light Photography, Inc.

CHAPTER 11

Heating the Atmosphere

FOCUS ON LEARNING

To assist you in learning the important concepts
in this chapter, you will find it helpful to focus
on the following questions:

1. What is weather? How is it different from climate?
2. What are the most important elements of weather and climate?
3. What are the major components of clean, dry air?
4. What is the extent and structure of the atmosphere?

5. What causes the seasons?
6. How is the atmosphere heated?
7. What causes temperature to vary from place to place?

Solar radiation provides more than 99.9 percent of the energy that heats Earth's surface.
Photo by John Sanford/Science Photo Library/Photo Researchers, Inc.

Figure 11.1 In January 1997, many parts of the northern Great Plains experienced paralyzing blizzards. This one struck Bismarck, North Dakota, on January 9, 1997. Probably no other aspect of our physical environment affects the daily lives of people more than the weather. *(Photo by Mike McCleary/*The Bismarck Tribune*)*

Weather influences our everyday activities, our jobs, and our health and comfort (Figure 11.1). Many of us pay little attention to the weather unless we are inconvenienced by it or when it adds to our enjoyment outdoors. Nevertheless, few other aspects of our physical environment affect our lives more than the phenomena we collectively call the weather.

The United States has the greatest variety of weather of any country in the world. Beyond its direct impact on the lives of individuals, the weather has a strong effect on the world economy by influencing agriculture, energy use, water resources, transportation, and industry.

Although the atmosphere influences our lives a great deal, it is also important to realize that the reverse is true: people influence the atmosphere and its behavior. There are, and will continue to be, significant political and scientific decisions that must be made involving these impacts. Important examples are air pollution control and the effects of human activities on global climate and the atmosphere's protective ozone layer. We all need increased awareness and understanding of our atmosphere and its behavior.

Weather and Climate

If you were to search through the library for books about the atmosphere, many of the titles would contain the term *weather,* whereas others would have the word *climate.* What is the difference between these two terms? **Weather** is a word used to denote the state of the atmosphere at a particular place for a *short period of time.* Weather is constantly changing—hourly, daily, and seasonally. **Climate,** on the other hand, might best be described as an aggregate or composite of weather. Stated another way, the climate of a place or region is a generalization of the weather conditions over a *long period of time.* Therefore, a climatic description is possible only after weather records have been kept for many years.

Although weather and climate are not identical, the nature of both is expressed in terms of the same **elements,** those quantities or properties that are measured regularly. The most important are (1) air temperature, (2) humidity, (3) type and amount of cloudiness, (4) type and amount of precipitation, (5) air pressure, and (6) the speed and direction of the wind. These elements are the major variables from which weather patterns and climate types are deciphered. Although we shall study these elements separately at first, keep in mind that they are very much interrelated. A change in any one of the elements will often bring about changes in the others.

Composition of the Atmosphere

Air is *not* a unique element or compound. Rather, **air** is a *mixture* of many discrete gases, each with its own physical properties, in which varying quantities of tiny solid and liquid particles are suspended.

The composition of air is not constant; it varies from time to time and from place to place. If the water vapor, dust, and other variable components were removed from the atmosphere, we would find that its makeup is very stable worldwide up to an altitude of about 80 kilometers (50 miles).

Major Components

Clean, dry air is composed almost entirely of two gases—78 percent nitrogen and 21 percent oxygen (Figure 11.2). Although these gases are the most plentiful components of air and are of great significance to life on Earth, they are of minor importance in affecting weather phenomena. The remaining 1 percent of dry air is mostly the inert gas argon (0.93 percent) plus tiny quantities of a number of other gases. Carbon dioxide, although present in only minute amounts (0.036 percent), is nevertheless an important constituent of air, because it has the ability to absorb heat energy radiated by Earth and thus helps keep the atmosphere warm.

Variable Components

Air includes many gases and particles that vary significantly from time to time and place to place. Important examples include water vapor, dust particles, and ozone. Although usually present in small percentages, they can have significant effects on weather and climate.

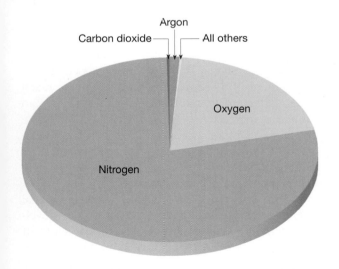

Figure 11.2 Proportional volume of gases composing dry air. Nitrogen and oxygen clearly dominate.

Water Vapor   The amount of water vapor in the air varies considerably, from practically none at all up to about 4 percent by volume. Why is such a small fraction of the atmosphere so significant? Certainly the fact that water vapor is the source of all clouds and precipitation would be enough to explain its importance. However, water vapor has other roles. Like carbon dioxide, it has the ability to absorb heat energy given off by Earth as well as some solar energy. It is therefore important when we examine the heating of the atmosphere.

When water changes from one state to another (see Figure 12.1, p. 300), it absorbs or releases heat. This energy is termed *latent heat,* which means "hidden" heat. As we shall see in later chapters, water vapor in the atmosphere transports this latent heat from one region to another, and it is the energy source that helps drive many storms.

Dust   Most of us probably think of dust as small, barely visible bits of dirt. However, from a meteorological standpoint, dust is much more than that. It includes many microscopic particles that are invisible to the naked eye, among them organic materials like pollen, spores, and seeds. Dust particles are most numerous in the lower atmosphere near their primary source, Earth's surface. Still, the upper atmosphere is not free of them, because some dust is carried to great heights by rising currents of air. Other dust particles are provided to the upper atmosphere by meteors that disintegrate as they pass through Earth's envelope of air.

Some particles are important because they act as surfaces upon which water vapor may condense. This function is essential to the formation of clouds and fog. In addition, dust may absorb or reflect incoming solar radiation. Thus, when the dust content of the atmosphere is high, as it may be following an explosive volcanic eruption, the amount of sunlight reaching Earth's surface can be measurably reduced. Finally, dust in the air contributes to a phenomenon we all have observed—the red and orange colors of sunrise and sunset.

Ozone   Another important component of the atmosphere is *ozone.* It is a form of oxygen that combines three oxygen atoms into each molecule (O_3). Ozone is not the same as oxygen we breathe, which has two atoms per molecule (O_2). There is very little of this gas in the atmosphere and its distribution is not uniform. It is concentrated well above the surface in a layer called the *stratosphere,* between 10 and 50 kilometers (6 and 31 miles).

In this altitude range, oxygen molecules (O_2) are split into single atoms of oxygen (O) when they absorb ultraviolet radiation emitted by the Sun. Ozone is then created when a single atom of oxygen (O) and a molecule of oxygen (O_2) collide. This must happen in the presence of a third, neutral molecule that acts as a *catalyst* by allowing the reaction to take place without itself being consumed in the process. Ozone is concentrated in the 10- to 50-kilometer height range because a crucial balance exists there: The ultraviolet radiation from the Sun is sufficient to produce single atoms of oxygen, and there are enough gas molecules to bring about the required collisions.

The presence of the ozone layer in our atmosphere is crucial to those of us who dwell on Earth. The reason is that ozone absorbs the potentially harmful ultraviolet (UV) radiation from the Sun. If ozone did not filter a great deal of the ultraviolet radiation, and if the Sun's UV rays reached the surface of Earth undiminished, our planet would be uninhabitable for most life as we know it. Thus, anything that reduces the amount of ozone in the atmosphere could affect the well-being of life on Earth. Just such a problem exists.

The Ozone Problem

Although stratospheric ozone is 10 to 50 kilometers up, it is vulnerable to human activities. Chemicals we produce are breaking up ozone molecules in the stratosphere, weakening our shield against UV rays. This loss of ozone is a serious global-scale environmental problem. Measurements over the past two decades confirm that ozone depletion is occurring worldwide and is especially pronounced above Earth's poles. You can see this effect over the South Pole in Figure 11.3.

Over the past half century, people have unintentionally placed the ozone layer in jeopardy by polluting the atmosphere. The offending chemicals are known as *chlorofluorocarbons* (*CFCs* for short). Over the decades many uses were developed for CFCs: coolants for air conditioning and refrigeration equipment, cleaning solvents for electronic components and computer chips, and propellants for aerosol sprays. CFCs are also used to produce certain plastic foams for products such as cups and home insulation.

Because CFCs are practically inert (that is, not chemically active) in the lower atmosphere, some of these gases gradually make their way to the ozone layer, where sunlight separates the chemicals into their constituent atoms. The chlorine atoms released this way break up some of the ozone molecules.

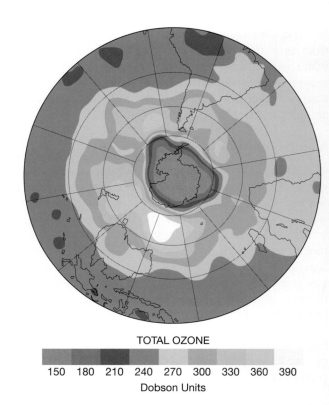

TOTAL OZONE

150 180 210 240 270 300 330 360 390
Dobson Units

Figure 11.3 Ozone distribution of the Southern Hemisphere for the month of October 1995. The image was produced with data acquired by the TIROS Observational Vertical Sounder (TOVS). The area of greatest depletion, called the ozone "hole," forms over Antarctica during the Southern Hemisphere spring. In October 1995, ozone levels reached a minimum of about 120 Dobson units, far below the 220 Dobson units typically seen over Antarctica before the hole forms. *(Data courtesy of NOAA)*

Because ozone filters out most of the UV radiation from the Sun, a decrease in its concentration permits more of these harmful wavelengths to reach Earth's surface. The most serious threat to human health is an increased risk of skin cancer. An increase in damaging UV radiation also can impair the human immune system as well as promote cataracts, a clouding of the eye lens that reduces vision and may cause blindness if not treated.

In response to this problem, an international agreement known as the *Montreal Protocol* was developed under the sponsorship of the United Nations. If ozone-depleting CFCs are eliminated from general use on the schedule established by the Montreal Protocol, it is estimated that the ozone layer will gradually recover and return to former levels by the middle of the twenty-first century. Positive action has been taken, yet the problem will remain for decades to come.

Height and Structure of the Atmosphere

To say that the atmosphere begins at Earth's surface and extends upward is obvious. However, where does the atmosphere end and outer space begin? There is no sharp boundary; the atmosphere rapidly thins as you travel away from Earth, until there are too few gas molecules to detect.

Pressure Changes

To understand the vertical extent of the atmosphere, let us examine the changes in the atmospheric pressure with height. Atmospheric pressure is simply the weight of the air above. At sea level, the average pressure is slightly more than 1000 millibars. This corresponds to a weight of slightly more than 1 kilogram per square centimeter (14.7 pounds per square inch). Obviously the pressure at higher altitudes is less (Figure 11.4).

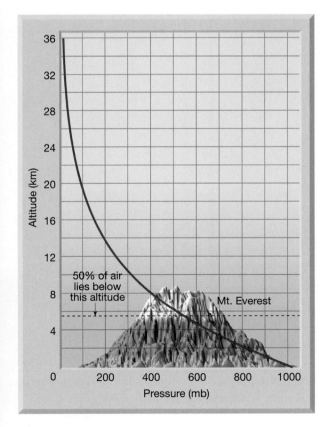

Figure 11.4 Atmospheric pressure variation with altitude. The rate of pressure decrease with an increase in altitude is not constant. Rather, pressure decreases rapidly near Earth's surface and more gradually at greater heights.

Figure 11.5 Temperatures drop with an increase in altitude in the troposphere. Therefore, it is possible to have snow on a mountaintop and warmer, snow-free lowlands below. Mount Kerkeslin, Jasper National Park, Alberta, Canada. *(Photo by Carr Clifton/Minden Pictures)*

One-half of the atmosphere lies below an altitude of 5.6 kilometers (3.5 miles). At about 16 kilometers (10 miles), 90 percent of the atmosphere has been transversed, and above 100 kilometers (62 miles), only 0.00003 percent of all the gases making up the atmosphere remain. Even so, traces of our atmosphere extend far beyond this altitude, gradually merging with the emptiness of space.

Temperature Changes

By the early twentieth century, much had been learned about the lower atmosphere. The upper atmosphere was partly known from indirect methods. Data from balloons and kites had revealed that the air temperature dropped with increasing height above Earth's surface. This phenomenon is felt by anyone who has climbed a high mountain, and is obvious in pictures of snowcapped mountaintops rising above snow-free lowlands (Figure 11.5). We divide the atmosphere vertically into four layers on the basis of temperature (Figure 11.6).

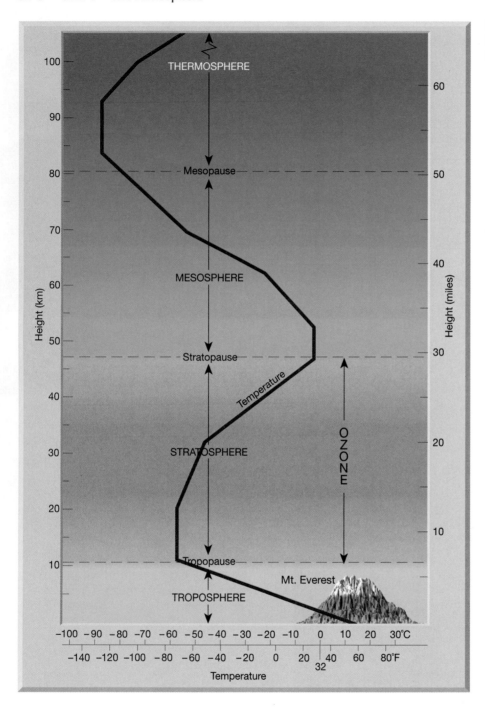

Figure 11.6 Thermal structure of the atmosphere to a height of about 110 kilometers (68 miles).

Troposphere The bottom layer in which we live, where temperature decreases with an increase in altitude, is the **troposphere.** The term literally means the region where air "turns over," a reference to the turbulent weather in this lowermost zone. The troposphere is the chief focus of meteorologists, because it is in this layer that essentially all important weather phenomena occur.

The temperature decrease in the troposphere is called the **environmental lapse rate.** Although its average value is 6.5°C per kilometer (3.5°F per 1000 feet), a figure known as the *normal lapse rate,* its

Figure 11.7 Atmospheric soundings using radiosondes supply data on vertical changes in temperature, pressure, and humidity. *(Courtesy of NOAA)*

value is quite variable. Thus, to determine the actual environmental lapse rate for any particular time and place, as well as gather information about vertical changes in pressure, wind, and humidity, radiosondes are used. The *radiosonde* is an instrument package that is attached to a balloon and transmits data by radio as it ascends through the atmosphere (Figure 11.7).

The thickness of the troposphere is not the same everywhere; it varies with latitude and the season. On the average, the temperature drop continues to a height of about 12 kilometers (7.4 miles). The outer boundary of the troposphere' is the *tropopause.*

Stratosphere Beyond the tropopause is the **stratosphere.** In the stratosphere, the temperature

remains constant to a height of about 20 kilometers and then begins a gradual increase that continues until the *stratopause,* at a height of about 50 kilometers above Earth's surface. Below the tropopause, atmospheric properties like temperature and humidity are readily transferred by large-scale turbulence and mixing. Above the tropopause, in the stratosphere, they are not. The reason for the increased temperatures in the stratosphere is that the atmosphere's ozone is concentrated in this layer. (Recall that ozone absorbs ultraviolet radiation from the sun.) As a consequence, the stratosphere is heated.

Mesosphere In the third layer, the **mesosphere,** temperatures again decrease with height until, at the *mesopause,* approximately 80 kilometers (50 miles) above the surface, the temperature approaches −90°C.

Thermosphere The fourth layer extends outward from the mesopause and has no well-defined upper limit. It is the **thermosphere,** a layer that contains only a minute fraction of the atmosphere's mass. In the extremely rarefied air of this outermost layer, temperatures again increase owing to the absorption of very short-wave, high-energy solar radiation by atoms of oxygen and nitrogen.

Temperatures rise to extremely high values of more than 1000°C in the thermosphere. But such temperatures are not comparable to those experienced near Earth's surface. Temperature is defined in terms of the average speed at which molecules move. Because the gases of the thermosphere are moving at very high speeds, the temperature is very high. But the gases are so sparse that, collectively, they possess only an insignificant quantity of heat. For this reason, the temperature of a satellite orbiting Earth in the thermosphere is determined chiefly by the amount of solar radiation it absorbs and not by the high temperature of the almost nonexistent surrounding air. If an astronaut inside were to expose his or her hand, it would not feel hot.

Earth–Sun Relationships

Always remember that nearly all of the energy that drives Earth's variable weather and climate comes from the Sun. Earth intercepts only a minute percentage of the energy given off by the Sun—less than one two-billionth. This may seem to be an insignificant amount until we realize that it is several hundred thousand times the electrical-generating capacity of the United States.

If solar energy were distributed uniformly over Earth's surface, our planet would have more uniform temperatures. But solar energy is not distributed evenly over Earth's surface. It is this unequal heating that creates winds and ocean currents, which, in turn, transport heat from the tropics to the poles in an unending attempt to reach an energy balance. If the Sun were to be "turned off," global winds would quickly subside. Yet as long as the Sun shines, the winds will blow and the phenomena we know as weather will persist.

Energy from the Sun is the most important control of our weather and climate. To understand atmospheric processes, we must understand what causes the variations in the amount of solar energy reaching Earth. The principal causes are variations in Sun angle and exposure time.

Earth's Rotation and Revolution

Earth has two principal motions—rotation and revolution. **Rotation** is the spinning of Earth about its axis. The axis is an imaginary line running through the poles. Our planet rotates once every 24 hours, producing the daily cycle of daylight and darkness. At any moment, half of Earth is experiencing daylight, and the other half darkness. The line separating the dark half of Earth from the lighted half is called the **circle of illumination.**

Revolution refers to the movement of Earth in its orbit around the Sun. Hundreds of years ago, most people believed that Earth was stationary in space and that the Sun and stars revolved around our planet. Today, we know that Earth is traveling at more than 107,000 kilometers per hour in its orbit about the Sun.

Seasons

We know that it is colder in winter than in summer. But why? Length of daylight certainly accounts for some of the difference. Long summer days expose us to more solar radiation, whereas short winter days expose us to less. But an even bigger factor is the angle of the Sun above the horizon.

During mid-summer the Sun is high above the horizon as it makes its daily journey across the sky. But as summer gives way to autumn, the noon Sun appears lower and lower in the sky. What we observe here is the annual shifting of the solar angle or *altitude* of the Sun.

The seasonal variation in the altitude of the Sun affects the amount of energy received at Earth's surface in two ways. First, when the Sun is high in the sky, the solar rays are most concentrated (you can see this in Figure 11.8A). The lower the angle, the more spread out and less intense is the solar radiation that reaches the surface (Figure 11.8B,C).

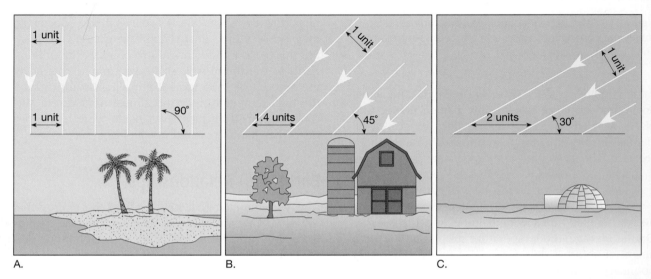

Figure 11.8 Changes in the Sun angle cause variations in the amount of solar energy reaching Earth's surface. The higher the angle, the more intense the solar radiation. Notice that one unit of solar energy striking at a 30° angle (C) spreads over twice as much area as that striking at a 90° angle (A).

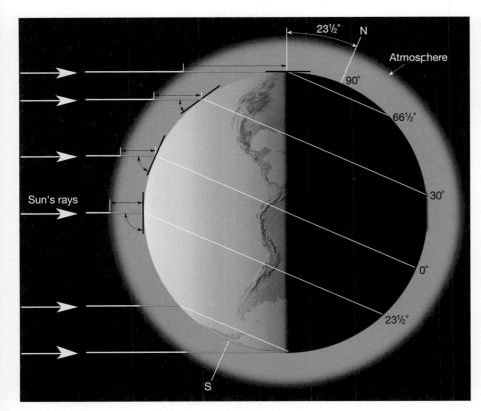

Figure 11.9 Rays striking at a low angle (toward the poles) must travel through more of the atmosphere than rays striking at a higher angle (around the equator) and thus are subject to greater depletion by reflection and absorption.

Second, and of lesser importance, the angle of the Sun determines the amount of atmosphere the rays must penetrate (Figure 11.9). When the Sun is directly overhead, the rays pass through a thickness of only 1 atmosphere, whereas rays entering at a 30-degree angle travel through twice this amount, and 5-degree rays travel through a thickness roughly equal to 11 atmospheres. The longer the path, the greater are the chances for absorption, reflection, and scattering by the atmosphere, all of which reduce the intensity at the surface.

If Earth were a flat surface, oriented at a right angle to the Sun, all places would receive the same amount of solar radiation. But Earth is spherical. Hence, on any given day only places located at a particular latitude receive vertical (90-degree) rays from the Sun. As we move either north or south of this location, the sun's rays strike at an ever-decreasing angle. Thus, the nearer a place is to the latitude receiving vertical rays of the Sun, the higher will be its noon Sun (Figure 11.9).

Causes of Seasons What causes yearly fluctuations in Sun angle and length of daylight? Variations occur because Earth's orientation to the Sun continually changes as it travels along its orbit. Earth's axis is not perpendicular to the plane of its orbit around the Sun, but instead is tilted 23 ½ degrees from the perpendicular. This is termed the **inclination of the axis,** and as we shall see, if the axis were not inclined, we would have no seasonal changes. In addition, because the axis remains pointed in the same direction (toward the North Star) as Earth journeys around the Sun, the orientation of Earth's axis to the sun's rays is constantly changing (Figure 11.10).

On one day each year the axis is such that the Northern Hemisphere is "leaning" 23 ½ degrees *toward* the Sun. Six months later, when Earth has moved to the opposite side of its orbit, the Northern Hemisphere "leans" 23 ½ degrees *away* from the Sun. On days between these extremes, Earth's axis "leans" at amounts less than 23 ½ degrees to the rays of the Sun. This change in orientation causes the vertical rays of the Sun to make a yearly migration from 23 ½ degrees north of the equator to 23 ½ degrees south of the equator.

This migration in turn causes the altitude angle of the noon Sun to vary by as much as 47 degrees (23 ½ + 23 ½) during the year at places located poleward of latitude 23 ½ degrees. For example, a mid-

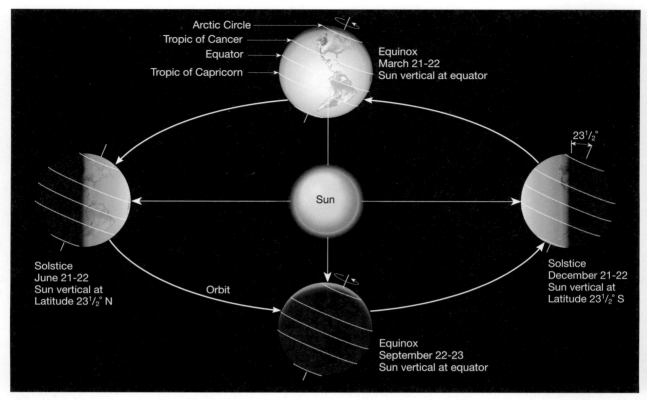

Figure 11.10 Earth–Sun relationships.

latitude city like New York (about 40 degrees north latitude) has a maximum noon Sun angle of 73 ½ degrees when the Sun's vertical rays reach their farthest northward location and a minimum noon Sun angle of 26 ½ degrees 6 months later.

Solstices and Equinoxes Historically, 4 days each year have been given special significance based on the annual migration of the direct rays of the Sun and its importance to the yearly weather cycle. On June 21 or 22, Earth is in a position such that the north end of its axis is tilted 23 ½ degrees *toward* the Sun (Figure 11.11A). At this time the vertical rays of the Sun strike 23 ½ degrees north latitude (23 ½ degrees north of the equator), a latitude known as the **Tropic of Cancer.** For people in the Northern Hemisphere, June 21 or 22 is known as the **summer solstice.**

Six months later, on about December 21 or 22, Earth is in the opposite position, with the sun's vertical rays striking at 23 ½ degrees south latitude (Figure 11.11B). This parallel is known as the **Tropic of Capricorn.** For those in the Northern Hemisphere,

December 21 and 22 is the **winter solstice.** However, at the same time in the Southern Hemisphere, people are experiencing just the opposite—the summer solstice.

Midway between the solstices are the equinoxes. September 22 or 23 is the date of the **autumnal equinox** in the Northern Hemisphere, and March 21 or 22 is the date of the **spring equinox.** On these dates, the vertical rays of the Sun strike the equator (0 degrees latitude) because Earth is in such a position in its orbit that the axis is tilted neither toward nor away from the Sun (Figure 11.11C).

The length of daylight versus darkness is also determined by Earth's position in orbit. The length of daylight on June 21, the summer solstice in the Northern Hemisphere, is greater than the length of night. This fact can be established from Figure 11.11A by comparing the fraction of a given latitude that is on the "day" side of the circle of illumination with the fraction on the "night" side. The opposite is true for the winter solstice, when the nights are longer than the days. Again for comparison let us

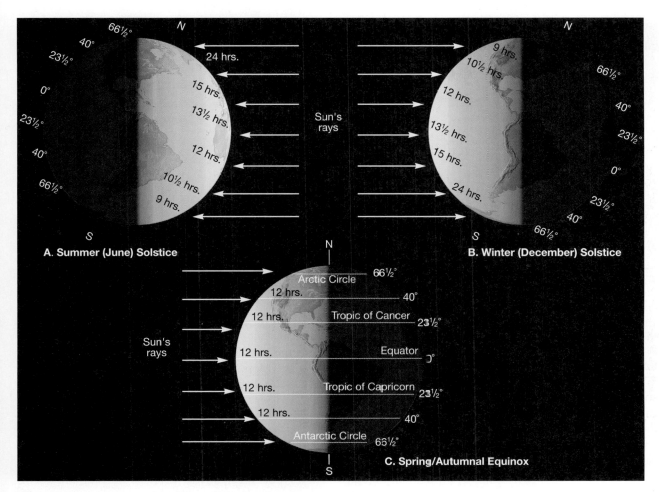

Figure 11.11 Characteristics of the solstices and equinoxes.

consider New York City, which has about 15 hours of daylight June 21 and only about 9 hours on December 21 (you can see this in Figure 11.11 and Table 11.1). Also note from Table 11.1 that on June 21 the farther you are north of the equator the longer is the period of daylight, until the Arctic Circle is reached, where the length of daylight is 24 hours.

During an equinox (meaning "equal night"), the length of daylight is 12 hours *everywhere* on Earth, because the circle of illumination passes directly through the poles, dividing the latitudes in half (Figure 11.11C).

As a review of the characteristics of the summer solstice for the Northern Hemisphere, examine Figure 11.11A and Table 11.1 and consider the following facts:

1. The solstice occurs on June 21 or 22.
2. The vertical rays of the Sun are striking the Tropic of Cancer (23 ½ degrees north latitude).
3. Locations in the Northern Hemisphere are experiencing their greatest length of daylight. (Opposite for the Southern Hemisphere.)
4. Locations north of the Tropic of Cancer are experiencing their highest noon Sun angles. (Opposite for places south of the Tropic of Capricorn.)
5. The farther you are north of the equator, the longer the period of daylight, until the Arctic Circle is reached, where daylight lasts for 24 hours. (Opposite for the Southern Hemisphere.)

Table 11.1 ■ Length of daylight			
Latitude (degrees)	Summer Solstice	Winter Solstice	Equinoxes
0	12 h	12 h	12 h
10	12 h 35 min	11 h 25 min	12
20	13 12	10 48	12
30	13 56	10 04	12
40	14 52	9 08	12
50	16 18	7 42	12
60	18 27	5 33	12
70	24 h (for 2 mo)	0 00	12
80	24 h (for 4 mo)	0 00	12
90	24 h (for 6 mo)	0 00	12

The facts about the winter solstice are just the opposite. It should now be apparent why a mid-latitude location is warmest in the summer, for it is then that days are longest and Sun's altitude is highest.

In summary, seasonal variations in the amount of solar energy reaching places on Earth's surface are caused by the migrating vertical rays of the Sun and the resulting variations in Sun angle and length of daylight. All places at the same latitude have identical Sun angles and lengths of daylight. If the Earth–sun relationships just described were the only controls of temperature, we would expect these places to have identical temperatures as well. Obviously this is not the case. Although the altitude of the Sun is a very important control of temperature, it is not the only control, as we shall see.

Mechanisms of Heat Transfer

All forms of matter, whether solid, liquid, or gas, consist of atoms or molecules that are in constant motion. Because of this vibratory motion, all matter is said to contain *thermal energy*. Whenever a substance is heated, its atoms move faster and faster. This leads to an increase in thermal energy. It is the average motion of the atoms or molecules in objects that we sense when we determine how hot or cold something is. We commonly describe the "hotness" or "coldness" of an object using the measurement called *temperature*. More specifically, temperature is a measure of the average motion of the atoms or molecules within a substance.

Although closely related, temperature and heat are different concepts (recall the situation in the thermosphere). Whereas temperature is a measure of the average motion of molecules, heat is the energy that flows because of temperature differences. In all situations, *heat is transferred from warmer to cooler objects*. Thus, if two objects of different temperature are in contact, the warmer object will become cooler and the cooler object will become warmer until they both reach the same temperature.

Three mechanisms of heat transfer are recognized: conduction, convection, and radiation. We experience all three in everyday life, as you shall see.

Conduction and Convection

Conduction is familiar to all of us. Anyone who has touched a metal spoon that was left in a hot pan has discovered that heat was conducted through the spoon. **Conduction** *is the transfer of heat through matter by molecular activity.* The energy of molecules is transferred through collisions from one molecule to another, with the heat flowing from the higher temperature to the lower temperature.

The ability of substances to conduct heat varies considerably. Metals are good conductors, as those of us who have touched hot metal have quickly learned (Figure 11.12). Air, on the other hand, is a very poor conductor of heat. Consequently, conduction is important only between Earth's surface and the air directly in contact with the surface. As a means of heat transfer for the atmosphere as a whole, conduction is the least significant.

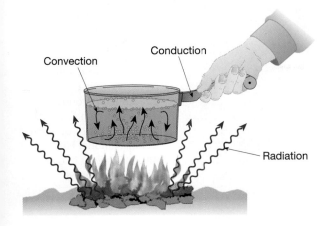

Figure 11.12 The three mechanisms of heat transfer: conduction, convection, and radiation.

Heat gained by the lowest layer of the atmosphere from radiation or conduction is most often transferred by convection. **Convection** *is the transfer of heat by the movement of a mass or substance from one place to another*. It can take place only in liquids and gases (Figure 11.12). Convective motions in the atmosphere transport heat from equatorial regions to the poles and from the surface upward. The term *advection* is usually reserved for horizontal convective motions such as winds, while *convection* is used to describe vertical motions in the atmosphere.

Radiation

The third mechanism of heat transfer is **radiation**. As shown in Figure 11.12, radiation travels out in all directions from its source. Unlike conduction and convection, which need a medium to travel through, radiant energy readily travels through the vacuum of space. Thus, radiation is the heat-transfer mechanism by which solar energy reaches our planet.

From our everyday experience we know that the Sun emits light and heat as well as the ultraviolet rays that cause suntan. Although these forms of energy comprise a major portion of the total energy that radiates from the Sun, they are only part of a large array of energy called **radiation** or **electromagnetic radiation.** This array, or spectrum, of electromagnetic energy is shown in Figure 11.13. All radiation, whether X rays, microwaves, or radio waves, transmit energy through the vacuum of space at 300,000 kilometers (186,000 miles) per second and only slightly slower through our atmosphere.

Nineteenth-century physicists were so puzzled by the seemingly impossible phenomenon of energy traveling through the vacuum of space, without a medium to transmit it, that they assumed that a material, which they named *ether,* existed between the Sun and Earth. This medium was thought to transmit radiant energy in much the same way that air transmits sound waves. Of course, this was incorrect.

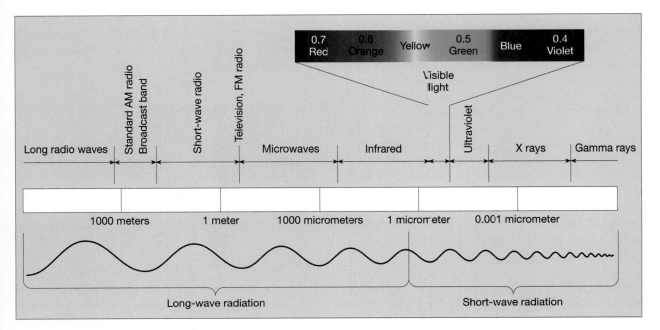

Figure 11.13 The electromagnetic spectrum, illustrating the wavelengths and names of various types of radiation.

Figure 11.14 Visible light consists of an array of colors we commonly call the "colors of the rainbow." Rainbows are relatively common optical phenomena produced by the bending and reflection of light by drops of water. Denali National Park, Alaska *(Photo © by Carr Clifton Photography)*

Today we know that, like gravity, radiation requires no material to transmit it.

In some respects, the transmission of radiant energy parallels the motion of the gentle swells in the open ocean. Like ocean swells, electromagnetic waves come in various sizes. For our purpose, the most important difference among electromagnetic waves is their wavelength, or the distance from one crest to the next. Radio waves have the longest wavelengths, ranging to tens of kilometers, whereas gamma waves are the very shortest, being less than one billionth of a centimeter long.

Visible light, as the name implies, is the only portion of the spectrum we can see. We often refer to visible light as "white" light since it appears "white" in color. However, it is easy to show that white light is really a mixture of colors, each corresponding to a particular wavelength (Figure 11.14). Using a prism, we can divide white light into a color array rainbow. Figure 11.13 shows that violet has the shortest wavelength—0.4 micrometer—and red has the longest—0.7 micrometer.

Located adjacent to red, and having a longer wavelength, is **infrared** radiation, which we cannot see but which we can detect as heat. The closest invisible waves to violet are called **ultraviolet** rays and are responsible for sunburn after an intense exposure to the Sun. Although we divide radiant energy into groups based on our ability to perceive them, all forms of radiation are basically the same. When any form of radiant energy is absorbed by an object, the result is an increase in molecular motion, which causes a corresponding increase in temperature.

To better understand how the atmosphere is heated, it is useful to have a general understanding of the basic laws governing radiation.

1. All objects, at whatever temperature, emit radiant energy. Hence, not only hot objects like the Sun but also Earth, including its polar ice caps, continually emit energy.
2. Hotter objects radiate more total energy per unit area than do colder objects.
3. The hotter the radiating body, the shorter the wavelength of maximum radiation. The Sun, with a surface temperature of about 5700°C, radiates maximum energy at 0.5 micrometer, which is in the visible range. The maximum radiation for Earth occurs at a wavelength of 10 micrometers, well within the infrared (heat) range. Because the maximum Earth radiation is

284

roughly 20 times longer than the maximum solar radiation, Earth radiation is often called *long-wave radiation,* and solar radiation is called *short-wave radiation.*

4. Objects that are good absorbers of radiation are good emitters as well. Earth's surface and the Sun approach being perfect radiators because they absorb and radiate with nearly 100 percent efficiency for their respective temperatures. On the other hand, *gases are selective absorbers and radiators.* Thus, the atmosphere, which is nearly transparent (does not absorb) to certain wavelengths of radiation, is nearly opaque (a good absorber) to others. Our experience tells us that the atmosphere is transparent to visible light; hence, it readily reaches Earth's surface. This is not the case for the longer wavelength radiation emitted by Earth.

Figure 11.12 summarizes the various mechanisms of heat transfer. A portion of the radiant energy generated by the campfire is absorbed by the pan. This energy is readily transferred through the metal container by the process of conduction. Conduction also increases the temperature of the water at the bottom. Once warmed, this layer of water moves upward and is replaced by cool water descending from above. Thus, convection currents that redistribute the newly acquired energy throughout the pan are established. Meanwhile, the camper is warmed by radiation emitted by the fire and the pan. Furthermore, because metals are good conductors, the camper's hands is likely to be burned if he (or she) does not use a potholder. Like this example, the heating of Earth's atmosphere involves the processes of conduction, convection, and radiation, all of which occur simultaneously.

Paths Taken by Incoming Solar Radiation

Although the atmosphere is largely transparent to incoming solar radiation, only about 25 percent penetrates directly to Earth's surface without some sort of interference by the atmosphere. The remainder is either *absorbed* by the atmosphere, *scattered* about until it reaches Earth's surface or returns to space, or is *reflected* back to space (Figure 11.15). What determines whether radiation will be absorbed, scattered, or reflected outward? It depends greatly on the wavelength of the energy being transmitted, as well as on the nature of the intervening material.

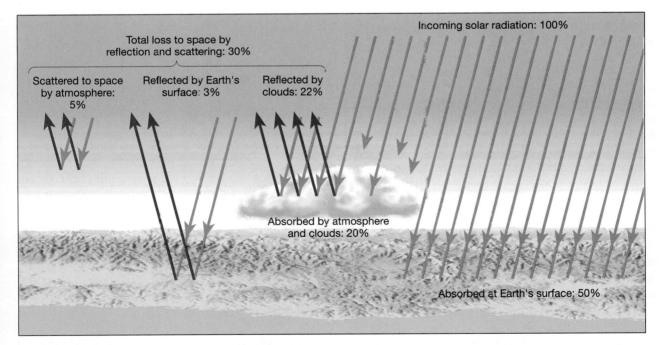

Figure 11.15 Distribution of incoming solar radiation by percentage. More solar energy is absorbed by Earth's surface than by the atmosphere. Consequently, the air is not heated directly by the Sun but is heated chiefly by reradiation from Earth's surface.

Scattering

Although solar radiation travels in a straight line, the gases and dust particles in the atmosphere can redirect this energy, a process called *scattering.* Some of the light is backscattered into space, while the remainder continues Earthward, where it interacts with other molecules that scatter it further by changing the direction of the radiation, but not its wavelength.

About 30 percent of the solar energy reaching the outer atmosphere is reflected back to space. Included in this figure is the amount sent skyward by backscattering. This energy is lost to Earth and does not play a role in heating the atmosphere.

Albedo

The fraction of the total radiation that is reflected by a surface is called its **albedo.** Thus, the albedo for Earth as a whole (the planetary albedo) is 30 percent. However, the albedo from place to place as well as from time to time in the same locale can vary greatly depending on the amount of cloud cover and dust in the air, as well as on the angle of the Sun's rays and the nature of the surface. A lower angle means that more atmosphere must be penetrated, thus making the "obstacle course" longer, and therefore the loss of solar radiation greater (see Figure 11.9, p. 279). Table 11.2 gives the albedo for various surfaces and clouds. Note that the angle at which the Sun's rays strike a water surface greatly affects its albedo.

Absorption

As stated earlier, gases are selective absorbers, meaning that they absorb strongly in some wavelengths, moderately in others, and only slightly in still others. When a gas molecule absorbs light waves, the energy is transformed into internal molecular motion, which is detectable as a rise in temperature. Nitrogen, the most abundant constituent in the atmosphere, is a poor absorber of all types of incoming solar radiation. Oxygen and ozone are efficient absorbers of ultraviolet radiation. Oxygen removes most of the shorter ultraviolet radiation high in the atmosphere, and ozone absorbs most of the remaining UV rays in the stratosphere. The absorption of UV radiation in the stratosphere accounts for the high temperatures experienced there. The only other significant absorber of incoming solar radiation is water vapor, which along with oxygen and ozone accounts for most of the solar radiation absorbed within the atmosphere.

Table 11.2 ■ Albedo of various surfaces	
Surface	Percent Reflected
Clouds, stratus	
<150 meters thick	25–63
150–300 meters thick	45–75
300–600 meters thick	59–84
Average of all types and thicknesses	50–55
Concrete	17–27
Crops, green	5–25
Forest, green	5–10
Meadows, green	5–25
Ploughed field, moist	14–17
Road, blacktop	5–10
Sand, white	30–60
Snow, fresh-fallen	80–90
Snow, old	45–70
Soil, dark	5–15
Soil, light (or desert)	25–30
Water	8*

*Typical albedo value for a water surface. The albedo of a water surface varies greatly depending upon the Sun angle. If the Sun angle is greater than 30 degrees, the albedo is less than 5 percent. When the Sun is near the horizon (Sun angle less than 3 degrees), the albedo is more than 60 percent.

For the atmosphere as a whole, none of the gases are effective absorbers of visible radiation. This explains why most visible radiation reaches Earth's surface and why we say that the atmosphere is transparent to incoming solar radiation. Thus, the atmosphere does not acquire the bulk of its energy directly from the Sun. Rather, it is heated chiefly by energy that is first absorbed by Earth's surface and then reradiated to the sky.

Heating the Atmosphere: The Greenhouse Effect

Approximately 50 percent of the solar energy that strikes the top of the atmosphere reaches Earth's surface and is absorbed. Most of this energy is then reradiated skyward. Because Earth has a much lower surface temperature than the Sun, the radiation that it emits has longer wavelengths than does solar radiation.

The atmosphere as a whole is an efficient absorber of the longer wavelengths emitted by Earth (*terrestrial radiation*). Water vapor and carbon dioxide are the principal absorbing gases. Water vapor absorbs roughly five times more terrestrial radiation than do all other gases combined and accounts for the warm temperatures found in the lower troposphere, where it is most highly concentrated. Because the atmosphere is quite transparent to shorter-wavelength solar radiation and more readily absorbs longer-wavelength terrestrial radiation, the atmosphere is heated from the ground up rather than vice versa. This explains the general drop in temperature with increasing altitude experienced in the troposphere. The farther from the "radiator," the colder it becomes.

When the gases in the atmosphere absorb terrestrial radiation, they warm, but they eventually radiate this energy away. Some travels skyward, where it may be reabsorbed by other gas molecules, a possibility less likely with increasing height because the concentration of water vapor decreases with altitude. The remainder travels Earthward and is again absorbed by Earth. For this reason, Earth's surface is continually being supplied with heat from the atmosphere as well as from the Sun.

This very important phenomenon has been termed the **greenhouse effect** because it was once thought that greenhouses were heated in a similar manner (Figure 11.16). The gases of our atmosphere, especially water vapor and carbon dioxide, act very much like the glass in the greenhouse. They allow shorter-wavelength solar radiation to enter, where it is absorbed by the objects inside. These objects in turn reradiate the heat, but at longer wavelengths, to which glass is nearly opaque. The heat therefore is trapped in the greenhouse. However, a more important factor in keeping a greenhouse warm is the fact that the greenhouse itself prevents mixing of air inside with cooler air outside. Nevertheless, the term *greenhouse effect* is still used.

Global Warming

In the preceding section you learned that carbon dioxide (CO_2) absorbs some of the radiation emitted by Earth and thus contributes to the greenhouse effect. Because CO_2 is an important heat absorber, it follows that a change in the atmosphere's CO_2 content could influence air temperature.

Paralleling the rapid growth of industrialization, which began in the 1800s, has been the burning of fossil fuels (coal, oil, and natural gas). Combustion of these fuels has added vast quantities of CO_2 to the atmosphere. Although not all of this CO_2 remains in the atmosphere, enough has lingered to increase atmospheric CO_2 about 70 parts per million since 1850 (Figure 11.17).

If we assume that the use of fossil fuels will increase at projected rates, current estimates indicate

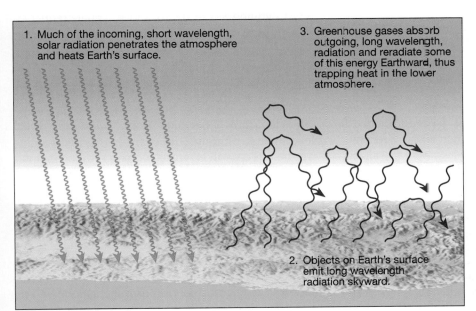

1. Much of the incoming, short wavelength, solar radiation penetrates the atmosphere and heats Earth's surface.

3. Greenhouse gases absorb outgoing, long wavelength, radiation and reradiate some of this energy Earthward, thus trapping heat in the lower atmosphere.

2. Objects on Earth's surface emit long wavelength radiation skyward.

Figure 11.16 The heating of the atmosphere. Most of the solar radiation that is not reflected back to space passes through the atmosphere and is absorbed at Earth's surface. Earth's surface, in turn, emits longer wavelength radiation. A portion of this energy is absorbed by certain gases in the atmosphere.

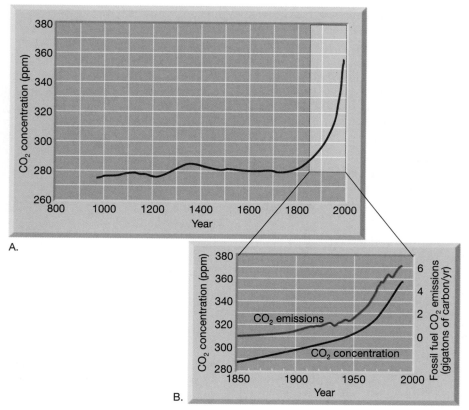

A.

B.

Figure 11.17 A. Carbon dioxide (CO_2) concentrations over the past 1000 years. Most of the record is based on data obtained from Antarctic ice cores. Bubbles of air trapped in the glacial ice provide samples of past atmospheres. The record since 1958 comes from direct measurements of atmospheric CO_2 taken at Mauna Loa Observatory, Hawaii. **B.** The rapid increase in CO_2 concentration since the onset of industrialization in the late 1700s is clear and has followed closely the rise in CO_2 emissions from fossil fuels.

that the atmosphere's carbon dioxide content will approach 600 parts per million by some time in the second half of the twenty-first century. If this occurs, the most realistic models predict an increase in the mean global surface temperature of 2.5°C. This may sound minor, but a change of this magnitude would be unprecedented in human history. Such an increase would come close to equaling the warming that has taken place since the peak of the most recent stage of the Ice Age 18,000 years ago, except that it would occur *much more rapidly.*

Carbon dioxide is not the only gas contributing to a future global temperature increase. In recent years, atmospheric scientists have come to realize that the industrial and agricultural activities of people are causing a buildup of certain trace gases that also play a significant role.

If global temperatures increase significantly, the possible consequences are potentially very serious. Potential changes include:

1. Shifts in temperature and rainfall patterns that could disrupt agriculture worldwide.

2. A gradual rise in sea level due to a warmer ocean (the warmed water expands, causing sea level to rise) and melting glaciers that would accelerate shoreline erosion in many coastal areas.

3. Changing storm tracks and a higher frequency and greater intensity of hurricanes.

4. Increases in the frequency and intensity of heat waves and droughts.

The changes that occur will probably take the form of environmental shifts that will be imperceptible to most people from year to year. Nevertheless, although the changes may seem gradual, the effects will clearly have powerful economic, social, and political consequences.

Temperature Measurement and Data

Changes in air temperature are probably noticed by people more often than changes in any other element of weather. At a weather station, the temperature is read on a regular basis from instruments mounted in

Figure 11.18 Standard instrument shelter. A shelter protects instruments from direct sunlight and allows for the free flow of air. *(Courtesy of Qualimetrics, Inc.)*

an instrument shelter (Figure 11.18). The shelter protects the instruments from direct sunlight and allows a free flow of air. In addition to a good-quality thermometer, the shelter is likely to contain a thermograph to continuously record temperature and a set of maximum-minimum thermometers. As their name implies, these thermometers record the highest and lowest temperatures during a measurement period, usually 24 hours.

The daily maximum and minimum temperatures are the bases for many of the temperature data compiled by meteorologists:

1. By adding the maximum and minimum temperatures and then dividing by two, the *daily mean temperature* is calculated.

2. The *daily range* of temperature is computed by finding the difference between the maximum and minimum temperatures for a given day.

3. The *monthly mean* is calculated by adding together the daily means for each day of the month and dividing by the number of days in the month.

4. The *annual mean* is an average of the 12 monthly means.

5. The *annual temperature range* is computed by finding the difference between the highest and lowest monthly means.

Mean temperatures are particularly useful for making comparisons, whether on a daily, monthly, or annual basis. It is quite common to hear a weather reporter state, "Last month was the hottest July on record," or "Today Chicago was ten degrees warmer than Miami." Temperature ranges are also useful statistics, because they give an indication of extremes.

Controls of Temperature

A "temperature control" is any factor that causes temperature to vary from place to place and from time to time. Earlier in this chapter we examined the single greatest cause for temperature variations—differences in the receipt of solar radiation. Because variations in Sun angle and length of daylight are a function of latitude, they are responsible for warm temperatures in the tropics and colder temperatures at more poleward locations.

However, latitude is not the only control of temperature; if it were, we would expect that all places along the same parallel of latitude would have identical temperatures. This is clearly not the case. For example, Eureka, California, and New York City are both coastal cities at about the same latitude and both have an average mean temperature of 11°C (52°F). However, New York City is 9°C (16°F) warmer than Eureka in July and 10°C cooler in January. Quito and Guayaquil, two cities in Ecuador, are relatively close to each other, yet the mean annual temperatures at these two cities differ by 12°C (21°F). To explain these situations and countless others, we must understand other factors that exert a strong influence upon temperatures. Among the most important are the differential heating of land and water, altitude, geographic position, and ocean currents.*

---

*For a discussion of the effects of ocean currents on temperatures, see Chapter 10.

Land and Water

The heating of Earth's surface directly influences the heating of the air above. Therefore, to understand variations in air temperature, we must examine the nature of the surface. Different land surfaces absorb varying amounts of incoming solar energy, which in turn cause variations in the temperature of the air above. The largest contrast, however, is not between different land surfaces, but between land and water. *Land heats more rapidly and to higher temperatures than water and cools more rapidly and to lower temperatures than water.* Temperature variations, therefore, are considerably greater over land than over water.

Among the reasons for the differential heating of land and water are the following:

1. The *specific heat* (amount of energy needed to raise 1 gram of a substance 1°C) is far greater for water than for land. Thus, water requires a great deal more heat to raise its temperature the same amount as an equal quantity of land.

2. Land surfaces are opaque, so heat is absorbed only at the surface. Water, being more transparent, allows heat to penetrate to a depth of many meters.

3. The water that is heated often mixes with water below, thus distributing the heat through an even larger mass.

4. Evaporation (a cooling process) from water bodies is greater than that from land surfaces.

Monthly temperature data for two cities will demonstrate the moderating influence of a large water body and the extremes associated with land (Figure 11.19). Vancouver, British Columbia, is located along a windward coast, whereas Winnipeg, Manitoba, is in a continental position far from the influence of water. Both cities are at about the same latitude and thus experience similar Sun angles and lengths of daylight. Winnipeg, however, has a mean January temperature that is 20°C lower than Vancouver's. Winnipeg's July mean is 2.6°C higher than Vancouver's. Although their latitudes are nearly the same, Winnipeg, which has no water influence, experiences much greater temperature extremes than Vancouver, which does. The key to Vancouver's moderate year-round climate is the Pacific Ocean.

On a much larger scale, the moderating influence of water may also be demonstrated when temperature variations in the Northern and Southern hemispheres

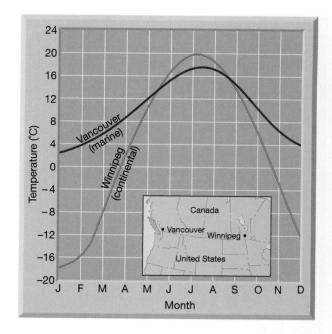

Figure 11.19 Mean monthly temperatures for Vancouver, British Columbia, and Winnipeg, Manitoba. Vancouver has a much smaller annual temperature range owing to the strong marine influence of the Pacific Ocean. The curve for Winnepeg illustrates the greater extremes associated with an interior location.

are compared. In the Northern Hemisphere, 61 percent is covered by water, and land accounts for the remaining 39 percent. However, in the Southern Hemisphere, 81 percent is covered by water and 19 percent by land. The Southern Hemisphere is correctly called the *water hemisphere* (see Figure 9.1, p. 228). Table 11.3 portrays the considerably smaller annual temperature variations in the water-dominated Southern Hemisphere as compared with the Northern Hemisphere.

Table 11.3 ■ Variation in mean annual temperature range (°C) with latitude

Latitude	Northern Hemisphere	Southern Hemisphere
0	0	0
15	3	4
30	13	7
45	23	6
60	30	11
75	32	26
90	40	31

Altitude

The two cities in Ecuador mentioned earlier—Quito and Guayaquil—demonstrate the influence of altitude upon mean temperatures. Although both cities are near the equator and not far apart, the annual mean at Guayaquil is 25°C (77°F) as compared to Quito's mean of 13°C (55°F). The difference is explained largely by the difference in the cities' elevations: Guayaquil is only 12 meters (40 feet) above sea level, whereas Quito is high in the Andes Mountains at 2800 meters (9200 feet). Recall that temperatures drop an average of 6.5°C per kilometer in the troposphere; thus, cooler temperatures are to be expected at greater heights (see Figure 11.5, p. 275).

Even so, the magnitude of the difference is not explained completely by the normal lapse rate. If the normal lapse rate is used, we would expect Quito to be about 18°C cooler than Guayaquil; the difference, however, is only 12°C. Places at high elevations, such as Quito, are warmer than the value calculated using the normal lapse rate because of the absorption and reradiation of solar energy by the ground surface.

Geographic Position

The geographic setting can greatly influence the temperatures experienced at a specific location. A coastal location where prevailing winds blow from the ocean onto the shore (a *windward* coast) experiences considerably different temperatures than does a coastal location where prevailing winds blow from the land toward the ocean (a *leeward* coast). In the first situation, the windward coast will experience the full moderating influence of the ocean—cool summers and mild winters—compared to an inland station at the same latitude.

A leeward coastal situation, however, will have a more continental temperature regime because the winds do not carry the ocean's influence onshore. Eureka, California, and New York City, the two cities mentioned earlier, illustrate this aspect of geographic position. The annual temperature range at New York City is 19°C greater than Eureka's.

Seattle and Spokane, both in Washington state, illustrate a second aspect of geographic position—mountains that act as barriers. Although Spokane is only about 360 kilometers (220 miles) east of Seattle, the towering Cascade Range separates the cities. Consequently, while the Pacific port city of Seattle has temperatures showing a moderating marine influence, Spokane's are more typically continental. Spokane is 7°C (13°F) cooler than Seattle in January and 4°C (7°F) warmer than Seattle in July. The annual range at Spokane is 11°C (20°F) greater than at Seattle. The Cascade Range effectively cuts Spokane off from the moderating influence of the Pacific Ocean.

Cloud Cover and Albedo

The extent of cloud cover is a factor that influences temperatures in the lower atmosphere. Cloud cover is important because many clouds have a high albedo; therefore, clouds reflect a significant portion of the sunlight that strikes them back into space (See Table 11.2, p. 286). By reducing the amount of incoming solar radiation, daytime temperatures will be lower than if the clouds were not present and the sky were clear (Figure 11.20A).

At night, clouds have the opposite effect as during daylight. They act as a blanket by absorbing outgoing terrestrial radiation and reradiating a portion of it back to the surface (Figure 11.20B). Consequently, some of the heat that otherwise would have been lost remains near the ground. Thus, nighttime air temperatures do not drop as low as they would on a clear night. The effect of cloud cover is to reduce the daily temperature range by lowering the daytime maximum and raising the nighttime minimum.

Clouds are not the only phenomenon that increase albedo and thereby reduce air temperatures. We also recognize that snow- and ice-covered surfaces have high albedos. This is one reason why mountain glaciers do not melt away in the summer and why snow may still be present on a mild spring day. In addition, during the winter, when snow covers the ground, daytime maximums on a sunny day are less than they otherwise would be because energy that the land would have absorbed and used to heat the air is reflected and lost.

World Distribution of Temperature

Temperature distribution is shown on a map by using **isotherms,** lines that connect places of equal temperature. On world maps that depict global patterns, temperatures are often adjusted to sea level to eliminate the complications caused by altitude variations. January and July are selected most often for analysis because, for most locations, they represent temperature extremes. You can see this in Figures 11.21 and 11.22.

On these figures the isotherms generally trend east–west and show a decrease in temperatures

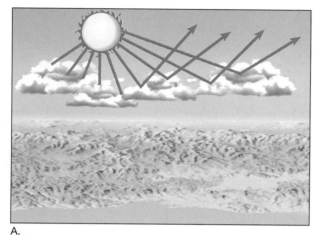

A.

B.

Figure 11.20 Clouds reduce daily temperature range. **A.** During daylight hours, clouds reflect solar radiation back to space. Therefore, the maximum temperature is lower than if the sky were clear. **B.** At night, the minimum temperature will not fall as low because clouds retard the loss of heat.

poleward from the tropics, illustrating one of the most fundamental and best-known aspects of the world distribution of temperature—that the effectiveness of incoming solar radiation in heating Earth's surface and the atmosphere above is largely a function of latitude. Further, there is a latitudinal shifting of temperatures caused by the seasonal migration of the Sun's vertical rays.

The added effect of the differential heating of land and water is also reflected on the January and July temperature maps. The warmest and coldest temperatures are found over land. Because temperatures do not fluctuate as much over water as over land, the seasonal north–south migration of isotherms is greater over the continents than over the oceans.

In addition, it is clear that the isotherms in the Southern Hemisphere, where there is little land and the oceans predominate, are much straighter and more stable than in the Northern Hemisphere, where they bend sharply northward in July and southward in January over the continents.

Isotherms also reveal the presence of ocean currents. Warm currents cause isotherms to be deflected poleward, whereas cold currents cause an equatorward bending. The horizontal transport of water poleward warms the overlying air and results in air temperatures that are higher than otherwise would be expected for the latitude. Conversely, currents moving toward the equator produce air temperatures cooler than expected.

Because Figures 11.21 and 11.22 show the seasonal extremes of temperature, they can be used to evaluate variations in the annual range of temperature from place to place. A comparison of the two maps shows that a station near the equator will record a very small annual range because it experiences little variation in the length of daylight and always has a relatively high Sun angle. By contrast, a station in the middle latitudes experiences much wider variations in Sun angle and length of daylight, and thus larger variations in temperature. Therefore, we can state that the annual temperature range increases with an increase in latitude.

Land and water also affect seasonal temperature variations, especially outside the tropics. A continental location must endure hotter summers and colder winters than a coastal location. Consequently, the annual range will increase with an increase in continentality.

A classic example of the effect of latitude and continentality on annual temperature range is Yakutsk, a city in Siberia, approximately 60 degrees north latitude and far from the influence of water. As a result, Yakutsk has an average annual temperature range of 62.2°C (112°F), one of the greatest in the world.

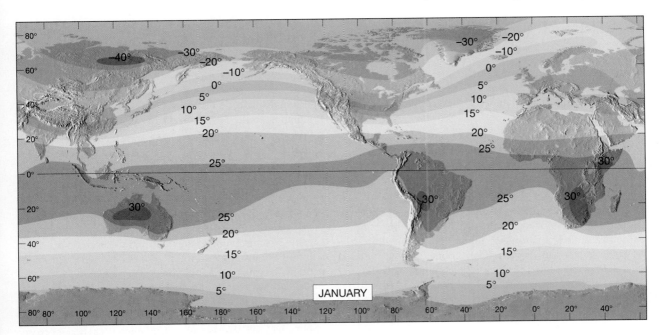

Figure 11.21 World distribution of mean temperatures (°C) for January.

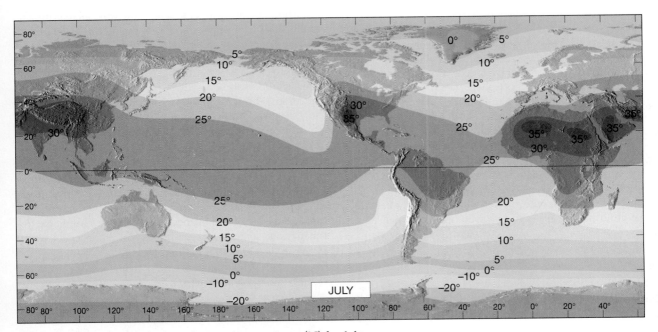

Figure 11.22 World distribution of mean temperatures (°C) for July.

The Chapter in Review

1. *Weather* is the state of the atmosphere at a particular place for a short period of time. *Climate,* on the other hand, is a generalization of the weather conditions of a place over a long period of time.

2. The most important *elements,* those quantities or properties that are measured regularly, of weather and climate are: (1) air *temperature,* (2) *humidity,* (3) type and amount of *cloudiness,* (4) type and amount of *precipitation,* (5) air *pressure,* and (6) the speed and direction of the *wind.*

3. If water vapor, dust, and other variable components of the atmosphere were removed, clean, dry air would be composed almost entirely of *nitrogen,* about 78% of the atmosphere by volume, and oxygen (O_2), about 21%. *Carbon dioxide* (CO_2), although present only in minute amounts (0.036%), is important because it has the ability to absorb heat radiated by Earth and thus helps keep the atmosphere warm. Among the variable components of air, *water vapor* is very important because it is the source of all clouds and precipitation and, like carbon dioxide, it is also a heat absorber.

4. *Ozone* (O_3), the triatomic form of oxygen, is concentrated in the 10- to 50-kilometer altitude range of the atmosphere, and is important to life because of its ability to absorb potentially harmful ultraviolet radiation from the Sun.

5. Because the atmosphere gradually thins with increasing altitude, it has no sharp upper boundary but simply blends into outer space. Based on temperature, the atmosphere is divided vertically into four layers. The *troposphere* is the lowermost layer. In the troposphere, temperature usually decreases with increasing altitude. This *environmental lapse rate* is variable, but averages about 6.5°C per kilometer (3.5°F per 1000 feet). Essentially all important weather phenomena occur in the troposphere. Beyond the troposphere is the *stratosphere,* which exhibits warming because of absorption of UV radiation by ozone. In the *mesophere,* temperatures again decrease. Upward from the mesosphere is the *thermosphere,* a layer with only a tiny fraction of the atmosphere's mass and no well-defined upper limit.

6. The two principal motions of Earth are (1) *rotation,* the spinning about its axis which produces the daily cycle of daylight and darkness; and (2) *revolution,* the movement in its orbit around the Sun.

7. Several factors acting together cause the seasons. Earth's axis is inclined 23 ½° from the perpendicular to the plane of its orbit around the Sun and remains pointed in the same direction (toward the North Star) as Earth journeys around the Sun. As a consequence, Earth's orientation to the Sun continually changes. Fluctuations in the Sun angle and length of daylight brought about by Earth's changing orientation to the Sun cause the seasons.

8. The three mechanisms of heat transfer are (1) *conduction,* the transfer of heat through matter by molecular activity; (2) *convection,* the transfer of heat by the movement of a mass or substance from one place to another; and (3) *radiation,* the transfer of heat by electromagnetic waves.

9. *Electromagnetic radiation* is energy emitted in the form of rays, or waves, called electromagnetic waves. All radiation is capable of transmitting energy through the vacuum of space. One of the most important differences between electromagnetic waves is their *wavelengths,* which range from very long *radio waves* to very short *gamma rays. Visible light* is the only portion of the electromagnetic spectrum we can see. Some of the basic laws that govern radiation as it heats the atmosphere are (1) all objects emit radiant energy, (2) hotter objects radiate more total energy than do colder objects, (3) the hotter the radiating body, the shorter the wavelengths of maximum radiation, and (4) objects that are good absorbers of radiation are good emitters as well.

10. The general drop in temperature with increasing altitude in the troposphere supports the fact that *the atmosphere is heated from the ground up.* Approximately 50 percent of the solar energy that strikes the top of the atmosphere reaches Earth's surface and is absorbed. Earth releases the absorbed radiation in the form of long-wave radiation. The atmospheric absorption of this long-wave *terrestrial radiation,* primarily by water vapor and carbon dioxide, is responsible for heating the atmosphere.

11. Carbon dioxide, an important heat absorber in the atmosphere, is one of several gases that influence *global warming.* Some consequences of global warming could be: (1) shifts in temperature and rainfall patterns, (2) a gradual rise in sea level, (3) changing storm tracks and a higher frequency and greater intensity of hurricanes, and (4) an increase in the frequency and intensity of heat waves and droughts.

12. The factors that cause temperature to vary from place to place, also called the *controls of temperature,* are (1) differences in the *receipt of solar radiation* due to latitude, (2) the unequal heating and cooling of *land and water,* (3) *altitude,* (4) *geographic position,* and (5) *ocean currents.*

13. Temperature distribution is shown on a map by using *isotherms,* which are lines that connect equal temperatures.

Key Terms

air (p. 272)

albedo (p. 286)

circle of illumination
  (p. 278)

climate (p. 272)

conduction (p. 282)

convection (p. 283)

element (of weather and
  climate) (p. 272)

environmental lapse rate
  (p. 276)

equinox (spring or autum-
  nal) (p. 280)

greenhouse effect (p. 287)

inclination of the axis
  (p. 279)

infrared (p. 284)

isotherm (p. 291)

mesosphere (p. 277)

radiation or electromag-
  netic radiation (p. 283)

revolution (p. 278)

rotation (p. 278)

solstice (summer or winter)
  (p. 280)

stratosphere (p. 277)

thermosphere (p. 277)

Tropic of Cancer (p. 280)

Tropic of Capricorn
  (p. 280)

troposphere (p. 276)

ultraviolet (UV) (p. 284)

visible light (p. 284)

weather (p. 272)

Questions for Review

1. Distinguish between weather and climate.
2. List the basic elements of weather and climate.
3. What are the two major components of clean, dry air (Figure 11.2, p. 273)? Are they important meteorologically?
4. Why are water vapor and dust important constituents of our atmosphere?
5. **a.** Why is ozone important to life on Earth?
   b. What are CFCs and what is their connection to the ozone problem?
   c. What is the most serious threat to human health of a decrease in the stratosphere's ozone?
6. What is the troposphere and why do some call it the "weather sphere"?
7. Why do temperatures rise in the stratosphere?
8. Briefly explain the primary cause of the seasons.
9. After examining Table 11.1 (p. 282), write a general statement that relates the season, the latitude, and the length of daylight.
10. Describe the relationship between the temperature of a radiating body and the wavelengths it emits.
11. Distinguish among the three basic mechanisms of heat transfer.
12. Figure 11.15 (p. 285) illustrates what happens to incoming solar radiation. The percentages shown, however, are only global averages. In particular, the amount of solar radiation reflected (albedo) may vary considerably. What factors might cause variations in albedo?
13. How does Earth's atmosphere act as a "greenhouse"?
14. How are temperatures in the lower atmosphere likely to change as carbon dioxide levels continue to increase? Why?
15. Quito, Ecuador, is located on the equator and is not a coastal city. It has an average annual temperature of only 13°C (55°F). What is the likely cause for this low average temperature?
16. In what ways can geographic position be considered a control of temperature?
17. Yakutsk is an inland city located in Siberia at about 60 degrees north latitude. This Russian city has one of the highest average annual temperature ranges in the world: 62.2°C (112°F). Explain the reasons for the very high annual temperature range.

Testing What You Have Learned

Multiple-Choice Questions

1. Which one of the following is NOT considered an element of weather and climate?
   - **a.** air temperature
   - **c.** altitude
   - **e.** air pressure
   - **b.** cloudiness
   - **d.** humidity

2. The most abundant gases in the atmosphere by volume are _____
   - **a.** ozone and oxygen.
   - **c.** oxygen and argon.
   - **e.** water vapor and oxygen.
   - **b.** helium and ozone.
   - **d.** nitrogen and oxygen.

3. Water vapor _____
   - **a.** makes clouds dark.
   - **c.** is invisible.
   - **e.** exists as tiny liquid droplets.
   - **b.** feels wet.
   - **d.** colors the sky blue.

4. Which atmospheric gas filters out most of the ultraviolet radiation in sunlight?
   - **a.** argon
   - **c.** helium
   - **e.** ozone
   - **b.** nitrogen
   - **d.** carbon dioxide

5. Earth's atmosphere is divided vertically into four layers on the basis of _____
   - **a.** temperature.
   - **c.** pressure.
   - **e.** color.
   - **b.** composition.
   - **d.** moisture.

6. Changes in which of the following are responsible for seasonal temperature changes?
   - **a.** Sun angle
   - **c.** Earth-sun distance
   - **e.** a., b., and c.
   - **b.** length of daylight
   - **d.** a. and b.

7. One of the most important differences between the various types of electromagnetic radiation is _____
   - **a.** color.
   - **c.** name.
   - **e.** wavelength.
   - **b.** velocity.
   - **d.** wave height.

8. Approximately _____ percent of the solar energy that strikes the top of the atmosphere reaches the surface and is absorbed.
   - **a.** 10
   - **c.** 50
   - **e.** 90
   - **b.** 30
   - **d.** 70

9. The most realistic models available predict an average global surface temperature increase of about _____ by the second half of the next century.
   - **a.** 0.1°C
   - **c.** 8.5°C
   - **e.** 23.0°C
   - **b.** 2.5°C
   - **d.** 12.0°C

10. Which one of the locations listed below should have the highest annual (yearly) temperature range?
    - **a.** equatorial island
    - **c.** tropical coast
    - **e.** mid-latitude continental center
    - **b.** mid-latitude coast
    - **d.** polar ice cap

Fill-In Questions

11. The word used to denote the state of the atmosphere at a particular place for a short period of time is ___weather___, while ___climate___ generalizes the state of the atmosphere over a long period of time.

12. The temperature decrease in the troposphere is called the _____ _____ _____.

13. The two principal motions of Earth are ___rotation___ about its axis and ___revolution___ in its orbit about the sun.

14. On June 21–22, the vertical rays of the Sun strike a latitude known as the Tropic of ___Cancer___.

15. The two gases _____ _____ and _____ _____, are the principal atmospheric absorbers of terrestrial radiation.

True/False Questions

16. The amount of water vapor in the air varies considerably, from almost none at all to about 4 percent by volume. _____

17. The line separating the dark half of Earth from the lighted half is called the circle of illumination. _____

18. During an equinox, the length of daylight is 12 hours everywhere on Earth. _____

19. When any form of radiant energy is absorbed by an object, the result is an increase in temperature. _____

20. Temperature distribution is often shown on a map by using isotherms. _____

Answers:

1. c; 2. d; 3. c; 4. e; 5. a; 6. d; 7. e; 8. c; 9. b; 10. e; 11. weather, climate; 12. environmental lapse rate; 13. rotation, revolution; 14. Cancer; 15. carbon dioxide, water vapor; 16. T; 17. T; 18. T; 19. T; 20. T.

Web   Work

The following are informative and interesting World Wide Web sites that address topics related to those presented in this chapter:

- Current Global Temperatures
  http://www.ssec.wisc.edu/data/comp/latest_cmobb.gif

- Earth's Radiation Energy Balance
  http://oldthunder.ssec.wisc.edu/wxwise/homerbe.html

For direct access to these sites and other relevant Web locations, contact the *Foundations of Earth Science* WWW home page at http://www.prenhall.com/lutgens.

CHAPTER 12
Clouds and Precipitation

FOCUS ON LEARNING

To assist you in learning the important concepts
in this chapter, you will find it helpful to focus
on the following questions:

1. What processes cause water to change from one state of matter to another?
2. What is humidity? What is the most common method used to express humidity?
3. What is the basic cloud-forming process?
4. What controls the stability of the atmosphere?
5. What are three mechanisms that initiate the vertical movement of air?
6. What are the necessary conditions for condensation?
7. What two criteria are used for cloud classification?
8. What is fog? How do fogs form?
9. How is precipitation produced in a cloud? What are the different forms of precipitation?

Storm clouds over Havana, Cuba. *Photo by Lorne Resnick/Tony Stone Images*

Scientists agree: when it comes to understanding atmospheric processes, *water vapor is the most important gas in the atmosphere.* Water vapor constitutes only a small fraction of the gases in the atmosphere, varying from nearly 0 to about 4 percent by volume. But the importance of water in the air vastly exceeds what these small percentages would indicate.

Recall from Chapter 11 that water vapor is an important heat-absorbing gas. This property is critical to the heating of the atmosphere. Water vapor is also the source of all condensation and precipitation. Clouds and fog, as well as rain, snow, sleet, and hail, are among the most conspicuous and observable of weather phenomena.

We all expect rain to fall when certain cloud types are present, yet most of us have no idea of the complex processes that must take place in a cloud to produce rain. The formation of a single average raindrop requires water from nearly a million microscopic cloud droplets. What mechanisms foster the creation of raindrops? The wide variation in the amount of precipitation from place to place, and the local differences from time to time, both have a significant impact on the nature of the physical landscape and on our lifestyles.

Changes of State

Water vapor is an odorless, colorless gas that mixes freely with the other gases of the atmosphere. Oxygen and nitrogen, the two gases that comprise 99 percent of the atmosphere, change state (solid to liquid to gas) at extremely low temperatures (−200°C or so). But water vapor changes state at the temperatures and pressures experienced near Earth's surface.

It is because of these changes of state at "ordinary" temperatures that water leaves the oceans as a gas and returns again as a liquid. The processes that involve such changes of state require that *heat* be absorbed or released (Figure 12.1). This heat energy is measured in calories. One *calorie* is the amount of heat required to raise the temperature of 1 gram of water 1°C. Thus, when 10 calories of heat are added to 1 gram of water, a 10°C temperature rise occurs.

Importance of Latent Heat

Under certain conditions, heat may be added to a substance without an accompanying temperature change. This may sound impossible, but consider what occurs during a change in state. To illustrate, let us apply heat to a container of ice water. Its temperature is 0°C. We keep adding heat, but the temperature remains constant until all the ice is melted. *Then* the temperature begins to rise.

Where did the heat go during melting? In this case, the energy was used to disrupt the internal crystalline structure of the ice cubes, freeing the water molecules to move around—in other words, to melt. *Because this heat energy is not associated with a temperature change,* it is referred to as **latent heat** (latent means hidden). This energy is not available as heat until the liquid returns to the solid state, when the heat is released.

As we shall examine later, latent heat plays a crucial role in many atmospheric processes. In particular, the release of latent heat aids the formation of the towering clouds often seen on warm summer days, and it is the major source of energy for violent thunderstorms, tropical storms, and hurricanes.

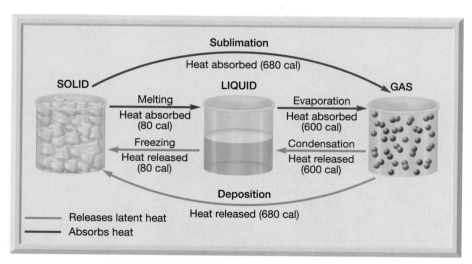

Figure 12.1 Changes of state.

Figure 12.2 Frost on a window pane. *(Photo by D. Cavagnaro/DRK Photo)*

Liquid/Gas Change of State The process of converting a liquid to a gas is termed **evaporation.** It takes approximately 600 calories of energy to convert 1 gram of water to water vapor (Figure 12.1). The energy absorbed by the water molecules during evaporation is used solely to give them the motion needed to escape the surface of the liquid and become a gas. This energy is referred to as *latent heat of vaporization.* During the process of evaporation, it is the higher-temperature (faster-moving) molecules that escape the surface. As a result, the average molecular motion (temperature) of the remaining water is reduced, causing a cooling effect—hence the common expression, "evaporation is a cooling process." You have undoubtedly experienced this cooling effect when stepping dripping wet from a swimming pool or bathtub.

Condensation is the process whereby water vapor changes to the liquid state. For condensation to occur, the water molecules must release their stored heat energy (*latent heat of condensation*), equal to what was absorbed during evaporation. This released energy plays an important role in producing violent weather and can transfer great quantities of heat from tropical oceans to more poleward locations.

Solid/Liquid Change of State. **Melting** is the process by which a solid is changed to a liquid. It requires absorption of approximately 80 calories of heat per gram of water (*latent heat of melting*). **Freezing,** the reverse process, releases these 80 calories per gram as *latent heat of fusion.*

The last of the processes illustrated in Figure 12.1 are sublimation and deposition. **Sublimation** is the conversion of a solid directly to a gas, without passing through the liquid state. You may have observed this change in watching the sublimation of "dry ice" (frozen carbon dioxide) into white, wispy vapor, sometimes used to generate "fog" or "smoke" in theatrical productions. Water ice sublimates, too, which is why unused ice cubes gradually shrink in a freezer.

The term **deposition** is used to denote the reverse process, the conversion of a vapor directly to a solid. This change occurs, for example, when water vapor is deposited as ice on cold objects such as grass or windows (Figure 12.2). These deposits are termed *white frost* or *hoar frost* but are generally just called *frost.* A household example of deposition is the "frost" that accumulates in a freezer compartment. As shown in Figure 12.1, sublimation and deposition involve an amount of energy equal to the total of the other two processes.

Humidity: Water Vapor in the Air

Humidity is the general term for the amount of water vapor in air. There are different ways of looking at humidity; we will examine two—specific humidity and relative humidity.

But, before we consider these humidity measures, you need to understand the concept of **saturation.** Imagine a closed jar half full of water and half full of dry air, both at the same temperature. As the water begins to evaporate from the water surface, a small increase in pressure can be detected in the air above. This increase is the result of the motion of the water vapor molecules that were added to the air through evaporation. In the open atmosphere, this pressure is termed **vapor pressure** and is defined as that part of the total atmospheric pressure that can be attributed to the water vapor content.

In the closed container, as more and more molecules escape from the water surface, the steadily increasing vapor pressure in the air above forces more and more of these molecules to return to the liquid. Eventually the number of vapor molecules returning to the surface will balance the number leaving. At that point, the air is said to be *saturated,* or filled to capacity. However, if we add heat to the container, increasing the temperature of the water and air, more water will evaporate before a balance is reached. Consequently, at higher temperatures, more moisture is required for saturation. Stated another way, the water vapor capacity of air is *temperature* dependent, with warm air having a much greater capacity than cold air. The amount of water vapor required for saturation at various temperatures varies surprisingly, as shown in Table 12.1.

Specific Humidity

Not all air is saturated, of course. Thus, we need ways to express how humid a parcel of air is. One method specifies the amount of water vapor contained in a unit of air. **Specific humidity** is the weight of water vapor per weight of a chosen mass of air, including the water vapor. Because it is measured in units of weight (usually in grams per kilogram), specific humidity is not affected by changes in pressure or temperature. *Specific humidity is a measure of the actual quantity of water vapor in a given mass of air.*

Relative Humidity

The most familiar, and perhaps the most misunderstood, term used to describe the moisture content of air is relative humidity. Stated in an admittedly oversimplified manner, **relative humidity** *is the ratio of the air's actual water vapor content to its potential water vapor capacity at a given temperature.* To illustrate, we see from Table 12.1 that at 25°C air is saturated when it contains 20 grams of water vapor per kilogram of air. Thus, if the air contains only 10 grams per kilogram on a 25°C day, the relative humidity is expressed as 10/20 or 50 percent. Further, if air with a temperature of 25°C had a water vapor content of 20 grams per kilogram, the relative humidity would be expressed at 20/20 or 100 percent. On those occasions when the relative humidity reaches 100 percent, the air is saturated.

Because relative humidity is a comparison of the air's water vapor content (specific humidity) and its capacity, relative humidity can be changed in either of two ways. First, if moisture is added to or subtracted from air, its relative humidity will change. This is illustrated in Figure 12.3.

The second condition that affects relative humidity is air temperature. Examine Figure 12.4 carefully, and note in Part A that, when air at 20°C contains 7 grams of water vapor per kilogram of air, it has a relative humidity of 50 percent. When the air is cooled from 20°C to 10°C as shown in Part B, it still contains 7 grams of water vapor, but the relative humidity increases from 50 percent to 100 percent. We can conclude from this that, when the water vapor content remains constant, *a decrease in temperature results in an increase in relative humidity.*

What happens when the air is cooled further, below the temperature at which saturation occurs? Part C of Figure 12.4 illustrates this situation. Notice from Table 12.1 that, when the flask is cooled to 0°C, the air is saturated at 3.5 grams of water vapor per kilogram of air. Because this flask originally contained 7 grams of water vapor, 3.5 grams of water vapor have to go somewhere, because there is no room for them in the cooler air. They will condense to form liquid droplets that collect on the walls of the

Table 12.1 ■ Amount of water vapor (grams) required to saturate a kilogram of air at various temperatures		
Temperature		Grams of water vapor per kg of air
(°C)	(°F)	
−40	−40	0.1
−30	−22	0.3
−20	−4	0.75
−10	14	2
0	32	3.5
5	41	5
10	50	7
15	59	10
20	68	14
25	77	20
30	86	26.5
35	95	35
40	104	47

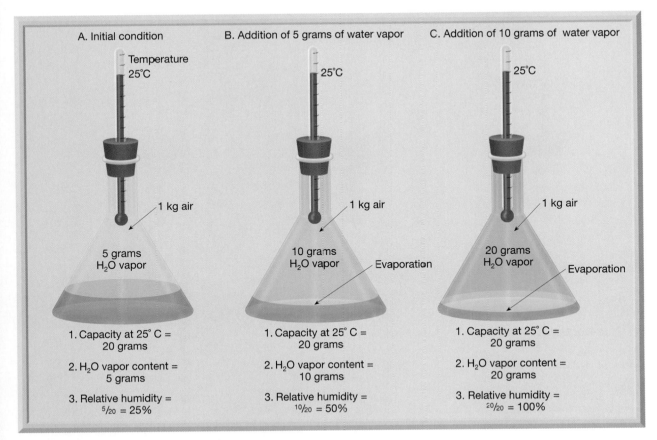

Figure 12.3 When the temperature remains constant, relative humidity will increase as water vapor is added to the air. Here the water vapor capacity remains constant at 20 grams per kilogram, while the relative humidity rises from 25 percent to 100 percent as the water vapor content (specific humidity) increases.

container. In the meantime, the relative humidity of the air inside remains at 100 percent.

We can generalize the effects of temperature on relative humidity as follows. When the water vapor content of air remains at a constant level, a decrease in air temperature results in an increase in relative humidity, and an increase in temperature causes a decrease in relative humidity. In Figure 12.5 the variations in temperature and relative humidity during a typical day demonstrate the relationship just described.

Dew Point

Another important idea related to relative humidity is the dew-point temperature. **Dew point** is the *temperature* to which air would have to be cooled to reach saturation. Note that in Figure 12.4 unsaturated air at 20°C is cooled to 10°C before saturation occurs. Therefore, 10°C would be the dew-point temperature for this air. If this same parcel of air

were cooled further, the air's capacity would be exceeded, and the excess vapor would condense.

This brings up an important concept. When air in the atmosphere is cooled below its dew point, some of the water vapor condenses to form clouds. Because clouds are made of *liquid droplets,* this moisture is no longer part of the *water vapor* content of the air. (Clouds are not water vapor; they are liquid water droplets too tiny to fall to Earth as rain.)

In summary, dew point is the temperature to which air would have to be cooled to reach saturation, and relative humidity indicates how near the air is to being saturated.

Measuring Humidity

Relative humidity is commonly measured using a **hygrometer**. One type of hygrometer, called a *sling psychrometer*, consists of two identical thermometers mounted side by side (Figure 12.6). One

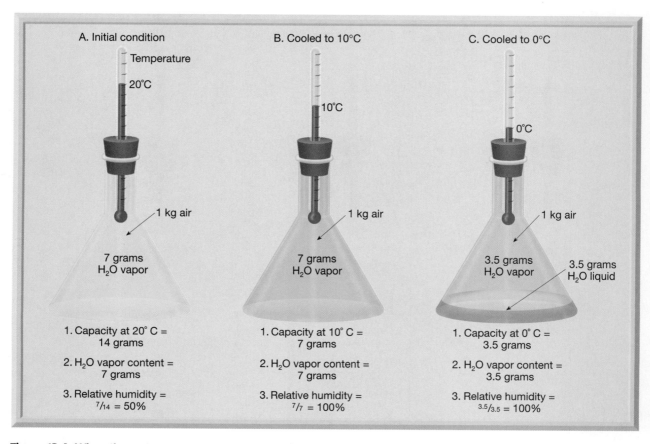

Figure 12.4 When the water vapor content remains constant, the relative humidity can be changed by increasing or decreasing the air temperature. In this example, when the temperature of the air in the flask was lowered from 20°C to 10°C, the relative humidity increased from 50 percent to 100 percent. Thus, 10°C is the dew point. Further cooling (from 10°C to 0°C) causes one-half of the water vapor to condense because colder air holds less moisture. In nature, cooling of air below its dew point generally causes condensation in the form of clouds, dew, or fog.

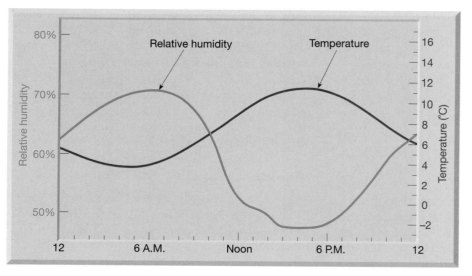

Figure 12.5 Typical daily variations in temperature and relative humidity during a spring day at Washington, D.C. When temperature increases, relative humidity drops (see mid-afternoon) and vice versa.

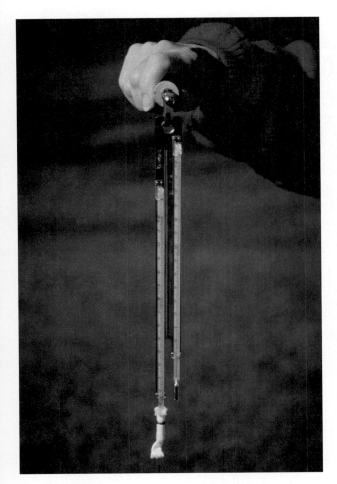

Figure 12.6 Sling psychrometer. This instrument is used to determine both relative humidity and dew point. The dry-bulb thermometer gives the current air temperature. The wet-bulb thermometer is covered with a cloth wick that is dipped in water. The thermometers are spun until the temperature of the wet-bulb thermometer stops declining. Then the thermometers are read and the data used in conjunction with Table 12.2. *(Photo by E. J. Tarbuck)*

thermometer, the *dry-bulb,* gives the present air temperature. The other, called the *wet-bulb* thermometer, has a thin muslin wick tied around the end (see end of thermometers in the photo).

To use the psychrometer, the cloth sleeve is saturated with water and a continuous current of air is passed over the wick. This is done either by swinging the instrument freely in the air or by fanning air past it. As a consequence, water evaporates from the wick, and the heat absorbed by the evaporating water makes the temperature of the wet bulb drop. The loss of heat that was required to evaporate water from the wet bulb lowers the thermometer reading.

The amount of cooling that takes place is directly proportional to the dryness of the air. The drier the air, the more moisture evaporates. The more heat the evaporating water absorbs, the greater the cooling. Therefore, the larger the difference that is observed between the thermometer readings, the lower the relative humidity; the smaller the difference, the higher the relative humidity. If the air is saturated, no evaporation will occur, and the two thermometers will have identical readings.

To determine the precise relative humidity from the thermometer readings, a standard table is used (Table 12.2). With the same information, but using a different table, the dew-point temperature may also be calculated.

The Basis of Cloud Formation: Adiabatic Cooling

Up to this point, we have considered basic properties of water vapor and how its variability is measured. We are now ready to examine some of the important roles that water vapor plays in weather, especially the formation of clouds.

Fog and Dew versus Cloud Formation

Recall that condensation occurs when water vapor changes to a liquid. This condensation may form dew, fog, or clouds. Although these three forms are different, all require saturated air to develop. As indicated earlier, saturation occurs either when water vapor is added to the air or, more commonly, when the air is cooled to its dew point. Near Earth's surface heat is readily exchanged between the ground and the air above. During evening hours, the surface radiates heat away, and the surface and adjacent air cool rapidly. This "radiation cooling" accounts for the formation of dew and some types of fog. At higher altitudes, however, clouds often form during the warmest part of the day, not during the cool evening or night. Consequently, some other mechanism must operate to form clouds.

Adiabatic Temperature Change

The process that is responsible for most cloud formation is easily visualized if you have ever pumped up a bicycle tire and noticed that the pump barrel became quite warm. The heat you felt was the consequence of the work you did on the air to compress it. When energy is used to compress air, the motion of the gas molecules increases

Table 12.2 ■ Relative humidity (percent)*

Dry bulb (°C)	Depression of Wet-Bulb Temperature (Dry-Bulb Temperature Minus Wet-Bulb Temperature = Depression of the Wet Bulb)																					
	1	2	3	4	5	6	7	8	9	10	11	12	13	14	15	16	17	18	19	20	21	22
−20	28																					
−18	40																					
−16	48	0																				
−14	55	11																				
−12	61	23																				
−10	66	33	0																			
−8	71	41	13																			
−6	73	48	20	0																		
−4	77	54	32	11																		
−2	79	58	37	20	1																	
0	81	63	45	28	11																	
2	83	67	51	36	20	6																
4	85	70	56	42	27	14																
6	86	72	59	46	35	22	10	0														
8	87	74	62	51	39	28	17	6														
10	88	76	65	54	43	33	24	13	4													
12	88	78	67	57	48	38	28	19	10	2												
14	89	79	69	60	50	41	33	25	16	8	1											
16	90	80	71	62	54	45	37	29	21	14	7	1										
18	91	81	72	64	56	48	40	33	26	19	12	6	0									
20	91	82	74	66	58	51	44	36	30	23	17	11	5									
22	92	83	75	68	60	53	46	40	33	27	21	15	10	4	0							
24	92	84	76	69	62	55	49	42	36	30	25	20	14	9	4	0						
26	92	85	77	70	64	57	51	45	39	34	28	23	18	13	9	5						
28	93	86	78	71	65	59	53	45	42	36	31	26	21	17	12	8	4					
30	93	86	79	72	66	61	55	49	44	39	34	29	25	20	16	12	8	4				
32	93	86	80	73	68	62	56	51	46	41	36	32	27	22	19	14	11	8	4			
34	93	86	81	74	69	63	58	52	48	43	38	34	30	26	22	18	14	11	8	5		
36	94	87	81	75	69	64	59	54	50	44	40	36	32	28	24	21	17	13	10	7	4	
38	94	87	82	76	70	66	60	55	51	46	42	38	34	30	26	23	20	16	13	10	7	5
40	94	89	82	76	71	67	61	57	52	48	44	40	36	33	29	25	22	19	16	13	10	7

Relative Humidity Values

*To determine the relative humidity, find the air (dry-bulb) temperature on the vertical axis (far left) and the depression of the wet-bulb on the horizontal axis (top). Where the two meet, the relative humidity is found. For example, when the dry-bulb temperature is 20°C and a wet-bulb temperature is 14°C, then the depression of the wet-bulb is 6°C (20°C – 14°C). From Table 12.2, the relative humidity is 51 percent.

and therefore the temperature of the air rises. Conversely, air that is allowed to escape from a bicycle tire *expands and cools*. This results because the expanding air pushes (does work on) the surrounding air and must cool by an amount equivalent to the energy expended.

You may have experienced the cooling effect of an expanding gas while applying a spray deodorant. As the compressed gas propellant in the aerosol can is released, it quickly expands and cools. This drop in temperature occurs *even though heat is neither added nor subtracted*. Such variations are known as adiabatic temperature changes and result when air is compressed or allowed to expand. In summary, when air is allowed to expand, it cools, and when it is compressed, it warms.

Dry Adiabatic Rate As you travel from Earth's surface upward through the atmosphere, the atmospheric pressure rapidly diminishes, because there are fewer and fewer gas molecules. Thus, any time a parcel of air moves upward, it passes through regions of successively lower pressure. As a result, the ascending air expands. As it expands, it cools adiabatically. Unsaturated air cools at the rather constant rate of 10°C for every 1000 meters of ascent (1°C per 100 meters).

Conversely, descending air comes under increasingly higher pressures, compresses, and is heated 10°C for every 1000 meters of descent. This rate of cooling or heating applies only to unsaturated air and is known as the **dry adiabatic rate.**

Wet Adiabatic Rate If air rises high enough, it will cool sufficiently to reach the dew point, and cause condensation. From this point on along its ascent, *latent heat of condensation* stored in the water vapor will be liberated. Although the air will continue to cool after condensation begins, the released latent heat works against the adiabatic process, thereby reducing the rate at which the air cools. This slower rate of cooling caused by the addition of latent heat is called the **wet adiabatic rate** of cooling. Because the amount

of latent heat released depends on the quantity of moisture present in the air, the wet adiabatic rate varies from 5°C per 1000 meters for air with a high moisture content to 9°C per 1000 meters for dry air.

Figure 12.7 illustrates the role of adiabatic cooling in the formation of clouds. Note that from the surface up to the condensation level the air cools at the dry adiabatic rate. The wet adiabatic rate commences at the condensation level.

Stability of Air

As you have learned, if air rises, it will cool adiabatically and eventually produce clouds. But why does air rise on some occasions and not on others? Why do the size of clouds and the amount of precipitation vary so much when air does rise? The answers to these questions are closely related to the stability of the air.

Imagine if you will, a large bubble of air with a thin flexible cover that allows it to expand but prevents it from mixing with the surrounding air. If the imaginary bubble were forced to rise, its temperature would *decrease because of expansion*. By comparing the bubble's temperature to that of the surrounding air, we can determine the stability of the bubble.

If the bubble's temperature is lower than that of its environment, it will be denser, and if allowed to move freely, it would sink to its original position. Air of this type is termed *stable air* because it resists vertical displacement.

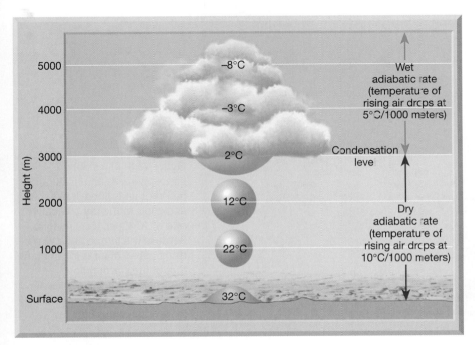

Figure 12.7 Rising air cools at the dry adiabatic rate of 10°C per 1000 meters, until the air reaches the dew point and condensation (cloud formation) begins. As air continues to rise, the latent heat released by condensation reduces the rate of cooling. The wet adiabatic rate is therefore always less than the dry adiabatic rate.

On the other hand, if our imaginary bubble were warmer, and therefore less dense than the surrounding air, it would continue to rise until it reached an altitude having the same temperature, much as a hot-air balloon rises as long as it is less dense than the surrounding air. This type of air is called *unstable air.*

Determining Stability

The stability of air is determined by examining the temperature of the atmosphere at various heights. Recall from Chapter 11 that this measure is termed the *environmental lapse rate.* The environmental lapse rate is the temperature of the atmosphere as determined from observations made by radiosondes and aircraft. It is important not to confuse this with *adiabatic temperature changes,* which are changes in temperature due to expansion or compression as a parcel of air rises or descends.

Stable Air For illustration, we will examine a situation where the environmental lapse rate is 5°C per 1000 meters (Figure 12.8). Under this condition, when air at the surface has a temperature of 25°C, the air at 1000 meters will be 5 degrees cooler, or 20°C, the air at 2000 meters will have a temperature of 15°C, and so forth. At first glance it appears that the air at the surface is less dense than the air at 1000 meters,

because it is 5 degrees warmer. However, if the air near the surface is unsaturated and were to rise to 1000 meters, it would expand and cool at the dry adiabatic rate of 10°C per 1000 meters. Therefore, upon reaching 1000 meters, its temperature would have dropped 10°C. Being 5 degrees cooler than its environment, it would be denser and tend to sink to its original position. Hence, we say that the air near the surface is potentially cooler than the air aloft and therefore will not rise on its own. The air just described is *stable* and resists vertical movement.

Stated quantitatively, *stability* prevails when the environmental lapse rate is less than the wet adiabatic rate. Figure 12.9 depicts this situation using an environmental lapse rate of 5°C per 1000 meters and a wet adiabatic rate of 6°C per 1000 meters. Note that at 1000 meters the temperature of the surrounding air is 15°C, while the rising parcel of air has cooled to 10°C and is therefore the denser air. Even if this stable air were to be forced above the condensation level, it would remain cooler and denser than its environment, and thus it would tend to return to the surface.

Unstable Air At the other extreme, air is said to be *unstable* when the environmental lapse rate is greater than the dry adiabatic rate. As shown in Figure 12.10, the ascending parcel of air is always warmer than its environment and will continue to rise because of its

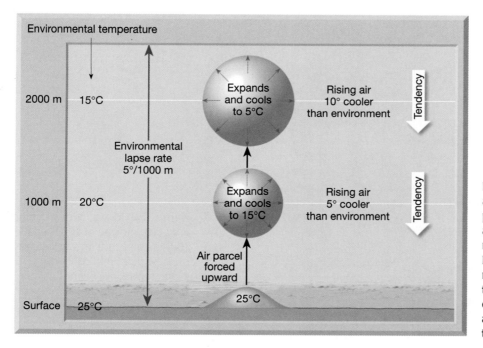

Figure 12.8 In a stable atmosphere, as an unsaturated parcel of air is lifted, it expands and cools at the dry adiabatic rate of 10°C per 1000 meters. Because the temperature of the rising parcel of air is lower than that of the surrounding environment, it will be heavier and, if allowed to do so, will sink to its original position.

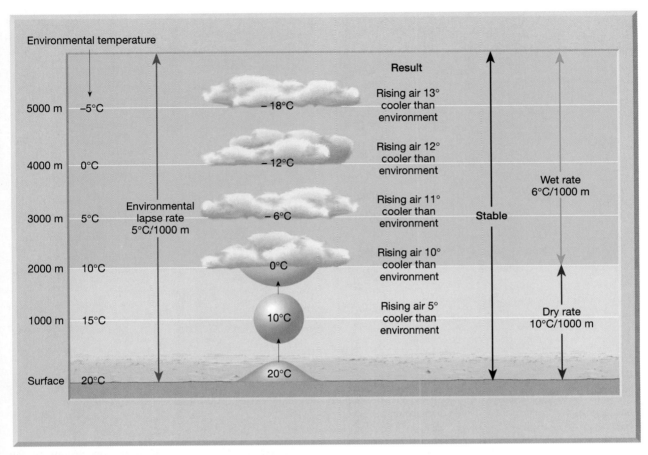

Environmental temperature

		Result
5000 m	−5°C	−18°C — Rising air 13° cooler than environment
4000 m	0°C	−12°C — Rising air 12° cooler than environment
3000 m	5°C	Environmental lapse rate 5°C/1000 m — −6°C — Rising air 11° cooler than environment
2000 m	10°C	0°C — Rising air 10° cooler than environment
1000 m	15°C	10°C — Rising air 5° cooler than environment
Surface	20°C	20°C

Stable

Wet rate 6°C/1000 m

Dry rate 10°C/1000 m

Figure 12.9 Stability prevails when the environmental lapse rate is less than the wet adiabatic rate. The rising parcel of air is therefore always cooler and denser than the surrounding air. When stable air is forced to rise, it produces flat, layered clouds.

own buoyancy. However, such instability is generally limited to near Earth's surface. On hot, sunny days the air above some surfaces, like paved shopping centers, is heated more than the air over adjacent surfaces. These invisible pockets of more intensely heated air, being less dense than the air aloft, will rise like a hot-air balloon. This phenomenon produces the small fluffy clouds we associate with fair weather. Occasionally, when the surface air is considerably warmer than the air aloft, clouds with great vertical development can form, as shown in Figure 12.10. Clouds of this type are associated with heavy precipitation.

In summary, the stability of air is determined by examining the temperature of the atmosphere at various heights. In simple terms, a column of air is deemed unstable when the air near the bottom of this layer is significantly warmer (less dense) than the air aloft, indicating a steep environmental lapse rate.

Under these conditions the air actually turns over; the warm air below rises and displaces the colder air aloft. Conversely, the air is considered to be stable when the temperature drops gradually with increasing altitude.

Stability and Daily Weather

From the previous discussion we can conclude that stable air resists vertical movement, whereas unstable air ascends freely because of its own buoyancy. But how do these facts manifest themselves in our daily weather?

Because stable air resists upward movement, we might conclude that clouds will not form when stable conditions prevail in the atmosphere. Although this seems reasonable, processes exist that *force* air aloft. These will be discussed in the following section, "Processes That Lift Air." On occasions when stable air is forced aloft, the clouds that form are

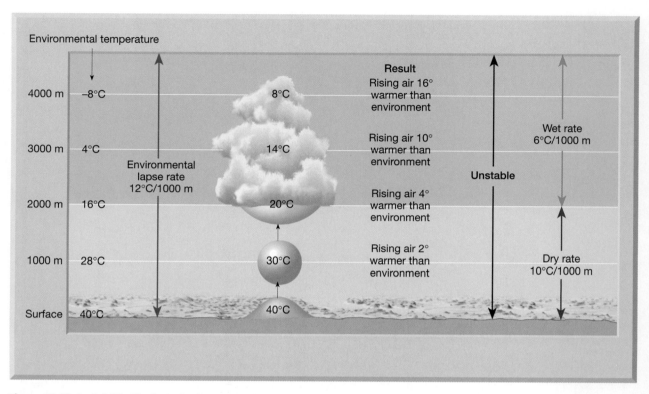

Figure 12.10 Instability illustrated using an environmental lapse rate of 12°C per 1000 meters. The rising air is always warmer and therefore lighter than the surrounding air.

widespread and have little vertical thickness when compared to their horizontal dimension, and precipitation, if any, is light to moderate.

By contrast, clouds associated with unstable air are towering and are usually accompanied by heavy precipitation. For this reason, we can conclude that on a dreary, overcast day with light drizzle, stable air has been forced aloft. On the other hand, during a day when cauliflower-shaped clouds appear to be growing as if bubbles of hot air are surging upward, we can be fairly certain that the ascending air is unstable.

In summary, the role of stability in determining our daily weather is very important. To a large degree, stability determines the type of clouds that develop and whether precipitation will come as a gentle shower or a heavy downpour.

Processes That Lift Air

Earlier you learned that on some summer days, surface heating can be sufficient to cause pockets of air to be warmed more than the surrounding air. If the temperature of this air is sufficiently high and ample moisture is present, these unstable parcels will rise to produce clouds and occasionally precipitation. By contrast, air that is stable will not rise on its own. It requires some mechanism to initiate the vertical movement. Three such mechanisms are orographic lifting, frontal wedging, and convergence.

Orographic Lifting

Orographic lifting occurs when elevated terrains, such as mountains, act as barriers to flowing air (Figure 12.11). As air is pushed up a mountain slope, adiabatic cooling often generates clouds and copious precipitation. In fact, many of the rainiest places in the world are located on windward mountain slopes, such as the Olympic Mountains in Washington State.

In addition to providing the lift to render air unstable, mountains remove additional moisture in other ways. By slowing the horizontal flow of air, they retard the passage of storm systems. Moreover, the irregular topography of mountains enhances differential heating and instability. These combined effects account for the generally higher precipitation associated with mountainous regions compared with surrounding lowlands.

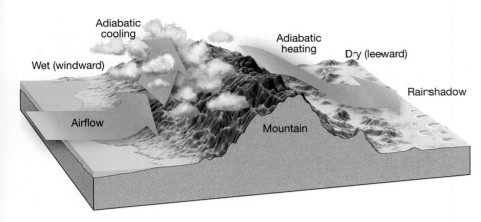

Figure 12.11 Orographic lifting. The name comes from *oro-*, which means "mountain."

By the time air reaches the leeward side of a mountain, much of its moisture has been lost. And, if the air descends, it becomes compressed and warms adiabatically, making condensation and precipitation even less likely. As shown in Figure 12.11, the result can be a **rainshadow desert.** The Great Basin desert of the western United States lies only a few hundred kilometers from the Pacific but is effectively cut off from the ocean's moisture by the imposing Sierra Nevada range (Figure 12.12). The Gobi desert of Mongolia, the Takla Makan of China, and the Patagonia desert of Argentina are other examples of rainshadow deserts found on the leeward sides of mountains.

Frontal Wedging

If orographic lifting were the only mechanism that forced air aloft, the relatively flat central portion of

Figure 12.12 Rainshadow desert. The arid conditions in Death Valley can be partially attributed to the adjacent mountains, which effectively remove the moisture from air originating over the Pacific. *(Photo by Scott T. Smith)*

North America would be an expansive desert. Fortunately, this is not the case. **Frontal wedging** occurs when cool air acts as a barrier over which warmer, less dense air rises. Figure 12.13 illustrates frontal wedging of both stable and unstable air. As this figure shows, forceful lifting is important in producing clouds. The stability of the air, however, determines to a great extent the type of clouds formed and the amount of precipitation that may be expected.

These weather-producing fronts are part of the storm systems called *middle-latitude cyclones*. These cyclones, which show up as the "lows" you see on weather reports, are responsible for producing a high proportion of the precipitation in the middle latitudes. We will examine them closely in Chapter 14.

Convergence

Whenever air masses flow together, **convergence** is said to occur. Such flow results in general upward movement, because when air converges, it has to go somewhere so the height of the air column increases. Consequently, air within the column must move upward.

The Florida peninsula provides an excellent example of the role that convergence can play in initiating cloud development and precipitation. On warm days, the airflow is from the ocean to the land along both coasts of Florida. This leads to general convergence over the peninsula. Such a pattern of air movement and the uplift that results is aided by intense solar heating. Rising parcels of humid air expand and cool adiabatically, the water vapor condenses to form clouds, and the result is the greatest frequency of mid-afternoon thunderstorms in the United States. More important, convergence as a mechanism of forceful lifting is a major contributor to

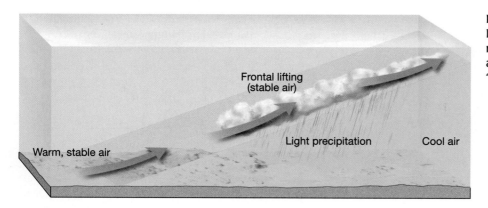

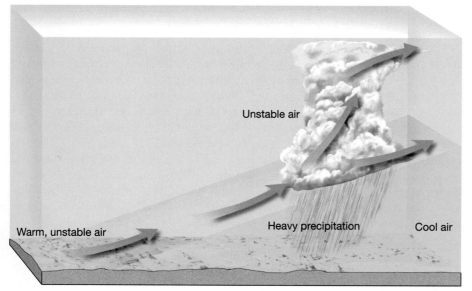

Figure 12.13 A. When stable air is lifted, layered clouds usually result. **B.** When warm, unstable air is forced to rise over cooler air, "towering" clouds develop.

the stormy weather associated with middle-latitude cyclones and hurricanes, discussed in Chapter 14.

Condensation and Cloud Formation

To review briefly, condensation occurs when water vapor in the air changes to a liquid. The result of this process may be dew, fog, or clouds. For any of these forms of condensation to occur, the air must be saturated. Saturation occurs most commonly when air is cooled to its dew point, or less often when water vapor is added to the air.

Generally, there must be a surface on which the water vapor can condense. When dew occurs, objects at or near the ground serve this purpose, like grass and car windows. But when condensation occurs in the air above the ground, tiny bits of particulate matter, known as **condensation nuclei,** serve as surfaces for water vapor condensation. These nuclei are very important, for in their absence, a relative humidity well in excess of 100 percent is needed to produce clouds.

Condensation nuclei such as microscopic dust, smoke, and salt particles (from the ocean) are profuse in the lower atmosphere. Because of this abundance of particles, relative humidity rarely exceeds 101 percent. Some particles, such as ocean salt, are particularly good nuclei because they absorb water. These particles are termed **hygroscopic** ("water-seeking") **nuclei.**

When condensation takes place, the initial growth rate of cloud droplets is rapid. It diminishes quickly because the excess water vapor is quickly absorbed by the numerous competing particles. This results in the formation of a cloud consisting of millions upon millions of tiny water droplets, all so fine that they remain suspended in air.

When cloud formation occurs at below-freezing temperatures, tiny ice crystals form. Thus, a cloud can consist of water droplets, ice crystals, or both.

Cloud Classification

Clouds are among the most conspicuous and observable aspects of the atmosphere and its weather. **Clouds** are a form of condensation best described as *visible aggregates of minute droplets of water or tiny crystals of ice.* In addition to being prominent and sometimes spectacular features in the sky, clouds are of continual interest to meteorologists, because they provide a visible indication of what is going on in the atmosphere. Anyone who observes clouds with the hope of recognizing different types often finds that there is a bewildering variety of these familiar white and gray masses streaming across the sky. Still, once one comes to know the basic classification scheme for clouds, most of the confusion vanishes.

Clouds are classified on the basis of their *form* and *height* (Figure 12.14). Three basic forms are recognized: cirrus, cumulus, and stratus.

- **Cirrus** clouds are high, white, and thin. They can occur as patches composed of small cells or as delicate veil-like sheets or extended wispy fibers that often have a feathery appearance.
- **Cumulus** clouds consist of globular individual cloud masses. Normally they exhibit a flat base and have the appearance of rising domes or towers. Such clouds are frequently described as having a cauliflower structure.
- **Stratus** clouds are best described as sheets or layers that cover much or all of the sky. While there may be minor breaks, there are no distinct individual cloud units.

All other clouds reflect one of these three basic forms or are combinations or modifications of them.

Three levels of cloud heights are recognized: high, middle, and low (Figure 12.14). *High clouds* normally have bases above 6000 meters (20,000 feet), *middle clouds* generally occupy heights from 2000 to 6000 meters, and *low clouds* form below 2000 meters (6500 feet). The altitudes listed for each height category are not hard and fast. There is some seasonal as well as latitudinal variation. For example, at high latitudes or during cold winter months in the mid-latitudes, high clouds are often found at lower altitudes.

High Clouds Three cloud types make up the family of high clouds (above 6000 meters): cirrus, cirrostratus, and cirrocumulus. *Cirrus* clouds are thin and delicate and sometimes appear as hooked filaments called "mares' tails" (Figure 12.15A). As the names suggest, *cirrocumulus* clouds consist of fluffy masses (Figure 12.15B), whereas *cirrostratus* clouds are flat layers (Figure 12.15C). Because of the low temperatures and small quantities of water vapor present at high altitudes, all high clouds are thin and white and are made up of ice crystals. Further, these clouds are not considered precipitation makers. However, when cirrus clouds are followed by cirrocumulus clouds and increased sky coverage, they may warn of impending stormy weather.

Middle Clouds Clouds that appear in the middle range (2000 to 6000 meters) have the prefix *alto* as part of their name. *Altocumulus* clouds are composed of globular masses that differ from cirrocumulus clouds in that they are larger and denser (Figure 12.15D). *Altostratus* clouds create a uniform white to grayish sheet covering the sky with the sun or moon visible as a bright spot (Figure 12.15E). Infrequent light snow or drizzle may accompany these clouds.

Low Clouds There are three members in the family of low clouds: stratus, stratocumulus, and nimbostratus. *Stratus* are a uniform fog-like layer of clouds that frequently covers much of the sky. On occasions these clouds may produce light precipitation. When stratus clouds develop a scalloped bottom that appears as long parallel rolls or broken globular patches, they are called *stratocumulus* clouds.

Nimbostratus clouds derive their name from the Latin *nimbus,* which means "rainy cloud," and *stratus,* which means "to cover with a layer" (Figure 12.15F). As the name suggests, nimbostratus clouds are one of the chief precipitation producers. Nimbostratus clouds form in association with stable conditions. We might not expect clouds to grow or persist in stable air, yet cloud growth of this type is common when air is forced to rise, as occurs along a mountain range, a front, or near the center of a cyclone where converging winds cause air to ascend. Such forced ascent of stable air leads to the formation of a stratified cloud layer that is large horizontally compared to its depth.

Clouds of Vertical Development Some clouds do not fit into any one of the three height categories mentioned. Such clouds have their bases in the low height range but often extend upward into the middle or high altitudes. Consequently, clouds in this category are called *clouds of vertical development*. They are all related to one another and are associated with unstable air. Although *cumulus* clouds are often connected with "fair" weather (Figure 12.15G), they may grow dramatically under the proper circumstances. Once upward movement is triggered, acceleration is powerful and clouds with great vertical extent form. The end result is often a towering cloud, called a *cumulonimbus,* that may produce rain showers or a thunderstorm (Figure 12.15H).

Definite weather patterns can often be associated with particular clouds or certain combinations of cloud types, so it is important to become familiar with cloud descriptions and characteristics.

Fog

Fog is generally considered to be an atmospheric hazard. When is it light, visibility is reduced to 2 or 3 kilometers. However, when it is dense, visibility may be cut to a few dozen meters or less, making travel by any mode not only difficult but often dangerous. Officially, visibility must be reduced to 1 kilometer or less before fog is reported. While this figure is arbitrary, it does provide an objective criterion for comparing fog frequencies at different locations.

Fog is defined as a *cloud with its base at or very near the ground*. Physically, there is no difference between a fog and a cloud; their appearance and structure are the same. The essential difference is the method and place of formation. Whereas clouds result when air rises and cools adiabatically, most fogs are the consequence of radiation cooling or the movement of air over a cold surface. (The exception is upslope fog.) In other circumstances, fogs form when enough water vapor is added to the air to bring about saturation (evaporation fogs). We will examine both of these types.

Fogs Caused by Cooling

When warm, moist air moves over a cool surface, the result may be a blanket of fog called **advection fog** (Figure 12.16, p. 318). Examples of such fogs are very common. The foggiest location in the United States, and perhaps in the world, is Cape Disappointment, Washington. The name is indeed appropriate, because this station averages 2552 hours of fog each year (there are about 8760 hours in a year). The fog experienced at Cape Disappointment,

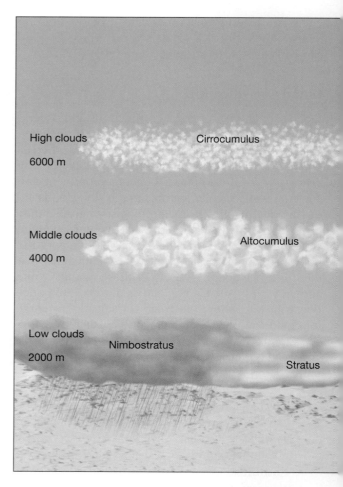

Figure 12.14 Classification of clouds according to height and form. *(After Ward's Natural Science Establishment, Inc., Rochester, N.Y.)*

and at other West Coast locations, is produced when warm, moist air from the Pacific Ocean moves over the cold California Current and is then carried onto shore by the prevailing winds. Advection fogs are also quite common in the winter season when warm air from the Gulf of Mexico blows across cold, often snow-covered surfaces of the Midwest and East.

Radiation fog forms on cool, clear, calm nights, when Earth's surface cools rapidly by radiation. As the night progresses, a thin layer of air in contact with the ground is cooled below its dew point. As the air cools and becomes denser, it drains into low areas, resulting in "pockets" of fog. The largest "pockets" are often river valleys, where thick accumulations may occur (Figure 12.17, p. 318).

As its name implies, **upslope fog** is created when relatively humid air moves up a gradually sloping plain or up the steep slopes of a mountain.

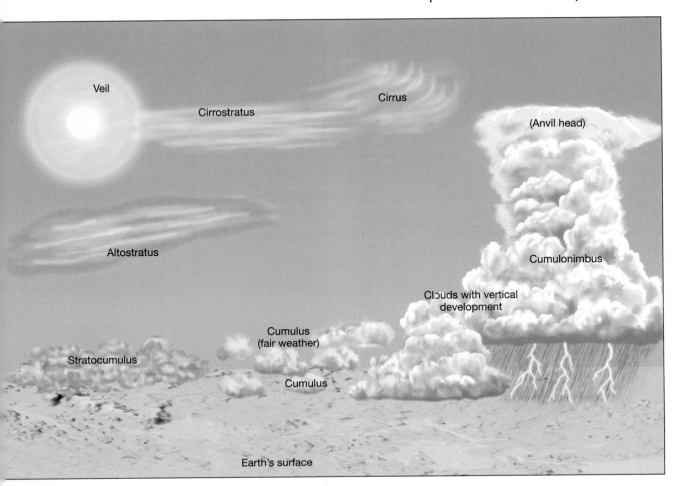

Because of this upward movement, air expands and cools adiabatically. If the dew point is reached, an extensive layer of fog may form. In the United States, the Great Plains offers an excellent example. When humid easterly or southeasterly winds blow westward from the Mississippi River upslope toward the Rocky Mountains, the air gradually rises, resulting in an adiabatic temperature decrease of about 13°C. When the difference between the air temperature and dew point of westward-moving air is less than 13°C, an extensive fog can result in the western plains.

Fogs Caused by Evaporation

When the saturation of air occurs primarily because of the addition of water vapor, the resulting fogs may be called *evaporation fogs*. Two types of evaporation fogs are recognized: steam fog and frontal or precipitation fog.

When cool air moves over warm water, enough moisture may evaporate from the water surface to produce saturation. As the rising water vapor meets the cold air, it immediately recondenses and rises with the air that is being warmed from below. Because the water has a steaming appearance, the phenomenon is called **steam fog** (Figure 12.18, p. 319). Steam fog is fairly common over lakes and rivers in the fall and early winter, when the water may still be relatively warm and the air is rather crisp. Steam fog is often shallow because as the steam rises it reevaporates in the unsaturated air above.

When frontal wedging occurs, warm air is lifted over colder air. If the resulting clouds yield rain, and the cold air below is near the dew point, enough rain will evaporate to produce fog. A fog formed in this manner is called **frontal fog,** or **precipitation fog.** The result is a more or less continuous zone of condensed water droplets reaching from the ground up through the clouds.

In summary, both steam fog and frontal fog result from the addition of moisture to a layer of air. As you saw, the air is usually cool or cold and

A. Cirrus

B. Cirrocumulus

C. Cirrostratus

D. Altocumulus

Figure 12.15 These photos depict common forms of several different cloud types. *(Photos A, B, D, E, F, and G by E. J. Tarbuck. Photo C by A. and J. Verkaik/The Stock Market. Photo H by Tom Bean)*

E. Altostratus

F. Nimbostratus

G. Cumulus

H. Cumulonimbus

Figure 12.16 Advection fog rolling into San Francisco Bay. *(Photo by Ed Pritchard/Tony Stone Worldwide Images)*

already near saturation. Because air's capacity to hold water vapor at low temperatures is small, only a relatively modest amount of evaporation is necessary to produce saturated conditions and fog.

Figure 12.17 Radiation fog in the Illinois River valley at Peoria, Illinois. *(Photo by David Zalaznik/Peoria Journal Star)*

Precipitation

All clouds contain water. But why do some produce precipitation while others drift placidly overhead? This simple question perplexed meteorologists for many years. First, cloud droplets are very small, averaging less than 10 micrometers in diameter (for comparison, a human hair is about 75 micrometers in diameter). Because of their small size, cloud droplets fall incredibly slowly. In addition, clouds are made up of many billions of these droplets, all competing for the available water vapor; thus their continued growth via condensation is very slow. So, what causes precipitation?

How Precipitation Forms

A raindrop large enough to reach the ground without completely evaporating contains roughly one million times more water than a single cloud droplet. Therefore, for precipitation to form, millions of cloud droplets must somehow coalesce (join together) into drops large enough to sustain themselves during their descent to the surface. Two mechanisms have been proposed to explain this phenomenon: the ice crystal process and the collision-coalescence process.

Figure 12.18 Early-morning steam fog on pond, Acadia, Maine. *(Photo by Nancy Rotenberg/Animals Animals/Earth Scenes)*

Ice Crystal Process Meteorologists discovered that, in the mid-to-high latitudes, precipitation often forms in clouds where the temperature is below 0°C. Although we would expect all cloud droplets to freeze in subzero temperatures, only those droplets that make contact with solid particles that have a particular structure (called *ice nuclei*) actually freeze. The remaining liquid droplets are said to be *supercooled* (below 0°C, but still liquid).

When ice crystals and supercooled water droplets coexist in a cloud, the stage is set to generate precipitation. It turns out that ice crystals collect the available water vapor at a much faster rate than liquid water. Thus, ice crystals grow larger at the expense of the water droplets. Eventually, this process generates ice crystals large enough to fall as snowflakes. During their descent, these ice crystals get bigger as they intercept supercooled cloud droplets that freeze on them. When the surface temperature is about 4°C (39°F) or higher, snowflakes usually melt before they reach the ground and continue their descent as rain.

Collision-Coalescence Process Precipitation can form in warm clouds that contain large hygroscopic ("water-seeking") condensation nuclei, such as salt particles. Relatively large droplets form on these large nuclei. Because the bigger droplets fall faster, they collide and join with smaller water droplets. After many collisions the droplets are large enough to fall to the ground as rain.

Forms of Precipitation

Because atmospheric conditions vary greatly from place to place as well as seasonally, several different forms of precipitation are possible. Rain and snow are the most common and familiar, but other forms of precipitation are important as well.

Rain and Drizzle We are familiar with rain. In meteorology, the term **rain** is restricted to drops of water that fall from a cloud and that have a diameter of at least 0.5 millimeter. Most rain originates in either nimbostratus clouds or in towering cumulonimbus clouds that are capable of producing unusually heavy

319

rainfalls known as *cloudbursts*. Raindrops rarely exceed about 5 millimeters. Larger drops do not survive because surface tension, which holds the drops together, is exceeded by the frictional drag of the air. Consequently, large raindrops regularly break apart into smaller ones.

Fine, uniform drops of water having a diameter less than 0.5 millimeter are called *drizzle*. Drizzle can be so fine that the tiny drops appear to float and their impact is almost imperceptible. Drizzle and small raindrops generally are produced in stratus or nimbostratus clouds where precipitation may be continuous for several hours, or on rare occasions for days.

Snow. Snow is precipitation in the form of ice crystals (snowflakes) or, more often, aggregates of crystals. The size, shape, and concentration of snowflakes depend to a great extent on the temperature at which they form.

Recall that at very low temperatures, the moisture content of air is small. The result is the generation of very light and fluffy snow made up of individual six-sided ice crystals. This is the "powder" that downhill skiers talk so much about. By contrast, at temperatures warmer than about –5°C, the ice crystals join together into larger clumps consisting of tangled aggregates of crystals. Snowfalls consisting of these composite snowflakes are generally heavy and have a high moisture content, which makes them ideal for making snowballs.

Sleet and Glaze. Sleet is a wintertime phenomenon and refers to the fall of small particles of ice that are clear to translucent. For sleet to be produced, a layer of air with temperatures above freezing must overlie a subfreezing layer near the ground. When raindrops, which are often melted snow, leave the warmer air and encounter the colder air below, they freeze and reach the ground as small pellets of ice the size of the raindrops from which they formed.

On some occasions, when the vertical distribution of temperatures is similar to that associated with the formation of sleet, **freezing rain** or **glaze** results instead. In such situations, the subfreezing air near the ground is not thick enough to allow the raindrops to freeze. The raindrops, however, do become supercooled as they fall through the cold air and turn to ice upon colliding with solid objects. The result can be a thick coating of ice having sufficient weight to break tree limbs, down power lines, and make walking or driving extremely hazardous (Figure 12.19).

Figure 12.19 Glaze forms when supercooled raindrops freeze on contact with objects. *(Photo by Annie Griffiths Belt/DRK Photo)*

Hail Hail is precipitation in the form of hard, rounded pellets or irregular lumps of ice. Moreover, large hailstones often consist of a series of nearly concentric shells of differing densities and degrees of opaqueness (Figure 12.20). Most hailstones have diameters between

Figure 12.20 A cross-section of the Coffeyville hailstone. This largest recorded hailstone fell over Kansas in 1970, and weighed 0.75 kilogram (1.67 pounds). *(Courtesy of the National Center for Atmospheric Research)*

1 centimeter (pea size) and 5 centimeters (golf ball size), although some can be as big as an orange or larger. Occasionally, hailstones weighing a pound or more have been reported. Many of these were probably composites of several stones frozen together.

The heaviest authenticated hailstone on record fell on Coffeyville, Kansas, September 3, 1970 (Figure 12.20). With a 14-centimeter (5 5-inch) diameter, this "giant" weighed 766 grams (1.67 pounds). It is estimated that this stone hit the ground at a speed in excess of 160 kilometers (100 miles) per hour.

The destructive effects of large hailstones are well known, especially to farmers whose crops have been devastated in a few minutes and to people whose windows and roofs have been damaged (Figure 12.21). In the United States, hail damage each year is significant. In 1994 alone, there was more than $450 million in property damage attributed to hail.

Hail is produced only in large cumulonimbus clouds where updrafts can sometimes reach speeds approaching 160 kilometers (100 miles) per hour, and where there is an abundant supply of supercooled water. Hailstones begin as small embryonic ice pellets that grow by collecting supercooled water droplets as they fall through the cloud. If they encounter a strong updraft, they may be carried upward again and begin the downward journey anew. Each trip through the supercooled portion of the cloud may be represented by an additional layer of ice. Hailstones can also form from a single descent through an updraft. Either way, the process continues until the hailstone encounters a downdraft or grows too heavy to remain suspended by the thunderstorm's updraft.

Rime. **Rime** is a deposit of ice crystals formed by the freezing of supercooled fog or cloud droplets on objects whose surface temperature is below freezing. When rime forms on trees, it adorns them with its characteristic ice feathers, which can be spectacular to behold (Figure 12.22). In these situations, objects such as pine needles act as freezing nuclei, causing the supercooled droplets to freeze on contact. On occasions when the wind is blowing, only the windward surfaces of objects will accumulate the layer of rime.

Measuring Precipitation

The most common form of precipitation, rain, is the easiest to measure. Any open container having a consistent cross section throughout can be a rain gauge. In general practice, however, more sophisticated devices are used so that small amounts of rainfall can

Figure 12.21 Hail damage to a corn field in central Illinois. *(University Corporation for Atmospheric Research/National Center for Atmospheric Research/National Science Foundation)*

Figure 12.22 Rime consists of delicate ice crystals that form when supercooled fog or cloud droplets freeze on contact with objects. Here, rime has formed on the needles of a pine tree. *(Photo by D. Cavagnaro/DRK Photo)*

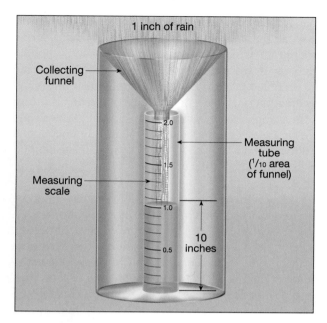

Figure 12.23 Precipitation measurement. The standard rain gauge allows for accurate rainfall measurement to the nearest 0.025 centimeter (0.01 inch). Because the cross-sectional area of the measuring tube is only one-tenth as large as the collector, rainfall is magnified ten times.

Labels in figure: 1 inch of rain; Collecting funnel; Measuring tube (¹/₁₀ area of funnel); Measuring scale; 10 inches

be measured more accurately, and to reduce losses from evaporation. The *standard rain gauge* (Figure 12.23) has a diameter of about 20 centimeters (8 inches) at the top. Once the water is caught, a funnel conducts the rain into a cylindrical measuring tube that has a cross-sectional area only one-tenth as large as the receiver. Consequently, rainfall depth is magnified ten times, which allows for accurate measurements to the nearest 0.025 centimeter (0.01 inch). The narrow opening also minimizes evaporation. When

the amount of rain is less than 0.025 centimeter, it is reported as a *trace* of precipitation.

In addition to the standard rain gauge, several types of recording gauges are routinely used. These instruments record not only the amount of rain but also its time of occurrence and intensity (amount per unit of time).

No matter which rain gauge is used, proper exposure is critical. Errors arise when the gauge is shielded from obliquely falling rain by buildings, trees, or other high objects. Hence, the instrument should be at least as far away from such obstructions as the objects are high. Another cause of error is the wind. It has been shown that with increasing wind and turbulence, it becomes more difficult to collect a representative quantity of rain.

When snow records are kept, two measurements are normally taken: depth and water equivalent. Usually the depth of snow is measured with a calibrated stick. The actual measurement is simple, but choosing a *representative* spot often poses a dilemma. Even when winds are light or moderate, snow drifts freely. As a rule, it is best to take several measurements in an open place away from trees and obstructions and then average them. To obtain the water equivalent, samples may be melted and then weighed or measured as rain.

The quantity of water in a given volume of snow is not constant. A general ratio of 10 units of snow to 1 unit of water is often used when exact information is not available, but the actual water content of snow may deviate widely from this figure. It may take as much as 30 centimeters of light and fluffy dry snow or as little as 4 centimeters of wet snow to produce 1 centimeter of water.

The Chapter in Review

1. *Water vapor,* an odorless, colorless gas, changes from one state of matter (solid, liquid, or gas) to another at the temperatures and pressures experienced near Earth's surface. The processes involved in *changing the state of matter* of water are: *evaporation, condensation, melting, freezing, sublimation,* and *deposition.*

2. *Humidity* is the general term used to describe the amount of water vapor in the air. *Relative humidity* is the most familiar term used to describe humidity. It is the *ratio* (expressed as a percent) of the *actual* water vapor content of an air sample to its *capacity* to hold water vapor at a given temperature. The water vapor capacity of air is temperature dependent, with warm air having a much greater capacity than cold air. *Relative humidity can be changed in*

two ways. One is by *adding or subtracting water vapor.* The second is by *changing air temperature.* When air is cooled, its relative humidity increases. Air is *saturated* when it contains the maximum quantity of water vapor possible for its temperature and pressure. *Dew point* is the temperature to which air must be cooled to become saturated.

3. The basic cloud-forming process is the cooling of air as it rises. The cooling results from expansion because the pressure becomes progressively lower with height. Air temperature changes brought about by compressing or expanding the air are called *adiabatic temperature changes.* Unsaturated air warms by compression and cools by expansion at the rather constant rate of 10°C per 1000 meters of altitude change, a figure called the *dry adiabatic* *rate.* If air rises high enough, it will cool sufficiently to cause condensation and form a cloud. From this point on, air that continues to rise will cool at the *wet adiabatic rate,* which varies from 5°C to 9°C per 1000 meters of ascent. The difference in the two rates results because the condensing water vapor releases *latent heat,* thereby slowing the rate at which the rising air cools.

4. The *stability of air* is determined by the temperature of the atmosphere at various altitudes. Air is said to be *unstable* when the *environmental lapse rate* (the rate of temperature decrease with increasing altitude in the troposphere) is greater than the *dry adiabatic rate.* Stated differently, a column of air is unstable when the air near the bottom is significantly warmer (less dense) than the air aloft.

5. Three mechanisms that can initiate the vertical movement of air are: (1) *orographic lifting,* which occurs when flowing air encounters elevated terrains, such as mountains, and the air rises to go over the barrier; (2) *frontal wedging,* when cool air acts as a barrier over which warmer, less dense air rises; and (3) *convergence,* which happens when volumes of air flow together and a general upward movement of air occurs.

6. For condensation to occur, air must be saturated. Saturation takes place either when air is cooled to its dew point, which most commonly happens, or when water vapor is added to the air. There must also be a surface on which the water vapor may condense. In cloud and fog formation, tiny particles called *condensation nuclei* serve this purpose.

7. *Clouds* are classified on the basis of their *appearance* and *height.* The three basic forms are *cirrus* (high, white, thin, wispy fibers), *cumulus* (globular, individual cloud masses), and *stratus* (sheets or layers that cover much or all of the sky). The four categories based on height are *high clouds* (bases normally above 6000 meters), *middle clouds* (from 2000 to 6000 meters), *low clouds* (below 2000 meters), and *clouds of vertical development.*

8. *Fog* is a cloud with its base at or very near the ground. Fogs form when air is cooled below its dew point or when enough water vapor is added to the air to bring about saturation. Various types of fog include *advection fog, radiation fog, upslope fog, steam fog,* and *frontal fog* (or *precipitation fog*).

9. For *precipitation* to form, millions of cloud droplets must somehow join together into large drops. Two mechanisms for the formation of precipitation have been proposed: One, in clouds where the temperatures are below freezing, ice crystals form and fall as snowflakes. At lower altitudes the snowflakes melt and become raindrops before they reach the ground. Two, large droplets form in warm clouds that contain large hygroscopic ("water seeking") nuclei, such as salt particles. As these big droplets descend, they collide and join with smaller water droplets. After many collisions, the droplets are large enough to fall to the ground as rain. The forms of precipitation include *rain, snow, sleet, hail,* and *rime.*

Key Terms

adiabatic temperature
  change (p. 306)
advection fog (p. 314)
cirrus (p. 313)
cloud (p. 313)
condensation (p. 301)
condensation nuclei
  (p. 312)
convergence (p. 311)
cumulus (p. 313)
deposition (p. 301)
dew point (p. 303)

dry adiabatic rate (p. 306)
evaporation (p. 301)
fog (p. 314)
freezing (p. 301)
freezing rain (p. 320)
frontal fog (p. 315)
frontal wedging (p. 311)
glaze (p. 320)
hail (p. 320)
humidity (p. 301)
hygrometer (p. 303)

hygroscopic nuclei (p. 311)
latent heat (p. 300)
melting (p. 301)
orographic lifting (p. 310)
precipitation fog (p. 315)
radiation fog (p. 314)
rain (p. 319)
rainshadow desert (p. 311)
relative humidity (p. 302)
rime (p. 321)
saturation (p. 301)

sleet (p. 320)
snow (p. 320)
specific humidity (p. 302)
steam fog (p. 315)
stratus (p. 313)
sublimation (p. 301)
upslope fog (p. 314)
vapor pressure (p. 302)
wet adiabatic rate (p. 306)

Questions for Review

1. Summarize the processes by which water changes from one state to another. Indicate whether heat energy is absorbed or liberated.
2. After studying Table 12.1 (p. 302), write a generalization relating temperature and the capacity of air to hold water vapor.
3. Define specific humidity.

4. Explain what 50 percent relative humidity means.
5. Referring to Figure 12.5 (p. 304), answer the following questions:
    a. During a typical day, when is the relative humidity highest? Lowest?
    b. At what time of day would dew most likely form?
    c. Write a generalization relating changes in air temperature to changes in relative humidity.
6. If the air temperature remains unchanged and the quantity of water vapor (specific humidity) decreases, how will relative humidity change?
7. Explain the principle of the sling psychrometer.
8. On a warm summer day when the relative humidity is high, it may seem even warmer than the thermometer indicates. Why do we feel so uncomfortable on a "muggy" day?
9. Why does air cool when it rises through the atmosphere?
10. Explain the difference between environmental lapse rate and adiabatic cooling.
11. How do orographic lifting and frontal wedging act to force air to rise?
12. Explain why the Great Basin area of the western United States is so dry. What term is applied to such a situation?
13. How does stable air differ from unstable air? Describe the general nature of the clouds and precipitation expected with each.
14. What is the function of condensation nuclei in cloud formation?
15. As you drink an ice-cold beverage on a warm day, the outside of the glass or bottle becomes wet. Explain.
16. What is the basis for the classification of clouds?
17. Why are high clouds thin?
18. List five types of fog and discuss the details of their formation.
19. What is the difference between precipitation and condensation?
20. List the forms of precipitation and the circumstances of their formation.

Testing What You Have Learned

Multiple-Choice Questions

1. The amount of heat required to raise the temperature of one gram of water 1°C is a _____
    a. degree.
    c. langley.
    e. thermogram.
    b. calorie.
    d. isorad.
2. What term is used to describe the conversion of a solid directly to a gas, without passing through the liquid state?
    a. evaporation
    c. freezing
    e. sublimation
    b. condensation
    d. deposition
3. The weight of water vapor per weight of a chosen mass of air, including the water vapor, is the air's _____ humidity.
    a. specific
    c. relative
    e. maximum
    b. temporary
    d. latent
4. When air is saturated, its relative humidity is _____
    a. 50%.
    c. 100%.
    e. decreasing.
    b. 75%.
    d. increasing.
5. If both the dry-bulb and wet-bulb thermometers of a psychrometer record the same temperature, the air is _____
    a. saturated.
    c. warming.
    e. a., b., and c.
    b. at the dew point.
    d. a. and b.
6. Cooling of air by expansion and warming of air by compression are known as _____ temperature changes.
    a. environmental
    c. latent
    e. atmospheric
    b. lapse rate
    d. adiabatic

7. A high proportion of the precipitation in the middle latitudes is produced when air is lifted during a process called _____

   a. monsoon flow.   **c.** thrusting.   **e.** rifting.

   b. dehydration.   **d.** frontal wedging.

8. Particles, such as salt from the ocean, that absorb water are called _____ nuclei.

   a. hydro   **c.** latent   **e.** hygroscopic

   b. aqueous   **d.** central

9. Towering clouds of great vertical development are called _____ clouds.

   a. cumulonimbus   **c.** nimbostratus   **e.** none of the above

   b. stratonimbus   **d.** cirronimbus

10. What type of fog is likely to form when warm, moist air is blown over a cool surface?

   a. steam fog   **c.** radiation fog   **e.** upslope fog

   b. advection fog   **d.** frontal fog

Fill-In Questions

11. "Hidden" energy that is not associated with a temperature change but is absorbed or released during a change of state is called _____ _____.

12. The temperature to which air would have to be cooled in order to reach saturation is called the _____ _____.

13. Three mechanisms that initiate the vertical movement of stable air are _____ _____, _____ _____, and _____.

14. On the basis of their appearance, the three basic forms of clouds are _____, _____, and _____.

15. Deposits of ice crystals formed by the freezing of supercooled fog or cloud droplets are called _____.

True/False Questions

16. Vapor pressure is that part of the total atmospheric pressure that can be attributed to the water vapor content. _____

17. Lowering the temperature of air below its dew point causes evaporation. _____

18. Rainshadow deserts are often found on the windward sides of mountains in the middle latitudes. _____

19. For condensation to occur, the air must be saturated and there generally must be a surface on which the water vapor may condense. _____

20. In meteorology, the term *rain* is restricted to drops of water that fall from a cloud and have a diameter of less than 0.5 millimeter. _____

Answers:

1. b; 2. e; 3. a; 4. c; 5. d; 6. d; 7. d; 8. e; 9. a; 10. b; 11. latent heat; 12. dew point; 13. orographic lifting, frontal wedging, convergence; 14. cirrus, cumulus, stratus; 15. rime; 16. T; 17. F; 18. F; 19. T; 20. F.

Web   Work

The following are informative and interesting World Wide Web sites that address topics related to those presented in this chapter:

- Clouds and Precipitation Online Tutorial
  http://www.2010.atmos.uiuc.edu/(Gh)/guides/mtr/cld/home.rxml

- Current United States Cloud Cover and Precipitation
  http://www.ems.psu.edu/cgi-bin/wx/cloudstats.cgi

For direct access to these sites and other relevant Web locations, contact the *Foundations of Earth Science* WWW home page at http://www.prenhall.com/lutgens.

CHAPTER 13

The Atmosphere in Motion

FOCUS ON LEARNING

To assist you in learning the important concepts
in this chapter, you will find it helpful to focus
on the following questions:

1. What is air pressure and how does it change with altitude?

2. What force creates wind and what other factors influence it?

3. What are the two types of pressure centers? What is the air movement and general weather associated with each type?

4. What is the idealized global circulation?

5. What is the general atmospheric circulation in the mid-latitudes?

6. What are the names and causes of some local winds?

A brisk breeze propels these sailboats. *Photo by Alastair Black/Tony Stone Images*

Of the various elements of weather and climate, changes in air pressure are the least noticeable. When listening to a weather report, we are generally interested in moisture conditions (humidity and precipitation), temperature, and perhaps wind. It is the rare individual, however, who wonders about air pressure. Although the hour-to-hour and day-to-day variations in air pressure are generally not noticed by people, they are very important in producing changes in our weather. Variations in air pressure from place to place cause the movement of air we call wind, and are a significant factor in weather forecasting (Figure 13.1). As we shall see, air pressure is tied very closely to the other elements of weather in a cause-and-effect relationship.

Measuring Air Pressure

In Chapter 11 you saw that air has weight; at sea level, it exerts a pressure of 1 kilogram per square centimeter (14.7 pounds per square inch). The **air pressure** at a particular place is simply the force exerted by the weight of the air above. With an increase in altitude, there is less air, so the weight of the air, and thus the pressure, decreases. It decreases rapidly at first, then much more slowly (see Figure 11.4, p. 275).

When meteorologists measure atmospheric pressure, they employ a unit called the *millibar*. Standard

sea level pressure is 1013.2 millibars. Although the millibar has been the unit of measure on all U.S. weather maps since January 1940, the media use "inches of mercury" to describe atmospheric pressure. In the United States, the National Weather Service converts millibar values to inches of mercury for public and aviation use.

"Inches of mercury" is easy to understand. The use of mercury for measuring air pressure dates from 1643, when Torricelli, a student of the famous Italian scientist Galileo, invented the **mercury barometer.** Torricelli correctly described the atmosphere as a vast ocean of air that exerts pressure on us and all objects about us. To measure this force he filled a glass tube, which was closed at one end, with mercury. The tube was then inverted into a dish of mercury (Figure 13.2A). Torricelli found that the mercury flowed out of the tube until the weight of the column was balanced by the pressure that the atmosphere exerted on the surface of the mercury in the dish. In other words, the weight of mercury in the column equaled the weight of the same diameter column of air that extended from the ground to the top of the atmosphere.

When air pressure increases, the mercury in the tube rises. Conversely, when air pressure decreases, so does the height of the mercury column. With some refinements the mercurial barometer invented by Torricelli is still the standard pressure-measuring instrument used today. Standard atmospheric pressure at sea level equals 29.92 inches of mercury.

The need for a smaller and more portable instrument for measuring air pressure led to the development of the **aneroid** ("without liquid") **barometer.** Based on a different principle from the mercury barometer, this instrument consists of partially evacuated metal chambers that have a spring inside to keep them from collapsing. The metal chambers, being very sensitive to air pressure variations, change shape, compressing as the pressure increases and expanding as the pressure decreases. Aneroids are often used in making **barographs,** instruments that continuously record pressure changes (Figure 13.2B).

Factors Affecting Wind

In Chapter 12 we examined the upward movement of air and its role in cloud formation. As important as vertical motion is, far more air moves horizontally, the

Figure 13.1 Strong winds from a tropical storm in Jamaica. Wind is simply air that flows horizontally with respect to Earth's surface. Air is set in motion by variations in air pressure from place to place. *(Photo by Denis Valentine/The Stock Market)*

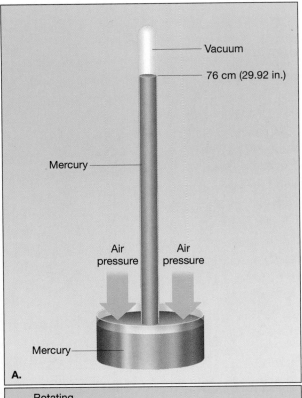

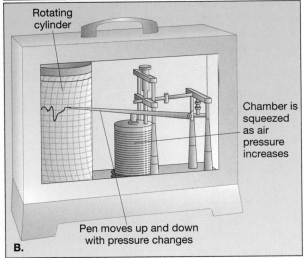

Figure 13.2 **A.** Simple mercury barometer. The weight of the column of mercury is balanced by the pressure exerted on the dish of mercury by the air above. If the pressure decreases, the column of mercury falls; if the pressure increases, the column rises. **B.** An aneroid barograph makes a continuous record of pressure changes. One important advantage of the aneroid barometer is that it is easily adapted to a recording mechanism.

phenomenon we call **wind.** What causes wind? Simply stated, wind is the result of horizontal differences in air pressure. *Air flows from areas of higher pressure to areas of lower pressure.* You may have experienced this when opening a vacuum-packed can of coffee. The noise you hear is caused by air rushing from the higher pressure outside the can to the lower pressure inside. Wind is nature's attempt to balance such inequalities in air pressure. Because unequal heating of Earth's surface generates these pressure differences, *solar energy is the ultimate driving force of wind.*

If Earth did not rotate, and if there were no friction between moving air and Earth's surface, air would flow in a straight line from areas of higher pressure to areas of lower pressure. But because Earth does rotate and friction exists, wind is controlled by the following combination of forces: (1) the pressure gradient force, (2) Coriolis effect, and (3) friction. We will now examine each of these factors.

Pressure Gradient Force

Pressure differences create wind, and the greater these differences, the greater the wind speed. Over Earth's surface, variations in air pressure are determined from barometric readings taken at hundreds of weather stations. These pressure data are shown on a weather map using **isobars,** lines that connect places of equal air pressure (Figure 13.3). The spacing of isobars indicates the amount of pressure change occurring over a given distance and is expressed as the **pressure gradient.**

You might find it easier to visualize a pressure gradient if you think of it like the slope of a hill. A steep pressure gradient, like a steep hill, causes greater acceleration of an air parcel than does a weak pressure gradient. Thus, the relationship between wind speed and the pressure gradient is straightforward: *Closely spaced isobars indicate a steep pressure gradient and high winds, whereas widely spaced isobars indicate a weak pressure gradient and light winds.* Figure 13.3 illustrates the relationship between the spacing of isobars and wind speed. Notice that wind speeds are greater in Ohio, Kentucky, Michigan, and Illinois, where isobars are more closely spaced, than in the western states where isobars are more widely spaced.

The pressure gradient is the driving force of wind, and it has both magnitude and direction. Its magnitude is determined from the spacing of isobars. The direction of force is always from areas of higher

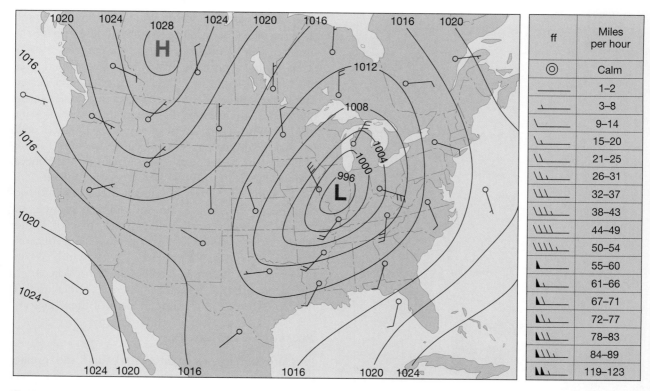

ff	Miles per hour
◎	Calm
──	1–2
──╮	3–8
──╲	9–14
──╲╮	15–20
──╲╲	21–25
──╲╲╮	26–31
──╲╲╲	32–37
──╲╲╲╮	38–43
──╲╲╲╲	44–49
──╲╲╲╲╮	50–54
▟──	55–60
▟──╮	61–66
▟──╲	67–71
▟──╲╮	72–77
▟──╲╲	78–83
▟──╲╲╮	84–89
▟▟──	119–123

Figure 13.3 The black lines are isobars that connect places of equal barometric pressure. They show the distribution of pressure on weather maps. The lines usually curve and often join around cells of high and low pressure. Flags indicate the expected airflow surrounding cells of high and low pressure and are plotted like flags flying with the wind. Wind speed is indicated by flags and "feathers" as shown along the right-hand side of this drawing.

pressure to areas of lower pressure and at right angles to the isobars. Once the air starts to move, the Coriolis effect and friction come into play, but then only to modify the movement, not to produce it.

Coriolis Effect

Figure 13.3 shows the typical air movements associated with high- and low-pressure systems. As expected, the air moves out of the regions of higher pressure and into the regions of lower pressure. However, the wind does not cross the isobars at right angles as the pressure gradient force directs it to do. The direction deviates as a result of Earth's rotation. This has been named the **Coriolis effect** after the French scientist who first thoroughly described it. All free-moving objects or fluids, including the wind, are deflected to the *right* of their path of motion in the Northern Hemisphere and to the *left* in the Southern Hemisphere.

The reason for this deflection can be illustrated by imagining the path of a rocket launched from the North Pole toward a target located on the equator

(Figure 13.4). If the rocket took an hour to reach its target, during its flight Earth would have rotated 15 degrees to the east. To someone standing on Earth it would look as if the rocket veered off its path and hit Earth 15 degrees west of its target. The true path of the rocket is straight and would appear so to someone out in space looking at Earth. It was Earth turning under the rocket that gave it its *apparent* deflection.

Note that the rocket was deflected to the right of its path of motion because of the counterclockwise rotation of the Northern Hemisphere. In the Southern Hemisphere, the effect is reversed. Clockwise rotation produces a similar deflection, but to the *left* of path of motion. The same deflection is experienced by wind regardless of the direction it is moving.

We attribute the apparent shift in wind direction to the Coriolis effect. This deflection: (1) is always directed at right angles to the direction of airflow; (2) affects only wind direction, not wind speed; (3) is affected by wind speed (the stronger the wind, the greater the deflection); and (4) is strongest at the poles and weakens equatorward, becoming nonexistent at the equator.

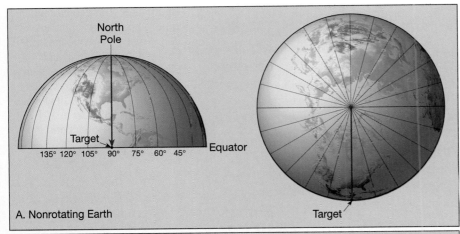

North Pole

Target

135° 120° 105° 90° 75° 60° 45°

Equator

A. Nonrotating Earth

Target

Figure 13.4 The Coriolis effect illustrated using a 1-hour flight of a rocket traveling from the North Pole to a location on the equator. **A.** On a nonrotating Earth, the rocket would travel straight to its target. **B.** However, Earth rotates 15° each hour. Thus, although the rocket travels in a straight line, when we plot the path of the rocket on Earth's surface, it follows a curved path that veers to the right of the target.

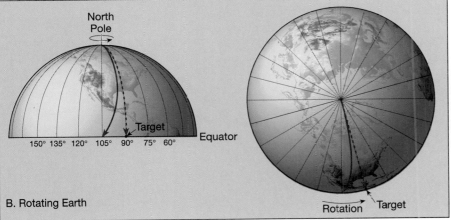

North Pole

Target

150° 135° 120° 105° 90° 75° 60°

Equator

B. Rotating Earth

Rotation   Target

Friction with Earth's Surface

The effect of friction on wind is important only within a few kilometers of Earth's surface. Friction acts to slow air movement and, as a consequence, alters wind direction. To illustrate friction's effect on wind direction, let us look at a situation in which it has no role. Above the friction layer, the pressure gradient force and Coriolis effect work together to direct the flow of air. Under these conditions, the pressure gradient force causes air to start moving across the isobars. As soon as the air starts to move, the Coriolis effect acts at right angles to this motion. The faster the wind speed, the greater the deflection.

Eventually, the Coriolis effect will balance the pressure gradient force and the wind will blow parallel to the isobars (Figure 13.5). Upper-air winds generally take this path and are called **geostrophic winds.** Owing to the lack of friction with Earth's surface, geostrophic winds travel at higher speeds than do surface winds. This can be observed in Figure 13.6 by noting the wind flags, many of which indicate winds of 50 to 100 miles per hour.

The most prominent features of upper-level flow are the **jet streams.** First encountered by high-flying bombers during World War II, these fast-moving "rivers" of air travel between 120 and 240 kilometers (75 and 150 miles) per hour in a west-to-east direction. One such stream is situated over the polar front, which is the zone separating cool polar air from warm subtropical air.

Below 600 meters (2000 feet), friction complicates the airflow just described. Recall that the Coriolis effect is proportional to wind speed. Friction lowers the wind speed, so it reduces the Coriolis effect. Because the pressure gradient force is not affected by wind speed, it wins the tug of war shown in Figure 13.7. The result is a movement of air at an angle across the isobars toward the area of lower pressure.

The roughness of the terrain determines the angle of airflow across the isobars. Over the smooth

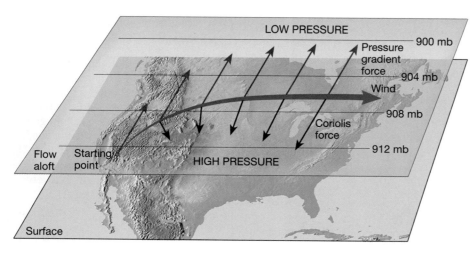

Figure 13.5 The geostrophic wind. Upper-level winds are deflected by the Coriolis effect until the Coriolis effect just balances the pressure gradient force. Above 600 meters (1970 feet), where friction is negligible, these winds will flow nearly parallel to the isobars and are called *geostrophic winds.*

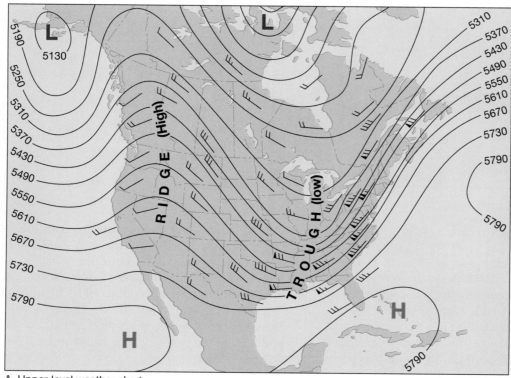

A. Upper-level weather chart

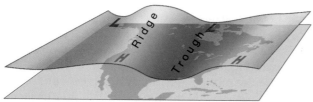

B. Representation of upper-level chart

Figure 13.6 Upper-air winds. This map shows the direction and speed of the upper-air wind for a particular day. Note that the airflow is nearly parallel to the contours. These isolines are height contours for the 500-millibar level.

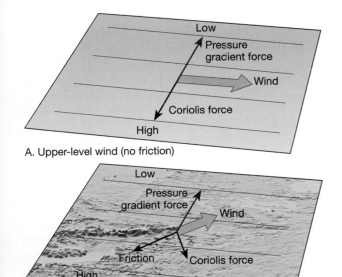

A. Upper-level wind (no friction)

B. Surface wind (effect of friction)

Figure 13.7 Comparison between upper-level winds and surface winds showing the effects of friction on airflow. Friction slows surface wind speed, which weakens the Coriolis effect, causing the winds to cross the isobars and move toward the lower pressure.

ocean surface, friction is low and the angle is small. Over rugged terrain, where friction is higher, the angle that air makes as it flows across the isobars can be as great as 45 degrees.

In summary, upper airflow is nearly parallel to the isobars, while the effect of friction causes the surface winds to move more slowly and cross the isobars at an angle.

Cyclones and Anticyclones

Among the most common features on any weather map are areas designated as pressure centers. **Cyclones,** or **lows,** are centers of low pressure, and **anticyclones,** or **highs,** are high-pressure centers. As Figure 13.8 illustrates, the pressure decreases from the outer isobars toward the center in a cyclone. In an anticyclone, just the opposite is the case—the values of the isobars increase from the outside toward the center. By knowing just a few basic facts about centers of high and low pressure, you can greatly increase your understanding of current and forthcoming weather.

Cyclonic and Anticyclonic Winds

From the preceding section, you learned that the two most significant factors that affect wind are the pressure gradient force and the Coriolis effect. Winds move from higher pressure to lower pressure and are deflected to the right or left by Earth's rotation. When these controls of airflow are applied to pressure centers in the Northern Hemisphere, the result is that winds blow inward and counterclockwise around a low. Around a high, they blow outward and clockwise (Figure 13.8).

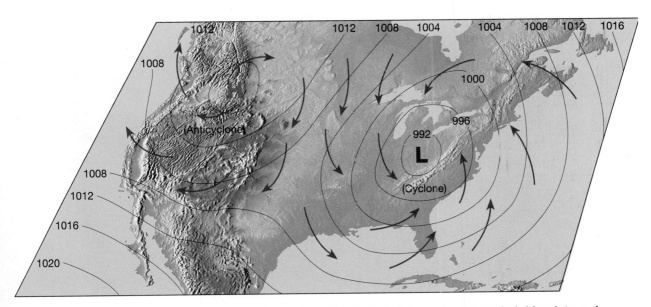

Figure 13.8 Cyclonic and anticyclonic winds in the Northern Hemisphere. Arrows show that winds blow into and counterclockwise around a low. By contrast, around a high, winds blow outward and clockwise.

In the Southern Hemisphere the Coriolis effect deflects the winds to the left, and therefore winds around a low are blowing clockwise, and winds around a high are moving counterclockwise. In either hemisphere, friction causes a net inflow (**convergence**) around a cyclone and a net outflow (**divergence**) around an anticyclone.

Weather Generalizations About Highs and Lows

Rising air is associated with cloud formation and precipitation, whereas subsidence produces clear skies. In this section we will learn how the movement of air can itself create pressure change and hence generate winds. After doing so, we will examine the relation between horizontal and vertical flow, and their effects on the weather.

Let us first consider the situation around a surface low-pressure system where the air is spiraling inward. Here the net inward transport of air causes a shrinking of the area occupied by the air mass, a process that is termed *horizontal convergence*. Whenever air converges horizontally, it must pile up, that is, increase in height to allow for the decreased

area it now occupies. This generates a "taller" and therefore heavier air column. Yet a surface low can exist only as long as the column of air exerts less pressure than that occurring in surrounding regions. We seem to have encountered a paradox—a low-pressure center causes a net accumulation of air, which increases its pressure. Consequently, a surface cyclone should quickly eradicate itself in a manner not unlike what happens when a vacuum-packed can is opened.

It should be apparent that for a surface low to exist for very long, compensation must occur at some layer aloft. For example, surface convergence could be maintained if divergence (spreading out) aloft occurred at a rate equal to the inflow below. Figure 13.9 shows the relationship between surface convergence (inflow) and divergence (outflow) aloft that is needed to maintain a low-pressure center. You can see that surface convergence around a cyclone causes a net upward movement.

The rate of this vertical movement is quite slow, generally less than 1 kilometer per day. Nevertheless, because rising air cools adiabatically, a low-pressure system often brings cloudiness and precipitation (Figure 13.10A). On occasion, divergence aloft may

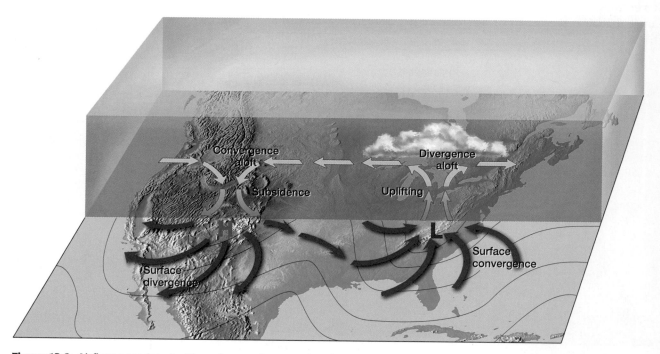

Figure 13.9 Airflow associated with surface cyclones and anticyclones. A low, or cyclone, has converging surface winds and rising air causing cloudy conditions. A high, or anticyclone, has diverging surface winds and descending air, which leads to clear skies and fair weather.

A.

B.

Figure 13.10 These two photographs illustrate the basic weather generalizations associated with pressure centers. **A.** New Yorkers struggle during a storm. Centers of low pressure are frequently associated with cloudy conditions and precipitation. *(Photo by Alex Brandon/AP/Wide World Photos)* **B.** By contrast, clear skies and "fair" weather may be expected when an area is under the influence of high pressure. Sunbathers on a beach in Hawaii. *(Photo by E. J. Tarbuck)*

even exceed surface convergence, resulting in intensified surface inflow (faster winds) and increased vertical motion. For this reason, divergence aloft can both intensify these storm centers and maintain them.

Diverging surface winds are associated with high-pressure centers. The outflow near the surface is accompanied by convergence aloft and general subsidence of the air column (Figure 13.9). Because descending air is compressed and warmed, cloud formation and precipitation are unlikely in an anticyclone, and "fair" weather can usually be expected with the approach of a high (Figure 13.10B).

For reasons that should now be obvious, it has been common practice to print on household barometers the words "stormy" at the low-pressure end and "fair" on the high-pressure end. By noting whether the pressure is rising, falling, or steady, we have a good indication of what the forthcoming weather will be. Such a determination, called the **pressure,** or **barometric tendency,** is a very useful aid in short-range weather prediction.

You should now be better able to understand why television weather reporters emphasize the positions and projected paths of cyclones and anticyclones. The "villain" on these weather programs is always the low-pressure center, which produces "bad" weather in any season. Lows move in roughly a west-to-east direction across the United States and require a few days to more than a week for the journey. Their paths can be somewhat erratic; thus, accurate prediction of their migration is difficult, although essential, for short-range forecasting.

Meteorologists must also determine if the flow aloft will intensify an embryo storm or act to suppress its development. Because the close tie between conditions at the surface and those aloft, a great deal of emphasis has been placed on the importance and understanding of the total atmospheric circulation, particularly in the mid-latitudes.

We will now examine the workings of Earth's general atmospheric circulation, and then again consider the structure of the cyclone in light of this knowledge.

General Circulation of the Atmosphere

As noted, the underlying cause of wind is unequal heating of Earth's surface. In tropical regions, more solar radiation is received than is radiated back to space. In polar regions the opposite is true—less solar energy is received than is lost. Attempting to balance these differences, the atmosphere acts as a giant heat transfer system, moving warm air poleward and cool air equatorward. On a smaller scale, but for the same reason, ocean currents also contribute to this global heat transfer. The general circulation is very complex, and there is a great deal that has yet to be explained. We can, however, develop a general understanding by first considering the circulation that would occur on a nonrotating Earth having a uniform surface. We will then modify this system to fit observed patterns.

Circulation on a Nonrotating Earth

On a hypothetical nonrotating planet with a smooth surface of either all land or all water, two large thermally produced cells would form (Figure 13.11). The heated equatorial air would rise until it reached the tropopause, which, acting like a lid, would deflect the air poleward. Eventually, this upper-level airflow would reach the poles, sink, spread out in all directions at the surface, and move back toward the equator. Once there, it would be reheated and start its journey over again. This hypothetical circulation system has upper-level air flowing poleward and surface air flowing equatorward.

If we add the effect of rotation, this simple convection system will break down into smaller cells. Figure 13.12 illustrates the three pairs of cells proposed to carry on the task of heat redistribution on a rotating planet. The polar and tropical cells retain the characteristics of the thermally generated convection described earlier. The nature of the mid-latitude circulation is more complex and will be discussed in more detail in a later section.

Idealized Global Circulation

Near the equator, the rising air is associated with the pressure zone known as the **equatorial low—** a region marked by abundant precipitation. As the upper-level flow from the equatorial low reaches 20 to 30 degrees latitude, north or south, it sinks back toward the surface. This subsidence and associated adiabatic heating produce hot, arid conditions. The center of this zone of subsiding dry air is the **subtropical high,** which encircles the globe near 30 degrees latitude, north and south (Figure 13.12). The great deserts of Australia, Arabia, and Africa exist because of the stable dry condition caused by the subtropical highs.

At the surface, airflow is outward from the center of the subtropical high. Some of the air travels equatorward and is deflected by the Coriolis effect, producing the reliable **trade winds.** The remainder travels poleward and is also deflected, generating the prevailing **westerlies** of the mid-latitudes. As the westerlies move poleward, they encounter the cool **polar easterlies** in the region of the **subpolar low.** The interaction of these warm and cool winds produces the stormy belt known as the **polar front.** The source region for the variable polar easterlies is the **polar high.** Here, cold polar air is subsiding and spreading equatorward.

In summary, this simplified global circulation is dominated by four pressure zones. The subtropical and polar highs are areas of dry subsiding air that flows outward at the surface, producing the prevailing winds. The low-pressure zones of the equatorial

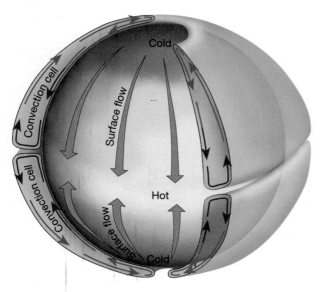

Figure 13.11 Global circulation on a nonrotating Earth. A simple convection system is produced by unequal heating of the atmosphere on a nonrotating Earth.

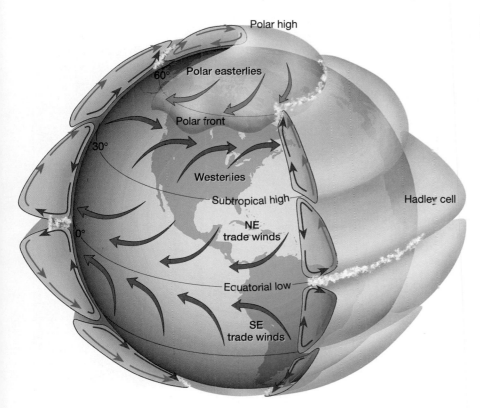

Figure 13.12 Idealized global circulation.

and subpolar regions are associated with inward and upward airflow accompanied by clouds and precipitation.

Influence of Continents

Up to this point, we have described the surface pressure and associated winds as continuous belts around Earth. However, the only truly continuous pressure belt is the subpolar low in the Southern Hemisphere. Here the ocean is uninterrupted by landmasses. At other latitudes, particularly in the Northern Hemisphere where landmasses break up the ocean surface, large seasonal temperature differences disrupt the pattern. Figure 13.13 shows the resulting pressure and wind patterns for January and July. The circulation over the oceans is dominated by semipermanent cells of high pressure in the subtropics and cells of low pressure over the subpolar regions. The subtropical highs are responsible for the trade winds and westerlies, as mentioned earlier.

The large landmasses, on the other hand, particularly Asia, become cold in the winter and develop a seasonal high-pressure system from which surface flow is directed off the land (Figure 13.13). In the summer, the opposite occurs; the landmasses are heated and develop a low-pressure cell, which permits air to flow onto the land. These seasonal changes in wind direction are known as the **monsoons.** During warm months, areas such as India experience a flow of warm, water-laden air from the Indian Ocean, which produces the rainy summer monsoon. The winter monsoon is dominated by dry continental air. A similar situation exists, but to a lesser extent, over North America.

In summary, the general circulation is produced by semipermanent cells of high and low pressure over the oceans and is complicated by seasonal pressure changes over land.

Circulation in the Mid-Latitudes

The circulation in the mid-latitudes, the zone of the westerlies, is complex and does not fit the convection system proposed for the tropics. Between 30 and 60 degrees latitude, the general west-to-east flow is interrupted by the migration of cyclones and anticyclones. In the Northern Hemisphere these cells move from west to east around the globe, creating an anticyclonic

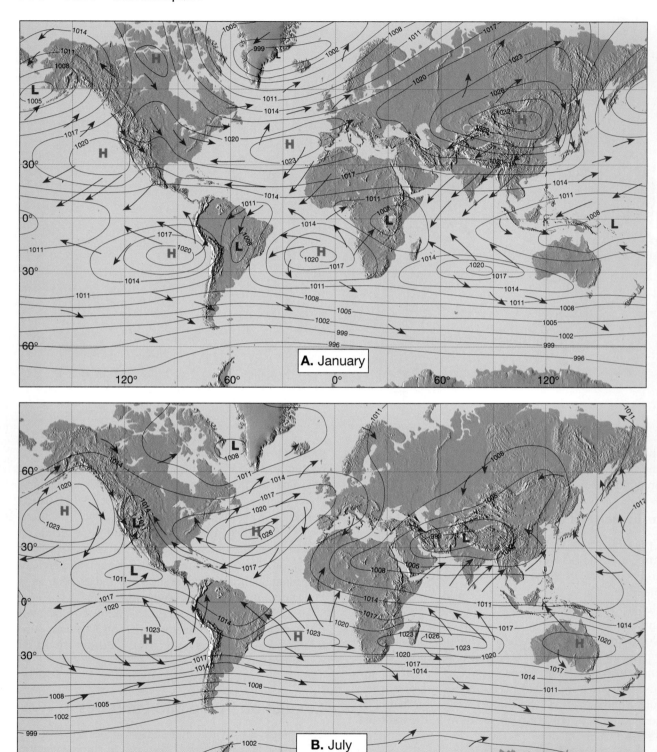

Figure 13.13 Average surface barometric pressure in millibars for **A.** January and **B.** July, with associated winds.

(clockwise) flow or a cyclonic (counterclockwise) flow in their area of influence. A close correlation exists between the paths taken by these surface pressure systems and the position of the upper-level airflow, indicating that the upper air steers the movement of cyclonic and anticyclonic systems.

Among the most obvious features of the flow aloft are the seasonal changes. The steep temperature gradient across the middle latitudes in the winter months corresponds to a stronger flow aloft. In addition, the polar jet stream fluctuates seasonally such that its average position migrates southward with the approach of winter and northward as summer nears. By midwinter, the jet core may penetrate as far south as central Florida.

Because the paths of low-pressure centers are guided by the flow aloft, we can expect the southern tier of states to experience more of their stormy weather in the winter season. During the hot summer months, the storm track is across the northern states, and some cyclones never leave Canada. The northerly storm track associated with summer applies also to Pacific storms, which move toward Alaska during the warm months, thus producing an extended dry season for much of our western coast. The number of cyclones generated is seasonal as well, with the largest number occurring in the cooler months when the temperature gradients are greatest. This fact is in agreement with the role of cyclonic storms in the distribution of heat across the mid-latitudes.

Local Winds

Having examined Earth's large-scale circulation, let us turn briefly to winds that influence much smaller areas. Remember that all winds are produced for the same reason: temperature differences that arise because of unequal heating of Earth's surface. In turn, these differences generate pressure variations. *Local winds* are simply small-scale winds produced by a locally generated pressure gradient. Those described here are caused either by topographic effects or variations in surface composition in the immediate area.

Sea and Land Breezes

In coastal areas during the warm summer months, the land is heated more intensely during the daylight hours than the adjacent body of water. As a result, the air above the land surface heats, expands, and rises, creating an area of lower pressure. A **sea**

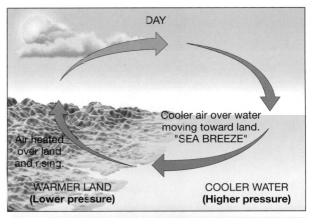

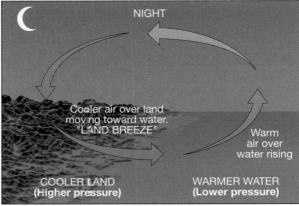

Figure 13.14 Sea and land breezes. During the daytime, land heats more intensely than does water. A low is created over the land, and the air moves from the water (higher pressure) to the land (lower pressure). At night, the land cools more rapidly, resulting in higher pressure over land and a reversal of the wind.

breeze then develops, because cooler air over the water (higher pressure) moves toward the warmer land (lower pressure) (Figure 13.14, top). The sea breeze begins to develop shortly before noon and generally reaches its greatest intensity during the mid- to late afternoon. These relatively cool winds can be a significant moderating influence on afternoon temperatures in coastal areas.

At night, the reverse may take place. The land cools more rapidly than the sea and the **land breeze** develops (Figure 13.14, bottom). Small-scale sea breezes can also develop along the shores of large lakes. People who live in a city near the Great Lakes, such as Chicago, recognize this lake effect, especially in the summer. They are reminded daily by weather reports of the cool temperatures near the lake as compared to warmer outlying areas.

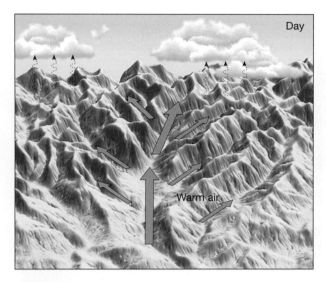
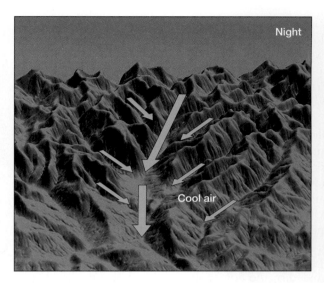

Figure 13.15 Mountain and valley breezes. Heating during the daylight hours warms the air along the mountain slopes. This warm air rises, generating a valley breeze. After sunset, cooling of the air near the mountain can result in cool air drainage into the valley, producing the mountain breeze.

Valley and Mountain Breezes

A daily wind similar to land and sea breezes occurs in many mountainous regions. During daylight hours, the air along the slopes of the mountains is heated more intensely than the air at the same elevation over the valley floor. Because this warmer air is less dense, it glides up along the slope and generates a **valley breeze** (Figure 13.15, left). The occurrence of these daytime upslope breezes can often be identified by the cumulus clouds that develop on adjacent mountain peaks.

After sunset, the pattern may reverse. Rapid radiation cooling along the mountain slopes produces a layer of cooler air next to the ground. Because cool air is denser than warm air, it drains downslope into the valley. This movement of air is called the **mountain breeze** (Figure 13.15, right). The same type of cool air drainage can occur in places that have very modest slopes. The result is that the coldest pockets of air are usually found in the lowest spots. Like many other winds, mountain and valley breezes have seasonal preferences. Although valley breezes are most common during the warm season when solar heating is most intense, mountain breezes tend to be more dominant in the cold season.

Chinook and Santa Ana Winds

Warm, dry winds are common on the eastern slopes of the Rockies, where they are called **chinooks.** Such winds are created when air descends the leeward (sheltered) side of a mountain and warms by compression. Because condensation may have occurred as the air ascended the windward side, releasing latent heat, the air descending the leeward slope will be warmer and drier than it was at a similar elevation on the windward side. Although the temperature of these winds is generally less than 10°C (50°F), which is not particularly warm, they occur mostly in the winter and spring when the affected areas may be experiencing below-freezing temperatures. Thus, by comparison, these dry, warm winds often bring a drastic change. When the ground has a snow cover, these winds are known to melt it in short order. The word chinook literally means "snow-eater."

A chinooklike wind that occurs in southern California is the **Santa Ana.** These hot, desiccating winds greatly increase the threat of fire in this already dry area.

How Wind Is Measured

Two basic wind measurements, direction and speed, are particularly significant to the weather observer. *Winds are always labeled by the direction from which they blow.* A north wind blows *from* the north *toward* the south, an east wind *from* the east *toward* the west. The instrument most commonly used to determine wind direction is the **wind vane** (Figure

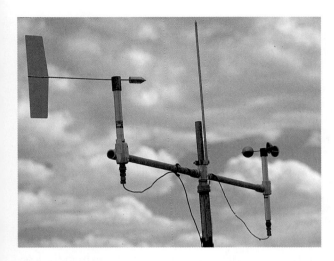

Figure 13.16 Wind vane and cup anemometer. The wind vane shows wind direction and the anemometer measures wind speed. *(Photo by Scott T. Smith)*

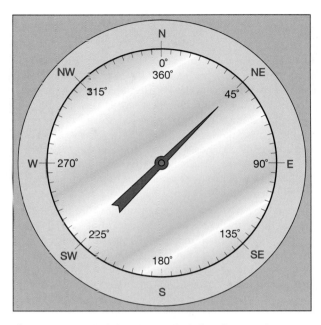

Figure 13.17 Wind direction. Wind direction can be expressed using the points of the compass or a scale of 0–360 degrees. Winds are always labeled according to the direction *from* which they are blowing.

13.16, left). This instrument, a common sight on many buildings, always points *into* the wind. Often the wind direction is shown on a dial that is connected to the wind vane (Figure 13.17). The dial indicates wind direction, either by points of the compass, (N, NE, E, SE, etc.) or by a scale of 0° to 360°. On the latter scale, 0 degrees or 360 degrees are both north, 90 degrees is east, 180 degrees is south, and 270 degrees is west. When the wind consistently blows more often from one direction than from any other, it is termed a **prevailing wind.**

Wind speed is commonly measured using a **cup anemometer** (Figure 13.16, right). The wind speed is read from a dial much like the speedometer of an automobile.

By knowing the locations of cyclones and anticyclones in relation to where you are, you can predict the changes in wind direction that will be experienced as a pressure center moves past. Because changes in wind direction often bring changes in temperature and moisture conditions, the ability to predict the winds can be very useful. In the Midwest, for example, a north wind may bring cool, dry air from Canada, whereas a south wind may bring warm, humid air from the Gulf of Mexico. Sir Francis Bacon summed it up nicely when he wrote, "Every wind has its weather."

The Chapter in Review

1. Air has weight: At sea level it exerts a pressure of 1 kilogram per square centimeter (14.7 pounds per square inch). *Air pressure* is the force exerted by the weight of air above. With increasing altitude there is less air above to exert a force, and thus air pressure decreases with altitude, rapidly at first, then much more slowly. The unit used by meteorologists to measure atmospheric pressure is the *millibar. Standard sea level pressure* is expressed as 1013.2 millibars. *Isobars* are lines on a weather map that connect places of equal air pressures.

2. A *mercury barometer* measures air pressure using a column of mercury in a glass tube sealed at one end and inverted in a dish of mercury. It measures atmospheric pressure in *"inches of mercury,"* the height of the column of mercury in the barometer. Standard atmospheric pressure at sea level equals 29.92 inches of mercury. As air pressure increases, the mercury in the tube rises and when air pressure decreases, so does the height of the column of mercury. *Aneroid* ("without liquid") *barometers* consist of partially evacuated metal chambers that compress as air pressure increases and expand as pressure decreases.

3. *Wind* is the horizontal flow of air from areas of higher pressure toward areas of lower pressure. Winds are controlled by the following combination of forces: (1) the *pressure gradient force* (amount of pressure change over a given distance), (2) *Coriolis effect* (deflective effect of Earth's rotation, to the right in the Northern Hemisphere and to the left in the Southern Hemisphere), and (3) *friction* with Earth's surface (slows the movement of air and alters wind direction).

4. The two types of pressure centers are: (1) *cyclones,* or *lows* (centers of low pressure), and (2) *anticyclones,* or *highs* (high-pressure centers). In the Northern Hemisphere, winds around a low (cyclone) are counterclockwise and inward. Around a high (anticyclone), they are clockwise and outward. In the Southern Hemisphere, the Coriolis effect causes winds to be clockwise around a low and counterclockwise around a high. Because air rises and cools adiabatically in a low-pressure center, cloudy conditions and precipitation are often associated with their passage. In a high-pressure center, descending air is compressed and warmed; therefore, cloud formation and precipitation are unlikely in an anticyclone, and "fair" weather is usually expected.

5. Earth's *global pressure zones* include the *equatorial low, subtropical high, subpolar low,* and *polar high.* The *global surface winds* associated with these pressure zones are the *trade winds, westerlies,* and *polar easterlies.*

6. Particularly in the Northern Hemisphere, large seasonal temperature differences over continents disrupt the idealized, or zonal, global patterns of pressure and wind. In winter, large, cold landmasses develop a seasonal high-pressure system from which surface airflow is directed off the land. In summer, landmasses are heated and a low-pressure system develops over them, which permits air to flow onto the land. These seasonal changes in wind direction are known as *monsoons.*

7. In the middle latitudes, between 30 and 60 degrees latitude, the general west-to-east flow of the westerlies is interrupted by the migration of cyclones and anticyclones. The paths taken by these cyclonic and anticyclonic systems is closely correlated to upper-level airflow and the polar jet stream. The average position of the polar *jet stream,* and hence the paths followed by cyclones, migrate southward with the approach of winter and northward as summer nears.

8. *Local winds* are small-scale winds produced by a locally generated pressure gradient. Local winds include: *sea and land breezes* (formed along a coast because of daily pressure differences over land and water), *valley and mountain breezes* (in mountainous areas where the air along slopes heats differently than does the air at the same elevation over the valley floor), *chinook and Santa Ana winds* (warm, dry winds created when air descends the leeward side of a mountain and warms by compression).

9. The two basic wind measurements are *direction* and *speed.* Winds are always labeled by the direction *from* which they blow. Wind direction is measured with a *wind vane* and wind speed is measured using a *cup anemometer.*

Key Terms

air pressure (p. 328)

aneroid barometer (p. 328)

anticyclone (high) (p. 333)

barograph (p. 328)

barometric tendency
  (p. 335)

chinook (p. 340)

convergence (p. 334)

Coriolis effect (p. 330)

cup anemometer (p. 341)

cyclone (low) (p. 333)

divergence (p. 334)

equatorial low (p. 336)

geostrophic wind (p. 331)

isobar (p. 329)

jet stream (p. 331)

land breeze (p. 339)

mercury barometer (p. 328)

monsoon (p. 337)

mountain breeze (p. 340)

polar easterlies (p. 336)

polar front (p. 336)

polar high (p. 336)

pressure gradient (p. 329)

pressure tendency (p. 335)

prevailing wind (p. 341)

Santa Ana (p. 340)

sea breeze (p. 339)

subpolar low (p. 336)

subtropical high (p. 336)

trade winds (p. 336)

valley breeze (p. 340)

westerlies (p. 336)

wind (p. 329)

wind vane (p. 340)

Questions for Review

1. What is standard sea level pressure in millibars? In inches of mercury? In pounds per square inch?
2. Mercury is 13 times heavier than water. If you built a barometer using water rather than mercury, how tall would it have to be to record standard sea level pressure (in inches of water)?
3. Describe the principle of the aneroid barometer.
4. What force is responsible for generating wind?
5. Write a generalization relating the spacing of isobars to the speed of wind.
6. How does the Coriolis effect modify air movement?
7. Contrast surface winds and upper-air winds in terms of speed and direction
8. Describe the weather that usually accompanies a drop in barometric pressure and a rise in barometric pressure.
9. Sketch a diagram (isobars and wind arrows) showing the winds associated with surface cyclones and anticyclones in both the Northern and Southern hemispheres.
10. If you live in the Northern Hemisphere and are directly west of a cyclone, what most probably will be the wind direction? What will the wind direction be if you are west of an anticyclone?
11. The following questions relate to the global pattern of air pressure and winds.
    a. The trade winds diverge from which pressure zone?
    b. Which prevailing wind belts converge in the stormy region known as the polar front?
    c. Which pressure belt is associated with the equator?
12. What influence does upper-level airflow seem to have on surface pressure systems?
13. Describe the monsoon circulation of India.
14. What is a local wind? List three examples.
15. A northeast wind is blowing *from* the _____ (direction) toward the _____ (direction).
16. The wind direction is 225 degrees. From what compass direction is the wind blowing (Figure 13.17, p. 341)?

Testing What You Have Learned

Multiple-Choice Questions

1. Which of the following instruments measures atmospheric pressure?
   - **a.** anemometer
   - **b.** aneroid barometer
   - **c.** mercury barometer
   - **d.** a. and b.
   - **e.** b. and c.

2. Which one of the following is NOT a factor that affects wind?
   - **a.** pressure gradient
   - **b.** friction
   - **c.** magnetism
   - **d.** Coriolis effect
   - **e.** both b. and d.

3. Fast-moving, upper-level, "rivers" of air that travel in a west-to-east direction are _____
   - **a.** doldrums.
   - **b.** easterlies.
   - **c.** gulf streams.
   - **d.** zonal flows.
   - **e.** jet streams.

4. Which one of the following best describes surface air circulation in a Northern Hemisphere anticyclone?
   - **a.** divergent-clockwise
   - **b.** stationary
   - **c.** inward-clockwise
   - **d.** rising
   - **e.** convergent-counterclockwise

5. When the barometric pressure is rising it is a good indication that the forthcoming weather will be _____
   - **a.** fair.
   - **b.** stormy.
   - **c.** snow.
   - **d.** hail.
   - **e.** overcast.

6. Near the equator, rising air is associated with the global pressure zone known as the _____
   - **a.** subtropical low.
   - **b.** equatorial high.
   - **c.** equatorial front.
   - **d.** equatorial rise.
   - **e.** equatorial low.

7. Winds that blow from the subtropical high toward the equator are the _____
   - **a.** trade winds.
   - **b.** polar easterlies.
   - **c.** westerlies.
   - **d.** doldrums.
   - **e.** monsoons.

8. The seasonal reversal of wind direction associated with continents (especially Asia) is known as the _____
   - **a.** doldrums.
   - **b.** westerlies.
   - **c.** monsoons.
   - **d.** chinooks.
   - **e.** coastal winds.

9. In the warm summer months, during the daylight hours a _____ breeze often develops along coastal areas.
   - **a.** monsoon
   - **b.** sea
   - **c.** chinook
   - **d.** land
   - **e.** solar

10. A _____ wind blows from west to east.
    - **a.** monsoon
    - **b.** south
    - **c.** east
    - **d.** west
    - **e.** sea

Fill-In Questions

11. The unit that is commonly used to express atmospheric pressure on a weather map is the _____.

12. Lines on a weather map that connect places of equal air pressure are called _____.

13. The Coriolis effect deflects winds to the right of their path of motion in the _____ Hemisphere.

14. Cyclones are _____ pressure centers and anticyclones are _____ pressure centers.

15. In the mid-latitudes, between 30 and 60 degrees latitude, the general _____ to _____ flow of air is interrupted by the migration of cyclones and anticyclones.

True/False Questions

16. Standard sea level pressure is 1013.2 millibars, which equals 29.92 inches of mercury. _____

17. With increasing altitude, air pressure *increases* rapidly at first, then much more slowly. _____

18. Winds blow from areas of higher pressure toward areas of lower pressure. _____

19. Cloudy conditions and precipitation are often associated with the passage of a low-pressure center. _____

20. Warm, dry winds called chinooks are common on the eastern slopes of the Rocky Mountains. _____

Answers:

1. e; 2. c; 3. e; 4. a; 5. a; 6. e; 7. a; 8. c; 9. b; 10. d; 11. millibar; 12. isobars; 13. Northern; 14. low, high; 15. west, east; 16. T; 17. F; 18. T; 19. T; 20. T.

Web Work

The following are informative and interesting World Wide Web sites that address topics related to those presented in this chapter:

- General Circulation of the Atmosphere
  http://oldthunder.ssec.wisc.edu/wxwise/class/gencirc.html

- Jet Stream Analyses and Forecasts
  http://squall.sfsu.edu/crws/jetstream.html

For direct access to these sites and other relevant Web locations, contact the *Foundations of Earth Science* WWW home page at http://www.prenhall.com/lutgens.

CHAPTER 14

Weather Patterns
and Severe Weather

FOCUS ON LEARNING

To assist you in learning the important concepts
in this chapter, you will find it helpful to focus
on the following questions:

1. What is an air mass?

2. How are air masses classified?
   What is the general weather associated with each air mass type?

3. What are fronts? How do warm
   fronts and cold fronts differ?

4. What are the primary weather
   producers in the middle latitudes?
   What are the weather patterns
   associated with these systems?

5. What atmospheric conditions produce thunderstorms, tornadoes,
   and hurricanes?

Tornadoes and hurricanes rank among nature's most destructive forces. Each spring, newspapers report the death and destruction left in the wake of a "band" of tornadoes. During late summer and fall we hear occasional reports about hurricanes in the news. Storms with names like Andrew, Camille, Fran, and Hugo make front-page headlines. Thunderstorms, although less intense and far more common than tornadoes and hurricanes, will also be part of our discussion on severe weather in this chapter. Before looking at violent weather, however, we will study those atmospheric phenomena that most often affect our day-to-day weather: air masses, fronts, and traveling middle-latitude cyclones. Here we will see the interplay of the elements of weather discussed in Chapters 11, 12, and 13.

Air Masses

For many people who live in the middle latitudes, including much of the United States, summer heat waves and winter cold spells are familiar experiences. In the first instance, several days of high temperatures and oppressive humidities may finally end when a series of thunderstorms pass through the area, followed by a few days of relatively cool relief. By contrast, the clear skies that often accompany a span of frigid subzero days may be replaced by thick gray clouds and a period of snow as temperatures rise to levels that seem mild when compared to those that existed just a day earlier. In both examples, what was experienced was a period of generally constant weather conditions followed by a relatively short period of change, and then the subsequent reestablishment of a new set of weather conditions that remained for perhaps several days before changing again.

The weather patterns just described result from movements of large bodies of air, called **air masses.** An air mass, as the term implies, is an immense body of air, typically 1600 kilometers (1000 miles) or more across, and perhaps several kilometers thick, that is characterized by a similarity of temperature and moisture at any given altitude. When this air moves out of its region of origin, it will carry these temperatures and moisture conditions with it, eventually affecting a large portion of a continent.

The horizontal uniformity of an air mass is not perfect, of course. Because air masses extend over large areas, small differences in temperature and humidity from place to place are to be expected. Nevertheless, the differences observed within an air mass are small compared to the rapid changes experienced across boundaries between air masses. Because it may take several days for an air mass to traverse an area, the region under its influence will probably experience fairly constant weather, a situation called **air-mass weather.** Certainly there may be some day-to-day variations, but the events will be quite unlike those in adjacent air masses that have different characteristics. For this reason the boundary between two adjoining air masses that have constrasting characteristics, called a *front,* marks a change in weather.

Source Regions

When a portion of the lower atmosphere moves slowly or stagnates over a relatively uniform surface, the air will assume the distinguishing features of that area, particularly with regard to temperature and moisture conditions (Figure 14.1).

The area where an air mass acquires its characteristic properties of temperature and moisture is called its **source region.** The source regions that produce air masses which influence North America are shown in Figure 14.2.

Air masses are classified according to their source region. **Polar (P) air masses** originate in high latitudes toward Earth's poles, whereas those that form in low latitudes are called **tropical (T) air masses.** The designation *polar* or *tropical* gives an indication of the temperature characteristics of the air masses. *Polar* indicates cold, and *tropical* indicates warm.

In addition, air masses are classified according to the nature of the surface in the source region. **Continental (c) air masses** form over land, and **maritime (m) air masses** originate over water. The designation *continental* or *maritime* thus suggests the moisture characteristics of the air mass. Continental air is likely to be dry, and maritime air, humid.

The four basic types of air masses according to this scheme of classification are continental polar (cP), continental tropical (cT), maritime polar (mP), and maritime tropical (mT).

Weather Associated with Air Masses

Continental polar and maritime tropical air masses influence the weather of North America most, especially east of the Rocky Mountains. Continental polar air masses originate in northern Canada, interior Alaska, and the Arctic—areas that are uniformly cold and dry in winter and cool and dry in summer. In winter, an invasion of continental polar air brings the

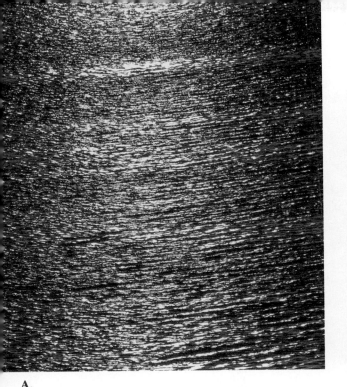

A.

B.

Figure 14.1 An air mass forms when air stagnates or moves slowly over a relatively uniform surface. It assumes the distinguishing features of that region, especially temperature and moisture conditions. **A.** Air masses that form over the ocean are humid. *(Photo by Brian Parker/Tom Stack & Associates)* **B.** Those that form over land are dryer. *(Photo by M. L. Sinibaldi/The Stock Market)*

clear skies and cold temperatures we associate with a cold wave as it moves southward from Canada into the United States. In summer, this air mass may bring a few days of cooling relief.

Although cP air masses are not normally associated with heavy precipitation, those that cross the Great Lakes in winter pick up moisture from the lakes and sometimes bring snow to the leeward shores. These localized storms often form when the surface weather map indicated no apparent cause for a snowstorm to occur. These are known as *lake-effect* snows and they make Buffalo and Rochester, New York, among the snowiest cities in the United States.

Maritime tropical air masses affecting North America most often originate in the Gulf of Mexico, the Caribbean Sea, or the adjacent Atlantic Ocean. As we might expect, these air masses are warm, moisture-laden, and usually unstable. Maritime tropical air is the source of much, if not most, of the precipitation received in the eastern two-thirds of the United States. In summer, when an mT air mass invades the central and eastern United States, and occasionally southern Canada, it brings the high temperatures and oppressive humidity typically associated with its source region.

Of the two remaining air masses, maritime polar and continental tropical, the latter has the least influence on the weather of North America. Hot, dry continental tropical air, originating in the Southwest and Mexico during the summer, only occasionally affects the weather outside its source region.

During the winter, maritime polar air masses coming from the North Pacific often originate as continental polar air masses in Siberia. As these air masses move across the Pacific, they become unstable as they warm and accumulate moisture. Upon entering North America, mP air masses drop much of their moisture because of orographic lifting in the western mountains. Maritime polar air also originates in the North Atlantic off the coast of eastern Canada and occasionally influences the weather of the northeastern United States. In winter, when New England is on the northern or northwestern side of a passing low, the counterclockwise cyclonic winds draw in maritime polar air. The result is a storm characterized by snow and cold temperatures, known locally as a *nor'easter*.

Fronts

Fronts are boundaries that separate different air masses, one warmer than the other and often higher in moisture content. Fronts can form between any two contrasting air masses. Considering the vast size

349

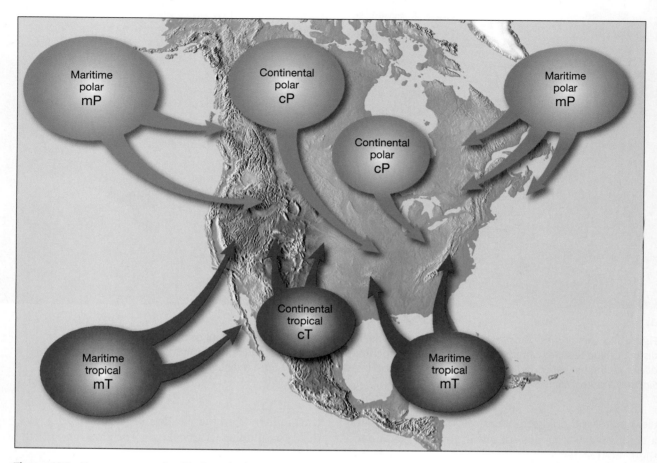

Figure 14.2 Air masses are classified on the basis of their source region. The designation continental (c) or maritime (m) gives an indication of moisture content, whereas polar (P) and tropical (T) indicate temperature conditions.

of the air masses involved, fronts are relatively narrow, being 15- to 200-kilometer-wide bands of discontinuity. On the scale of a weather map, they are generally narrow enough to be represented by a broad line (as in Figure 14.6, p. 354).

Above Earth's surface, the front slopes at a low angle so that warmer air overlies cooler air (Figure 14.3). In the ideal case, the air masses on both sides of the front move in the same direction and at the same speed. Under this condition, the front acts simply as a barrier, moving along between the two air masses, and through which they cannot penetrate. Generally, however, an air mass on one side of a front is moving faster in the direction perpendicular to the front than the air mass on the other side. Thus, one air mass actively advances into another and "clashes" with it. In fact, the boundaries were tagged *fronts* during World War I by Norwegian meteorologists, who visualized them as analogous to battle lines between two armies.

As one air mass moves into another, some mixing does occur along the frontal surface, but for the most part the air masses retain their distinct identities as one is displaced upward over the other. No matter which air mass is advancing, *it is always the warmer, less dense air that is forced aloft,* while *the cooler, denser air acts as the wedge upon which lifting takes place.*

Warm Fronts

When the surface position of a front moves so that warm air occupies territory formerly covered by cooler air, it is called a **warm front** (Figure 14.3). On a weather map, the surface position of a warm front is shown by a line with semicircles extending into the cooler air (Figure 14.3). East of the Rockies, warm tropical air often enters the United States from the Gulf of Mexico and overruns receding cool air. As the cold wedge retreats, friction slows the advance of the surface position of the front more so

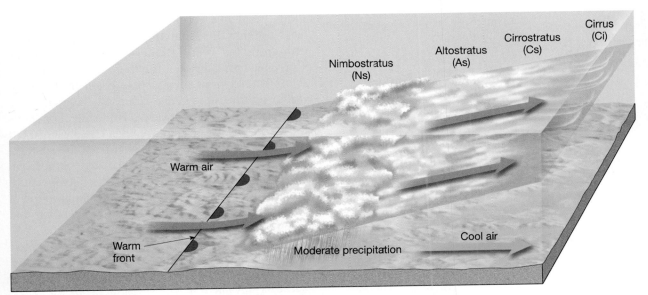

Figure 14.3 Warm front produced as warm air glides up over a cold air mass. Precipitation is moderate and occurs within a few hundred kilometers of the surface front.

than its position aloft; for this reason, the boundary separating these air masses acquires a small slope. The average slope of a warm front is about 1:200, which means that if you are 200 kilometers ahead of the surface location of a warm front, you will find the frontal surface at a height of 1 kilometer.

As warm air ascends the retreating wedge of cold air, it cools adiabatically to produce clouds and frequently, precipitation. The sequence of clouds shown in Figure 14.3 typically precedes a warm front. The first sign of the approach of a warm front is the appearance of cirrus clouds overhead (Figure 14.4). These high clouds form 1000 kilometers or more ahead of the surface front where the overrunning warm air has ascended high up the wedge of cold air.

As the front nears, cirrus clouds grade into cirrostratus, which blend into denser sheets of altostratus. About 300 kilometers ahead of the front, thicker stratus and nimbostratus clouds appear and rain or snow begins. Because of their slow rate of advance and very low slope, warm fronts usually produce light-to-moderate precipitation over a large area for an extended period. Warm fronts, however, are occasionally associated with cumulonimbus clouds and thunderstorms. This occurs when the overrunning air is unstable and the temperatures on opposite sides of the front contrast sharply. At the other extreme, a warm front associated with a dry air mass could pass unnoticed at the surface.

A gradual increase in temperature occurs with the passage of a warm front. The increase is most noticeable when there is a large temperature difference between the adjacent air masses. The moisture content and stability of the encroaching warm air mass largely determine when clear skies will return. During the summer, cumulus, and occasionally cumulonimbus, clouds are embedded in the warm unstable air mass that follows the front. Precipitation from these clouds is usually sporadic and not extensive.

Cold Fronts

When cold air is actively advancing into a region occupied by warmer air, the boundary is called a **cold front** (Figure 14.5). As with warm fronts, friction tends to slow the surface position of a cold front more so than its position aloft. However, because of the relative positions of the adjacent air masses, the cold front steepens as it moves. On the average, cold fronts are about twice as steep as warm fronts, having a slope of perhaps 1:100. In addition, cold fronts advance more rapidly than do warm fronts. These two differences—rate of movement and steepness of slope—largely account for the more violent nature of cold-front weather.

The forceful lifting of air along a cold front is often so rapid that the latent heat released when water vapor condenses will increase the air's buoyancy appreciably. The heavy downpours and vigorous wind gusts associated with mature cumulonimbus

Figure 14.4 Cirrus clouds are high, thin clouds that are often the first sign that a warm front is approaching. As warm air is wedged aloft, clouds form at increasingly greater heights. *(Photo by E. J. Tarbuck)*

clouds frequently result. Because a cold front produces roughly the same amount of lifting as does a warm front, but over a shorter distance, the intensity of precipitation is greater, but the duration is shorter.

As a cold front approaches, generally from the west or northwest, towering clouds often can be seen in the distance. Near the front, a dark band of ominous clouds foretells the ensuing weather. Usually a marked temperature drop and a wind shift from the south to west or northwest accompany the passage of the front. The sometimes violent weather and sharp temperature contrast along the cold front are symbolized on a weather map by a line with triangle-shaped points that extend into the warmer air mass (Figure 14.5).

The weather behind a cold front is dominated by a subsiding and relatively cold air mass. Thus, clearing begins relatively soon after the front passes. Although the compression of air by subsidence causes some adiabatic heating, the effect on surface temperatures is minor. In winter, the clear skies that follow the passing of a cold front further reduce surface temperatures because of the high rate of radiation cooling that occurs during the night. If the continental polar air mass, which most frequently accompanies a cold front, moves into a relatively warm and humid area, surface heating can produce shallow convection. This, in turn, may generate low cumulus or stratocumulus clouds behind the front.

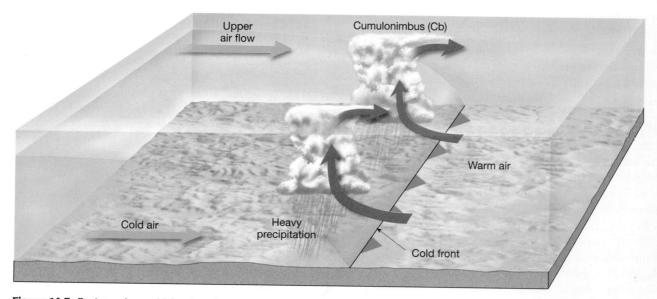

Figure 14.5 Fast-moving cold front and cumulonimbus clouds. Often thunderstorms occur if the warm air is unstable.

The Middle-Latitude Cyclone

So far we have examined the basic elements of weather as well as the dynamics of atmospheric motions. We are now ready to apply our knowledge of these diverse phenomena to an understanding of day-to-day weather patterns in the middle latitudes. For our purposes, *middle latitudes* refers to the region between southern Florida and Alaska.

The primary weather producers here are **middle-latitude cyclones.** They are large centers of low pressure that generally travel from west to east (Figure 14.6). Lasting from a few days to more than a week, these weather systems have a counterclockwise circulation with an air flow inward toward their centers. Most middle-latitude cyclones also have a cold front extending from the central area of low pressure, and frequently a warm front as well. Convergence and forceful lifting initiate cloud development and frequently cause abundant precipitation.

As early as the 1800s it was known that cyclones were the bearers of precipitation and severe weather. But it was not until the early part of the twentieth century that a model was developed to explain how cyclones form. It was formulated by a group of Norwegian scientists and published in 1918. The model was created primarily from near-surface observations.

Years later, as data from the middle and upper troposphere and from satellite images became available, some modifications were necessary. Yet this model is still an accepted working tool in interpreting the weather. If you keep this model in mind when you observe changes in the weather, the changes will no longer come as a surprise. You should begin to see some order in what had appeared to be disorder, and you might even occasionally "predict" the impending weather.

Idealized Weather

The middle-latitude cyclone model provides a useful tool for examining the weather patterns of the middle latitudes. Figure 14.6 illustrates the distribution of clouds and thus the regions of possible precipitation associated with a mature system. Compare this drawing to the satellite image of a cyclone shown in Figure 14.7.

Guided by the westerlies aloft, cyclones generally move eastward across the United States, so we can expect the first signs of their arrival in the west. However, often in the region of the Mississippi valley, cyclones begin a more northeasterly path and

occasionally move directly northward. A mid-latitude cyclone typically requires two to four days to move across a given region. During that short time, rather abrupt changes in atmospheric conditions may be experienced. This is particularly true in the spring, when the largest temperature contrasts occur across the mid-latitudes.

Using Figure 14.6 as a guide, we will now consider these weather producers and what we should expect from them as they move over an area during the spring. To facilitate our discussion, Figure 14.6 includes two profiles along lines A–E and F–G.

- First, imagine the change in weather as you move along profile A–E. At point A the sighting of high cirrus clouds would be the first sign of the approaching cyclone. These high clouds can precede the surface front by 1000 kilometers or more and they generally will be accompanied by falling pressure. As the warm front advances, a lowering and thickening of the cloud deck is noticed.
- Usually within 12 to 24 hours after the first sighting of cirrus clouds, light precipitation begins (point B). As the front nears, the rate of precipitation increases, a rise in temperature is noticed, and winds begin to change from east or southeast to south or southwest.
- With the passage of the warm front, an area is under the influence of a maritime tropical air mass (point C). Generally the region affected by this sector of the cyclone experiences warm temperatures, south or southwest winds, and generally clear skies, although fair-weather cumulus or altocumulus are not uncommon here.
- The relatively warm, humid weather of the warm sector passes quickly and is replaced by gusty winds and precipitation generated along the cold front. The approach of a rapidly advancing cold front is marked by a wall of dark clouds (point D). Severe weather accompanied by heavy precipitation, hail, and an occasional tornado is a definite possibility at this time of year. The passage of the cold front is easily detected by a wind shift; the southwest winds are replaced by winds from the west to northwest and by a pronounced drop in temperature. Also, rising pressure hints of the subsiding cool, dry air behind the front.
- Once the front passes, skies clear as cooler air invades the region (point E). Often a day or two

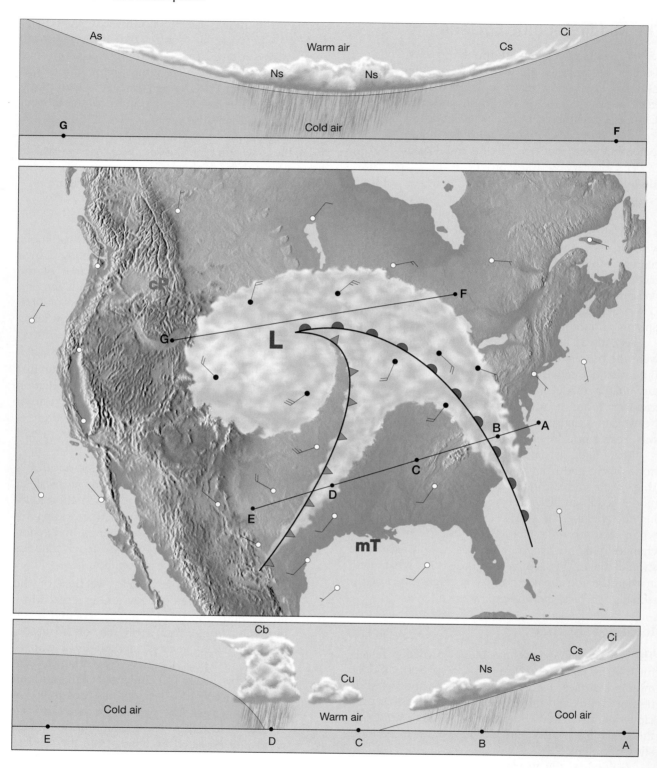

Figure 14.6 Distribution of clouds associated with an idealized middle-latitude cyclone.

Figure 14.7 Satellite view of a mature cyclone over the eastern United States. It is easy to see why we often refer to the cloud pattern of a cyclone as having a "comma" shape. *(Photo courtesy of John Jensenius/ National Weather Service)*

of almost cloudless deep-blue skies occurs unless another cyclone is edging into the region.

A very different set of weather conditions will prevail in those regions north of the storm's center along profile *F–G*. Often the storm reaches its greatest intensity in this zone and the area along profile *F-G* receives the brunt of the storm's fury. Here temperatures remain cold during the passage of the system and heavy snow, sleet, and/or freezing rain may develop during the winter months.

The Role of Airflow Aloft

When the earliest studies of cyclones were made, little was known about the nature of the airflow in the middle and upper troposphere. Since then, a close relationship has been established between surface disturbances and the flow aloft. Airflow aloft plays an important role in maintaining cyclonic and anticyclonic circulation. In fact, more often than not, these rotating surface wind systems are actually generated by upper-level flow.

Recall that the airflow around a surface low is inward, a fact that leads to mass convergence, or coming together (Figure 14.8). The resulting accumulation of air must be accompanied by a corresponding increase in surface pressure. Consequently, we might expect a low-pressure system to "fill"

rapidly and be eliminated, just as the vacuum in a coffee can is quickly dissipated when we open it. However, this does not occur. On the contrary, cyclones often exist for a week or longer. For this to happen, surface convergence must be offset by a mass outflow at some level aloft (Figure 14.8). As long as divergence (spreading out) aloft is equal to, or greater than, the surface inflow, the low pressure and its accompanying convergence can be sustained.

Because cyclones are bearers of stormy weather, they have received far more attention than anticyclones. Nevertheless, a close relation exists, which makes it difficult to separate any discussion of these two types of pressure systems. The surface air that feeds a cyclone, for example, generally originates as air flowing out of an anticyclone. Consequently, cyclones and anticyclones typically are found adjacent to one another. Like the cyclone, an anticyclone depends on the flow far above to maintain its circulation. In this instance, divergence at the surface is balanced by convergence aloft and general subsidence of the air column (Figure 14.8).

Thunderstorms

This is the first of three severe weather types we will examine in this chapter. Sections on tornadoes and hurricanes follow this look at thunderstorms.

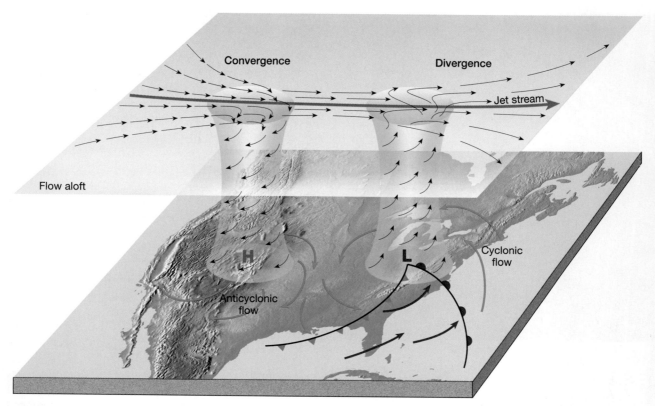

Figure 14.8 Idealized diagram depicting the support that divergence and convergence aloft provide to cyclonic and anticyclonic circulation at the surface. Divergence aloft initiates upward air movement, reduced surface pressure, and cyclonic flow. On the other hand, convergence along the jet stream results in general subsidence of the air column, increased surface pressure, and anticyclonic surface winds.

Severe weather has a fascination that everyday weather phenomena cannot provide. The **lightning** display and booming thunder generated by a severe thunderstorm can be a spectacular event that elicits both awe and fear (Figure 14.9). Of course, hurricanes and tornadoes also attract a great deal of much-deserved attention. A single tornado outbreak or hurricane can cause billions of dollars in property damage as well as many deaths. During a recent 30-year span (1966–1995), the number of fatalities in the United States caused by these meteorological hazards was nearly 5500 persons. It may surprise you that the number of deaths attributed to lightning (more than 2500) was greater than those caused by either tornadoes (almost 2200) or hurricanes (750). However, this may not necessarily be the case in the future.

Thunderstorm Occurrence

Most everyone has observed a small-scale phenomenon that is caused by the vertical motion of warm,

unstable air. Perhaps on a hot day you have seen a small dust devil that formed over an open field and whirled its dusty load to great heights. Or maybe you have noticed a bird glide skyward effortlessly upon an invisible thermal of hot air. These examples illustrate the dynamic thermal instability that occurs during the development of a **thunderstorm.** Thunderstorm activity is associated with cumulonimbus clouds that generate heavy rainfall, thunder, lightning, and occasionally hail and tornadoes.

At any given time there are an estimated 2000 thunderstorms in progress on Earth. As we would expect, the greatest number occur in the tropics where warmth, plentiful moisture, and instability are always present. About 45,000 thunderstorms take place each day and more than 16 million occur annually around the world. The lightning from these storms strikes Earth 100 times each second! Annually, the United States experiences about 100,000 thunderstorms and millions of lightning strikes. A glance at Figure 14.10 shows that thunderstorms are most frequent in Florida

Figure 14.9 Lightning near Sterling, Colorado. Cumulonimbus clouds can produce lightning, thunderstorms, and other forms of severe weather. *(Photo by Keith Kent/Science Photo Library/Photo Researchers, Inc.)*

Stages of Thunderstorm Development

All thunderstorms require warm, moist air, which, when lifted, will release sufficient latent heat to provide the buoyancy necessary to maintain its upward flight. Although this instability and associated buoyancy are triggered by a number of different processes, all thunderstorms have a similar life history.

Because instability and buoyancy are enhanced by high surface temperatures, thunderstorms are most common in the afternoon and early evening. However, surface heating alone is not sufficient for the growth of towering cumulonimbus clouds. A solitary cell of rising hot air produced by surface heating could, at best, produce a small cumulus cloud, which would evaporate within 10 to 15 minutes.

The development of 12,000-meter (or on rare occasions 18,000-meter) cumulonimbus towers requires a continual supply of moist air (Figure 14.11). Each new surge of warm air rises higher than the last, adding to the height of the cloud (Figure 14.12). These updrafts must occasionally reach speeds over 100 kilometers (60 miles) per hour to accommodate the size of hailstones they are capable of carrying upward. Usually within an hour the amount and size of precipitation that has accumulated is too much for the updrafts to support, and in one part of the cloud downdrafts develop, releasing heavy precipitation. This represents the most active stage of the thunderstorm. Gusty winds, lightning, heavy precipitation, and sometimes hail are experienced.

Eventually the warm, moist air supplied by updrafts ceases as downdrafts dominate throughout the cloud. The cooling effect of falling precipitation, coupled with the influx of colder air aloft, marks the end of the thunderstorm activity. The life span of a cumulonimbus cell within a thunderstorm complex is only about an hour, but as the storm moves, fresh supplies of warm, water-laden air generate new cells to replace those that are dissipating.

Tornadoes

Tornadoes are local storms of short duration that must be ranked high among nature's most destructive forces. Their sporadic occurrence and violent winds cause many deaths each year. Tornadoes are violent windstorms that take the form of a rotating column of air or *vortex* that extends downward from a cumulonimbus cloud (Figure 14.13).

and the eastern Gulf Coast region, where such activity is recorded between 70 and 100 days each year.

The greatest number of thunderstorms occur in association with relatively short-lived cumulonimbus clouds that produce local precipitation. In the United States these cells typically form within warm, humid (maritime tropical) air masses that originate over the Gulf of Mexico and migrate northward.

Occasionally, however, thunderstorms grow very large and remain active for hours. These *severe thunderstorms* produce frequent lightning and are accompanied by locally damaging winds or hail. Most severe thunderstorms in the middle latitudes form along or ahead of cold fronts. Here, forceful lifting of unstable mT air masses triggers thunderstorm development.

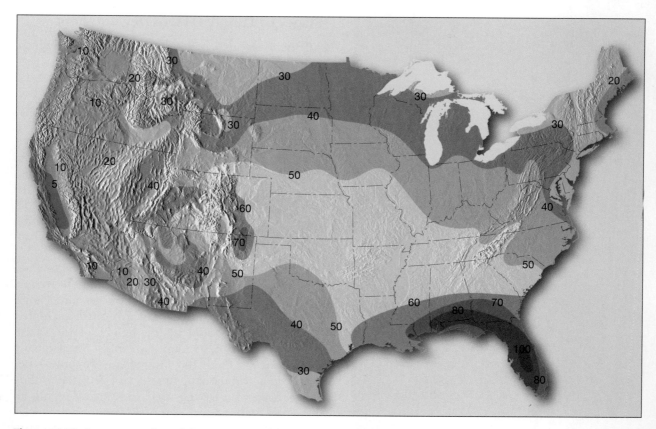

Figure 14.10 Average number of days per year with thunderstorms. Because of its close proximity to the source region for warm, humid, and unstable air masses, the Gulf Coast receives much of its precipitation from thunderstorms. *(Source: Environmental Data Service, NOAA)*

Figure 14.11 This developing cumulonimbus cloud became a towering August thunderstorm over the high plains near Boulder, Colorado. *(Photo by Henry Lansford)*

Pressures within some tornadoes have been estimated to be as much as 10 percent lower than immediately outside the storm. Drawn by the much lower pressure in the center of the vortex, air near the ground rushes into the tornado from all directions. As the air streams inward, it is spiraled upward around the core until it eventually merges with the airflow of the parent thunderstorm deep in the cumulonimbus tower. Because of the tremendous pressure gradient associated with a strong tornado, maximum winds can sometimes approach 480 kilometers (300 miles) per hour.

Tornado Occurrence and Development

An average of about 770 tornadoes are reported each year in the United States. Still, the actual number occurring from one year to the next varies greatly. During a recent 40-year span, for example, yearly totals ranged from a low of 421 to a high of 1126.

Tornadoes occur during every month of the year. April through June is the period of greatest tornado

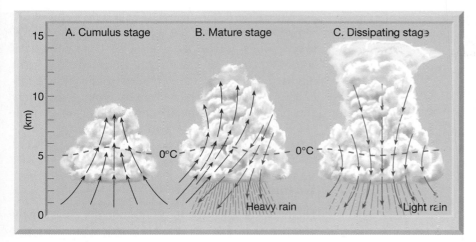

Figure 14.12 Stages in the development of a thunderstorm. During the cumulus stage, strong updrafts act to build the storm. The mature stage is marked by heavy precipitation and cool downdrafts in part of the storm. When the warm updrafts disappear completely, precipitation becomes light, and the cloud begins to evaporate.

frequency in the United States, whereas the number is lowest during December and January (Figure 14.14). Of the more than 21,000 confirmed tornadoes reported over the contiguous 48 states during the period depicted in Figure 14.14, an average of almost five per day occurred during May. At the other extreme, a tornado was reported only every other day in January.

Tornadoes form in association with severe thunderstorms that produce high winds, heavy (sometimes torrential) rainfall, and often damaging hail. Fortunately, less than 1 percent of all thunderstorms produce tornadoes. Although meteorologists are still not sure what triggers tornado formation, it has become apparent that they are products of the interaction between strong updrafts in a thunderstorm and winds in the troposphere. Despite recent advances in modeling the many variables that eventually produce a strong tornado, our knowledge is still limited. Nevertheless, the general atmospheric conditions that are most likely to develop into tornado activity are known.

General Atmospheric Conditions   Severe thunderstorms—and hence tornadoes—are most often spawned along the cold front of a middle-latitude cyclone. Throughout spring, air masses associated with middle-latitude cyclones are most likely to have greatly contrasting conditions. Continental polar air from Canada may still be very cold and dry, whereas maritime tropical air from the Gulf of Mexico is warm, humid, and unstable. The greater the contrast when these air masses meet, the more intense the storm. These two contrasting air masses are most likely to meet in the central United States, because there is no significant natural barrier separating the center of the country from

Figure 14.13 A tornado is a violently rotating column of air in contact with the ground. The air column is visible when it contains condensation or dust and debris. Often the appearance is the result of both. When the column of air is aloft and does not produce damage, the visible portion is properly called a funnel cloud. *(Photo by A. and J. Verkaik/The Stock Market)*

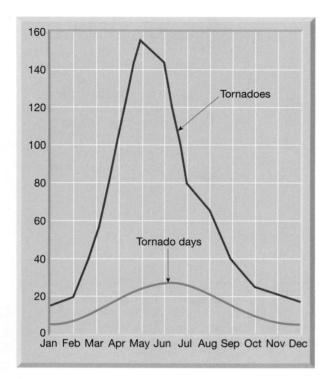

Figure 14.14 Average number of tornadoes and tornado days each month in the United States for a 27-year period. *(After NOAA)*

the arctic or the Gulf of Mexico. Consequently, this region generates more tornadoes than any other area of the country or, in fact, the world. Figure 14.15, which depicts tornado incidence in the United States for a 27-year period, readily substantiates this fact.

Profile of a Tornado An average tornado has a diameter of between 150 and 600 meters (500 and 2000 feet), travels across the landscape at approximately 45 kilometers (30 miles) per hour, and cuts a path about 10 kilometers (6 miles) long.* Because tornadoes usually occur slightly ahead of a cold front, in the zone of southwest winds, most move toward the northeast.

Of the hundreds of tornadoes reported in the United States annually, over half are comparatively weak and short-lived. Most of these small tornadoes have lifetimes of three minutes or less and paths that seldom exceed 1 kilometer in length and 100 meters wide. Typical wind speeds are on the order of 150

---

*The 10-kilometer figure applies to documented tornadoes. Because many small tornadoes go undocumented, the real average path of all tornadoes is unknown, but is shorter than 10 kilometers.

kilometers per hour or less. On the other end of the tornado spectrum are the infrequent and often long-lived violent tornadoes. Although large tornadoes constitute only a small percentage of the total reported, their effects are often devastating. Such tornadoes may exist for periods in excess of three hours and produce an essentially continuous damage path more than 150 kilometers long and perhaps a kilometer or more wide. Maximum winds range upward to 450 kilometers per hour.

Tornado Destruction

The potential for tornado destruction depends largely upon the strength of the winds generated by the storm. One commonly used guide to tornado intensity is the *Fujita intensity scale,* or simply the *F-scale* (Table 14.1). Because tornado winds cannot be measured directly, a rating on the F-scale is determined by assessing the worst damage produced by a storm.

Tornadoes take many lives each year, sometimes hundreds in a single day. When tornadoes struck an area stretching from Canada to Georgia on April 3, 1974, the death toll exceeded 300, the worst in half a century. Most tornadoes, however, do not result in a loss of life. In one statistical study that examined a 29-year period, there were 689 tornadoes that resulted in deaths. This figure represents slightly less than 4 percent of the total 19,312 reported storms.

Although the percentage of tornadoes resulting in death is small, each tornado is potentially lethal. When tornado fatalities and storm intensities are compared, the results are quite interesting: The majority (63 percent) of tornadoes are weak (F0 and F1), and the number of storms decreases as tornado intensity increases. The distribution of tornado fatalities, however, is just the opposite. Although only 2

Table 14.1 ■ Fujita Intensity Scale

Scale	Wind Speed		Damage
	Km/Hr	Mi/Hr	
F0	<116	<72	Light damage
F1	116–180	72–112	Moderate damage
F2	181–253	113–157	Considerable damage
F3	254–332	158–206	Severe damage
F4	333–419	207–260	Devastating damage
F5	>419	>260	Incredible damage

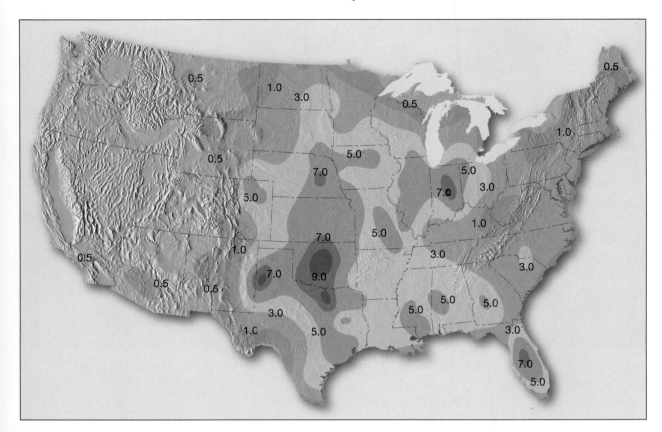

Figure 14.15 Average annual tornado incidence per 10,000 square miles (26,000 square kilometers) for a 27-year period.

percent of tornadoes are classified as violent (F4 and F5), they account for nearly 70 percent of the deaths. If there is some question as to the causes of tornadoes, there certainly is no question about the destructive effects of these violent storms (Figure 14.16).

Predicting Tornadoes

Because severe thunderstorms and tornadoes are small and relatively short-lived phenomena, they are among the most difficult weather features to forecast precisely.

Tornado Watch   When conditions appear favorable for tornado formation, a **tornado watch** is issued for areas covering about 65,000 square kilometers (25,000 square miles). Between 35 and 40 percent of these predictions are correct; that is, one or more tornadoes occur somewhere in the specified region. The incorrect forecasts are about evenly divided between cases in which no tornadoes are sighted and cases when tornadoes occur outside, but near, the watch area.

Tornado Warning   Whereas a tornado watch is designed to alert people to the possibility of tornadoes, a **tornado warning** is issued when a funnel cloud has actually been sighted or is indicated by radar. It warns of high probability of imminent danger.

When severe weather threatens, radar screens are monitored for very intense echoes, which in turn are associated with heavy precipitation and the greater likelihood of hail, strong winds, and tornadoes. In addition, the echo from a tornadic storm sometimes displays a hook-shaped appendage. If the direction and approximate speed of the storm are known, the storm's most probable path can be estimated. Because tornadoes often move erratically, the warning area is fan-shaped downwind from the point where the tornado has been spotted.

Since the late 1960s, warnings have been given for most major tornadoes. It is believed that such warnings substantially reduce the number of deaths and serious injuries that might otherwise occur.

Figure 14.16 This home was completely destroyed by a tornado. *(Photo by Michael Meinhardt/Sygma)*

Doppler Radar The tornado warning system that has been used throughout the United States for many years relies heavily on visual sightings by a few trained observers as well as the general public. Unfortunately, such a system is prone to incomplete coverage and mistakes. The errors are most likely to occur at night when tornadoes may go unnoticed or harmless clouds may be mistaken for funnel clouds. Hence, there may be a lack of adequate warning on the one hand, or unnecessary warnings on the other.

Many of the difficulties that once limited the accuracy of tornado warnings have been reduced or eliminated by an advancement in radar technology called **Doppler radar.** Doppler radar not only performs the same tasks as conventional radar, but also has the ability to detect motion directly. Doppler radar can detect the initial formation and subsequent development of a *mesocyclone,* an intense rotating wind system in the lower part of a thunderstorm that frequently precedes tornado development. Almost all mesocyclones produce damaging hail, severe winds, or tornadoes. Those that produce tornadoes (about 60 percent) can be distinguished by their stronger wind speeds and their sharper gradients of wind speeds.

It should also be pointed out that not all tornado bearing storms have clear-cut radar signatures and that other storms can give false signatures. Detection, therefore, is sometimes a subjective process and a given display could be interpreted in several ways. Consequently, trained observers will continue to form an important part of the warning system in the foreseeable future.

Although some operational problems exist, the benefits of Doppler radar are many. As a research tool, it is not only providing data on the formation of tornadoes, but is also helping meteorologists gain new insights into thunderstorm development, the structure and dynamics of hurricanes, and air turbulence hazards that plague aircraft. As a practical tool for tornado detection, it has significant advantages over a system that uses conventional radar.

Hurricanes

Most of us view the weather in the tropics with favor. Places like Hawaii and the islands of the Caribbean are known for their lack of significant day-to-day variations. Warm breezes, steady temperatures, and rains that come as heavy but brief tropical showers are expected. It is ironic that these relatively tranquil regions sometimes produce the most violent storms on Earth.

The whirling tropical cyclones that on occasion have wind speeds attaining 300 kilometers (185 miles) per hour are known in the United States as **hurricanes** and are the greatest storms on Earth. Out at sea, they can generate 15-meter (50-foot) waves capable of inflicting destruction hundreds of kilometers from their source. Should a hurricane smash onto land, strong winds coupled with extensive flooding can impose billions of dollars in damage and great loss of life (Figure 14.17).

The vast majority of hurricane-related deaths and damage are caused by relatively infrequent, yet powerful, storms. Hurricane Andrew was one such storm. On August 24, 1992, Andrew slammed into southern Florida, crossed the peninsula, and then went on across the Gulf of Mexico to strike the Louisiana coast two days later (Figure 14.18). The storm's destructive accomplishments were awesome. In all, the hurricane was responsible for at least 62 deaths and caused $20 billion to $30 billion in damages. It was the costliest natural disaster in U.S. history.

Hurricanes are becoming a growing threat because more and more people are living and working along and near coasts. In the mid-1990s, more than 50 percent of the U.S. population lived within 75 kilometers (45 miles) of a coast. This number is projected to increase substantially in the early decades of the twenty-first century. The concentration of such large numbers of people near the shoreline means that hurricanes and other large storms place millions at risk. Moreover, the potential costs of property damage are incredible.

Profile of a Hurricane

Hurricanes form in all tropical waters (except those of the South Atlantic and eastern South Pacific) between the latitudes of 5 degrees and 20 degrees. They are known by different names in various parts of the world. In the western Pacific, they are called *typhoons* and in the Indian Ocean, *cyclones*. The North Pacific has the greatest number of storms, averaging 20 each year. Fortunately for those living in the coastal regions of the southern and eastern United States, fewer than five hurricanes, on the

Figure 14.17 This destruction on North Carolina's Topsail Isle occurred when Hurricane Fran struck on September 5, 1996. The storm caused an estimated $3 billion in total damages. *(Photo by Cheryl Hogue/Photo Researchers, Inc.)*

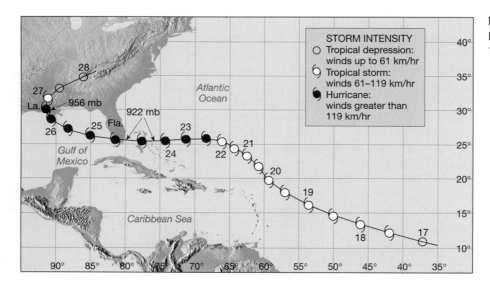

Figure 14.18 Positions for Hurricane Andrew, August 17–28, 1992. *(National Hurricane Center)*

average, develop annually in the warm sector of the North Atlantic.

Although many tropical disturbances develop each year, only a few reach hurricane status. By international agreement, a hurricane has wind speeds in excess of 119 kilometers (74 miles) per hour and a rotary circulation. Mature hurricanes average 600 kilometers (375 miles) in diameter and often extend 12,000 meters (40,000 feet) above the ocean surface. From the outer edge to the center, the barometric pressure has on occasion dropped 60 millibars, from 1010 millibars to 950 millibars. The lowest pressures ever recorded in the Western Hemisphere are associated with these storms. A steep pressure gradient generates the rapid, inward-spiraling winds of a hurricane.

As the inward rush of warm, moist surface air approaches the core of the storm, it turns upward and ascends in a ring of cumulonimbus towers. This doughnut-shaped wall of intense convective activity surrounding the center of the storm is called the **eye wall.** It is here that the greatest wind speeds and heaviest rainfall occur. Surrounding the eye wall are curved bands of clouds that trail away in a spiral fashion. Near the top of the hurricane the airflow is outward, carrying the rising air away from the storm center, thereby providing room for more inward flow at the surface.

At the very center of the storm is the **eye** of the hurricane (Figure 14.19). This well-known feature is a zone about 20 kilometers (12.5 miles) in diameter where precipitation ceases and winds subside. It offers a brief but deceptive break from the extreme

weather in the enormous curving wall clouds that surround it. The air within the eye gradually descends and heats by compression, making it the warmest part of the storm. Although many people believe that the eye is characterized by clear blue skies, such is usually not the case because the subsidence in the eye is seldom strong enough to produce cloudless conditions. Although the sky appears much brighter in this region, scattered clouds at various levels are common.

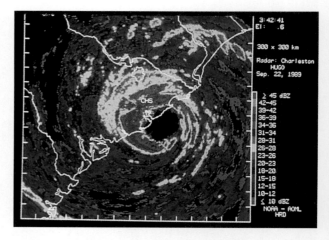

Figure 14.19 Radar image of Hurricane Hugo moving over the coast of South Carolina on September 22, 1989. Colors show rainfall intensity. Red is the most intense and blue is less intense. No rain is falling in the black areas. The rainless "hole" in the center is the eye. As expected, the heaviest rains are occurring in the eye wall. *(Courtesy of Peter Dodge, Hurricane Research Division/National Oceanic and Atmospheric Administration)*

Hurricane Formation and Decay

A hurricane is a heat engine that is fueled by the latent heat liberated when huge quantities of water vapor condense. The amount of energy produced by a typical hurricane in just a single day is truly immense—roughly equivalent to the entire electrical energy production of the United States in a year. The release of latent heat warms the air and provides buoyancy for its upward flight. The result is to reduce the pressure near the surface, which encourages a more rapid inward flow of air. To get this engine started, a large quantity of warm, moisture-laden air is required, and a continual supply is needed to keep it going.

Hurricanes develop most often in the late summer when water temperatures have reached 27°C (80°F) or higher and thus are able to provide the necessary heat and moisture to the air. This ocean-water temperature requirement accounts for the fact that hurricanes do not form over the relatively cool waters of the South Atlantic and the eastern South Pacific. For the same reason, few hurricanes form poleward of 20 degrees of latitude. Although water temperatures are sufficiently high, hurricanes do not form within 5 degrees of the equator, because the Coriolis effect is too weak to initiate the necessary rotary motion.

The initial stage of a tropical cyclone's life cycle is not well understood, owing in part to the complexity of these storms. Another important factor is the lack of observations in regions where storms form. However, we do know that smaller tropical cyclones initiate the process. These initial disturbances are regions of low-level convergence and lifting.

Many tropical disturbances of this type occur each year, but only a few develop into full-fledged hurricanes. By international agreement, lesser tropical cyclones are placed in different categories, based on wind strength. When a cyclone's strongest winds do not exceed 61 kilometers (38 miles) per hour, it is called a **tropical depression.** When winds are between 61 and 119 kilometers (38 and 74 miles) per hour, the cyclone is termed a **tropical storm.** It is during this phase that a name is given (Andrew, Fran, Opal, etc.). Should the tropical storm become a hurricane, the name remains the same. Each year between 80 and 100 tropical storms develop around the world. Of them, usually half or more eventually become hurricanes.

Hurricanes diminish in intensity whenever they (1) move over ocean waters that cannot supply warm, moist tropical air; (2) move onto land; or (3) reach a location where the large-scale flow aloft is unfavorable. Whenever a hurricane moves onto land, it loses its punch rapidly. The most important reason for this rapid demise is the fact that the storm's source of warm, moist air is cut off. When an adequate supply of water vapor does not exist, condensation and the release of latent heat must diminish. In addition, friction from the increased roughness of the land surface rapidly slows surface wind speeds. This factor causes the winds to move more directly into the center of the low, thus helping to eliminate the large pressure differences.

Hurricane Destruction

A location only a few hundred kilometers from a hurricane—just one day's striking distance away—may experience clear skies and virtually no wind. Prior to the age of weather satellites, such a situation made the task of warning people of impending storms very difficult.

The worst natural disaster in U.S. history resulted from a hurricane that struck an unprepared Galveston, Texas, on September 8, 1900. The strength of the storm, coupled with the lack of adequate warning, caught the population by surprise and cost 6000 people in the city their lives. At least 2000 more were killed elsewhere. Fortunately, hurricanes are no longer the unheralded killers they once were.

Once a storm develops cyclonic flow and the spiraling bands of clouds characteristic of a hurricane, it receives continuous monitoring. When hurricanes form, satellites are able to identify and track the storms long before they make landfall (Figure 14.20). In the United States, early warning systems have greatly reduced the number of deaths caused by hurricanes. At the same time, however, there has been an astronomical rise in the amount of property damage. The primary reason, of course, has been the extensive building of homes and businesses in coastal areas.

Although the amount of damage caused by a hurricane depends on several factors, including the size and population density of the area affected and the shape of the ocean bottom near the shore, certainly the most significant factor is the strength of the storm itself. By studying past storms, a scale has been established to rank the relative intensity of hurricanes (Table 14.2). As Table 14.2 indicates, a *category-5* storm is the worst possible, whereas a *category-1* hurricane is least severe. During the hurricane season it is common to hear scientists and reporters alike use the numbers from the *Saffir-Simpson Hurricane*

Figure 14.20 Weather satellites are invaluable tools for tracking storms and gathering atmospheric data. This enhanced infrared image shows Hurricane Andrew as the storm crossed the Gulf of Mexico on August 25, 1992. The eye is clearly visible in the center of the storm. The eye wall, the most intense part of the storm, is a doughnut-shaped wall of cumulonimbus development surrounding the eye. *(Photo by Photri/The Stock Market)*

Wind Damage For some structures, the force of the wind is sufficient to cause total destruction. This was demonstrated in southern Florida in 1992. The billions of dollars in property damages from Hurricane Andrew were largely the result of strong winds. However, wind damage is not necessarily responsible for a hurricane's greatest destructiveness.

Storm Surge The most devastating damage in the coastal zone is caused by the storm surge. It not only accounts for a large share of coastal property losses, but is also responsible for 90 percent of all hurricane-caused deaths. A **storm surge** is a dome of water 65 to 80 kilometers (40 to 50 miles) wide that sweeps across the coast near the point where the eye makes landfall. Ignoring wave activity, the storm surge is the height of the water above normal tide level. Thus, a storm surge commonly adds 2 to 3 meters (6 to 10 feet) to normal tide heights—to say nothing of tremendous wave activity superimposed atop the surge.

We can easily imagine the damage this surge of water can inflict on low-lying coastal areas. In the delta region of Bangladesh, for example, the land is mostly less than 2 meters above sea level. When a storm surge superimposed upon normal high tide inundated that area on November 13, 1970, the official death toll was 200,000. This was one of the worst disasters of modern times. In May 1991 a similar event struck Bangladesh again.

Scale. The famous Galveston hurricane just mentioned, with winds in excess of 209 kilometers (130 miles) per hour and a pressure of 931 millibars, would be placed in category 4. Storms that fall into category 5 are rare.

Damage caused by hurricanes can be divided into three categories: (1) wind damage, (2) storm surge, and (3) inland freshwater flooding.

Inland Flooding The torrential rains that accompany most hurricanes represent a third significant threat—flooding. Whereas the effects of storm surge and strong winds are concentrated in coastal areas, heavy rains may affect places hundreds of kilometers from the coast for several days after the storm has lost its hurricane-force winds.

Table 14.2 ■ Saffir-Simpson Hurricane Scale

Scale Number (category)	Central Pressure (millibars)	Winds (km/hr)	Storm Surge (meters)	Damage
1	≥980	119–153	1.2–1.5	Minimal
2	965–979	154–177	1.6–2.4	Moderate
3	945–964	178–209	2.5–3.6	Extensive
4	920–944	210–250	3.7–5.4	Extreme
5	>920	>250	>5.4	Catastrophic

Hurricanes weaken rapidly as they move inland, yet the remnants of the storm can still yield 15 to 30 centimeters (6 to 12 inches) or more of rain as they move inland. A good example of such destruction was Hurricane Agnes (1972). Although this was just a category-1 storm on the Saffir-Simpson scale, it was one of the costliest hurricanes of the century, creating more than $2 billion in damage and taking 122 lives. Most destruction was attributed to flooding caused by an inordinate amount of rainfall.

To summarize, extensive damage and loss of life in the coastal zone can result from storm surge, torrential rains, and strong winds. When loss of life occurs, it is commonly caused by the storm surge, which can devastate entire barrier islands and low-lying land along the coast. Although wind damage is usually not as catastrophic as the storm surge, it affects a much larger area. Where building codes are inadequate, economic losses can be especially severe. Because hurricanes weaken as they move inland, most wind damage occurs within 200 kilometers of the coast. Far from the coast a weakening storm can produce extensive flooding long after the winds have diminished below hurricane levels. Sometimes the damage from inland flooding exceeds storm-surge destruction.

The Chapter in Review

1. An *air mass* is a large body of air, usually 1600 kilometers (1000 miles) or more across, that is characterized by a *homogeneity of temperature and moisture at any given altitude*. When this air moves out of its region of origin, called the *source region,* it will carry these temperatures and moisture conditions elsewhere, perhaps eventually affecting a large portion of a continent.

2. Air masses are classified according to (1) the nature of the surface in the source region and (2) the latitude of the source region. *Continental (c)* designates an air mass of land origin, with the air likely to be dry; whereas a *maritime (m)* air mass originates over water, and therefore will be humid. *Polar (P)* air masses originate in high latitudes and are cold. *Tropical (T)* air masses form in low latitudes and are warm. According to this classification scheme, the *four basic types of air masses are: continental polar (cP), continental tropical (cT), maritime polar (mP), and maritime tropical (mT).* Continental polar (cP) and maritime tropical (mT) air masses influence the weather of North America most, especially east of the Rocky Mountains. Maritime tropical air is the source of much, if not most, of the precipitation received in the eastern two-thirds of the United States.

3. *Fronts* are boundaries that separate air masses of different densities, one warmer and often higher in moisture content than the other. A *warm front* occurs when the surface position of the front moves so that warm air occupies territory formerly covered by cooler air. Along a warm front, a warm air mass overrides a retreating mass of cooler air. As the warm air ascends, it cools adiabatically to produce clouds and frequently, light-to-moderate precipitation over a large area. A *cold front* forms where cold air is actively advancing into a region occupied by warmer air. Cold fronts are about twice as steep and move more rapidly than do warm fronts. Because of these two differences, precipitation along a cold front is generally more intense and of shorter duration than precipitation associated with a warm front.

4. The primary weather producers in the middle latitudes are *large centers of low pressure* that generally travel from *west to east,* called *middle-latitude cyclones.* These *bearers of stormy weather,* which last from a few days to a week, have a *counterclockwise circulation* pattern in the Northern Hemisphere, with an *inward flow of air* toward their centers. Most middle-latitude cyclones have a *cold front and frequently a warm front* extending from the central area of low pressure. *Convergence* and *forceful lifting along the fronts* initiate cloud development and frequently cause precipitation. The particular weather experienced by an area depends on the path of the cyclone.

5. *Thunderstorms* are caused by the upward movement of warm, moist, unstable air. They are associated with cumulonimbus clouds that generate heavy rainfall, thunder, lightning, and occasionally hail and tornadoes.

6. *Tornadoes,* destructive, local storms of short duration, are violent windstorms associated with severe thunderstorms that take the form of a rotating column of air that extends downward from a cumulonimbus cloud. Tornadoes are most often spawned along the cold front of a middle-latitude cyclone, most frequently during the spring months.

7. *Hurricanes,* the greatest storms on Earth, are tropical cyclones with wind speeds in excess of 119 kilometers (74 miles) per hour. These complex tropical disturbances develop over tropical ocean waters and are fueled by the latent heat liberated when huge quantities of water vapor condense. Hurricanes form most often in late summer when ocean-surface temperatures reach 27°C (80°F) or higher and thus are able to provide the necessary heat and moisture to air. Hurricanes diminish in intensity whenever they (1) move over cool ocean water that cannot supply adequate heat and moisture; (2) move onto land; or (3) reach a location where large-scale flow aloft is unfavorable. Hurricane damage is of three main types: (1) wind damage, (2) storm surge, and (3) inland flooding.

Key Terms

air mass (p. 348)

air-mass weather (p. 348)

cold front (p. 351)

continental (c) air mass (p. 348)

Doppler radar (p. 362)

eye (p. 364)

eye wall (p. 364)

front (p. 349)

hurricane (p. 363)

lightning (p. 356)

maritime (m) air mass (p. 348)

middle-latitude cyclone (p. 353)

polar (P) air mass (p. 348)

source region (p. 348)

storm surge (p. 366)

thunderstorm (p. 356)

tornado (p. 357)

tornado warning (p. 361)

tornado watch (p. 361)

tropical (T) air mass (p. 348)

tropical depression (p. 365)

tropical storm (p. 365)

warm front (p. 350)

Questions for Review

1. Describe the weather associated with a continental polar air mass in the winter and in the summer. When would this air mass be most welcome in the United States?

2. What are the characteristics of a maritime tropical air mass? Where are the source regions for the maritime tropical air masses that affect North America? Where are the source regions for the maritime polar air masses?

3. Describe the weather along a cold front where very warm, moist air is being displaced.

4. For each weather element listed, describe the changes that an observer experiences when a middle-latitude cyclone passes with its center north of the observer: wind direction, pressure tendency, cloud type, cloud cover, precipitation, temperature.

5. Describe the weather conditions an observer would experience if the center of a middle-latitude cyclone passed to the south.

6. Briefly explain how the flow aloft aids the formation of cyclones at the surface.

7. What is the primary requirement for the formation of thunderstorms?

8. Based on your answer to Question 7, where would you expect thunderstorms to be most common on Earth? Where specifically in the United States?

9. Why do tornadoes have such high wind speeds?

10. What general atmospheric conditions are most conducive to the formation of tornadoes?

11. When is "tornado season"? That is, during what months is tornado activity most pronounced?

12. Distinguish between a tornado watch and a tornado warning.

13. What advantages does Doppler radar have over conventional radar?

14. Which has stronger winds, a tropical storm or a tropical depression?

15. Why does the intensity of a hurricane diminish rapidly when it moves onto land?

16. Hurricane damage can be divided into three broad categories. Name them. Which category is responsible for the highest percentage of hurricane-related deaths?

17. A hurricane has slower wind speeds than does a tornado, yet it inflicts more total damage. How might this be explained?

Testing What You Have Learned

Multiple-Choice Questions

1. A large body of air characterized by relatively uniform temperatures and moisture conditions at any given altitude is called a(n) _____

   a. cold front.　　　**c.** cyclone.　　　**e.** anemometer.

   b. air mass.　　　**d.** warm front.

2. Which air mass would be associated with afternoon thunderstorms along the Florida coast?

   a. cP　　　**c.** mP　　　**e.** none of the above

   b. cT　　　**d.** mT

3. Which pair of air masses most influence the weather east of the Rocky Mountains?

   a. cP and cT　　　**c.** cP and mT　　　**e.** cP and mP

   b. cT and mP　　　**d.** mT and mP

4. Maritime tropical air masses affecting North America can originate in the _____

   a. Caribbean Sea.　　　**c.** Gulf of Mexico.　　　**e.** a., b., and c.

   b. Atlantic Ocean.　　　**d.** a. and c.

5. The passage of a(n) _____ front occurs when warm air occupies the territory formerly covered by cooler air.

   a. warm　　　**c.** stationary　　　**e.** mixed

   b. cold　　　**d.** occluded

6. Which type of front has the steepest slope?

   a. warm　　　**c.** stationary　　　**e.** mixed

   b. cold　　　**d.** occluded

7. What is the general direction of middle-latitude cyclone movement?

   a. north to south　　　**c.** northeast to southwest　　　**e.** west to east

   b. south to north　　　**d.** east to west

8. As the center of a middle-latitude cyclone *approaches,* air pressure will _____

   a. rise.　　　**c.** fall.　　　**e.** rise then remain steady.

   b. remain steady.　　　**d.** fall then rise.

9. After the *passage* of a warm front, what air mass will most likely influence an area in the central United States?

   a. cP　　　**c.** mP　　　**e.** cA

   b. cT　　　**d.** mT

10. The lowest air pressures ever recorded in the Western Hemisphere are associated with these storms.

    a. thunderstorms　　　**c.** lightning　　　**e.** middle-latitude cyclones

    b. tornadoes　　　**d.** hurricanes

Fill-In Questions

11. The area in which an air mass acquires its characteristic properties of temperature and moisture is called its _____ _____.

12. Usually a rapid temperature drop and a wind shift from the south to west or northwest accompany the passage of a _____ front.

13. In the region between southern Florida and Alaska, the primary weather producers are _____-_____ _____.

14. Cyclones are centers of _____ air pressure that are typically found adjacent to centers of high air pressure, called _____.

15. When conditions appear favorable for tornado formation, a tornado _____ is issued; however, when a funnel cloud has actually been sighted or is indicated by radar, a tornado _____ is issued.

True/False Questions

16. Continental tropical air has a strong influence on the weather of southeastern states such as Alabama. _____

17. The first sign of an approaching warm front is often the appearance of high, cirrus clouds overhead. _____

18. Because of their slow rate of advance and very gentle slope, cold fronts usually produce light-to-moderate precipitation over a large area for an extended period. _____

19. The surface circulation pattern of a Northern Hemisphere middle-latitude cyclone is counter-clockwise and convergent. _____

20. In the United States, tornadoes occur *only* during the summer months. _____

Answers:

1. b, 2. d, 3. c, 4. e, 5. a, 6. b, 7. e, 8. c, 9. d, 10. d, 11. source region, 12. cold, 13. middle-latitude cyclones, 14. low, anticyclones, 15. watch, warning, 16. F, 17. T, 18. F, 19. T, 20. F.

Web Work

The following are informative and interesting World Wide Web sites that address topics related to those presented in this chapter:

- Current North American Frontal Locations
  http://wxp.atms.purdue.edu/maps//surface/sfc_front.gif

- The Weather Channel
  http://www.weather.com/twc/homepage.twc

For direct access to these sites and other relevant Web locations, contact the *Foundations of Earth Science* WWW home page at http://www.prenhall.com/lutgens.

Earth's Place in the Universe

Cygnus Loop. *Photo courtesy Space Telescope Science Institute*

CHAPTER 15
The Nature
of the Solar System

FOCUS ON LEARNING

To assist you in learning the important concepts
in this chapter, you will find it helpful to focus
on the following questions:

1. What was the geocentric theory of the universe held by many early Greeks?

2. What were the contributions to modern astronomy of Nicolaus Copernicus, Tycho Brahe, Johannes Kepler, Galileo Galilei, and Isaac Newton?

3. What are the two groups of planets in the solar system? What are the general characteristics of the planets in each group?

4. How is the solar system thought to have formed?

5. What are the major features of the lunar surface?

6. What are some distinguishing features of each planet in the solar system?

7. What are the minor members of the solar system?

Montage of Saturn and some of its satellites. *NASA Science Source/Photo Researchers, Inc.*

Earth is one of nine planets and numerous smaller bodies that orbit the Sun. The Sun is part of a much larger family of perhaps 100 billion stars that comprise the Milky Way, which in turn is only one of billions of galaxies in an incomprehensibly large universe. This view of Earth's position in space is considerably different from that held only a few hundred years ago, when our planet was thought to occupy a privileged position as the center of the universe. This chapter traces the fascinating events that led to modern astronomy. In addition, it examines the formation and structure of the solar system.

Ancient Astronomy

Long before recorded history, which began about 5000 years ago, people were aware of the close relationship between events on Earth and the positions of heavenly bodies, the Sun in particular. People noted that changes in the seasons and floods of great rivers like the Nile in Egypt occurred when the celestial bodies, including the Sun, Moon, planets, and stars, reached a particular place in the heavens. Early agrarian cultures, which were dependent on the weather, believed that if the heavenly objects could control the seasons, they must also strongly influence all Earthly events. This belief undoubtedly was the reason why early civilizations began keeping records of the positions of celestial objects. The Chinese, Egyptians, and Babylonians in particular are noted for this.

A study of Chinese archives shows that they recorded every appearance of the famous Halley's comet for at least ten centuries. However, because this comet appears only once in a lifetime—once every 76 years—they were unable to link these appearances to establish that what they saw was the same object each time. Thus, like most ancients, the Chinese considered comets to be mystical. Comets were seen as bad omens and were blamed for a variety of disasters, from wars to plagues (Figure 15.1).

Early Greeks

The "Golden Age" of early astronomy (600 B.C.–A.D. 150) was centered in Greece. The early Greeks have been criticized, and rightly so, for using philosophical arguments to explain natural phenomena. However, they did rely on observational data as well. The basics of geometry and trigonometry, which they had developed, were used to measure the sizes and distances of the largest-appearing bodies in the heavens—the Sun and the Moon.

Many astronomical discoveries have been credited to the Greeks. They held the **geocentric** ("Earth-centered") view, believing that Earth was a sphere that stayed motionless at the center of the universe. Orbiting Earth were the moon, sun, and the known planets—Mercury, Venus, Mars, Jupiter, and Saturn. Beyond the planets was a transparent, hollow sphere (**celestial sphere**) on which the stars traveled daily around Earth (this is how it looks, but, of

Figure 15.1 Comet Hale-Bopp. The two tails are about 10 to 15 million miles long. (*A Peoria Astronomical Society, Inc. photograph by Eric Clifton and Greg Neaveill*)

course, the effect is actually caused by Earth's rotation about its axis). Some early Greeks realized that the motion of the stars could be explained just as easily by a rotating Earth, but they rejected that idea, because Earth exhibited no sense of motion and seemed too large to be movable. In fact, proof of Earth's rotation was not demonstrated until 1851.

To the Greeks, all of the heavenly bodies, except seven, appeared to remain in the same relative position to one another. These seven "wanderers" (*planetai* in Greek) included the Sun, the Moon, Mercury, Venus, Mars, Jupiter, and Saturn. Each was thought to have a circular orbit around Earth. Although this system was incorrect, the Greeks refined it to the point that it explained the apparent movements of all celestial bodies.

Although many of the Greek discoveries were lost during the Middle Ages, the Earth-centered view that the Greeks proposed became established in Europe. Presented in its finest form by Claudius Ptolemy, this geocentric outlook became known as the **Ptolemaic system**.

Ptolemy's Model

Much of our knowledge of Greek astronomy comes from a 13-volume treatise, *Almagest* ("the great work"), which was compiled by Ptolemy in A.D. 141 and which survived thanks to the work of Arab scholars. In this work, Ptolemy is credited with developing a model of the universe that accounted for the observable motions of the planets (Figure 15.2). The precision with which his model was able to predict planetary motion is attested to by the fact that it went virtually unchallenged, in principle if not in detail, for nearly 13 centuries.

In the Greek tradition, the Ptolemaic model had the planets moving in circular orbits around a motionless Earth. (The circle was considered the pure and perfect shape by the Greeks.) However, the motion of the planets, as seen against the background of stars, is not so simple. Each planet, if watched night after night, moves slightly eastward among the stars. Periodically, each planet appears to stop, reverse direction for a period of time, and then resume an eastward motion. The apparent westward drift is called **retrograde motion**. This rather odd apparent motion results from the combination of the motion of Earth and the planet's own motion around the Sun.

Figure 15.3 illustrates the retrograde motion of Mars. Earth has a faster orbital speed than does Mars, so it overtakes its neighbor. While doing so, Mars

appears to be moving backward, in retrograde motion. This is analogous to what a race-car driver sees out the side window when passing a slower car. The slower planet, like the slower car, appears to be going backward, although its actual motion is in the same direction as the faster-moving body.

Figure 15.3 shows how retrograde motion works. It is much more difficult to represent retrograde motion accurately using the incorrect Earth-centered model, but Ptolemy was able to do so (Figure 15.2B). Rather than using a simple circle for each planet's orbit, he showed that the planets orbited on small circles (*epicycles*), revolving along large circles (*deferents*). By trial and error, he found the right combinations of circles to produce the amount of retrograde motion observed for each planet. (An interesting note is that almost any closed curve can be produced by the combination of two circular motions, a fact that can be verified by anyone who has used the Spirograph© design-drawing toy.)

It is a tribute to Ptolemy's genius that he was able to account for the planets' motions as well as he did, considering that he used an incorrect model. Some suggest that he did not mean his model to represent reality, but only to be used for calculating the positions of the heavenly bodies. We probably will never know his intentions. However, the Roman Catholic Church, which dominated European thought for centuries, accepted Ptolemy's theory as the correct representation of the heavens, and this created problems for those who found fault with it.

The Birth of Modern Astronomy

Modern astronomy was not born overnight. Its development involved a break from deeply entrenched philosophical and religious views and, during the 1500s and 1600s, the founding of a "new and greater universe" governed by discernible laws. Let us now look at the work of five noted scientists involved in this transition: Nicolaus Copernicus, Tycho Brahe, Johannes Kepler, Galileo Galilei, and Sir Isaac Newton.

Nicolaus Copernicus

For almost 13 centuries after the time of Ptolemy, very few astronomical advances were made in Europe. The first great astronomer to emerge after the Middle Ages was Nicolaus Copernicus (1473–1543) from Poland. Copernicus became convinced that

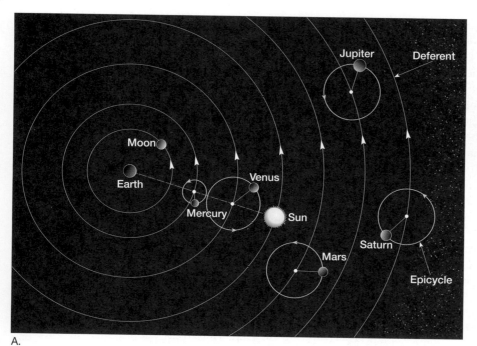

A.

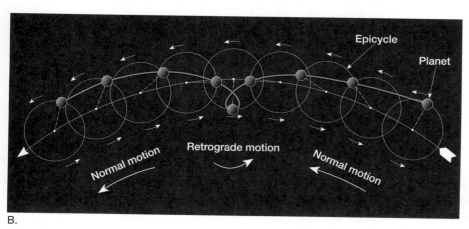

B.

Figure 15.2 View of the universe according to Ptolemy, second century A.D. **A.** Ptolemy believed that the star-studded celestial sphere made a daily trip around a motionless Earth. In addition he proposed that the Sun, Moon, and planets made trips of various lengths along individual orbits. **B.** Retrograde motion as explained by Ptolemy.

Earth is a planet, just like the other five then-known planets. The daily motions of the heavens, he reasoned, could be better explained by a rotating Earth.

Having concluded that Earth is a planet, Copernicus reconstructed the solar system with the Sun at the center and the planets Mercury, Venus, Earth, Mars, Jupiter, and Saturn orbiting around it. This was a major break from the ancient idea that a motionless Earth lies at the center of all movement. However, Copernicus retained a link to the past and used circles, which were considered to be the perfect geometric shape, to represent the orbits of the planets. Although these circular orbits were close to reality, they didn't quite match what people saw.

Unable to get satisfactory agreement between predicted locations of the planets and their observed positions, Copernicus found it necessary to add epicycles like those used by Ptolemy. The discovery that the planets have *elliptical* orbits would wait another century for the insights of Johannes Kepler.

Also, like his predecessors, Copernicus used philosophical justifications to support his point of view. Copernicus's monumental work, *De Revolutionibus, Orbium Coelestium (On the Revolution of the Heavenly Spheres)*, which set forth his controversial ideas, was published as he lay on his deathbed. Hence, he never suffered the criticism that befell many of his followers.

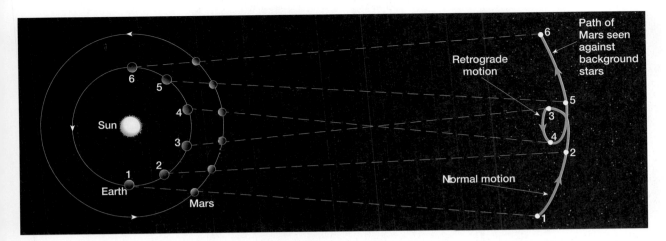

Figure 15.3 Retrograde (backward) motion of Mars as seen against the background of distant stars. When viewed from Earth, Mars moves eastward among the stars each day, then periodically appears to stop and reverse direction. This apparent westward drift is a result of the fact that Earth has a faster orbital speed than Mars and overtakes it. As this occurs, Mars appears to be moving backward; that is, it exhibits retrograde motion.

Tycho Brahe

Tycho Brahe (1546–1601) was born of Danish nobility three years after the death of Copernicus. Reportedly, Tycho became interested in astronomy while viewing a solar eclipse that had been predicted by astronomers. He persuaded King Fredrich II to establish an observatory, which Tycho headed, near Copenhagen. There he designed and built pointers (the telescope would not be invented for a few more decades), which he used for 20 years to systematically measure the locations of the heavenly bodies. These observations, particularly of Mars, were far more precise than any made previously and are his legacy to astronomy.

Tycho did not believe in the Copernican (Sun-centered) system, because he was unable to observe an apparent shift in the position of stars that would be caused by Earth's motion. His argument went like this: If Earth does revolve along an orbit around the Sun, the position of a nearby star, when observed from extreme points in Earth's orbit six months apart, should shift with respect to the more distant stars. His *idea* was correct, and this apparent shift of the stars is called *stellar parallax* (see Figure 16.2, p. 413).

The principle of parallax is easy to visualize: Close one eye, and with your index finger vertical, use your eye to line up your finger with some distant object. Now, without moving your finger, view the object with your other eye and notice that the object's position appears to shift. The farther away you hold your finger, the less the object's position

seems to shift. Herein lay the flaw in Tycho's argument. He was right about parallax, but, because the distance to even the nearest stars is enormous compared to the width of Earth's orbit, the shift that occurs is too small to be noticed by using the first primitive telescopes, let alone the unaided eye.

With the death of his patron, the King of Denmark, Tycho was forced to leave his observatory. It was probably his arrogant and extravagant nature that caused a conflict with the next ruler, so Tycho moved to Prague in what is now the Czech Republic. Here, in the last year of his life, he acquired an able assistant, Johannes Kepler. Kepler retained most of the observations made by Tycho and put them to exceptional use. Ironically, the data Tycho collected to refute the Copernican view would later be used by Kepler to support it.

Johannes Kepler

If Copernicus ushered out the old astronomy, Johannes Kepler (1571–1630) ushered in the new. Armed with Tycho's data, a good mathematical mind, and, of greater importance, a strong faith in the accuracy of Tycho's work, Kepler derived three basic laws of planetary motion. The first two laws resulted from his inability to fit Tycho's observations of Mars to a circular orbit. Unwilling to concede that the discrepancies were due to observational error, he searched for another solution. This endeavor led him to discover that the orbit of Mars is not a perfect circle but is elliptical (Figure 15.4). About the same time, he

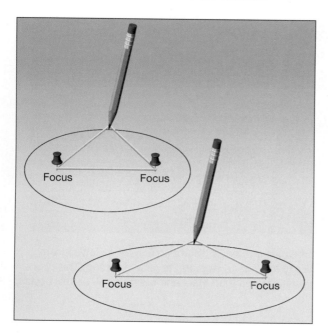

Figure 15.4 Drawing ellipses with various eccentricities. Using two straight pins for foci and a loop of string, trace out a curve while keeping the string taut, and you will have drawn an ellipse. The farther the pins (the foci) are moved apart, the more flattened (more eccentric) is the resulting ellipse.

realized that the orbital speed of Mars varies in a predictable way. As it approaches the Sun, it speeds up, and as it moves away from the Sun, it slows down.

In 1609, after almost a decade of work, Kepler proposed his first two laws of planetary motion:

1. The path of each planet around the Sun is an ellipse, with the Sun at one focus (Figure 15.4). The other focus is symmetrically located at the opposite end of the ellipse.

2. Each planet revolves so that an imaginary line connecting it to the Sun sweeps over equal areas in equal intervals of time (Figure 15.5). This law of equal areas expresses geometrically the variations in orbital speeds of the planets.

Figure 15.5 illustrates the second law. Note that in order for a planet to sweep equal areas in the same amount of time, it must travel more rapidly when it is nearer the Sun and more slowly when it is farther from the Sun.

Kepler was very religious and believed that the Creator made an orderly universe. The uniformity he tried to find eluded him for nearly a decade. Then, in 1619, he published his third law in *The Harmony of the Worlds*.

3. The orbital periods of the planets and their distances to the Sun are proportional.

In its simplest form, the orbital period of revolution is measured in Earth years, and the planet's distance to the Sun is expressed in terms of Earth's mean distance to the Sun. The latter "yardstick" is

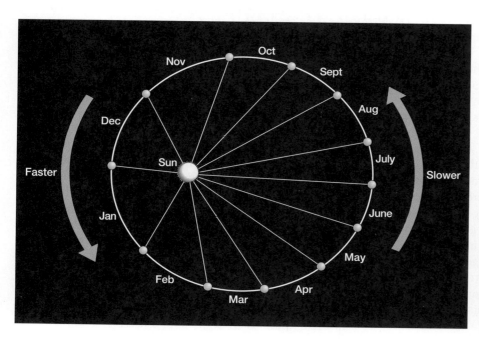

Figure 15.5 Kepler's law of equal areas. A line connecting a planet (Earth) to the Sun sweeps out an area in such a manner that equal areas are swept out in equal times. Thus, Earth revolves (moves in orbit) slower when it is farther from the Sun (aphelion) and faster when it is closest (perihelion). The eccentricity of Earth's orbit is greatly exaggerated in this diagram.

called the **astronomical unit (AU)** and averages about 150 million kilometers, or 93 million miles.

Kepler's laws assert that the planets revolve around the Sun and therefore support the Copernican view. Kepler, however, did fall short of determining the *forces* that act to produce the planetary motion he had so ably described. That task would remain for Galileo Galilei and Sir Isaac Newton.

Galileo Galilei

Galileo Galilei (1564–1642) was the greatest Italian scientist of the Renaissance. He was a contemporary of Kepler and, like Kepler, strongly supported the Copernican theory of a Sun-centered solar system. Galileo's greatest contributions to science were his descriptions of the behavior of moving objects. These he derived from experimentation. The method of using experiments to determine natural laws had essentially been lost since the time of the early Greeks.

All astronomical discoveries before Galileo's time were made without the aid of a telescope. In 1609, Galileo heard that a Dutch lens maker had devised a system of lenses that magnified objects. Apparently without ever seeing a telescope, Galileo constructed his own, which magnified distant objects to three times the size seen by the unaided eye. He immediately made others, the best having a magnification of about 30 times.

With the telescope, Galileo was able to view the universe in a new way. He made many important discoveries that supported the Copernican view of the universe, such as:

1. The discovery of four satellites, or moons, orbiting Jupiter. Galileo accurately determined their periods of revolution, which range from 2 to 17 days (Figure 15.6). This find dispelled the old idea that Earth was the only center of motion in the universe; for here, plainly visible, was another center of motion—Jupiter.

2. The discovery, through observation, that the planets are circular disks rather than just points of light, as was previously thought. This indicated that the planets might be Earthlike.

3. The discovery that Venus has phases just like the Moon, demonstrating that Venus orbits its source of light—the Sun. He saw that Venus appears smallest when it is in full phase, and thus is farthest from Earth (Figure 15.7). In the Ptolemaic system, the orbit of Venus lies between Earth

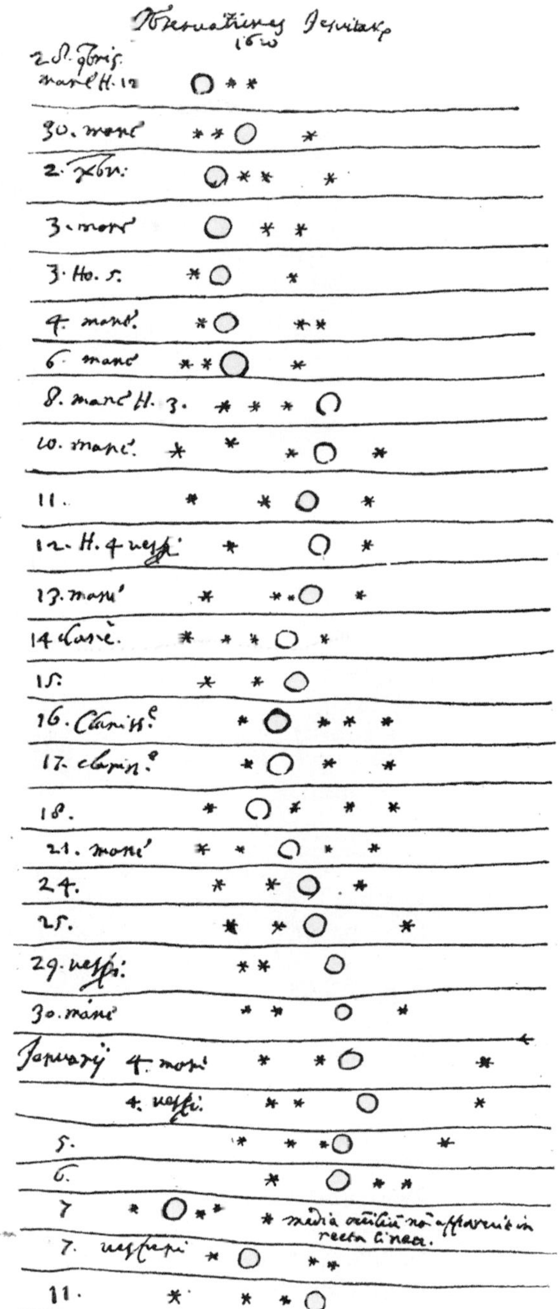

Figure 15.6 Sketch by Galileo of how he saw Jupiter and its four largest satellites through his telescope. The positions of Jupiter's four largest moons (drawn as stars) change nightly. You can observe these same changes with binoculars. *(Yerkes Observatory photograph)*

and the Sun, which means that only the crescent phase of Venus could be seen from Earth.

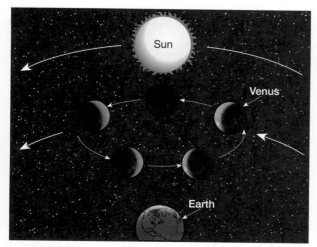

A. Phases of Venus as seen from
Earth in the Earth-centered model.

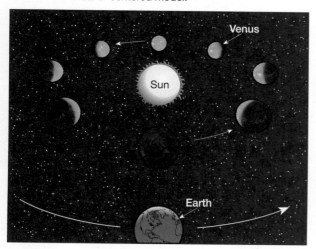

B. Phases of Venus as seen from
Earth in the sun-centered model.

C.

Figure 15.7 Using a telescope, Galileo discovered that Venus has phases just like the moon. **A.** In the Ptolemaic (Earth-centered) system, the orbit of Venus lies between the Sun and Earth as shown in Figure 15.2A. Thus, in an Earth-centered solar system, only the crescent phase of Venus would be visible from Earth. **B.** In the Copernican (sun-centered) system, Venus orbits the Sun and hence all of the phases of Venus should be visible from Earth. **C.** As Galileo observed, Venus goes through a series of moonlike phases. Further, Venus appears smallest during the full phase because it is farthest from Earth, and largest in the crescent phase because it is closest to Earth. This verified Galileo's belief that the Sun was the center of the solar system. *(Photo courtesy of Lowell Observatory)*

4. The discovery that the Moon's surface is not a smooth glass sphere, as the ancients had suspected and the Catholic Church had decreed. Rather, Galileo saw mountains, craters, and plains. He thought the plains might be bodies of water, and this idea was strongly promoted by others, as we can tell from the names given to these features (Sea of Tranquility, Sea of Storms, and so forth).

5. The discovery that the Sun (the viewing of which may have caused the eye damage that later blinded him) had sunspots (dark regions caused by slightly lower temperatures). He tracked the movement of these spots and estimated the rotational period of the Sun as just under a month.

Hence, another heavenly body was found to have both "blemishes" and rotational motion.

In 1616 the Catholic Church condemned the Copernican theory as contrary to Scripture, and Galileo was told to abandon it. Unwilling to accept this verdict, Galileo began writing his most famous work, *Dialogue of the Great World Systems.* Despite poor health, he completed the project and in 1630 went to Rome, seeking permission from Pope Urban VIII to publish. Because the book was a dialogue that expounded both the Ptolemaic and Copernican systems, publication was allowed. However, Galileo's enemies were quick to realize that he was promoting the Copernican view at the expense of the Ptolemaic system. Sale of the book was quickly halted, and Galileo was called before the Inquisition. Tried and convicted of proclaiming doctrines contrary to religious doctrine, he was sentenced to permanent house arrest, under which he remained for the last ten years of his life. It was not until 1992 that Galileo was finally exonerated by the Church.

Sir Isaac Newton

Sir Isaac Newton (1642–1727) was born in the year of Galileo's death. His many accomplishments in mathematics and physics led a successor to say that "Newton was the greatest genius that ever existed."

Although Kepler and those who followed attempted to explain the forces involved in planetary motion, their explanations were less than satisfactory. Kepler believed that some force pushed the planets along in their orbits. Galileo, however, correctly reasoned that no force is required to keep an object in motion. Galileo proposed that the natural tendency for a moving object (that is unaffected by an outside force) is to continue moving at a uniform speed and in a straight line. This concept, *inertia*, was later formalized by Newton as his first law of motion.

The problem, then, was not to explain the force that keeps the planets moving but rather to determine the force that *keeps them from going in a straight line out into space.* It was to this end that Newton conceptualized the force of *gravity.* At the early age of 23, he envisioned a force that extends from Earth into space and holds the Moon in orbit around Earth. Although others had theorized the

existence of such a force, he was the first to formulate and test the *law of universal gravitation*. It states:

Every body in the universe attracts every other body with a force that is directly proportional to their masses and inversely proportional to the square of the distance between them.

Thus, the gravitational force decreases with distance, so that two objects 3 kilometers apart have 3^2, or 9, times less gravitational attraction than if the same objects were 1 kilometer apart.

The law of gravitation also states that the greater the mass of the object, the greater its gravitational force. For example, the large mass of the Moon has a gravitational force strong enough to cause ocean tides on Earth, whereas the tiny mass of a communications satellite has no measurable effect on Earth.

With his laws of motion, Newton proved that the force of gravity, combined with the tendency of a planet to remain in straight-line motion, results in the elliptical orbits discovered by Kepler. Earth, for example, moves forward in its orbit about 30 kilometers (18.5 miles) each second, and during the same second, the force of gravity pulls it toward the Sun 0.5 centimeter (1/8 inch). Therefore, as Newton concluded, it is the combination of Earth's forward motion and its "falling" motion that defines its orbit (Figure 15.8). If gravity were somehow eliminated, Earth would move in a straight line out into space. On the other hand, if Earth's forward motion suddenly stopped, gravity would pull it directly toward the Sun.

With the acceptance of Newton's explanation of planetary motion, it became clear that the planets

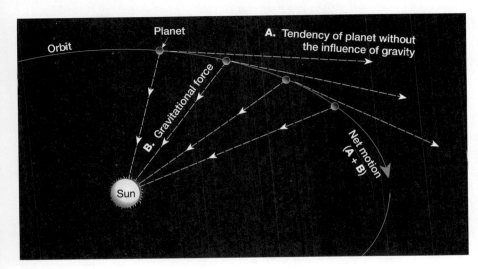

Figure 15.8 Orbital motion of Earth and other planets.

were more similar to Earth than to the stars. This realization stimulated a great deal of interest in observational astronomy. Interest in the planets was further enhanced by the possibility of discovering evidence of intelligent life elsewhere in the solar system.

The Planets: An Overview

The Sun is the hub of a huge rotating system consisting of nine planets, their satellites, and numerous small but interesting bodies, including asteroids, comets, and meteoroids. An estimated 99.85 percent of the mass of our solar system is contained within the Sun, while the planets collectively make up most of the remaining 0.15 percent. The planets, in order from the Sun, are Mercury, Venus, Earth, Mars, Jupiter, Saturn, Uranus, Neptune, and Pluto (Figure 15.9).

Under the control of the Sun's gravitational force, each planet maintains an elliptical orbit, and all of them travel in the same direction. The nearest planet to the Sun, Mercury, has the fastest orbital motion, 48 kilometers per second, and the shortest period of revolution, 88 days. By contrast, the most distant planet, Pluto, has an orbital speed of 5 kilometers per second and requires 248 years to complete one revolution.

Imagine a planet's orbit drawn on a flat sheet of paper. The paper represents the planet's *orbital plane*. The orbital planes of all nine planets lie within 3 degrees of the plane of the Sun's equator, except for those of Mercury and Pluto, which are inclined 7 and 17 degrees, respectively.

Careful examination of Table 15.1 shows that the planets fall quite nicely into two groups: the **terrestrial** (Earthlike) **planets** (Mercury, Venus, Earth, and Mars) and the **Jovian** (Jupiterlike) **planets** (Jupiter, Saturn, Uranus, and Neptune). Pluto is not included in either category, because its great distance from Earth and its small size make this planet's true nature a mystery.

The most obvious difference between the terrestrial and the Jovian planets is their size (Figure 15.10). The largest terrestrial planet (Earth) has a diameter only one-quarter as great as the diameter of the smallest Jovian planet (Neptune), and its mass is only one-seventeenth as great. Hence, the Jovian planets are often called giants. Also, because of their relative locations, the four Jovian planets are referred to as the *outer planets*, while the terrestrial planets are called the *inner planets*. As we shall see, there appears to be a correlation between the positions of these planets and their sizes.

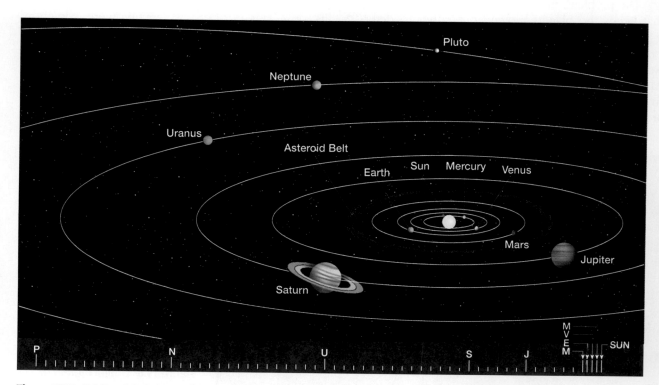

Figure 15.9 Orbits of the planets (not to scale). Orbits are shown to scale along bottom of diagram.

Table 15.1 ■ Planetary data

Planet	Symbol	Mean Distance from Sun: AU*	Mean Distance from Sun: Millions of Miles	Mean Distance from Sun: Millions of Kilometers	Period of Revolution	Inclination of Orbit	Orbital Velocity: mi/s	Orbital Velocity: km/s
Mercury	☿	0.39	36	58	88^d	7°00′	29.5	47.5
Venus	♀	0.72	67	108	225^d	3°24′	21.8	35.0
Earth	⊕	1.00	93	150	365.25^d	0°00″	18.5	29.8
Mars	♂	1.52	142	228	687^d	1°51′	14.9	24.1
Jupiter	♃	5.20	483	778	12^{yr}	1°18′	8.1	13.1
Saturn	♄	9.54	886	1427	29.5^{yr}	2°29′	6.0	9.6
Uranus	⛢	19.18	1783	2870	84^{yr}	0°46′	4.2	6.8
Neptune	♆	30.06	2794	4497	165^{yr}	1°46′	3.3	5.3
Pluto	♇	39.44	3666	5900	248^{yr}	17°12′	2.9	4.7

Planet	Period of Rotation	Diameter: Miles	Diameter: Kilometers	Relative Mass (Earth = 1)	Average Density (g/cm^3)	Polar Flattening (%)	Eccentricity	Number of Known Satellites
Mercury	59^d	3015	4878	0.06	5.4	0.0	0.206	0
Venus	244^d	7526	12,104	0.82	5.2	0.0	0.007	0
Earth	$23^h56^m04^s$	7920	12,756	1.00	5.5	0.3	0.017	1
Mars	$24^h37^m23^s$	4216	6794	0.11	3.9	0.5	0.093	2
Jupiter	9^h50^m	88,700	143,884	317.87	1.3	6.7	0.048	16
Saturn	10^h14^m	75,000	120,536	95.14	0.7	10.4	0.056	21
Uranus	17^h14^m	29,000	51,118	14.56	1.2	2.3	0.047	15
Neptune	16^h03^m	28,900	50,530	17.21	1.7	1.8	0.009	8
Pluto	6.4^d	~1500	2445	0.002	1.8	0.0	0.250	1

*AU = astronomical unit, Earth's mean distance from the Sun.

Other dimensions along which the two groups markedly differ include density, composition, and rate of rotation. The densities of the terrestrial planets average about 5 times the density of water, whereas the Jovian planets have densities that average only 1.5 times that of water. One of the outer planets, Saturn, has a density of only 0.7 that of water, which means that Saturn would float in water. Variations in the compositions of the planets are largely responsible for the density differences.

The substances that make up both groups of planets are divided into three groups—*gases, rocks,* and *ices*—based on their melting points.

1. The gases, hydrogen and helium, are those with melting points near absolute zero (0 Kelvin or –273°C), the lowest possible temperature.
2. The rocks are principally silicate minerals and metallic iron, which have melting points exceeding 700°C.
3. The ices have intermediate melting points (for example, H_2O has a melting point of 0°C) and include ammonia (NH_3), methane (CH_4), carbon dioxide (CO_2), and water (H_2O).

The terrestrial planets are mostly rocks: dense, rocky, and metallic material, with minor amounts of gases. The Jovian planets, on the other hand, contain a large percentage of gases (hydrogen and helium), with varying amounts of ices (mostly water, ammonia, and methane). This accounts for their low densities. (The outer planets may contain as much rocky and metallic material as the terrestrial planets, but this material would be concentrated in their central cores.)

Figure 15.10 The planets drawn to scale.

The Jovian planets have very thick atmospheres consisting of varying amounts of hydrogen, helium, methane, and ammonia. By comparison, the terrestrial planets have meager atmospheres at best. A planet's ability to retain an atmosphere depends on its temperature and mass. Simply stated, a gas molecule can "evaporate" from a planet if it reaches a speed known as the **escape velocity**. For Earth, this velocity is 11 kilometers (7 miles) per second. Any material, including a rocket, must reach this speed before it can leave Earth and go into space.

The Jovian planets, because of their greater surface gravities, have higher escape velocities (21–60 km/sec) than do the terrestrial planets. Consequently, it is more difficult for gases to "evaporate" from them. Also, because the molecular motion of a gas is temperature-dependent, at the low temperatures of the Jovian planets even the lightest gases are unlikely to acquire the speed needed to escape.

On the other hand, a comparatively warm body with a small surface gravity, like our Moon, is unable to hold even the heaviest gas and thus lacks an atmosphere. The slightly larger terrestrial planets of Earth, Venus, and Mars retain some heavy gases like carbon dioxide, but even their atmospheres make up only an infinitesimally small portion of their total mass.

Origin of the Solar System

The orderly nature of our solar system leads most astronomers to conclude that the planets formed at essentially the same time and from the same primordial material as the Sun. This material formed a vast cloud of dust and gases called a *nebula*. The **nebular hypothesis** suggests that all bodies of the solar system formed from an enormous nebular cloud consisting of approximately 80 percent hydrogen, 15 percent helium, and a few percent of all the other heavier elements known to exist. The heavier substances in this frigid cloud of dust and gases consisted mostly of elements such as silicon, aluminum, iron, and calcium—the substances of today's common rocky materials. Also prevalent were other familiar elements, including oxygen, carbon, and nitrogen.

About 5 billion years ago, and for reasons not yet fully understood, this huge cloud of minute rocky fragments and gases began to contract under its own gravitational influence (Figure 15.11A). The contracting clump of material apparently had some rotational motion. As this slowly rotating cloud gravitationally contracted, it rotated faster and faster for the same

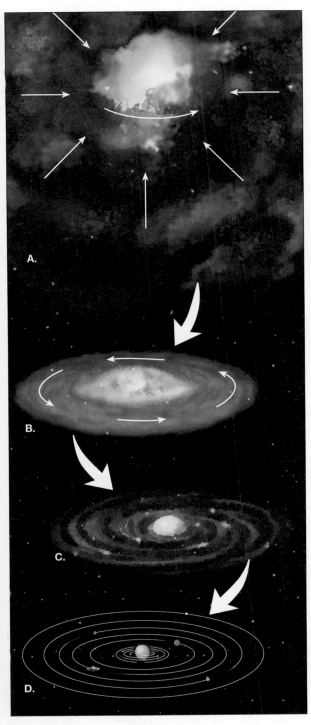

Figure 15.11 Nebular hypothesis for origin of the solar system. **A.** A huge rotating cloud of dust and gases (nebula) begins to contract. **B.** Most of the material is gravitationally pulled toward the center, producing the Sun. However, because of rotational motion, some dust and gases remain, orbiting the central body as a flattened disk. **C.** The planets begin to accrete from the material that is orbiting within the flattened disk. **D.** In time, most of the remaining debris collects into the nine planets and their moons, or is swept into space by the solar wind.

reason ice skaters do when they draw their arms toward their bodies. This rotation caused the nebular cloud to assume a disk shape (Figure 15.11B). Within this rotating disk, relatively small contractions, like eddies in a stream, formed the nuclei from which the planets would eventually develop (Figure 15.11C). However, the greatest concentration of material was gravitationally pulled toward the center, forming the *protosun*.

As this gaseous cloud collapsed, the temperature of the central mass continued to increase. Nebular material near the protosun reached temperatures of several thousand degrees and was completely vaporized

In a relatively short time after the protosun formed, the temperature in the inner portion of the nebula dropped significantly. This temperature decrease caused those substances with high melting points to condense into sand-sized particles. Materials such as iron and nickel solidified first. Next to condense were the elements of which the rock-forming minerals are composed—silicon, calcium, iron, and so forth. As these fragments collided, they joined into larger asteroid-sized objects, which, in a few tens of millions of years, accreted into the four inner planets we call Mercury, Venus, Earth, and Mars (Figure 15.11D).

As more and more of the nebular debris was swept up by these *protoplanets*, the inner solar system began to clear, allowing solar radiation to pass through to heat the planets' surfaces. Owing to their relatively high temperatures and weak gravitational fields, the inner planets were unable to accumulate much of the lighter components of this nebular cloud. These lighter components—hydrogen, ammonia, methane, and water—were eventually whisked from the inner solar system by the solar winds.

At the same time that the terrestrial planets were forming, the larger Jovian planets, along with their extensive satellite systems, were also developing. However, because of the frigid temperatures existing far from the Sun, the fragments from which these planets formed contained a high percentage of ices—water, carbon dioxide, ammonia, and methane. Perhaps by random chance, two of the outer planets, Jupiter and Saturn, grew many times larger (by mass) than did Uranus and Neptune. (For comparison, Jupiter is 318 and Saturn is 95 times more massive than Earth. However, Uranus and Neptune have masses only about 15 and 17 times greater, respectively, than Earth.)

In the remainder of this chapter, we will consider each planet in more detail, as well as some minor members of the solar system. First, however, a discussion of our Moon, Earth's companion in space, is appropriate.

Earth's Moon

Only one natural satellite, the Moon, accompanies Earth on its annual journey around the Sun. Although other planets have moons, our planet-satellite system is unique in the solar system, because Earth's Moon is unusually large compared to its parent planet. The diameter of the Moon is 3475 kilometers (2150 miles), about one-fourth of Earth's 12,751 kilometers. From the calculation of the Moon's mass, its density is 3.3 times that of water. This density is comparable to that of crustal rocks on Earth but is a fair amount less than Earth's average density, which is 5.5 times that of water. Geologists have suggested that this difference can be accounted for if the Moon's iron core is relatively small. The gravitational attraction at the lunar surface is one-sixth of that experienced on Earth's surface (a 100-pound person on Earth weighs only 17 pounds on the Moon). This difference allows an astronaut to lift a "heavy" life-support system with relative ease. If it were not necessary to carry such a load, an astronaut could jump six times higher than on Earth.

The Lunar Surface

When Galileo first pointed his telescope toward the Moon, he saw two different types of terrain. The dark areas he observed are now known to be fairly smooth lowlands, whereas the bright regions are densely cratered highlands (Figure 15.12). Because the dark regions resembled seas on Earth, they were later named **maria** (singular, *mare*: Latin for "sea").

Today we know that the Moon has no atmosphere and lacks water as well. Therefore, the processes of weathering and erosion that continually modify Earth's surface were virtually lacking on the Moon. In addition, tectonic forces are not active on the Moon, so events such as earthquakes and volcanic eruptions no longer occur. However, because the Moon is unprotected by an atmosphere, a different kind of erosion occurs: tiny particles from space (micrometeorites) continually bombard its surface and ever so gradually smooth the landscape. Rocks, for example, can become slightly rounded on top if exposed at the lunar surface for a long enough period. Nevertheless, it is unlikely that the Moon has changed appreciably in

the last 3 billion years, except for a few craters created by large meteorites.

Craters   The most obvious features of the lunar surface are craters. They are so profuse that craters-within-craters-within-craters is the rule! The larger ones in the lower portion of Figure 15.12 are about 250 kilometers (155 miles) in diameter, roughly the width of Indiana. Most craters were produced by the impact of rapidly moving debris (meteoroids), a phenomenon that was considerably more common in the early history of the solar system than it is today.

By contrast, Earth has only about a dozen easily recognized impact craters. This difference can be attributed to Earth's atmosphere. Friction with the air burns up small debris before it reaches the ground. In addition, evidence for most of the craters that formed early in Earth's history has been obliterated by erosion or tectonic processes.

The formation of an impact crater is illustrated in Figure 15.13. Upon impact, the high-speed meteoroid compresses the material it strikes, then almost instantaneously the compressed rock rebounds, ejecting material from the crater. This process is analogous to the splash that occurs when a rock is dropped into water, and it often results in the formation of a central peak, as seen in the large crater in Figure 15.14. Most of the ejected material (*ejecta*) lands near the crater, building a rim around it. The heat generated by the impact is sufficient to melt some of the impacted rock. Astronauts have brought back samples of glass beads produced in this manner, as well as rock formed when angular fragments and dust were welded together by the impact.

A meteoroid only 3 meters (10 feet) in diameter can blast out a 150-meter (500-foot)-wide crater. A few of the large craters such as Kepler and Copernicus, shown in Figure 15.12, formed from the impact of bodies 1 kilometer, or more, in diameter. These two large craters are thought to be relatively young because of the bright *rays* ("splash" marks) that radiate outward for hundreds of kilometers. These bright rays consist of fine debris ejected from the primary crater, including impact-generated glass beads, as well as material displaced during the formation of smaller, secondary craters.

Highlands   Densely pockmarked highland areas make up most of the lunar surface. In fact, all of the "back" side of the Moon is characterized by such topography. Within the highland regions are mountain ranges that have been named for mountainous

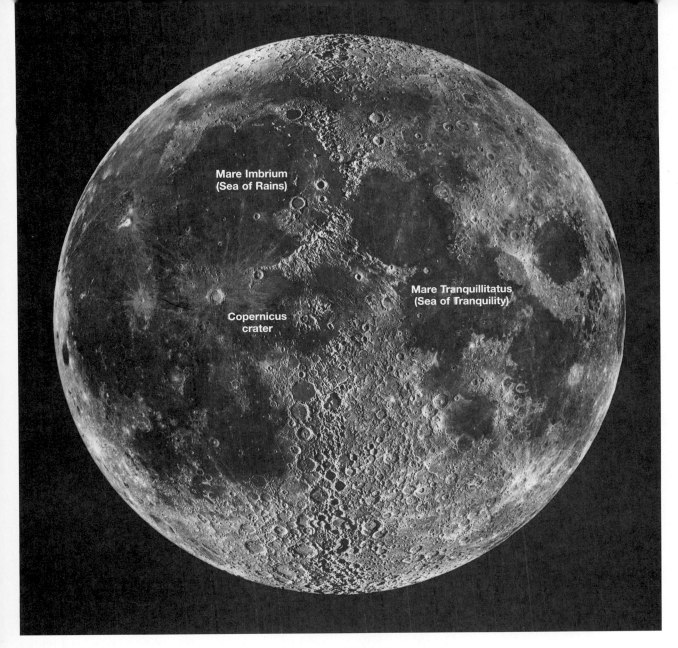

Figure 15.12 Telescopic view from Earth of the lunar surface. *(UCO/Lick Observatory Image)*

terrains on Earth. The highest lunar peaks reach elevations approaching 8 kilometers, only 1 kilometer lower than Mount Everest.

Maria Although highlands predominate, the less-rugged maria have attracted a great deal of interest. The origin of maria basins as enormous impact craters produced by the violent impact of at least a dozen asteroid-sized bodies was hypothesized before the turn of the century by the noted American geologist G. K. Gilbert (Figure 15.15A). However, it remained for the *Apollo* missions to determine what filled these depressions to produce the relatively flat topography.

Apparently the craters were flooded with layer upon layer of very fluid basaltic lava somewhat resembling the Columbia Plateau in the northwestern United States (Figure 15.15B). Astronauts have viewed and photographed the layered nature of maria. The lava flows are often over 30 meters (100 feet) thick, and the total thickness of the material that fills the maria must approach thousands of meters.

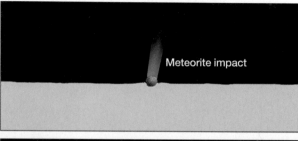

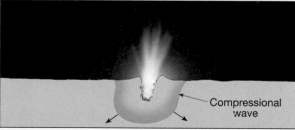

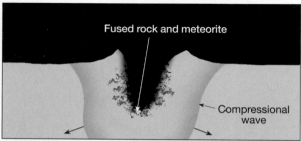

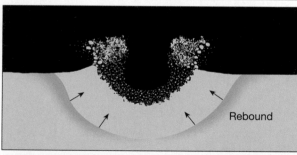

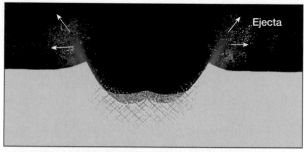

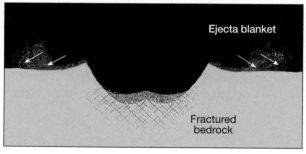

Figure 15.13 Formation of an impact crater. The energy of the rapidly moving meteoroid is transformed into heat energy and compressional waves. The rebound of the compressed rock causes debris to be ejected from the crater, and the heat melts some material, producing glass beads. Small secondary craters are formed by the material "splashed" from the impact crater. *(After E. M. Shoemaker)*

Regolith All lunar terrains are mantled with a layer of gray, unconsolidated debris derived from a few billion years of meteoric bombardment (Figure 15.16). This soil-like layer, properly called **lunar regolith**, is composed of igneous rocks, glass beads, and fine particles commonly called *lunar dust*. As meteoroid after meteoroid collided with the lunar surface, the thickness of the lunar regolith increased, while the size of the bombarded debris diminished. In the maria that have been explored by *Apollo* astronauts, the lunar regolith is apparently just over 3 meters (10 feet) thick, but it is believed to form a thicker mantle on the older highlands.

The Planets: A Short Tour

Mercury: Innermost, Hot, and Quick

Mercury, the innermost planet, has a diameter of 4878 kilometers. It is hardly larger than Earth's Moon and is smaller than three other moons in the solar system. Also, like our own Moon, it absorbs most of the sunlight that strikes it, reflecting only 6 percent into space. This is characteristic of terrestrial bodies that have no atmosphere. (By way of comparison, Earth reflects about 30 percent of the light that strikes it, most of it from clouds.)

Mercury's close proximity to the Sun makes viewing from Earthbound telescopes difficult at best. The first good glimpse of this planet came in the spring of 1974, when *Mariner 10* passed within 800 kilometers (500 miles) of its surface. Its striking resemblance to the Moon was immediately evident from the high-resolution images that were radioed back.

Mercury has cratered highlands, much like the Moon, and vast smooth terrains that resemble maria. However, unlike the Moon, Mercury is a very dense planet, which implies that it contains an iron core, perhaps larger than Earth's. Also, Mercury has very long cliffs that cut across the plains and craters alike. One proposal is that these

Figure 15.14 The 20-kilometer-wide (12 miles) lunar crater Euler in the southwestern part of Mare Imbrium. Clearly visible are the rays, central peak, secondary craters, and the large accumulation of ejecta near the crater rim. *(Courtesy of NASA)*

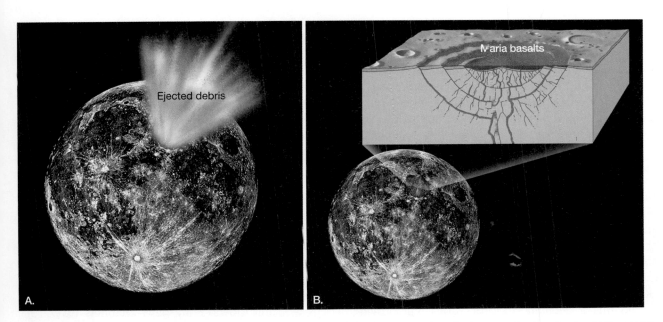

Figure 15.15 Formation of lunar maria. **A.** Impact of an asteroid-sized mass produced a huge crater hundreds of kilometers in diameter and disturbed the lunar crust far beyond the crater. **B.** Filling of the impact area with fluid basalts, perhaps derived from partial melting deep within the lunar mantle.

Figure 15.16 Astronaut Harrison Schmitt sampling the lunar surface. Notice the footprints (inset) in the lunar "soil." (Courtesy of NASA Headquarters)

cliffs resulted from crustal shortening as the planet cooled and shrank.

Mercury revolves quickly but rotates slowly. The time from one sunrise to the next on Mercury is 179 Earth days. Thus, a night on Mercury lasts for about three months and is followed by three months of daylight. The night temperatures drop as low as −173°C (−280°F) and the noontime temperatures exceed 427°C (800°F), hot enough to melt tin and lead. Mercury has the greatest temperature extremes of any planet. The odds of life as we know it existing on Mercury are nil.

Venus: Veiled in Clouds

Venus, second only to the Moon in brilliance in the night sky, is named for the Roman goddess of love and beauty. It orbits the Sun in a nearly perfect circle once every 225 days. Venus is similar to Earth in size, density, mass, and its location in the inner portion of the solar system. Thus, it has been referred to as "Earth's twin." Because of these similarities, it is hoped that a detailed study of Venus will provide geologists with a better understanding of Earth's evolutionary history.

Venus is shrouded in thick clouds impenetrable to visible light. Nevertheless, radar mapping by unmanned spacecraft, as well as by Earthbound instruments, has revealed a varied topography with features somewhat between those of Earth and Mars. Radar mapping is done by sending radar pulses in the microwave frequency to the Venusian surface, and the heights of features such as plateaus and mountains are measured by timing the return of the radar echo. In 1990, scientists began receiving crisp radar images from the *Magellan* space probe with a resolution that permits structures the size of Mount St. Helens to be clearly defined. These new data have confirmed earlier proposals that basaltic volcanism (Figure 15.17) and tectonic deformation are the dominant processes operating on Venus. Further, based on the low density of impact craters, researchers concluded that volcanism and tectonic deformation were also very active during the recent geologic past.

Over 80 percent of the Venusian surface consists of relatively subdued plains that are mantled by volcanic flows. Some lava channels extend for tens to hundreds of kilometers and have narrow widths from 1 to 2 kilometers. One meanders 6800

Figure 15.17 Sapas Mons, a volcanic cone, is seen in the center of this computer-generated view of Venus. Light-colored lava flows extending for hundreds of kilometers across the fractured plains are seen in the foreground, and Maat Mons, a large volcano, is located on the horizon. *(Courtesy of NASA Headquarters)*

kilometers across the planet. Thousands of small shield volcanoes have been found, a few hundred meters high with diameters of 2 to 8 kilometers.

In addition, over 1500 volcanoes with diameters exceeding 20 kilometers have been mapped. One of these, Sapas Mons, is approximately 400 kilometers (250 miles) across and 1.5 kilometers (0.9 mile) high. Many of the flows from this volcano erupted from the flanks rather than the summit. This eruptive style is shared by large shield volcanoes on Earth, such as those in Hawaii. Also like the Hawaiian volcanoes, these large structures appear to be associated with the upwelling of hot plumes within the planet's interior. Other volcanic structures discovered on Venus are circular, pancake-shaped domes about 25 kilometers (15 miles) in diameter and nearly 1 kilometer high (Figure 15.18). These domes are thought to be the result of outpourings of very viscous lava, much like volcanic domes on Earth.

Before the advent of space vehicles, Venus and Mars were considered the most hospitable other sites in the solar system for living organisms. However, evidence from *Mariner* fly-by space probes and Russian *Venera* landings on Venus indicates differently. The surface of Venus reaches temperatures of 475°C (900°F), and the Venusian atmosphere is 97 percent carbon dioxide. Only minor amounts of water vapor and nitrogen have been detected. Its atmosphere contains an opaque cloud deck about 25 kilometers thick, which begins approximately 70 kilometers from the surface. Although the unmanned *Venera 8* survived less than one hour on the Venusian surface, it determined that the atmospheric pressure on that planet is 90 times that on Earth's surface. This hostile environment makes it unlikely that life as we know it exists on Venus and makes manned space flights to Venus improbable.

Mars: The Red Planet

Mars has evoked greater interest than any other planet, for both astronomers and nonscientists. When we think of intelligent life on other worlds, "little green Martians" may come to mind. Interest in Mars stems mainly from this planet's accessibility to observation. Through the telescope, Mars appears as a reddish ball interrupted by some permanent dark regions that change intensity during the Martian year. The most prominent telescopic features of Mars are its brilliant white polar caps, resembling Earth's.

The Martian Atmosphere The Martian atmosphere is only 1 percent as dense as that of Earth. It is

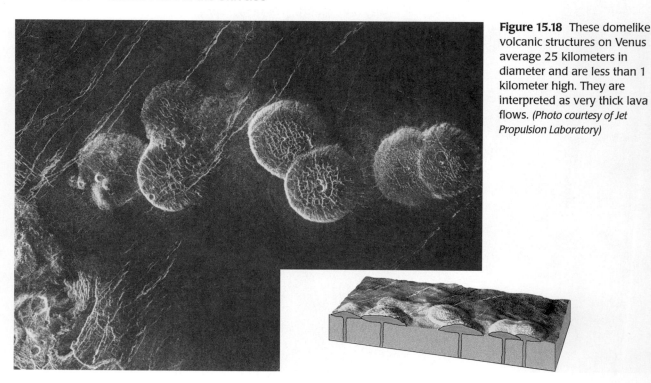

Figure 15.18 These domelike volcanic structures on Venus average 25 kilometers in diameter and are less than 1 kilometer high. They are interpreted as very thick lava flows. *(Photo courtesy of Jet Propulsion Laboratory)*

primarily carbon dioxide with very small amounts of water vapor. Data radioed back to Earth from Mars probes confirmed speculations that the polar caps of Mars are made of water ice, covered by a relatively thin layer of frozen carbon dioxide. As the winter nears in either the Martian northern or southern hemisphere, we see the equatorward growth of that hemisphere's ice cap as additional carbon dioxide is deposited (becomes frozen out of the atmosphere). This finding is compatible with the temperatures observed at the polar caps—frigid −125°C (−193°F), cold enough to solidify carbon dioxide.

Although the atmosphere of Mars is very thin, extensive dust storms do occur and may be responsible for the color changes observed from Earth-based telescopes. Winds up to 270 kilometers (170 miles) per hour can persist for weeks. Images radioed back by *Viking 1* and *Viking 2* revealed a Martian landscape remarkably similar to a rocky desert on Earth. Sand dunes are abundant, and many Martian impact craters have flat bottoms because they are partially filled with dust. Thus, unlike the craters of the Moon, the oldest craters on Mars might be completely obscured by deposits of windblown material.

Mars's Dramatic Surface *Mariner 9*, the first artificial satellite to orbit another planet, reached Mars in 1971. It witnessed a large dust storm raging, and only the ice caps were initially observable. When the dust cleared, images of Mars's northern hemisphere revealed numerous large volcanoes. The largest, Mons Olympus, covers an area the size of Ohio and is no less than 23 kilometers (75,000 feet) in elevation. This gigantic volcano and others resemble the shield volcanoes found on Earth, like those of Hawaii (Figure 15.19). Their extreme size is thought to result from the absence of tectonic plate movements on Mars. Therefore, rather than a chain of volcanoes forming as we find in Hawaii, single, larger cones developed.

Another surprising find made by *Mariner 9* was the existence of several canyons that dwarf even Earth's Grand Canyon of the Colorado River. One of the largest, Valles Marineris, is roughly 6 kilometers deep, up to 160 kilometers wide, and extends for almost 5000 kilometers along the Martian equator (Figure 15.20). This vast chasm is thought to have formed by slippage of crustal material along huge faults in the crustal layer. In this respect, it would be comparable to the rift valleys of Africa.

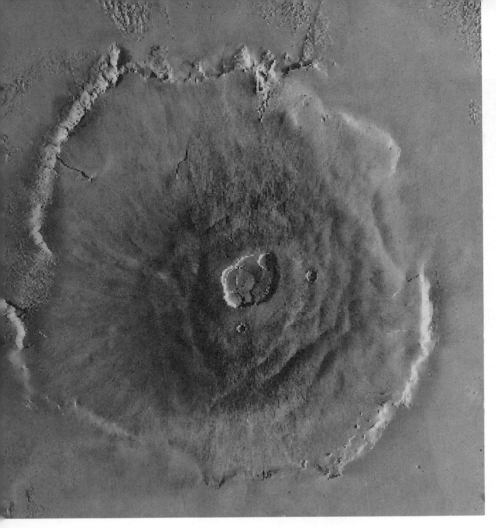

Figure 15.19 Image of Mons Olympus, an inactive shield volcano on Mars that covers an area about the size of the state of Ohio. *(Courtesy of the U.S. Geological Survey, U.S. Department of the Interior)*

Water on Mars?   Not all valleys on Mars have a tectonic origin. Many Martian valleys have tributaries exhibiting drainage patterns similar to those of stream valleys found on Earth. In addition, *Viking* orbiter images have revealed streamlined features that are unmistakably ancient islands located in what is now a dry stream bed. When these streamlike channels were first discovered, some observers speculated that a thick water-laden atmosphere capable of generating torrential downpours once existed on Mars. What happened to this water? The present Martian atmosphere contains only traces of water.

Pathfinder: The First Geologist on Mars   On July 4, 1997, Mars *Pathfinder* bounced onto the rock-littered surface of Mars and deployed its wheeled companion, Sojourner. For the next three months the lander sent back to Earth three gigabits of data, including 16,000 images and 20 chemical analyses. Sojourner carried an alpha photon X-ray spectrometer

(APXS) used to determine the chemical composition of rocks and Martian "soil" (regolith) at the landing site (Figure 15.21). In addition, the rover was able to take close-up images of the rocks. From these data, researchers concluded that the rocks were igneous.

During its first week on Mars, Sojourner's APXS obtained data for a patch of windblown soil and a medium-sized rock, known affectionately as Barnacle Bill. Preliminary evaluation of the APXS data on Barnacle Bill shows that it contains over 60 percent silica. If these data are confirmed, it could indicate that Mars contains the volcanic rock andesite. Researchers had expected that most volcanic rocks on Mars would be basalt, which is lower in silica (less than 50 percent). On Earth andesites are associated with tectonically active regions where oceanic crust is subducted into the mantle. Examples include the volcanoes of South America's Andes mountains and the Cascades of North America.

Figure 15.20 This image shows the entire Valles Marineris canyon system, over 2000 kilometers long and up to 8 kilometers deep. The dark red spots on the left edge of the image are huge volcanoes, each about 25 kilometers (15.5 miles) high. *(Courtesy of U.S. Geological Survey, U.S. Department of the Interior)*

Figure 15.21 Mars *Pathfinder* image of the Sojourner rover sampling a rock known as Yogi. *(NASA/Science Photo Library/Photo Researchers, Inc.)*

In January 2002, NASA plans to land another rover on Mars. It will be bigger than Sojourner and will explore for up to one year what is thought to be an ancient sedimentary environment. This rover will be capable of drilling a 5-centimeter bore into rocks and extracting a core for possible later return to Earth.

Jupiter: Lord of the Heavens

Jupiter, the largest planet in our solar system, has a mass 2.5 times greater than the combined mass of all the remaining planets, satellites, and asteroids. It is truly a giant among planets. In fact, had Jupiter been about ten times larger, it would have evolved into a small star. Despite its great size, however, it is only 1/800 as massive as the Sun. Jupiter also rotates more rapidly than any other planet, completing one rotation in slightly less than ten hours. The effect of Jupiter's rapid rotation causes its equatorial region to be slightly bulged and the polar dimension to be flattened (see "Polar Flattening" column in Table 15.1, p. 383).

When viewed through a telescope or binoculars, Jupiter appears to be covered with alternating bands of multicolored clouds aligned parallel to its equator (Figure 15.22). The most striking feature on the face of Jupiter is the *Great Red Spot* in its southern hemisphere (Figure 15.22). Although its color varies greatly in intensity, the Great Red Spot has been a prominent feature ever since it was first discovered more than three centuries ago. When *Voyager 2* swept by Jupiter in 1979, the size of the red spot was 11,000 kilometers by 22,000 kilometers, which equals two Earth-sized circles placed side by side. On occasion it has grown even larger. Although the Great Red Spot varies in size, it does remain the same distance from the Jovian equator.

The cause of the spot has been attributed to everything from volcanic activity to a large cyclonic storm. Images obtained by *Pioneer 11* as it moved to within 42,000 kilometers of Jupiter's cloud tops in 1974 support the cyclone view. The Great Red Spot apparently is a giant counterclockwise-rotating storm caught between two jetstream-like bands of atmosphere flowing in opposite directions. This huge hurricane-like storm rotates once every 12 Earth days. Although several smaller storms have been observed in other regions of Jupiter's atmosphere, none have survived for more than a few days.

One of the most interesting discoveries made by *Voyager 1* is the ring system of Jupiter. Thought to be less than 30 kilometers thick, the apparently continuous ring may extend outward from the surface of Jupiter to a distance equal to twice the diameter of the planet.

Jupiter's Moons Jupiter's satellite system, consisting of at least 16 moons, resembles a miniature solar system. The four largest satellites were discovered by Galileo and travel in nearly circular orbits around the parent with periods of from 2 to 17 days. These Galilean moons can be observed with a small telescope and are interesting in their own right. Because their orbits are along Jupiter's equatorial plane, and because they all have the same orbital direction, these moons

Figure 15.22 Jupiter with two of its largest satellites as seen from *Voyager I*. Io is seen against the Great Red Spot and Europa appears against a white oval. *(Courtesy of NASA Headquarters)*

A.

B.

C.

D.

Figure 15.23 Jupiter's four largest (Galilean) moons. These are the four moons discovered by Galileo. **A.** Callisto, the outermost of the Galilean satellites, is densely cratered, much like the moon. **B.** Europa, smallest of the Galilean moons, has an icy surface that is criss-crossed by many linear features. **C.** Ganymede, the largest Jovian satellite, contains cratered areas, smooth regions, and areas covered by numerous parallel grooves. **D.** The innermost moon, Io, is one of only three volcanically active bodies in the solar system. *(Courtesy of NASA Headquarters)*

most probably formed from "leftover" debris in much the same way the planets did.

By contrast, Jupiter's four outermost satellites are very small (20 kilometers in diameter), revolve in a direction that is opposite the other moons, and have orbits that are steeply inclined to the Jovian equator. These satellites appear to be asteroids that passed near enough to be captured gravitationally by Jupiter.

The images obtained by *Voyagers 1* and *2* in 1979 revealed, to the surprise of almost everyone, that each of the four Galilean satellites has a character all its own. The entire surface of *Callisto*, the outermost of the Galilean satellites, is densely cratered, much like the surfaces of Mercury and our own Moon (Figure 15.23A). However, on Callisto, the impacts occurred in a crust that appears to be a dirty, frozen ocean of water ice.

Europa, smallest of the Galilean satellites, has an icy surface that is crisscrossed by many linear features (Figure 15.23B). Further, this satellite is notably void of large impact craters. Therefore, the present surface of Europa must have formed sometime after the early period of bombardment, when rocky chunks were far more numerous in the solar system.

Ganymede, the largest Jovian satellite, contains the most diverse terrain. Like our own Moon, it has densely cratered regions and other very smooth areas where a younger icy layer covers the older cratered surface (Figure 15.23C). In addition, Ganymede has numerous parallel grooves. This suggests some type of tectonic activity in the distant past.

The innermost of the Galilean moons, *Io*, is the only volcanically active body discovered in our solar system, other than Earth and Neptune's moon Triton (Figure 15.23D). To date, eight active sulfurous volcanic centers have been seen rising from the surface of Io to heights approaching 200 kilometers (Figure 15.24). The surface of Io is very colorful, a result of its sulfurous "rocks" that change colors from red to yellow and from black to white, depending on temperature.

The heat source for Io's volcanic activity is thought to be tidal energy generated by a relentless gravitational "tug of war" between Jupiter and the Galilean satellites. Because Io is gravitationally locked to Jupiter, the same side always faces the giant planet. The gravitational influence of Jupiter and the other nearby satellites pulls and pushes on Io's tidal bulge as its slightly eccentric orbit takes it alternately closer to and farther from Jupiter. This gravitational flexing

Figure 15.24 A volcanic eruption on Io. This plume is rising over 100 kilometers (60 miles) above Io's surface. *(Courtesy of NASA Headquarters)*

of Io is transformed into heat energy, making the moon volcanically active.

Saturn: The Elegant Planet

Requiring 29.46 years to make one revolution, Saturn is almost twice as far from the Sun as Jupiter, yet its atmosphere, composition, and internal structure are thought to be remarkably similar to Jupiter's.

Saturn's Rings The most prominent feature of Saturn is its system of rings (Figure 15.25). Until the recent discovery that Jupiter, Uranus, and Neptune have very faint ring systems, this phenomenon was thought to be unique to Saturn.

When viewed from Earth, Saturn's rings appear as distinct bands, which classically have been called the A, B, and C rings (Figure 15.25). The A ring is the outermost of the bright rings and is separated from the brightest ring (B ring) by a large gap. This space, called the **Cassini division**, is easily seen in photographs. Having a width of 5000 kilometers, the Cassini division is large enough to accommodate Earth's Moon.

The origin of the rings is believed to be related to their distance from the surface of Saturn. Because they

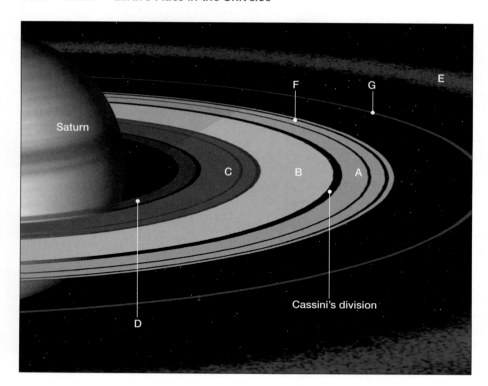

Saturn

F   G   E

C   B   A

Cassini's division

D

Figure 15.25 A view of the
dramatic ring system of Saturn.

are very close, the disruptive gravitational force of the parent prevented the individual particles from accreting into a larger satellite. Stated another way, objects cannot be held together by *self-gravitation* when they are within the influence of a stronger external gravitational field. In fact, should one of Saturn's large icy satellites approach closer than the outer edge of the bright rings, it would be destroyed and its remains distributed among the rings. The gravitational (tidal) force of Saturn pulling on the near side of such a satellite would be sufficiently greater than the force pulling on its far side, and it would tear the satellite apart.

From Earth, our view of Saturn slowly changes as both planets proceed along their orbits, continually changing relative positions. This changes our angle of view of Saturn's rings. Once every 15 years, we see them edge-on, and they appear as an extremely fine line.

Saturn Fly-By   In 1980 and 1981, fly-by missions of the nuclear-powered *Voyager 1* and *Voyager 2* space vehicles came within 100,000 kilometers of the surface of Saturn. More information on Saturn and its satellite system was gained in a few days than had been acquired since Galileo first viewed this elegant planet telescopically. Some of the information acquired by these space probes is as follows:

1. Saturn's atmosphere is very dynamic, with winds roaring at up to 1500 kilometers (930 miles) per hour.
2. Large cyclonic "storms" similar to Jupiter's Great Red Spot, although much smaller, occur in Saturn's atmosphere.
3. Eleven additional moons were discovered.
4. The icy rings of Saturn were discovered to be more complex than expected. Each of the seven rings is made of numerous ringlets, resembling the grooves on a phonograph record. The rings in the faint F ring are intertwined in stable, kinked, and braidlike configurations. Further, the B ring develops perplexing outwardly radiating spokes that survive for hours.
5. Satellite images reveal the thickness of the ring system to be no more than a few hundred meters, whereas its lateral extent exceeds 200,000 kilometers. We easily see the thin rings from more than a billion kilometers distance because they are highly reflective.

Although none of the images obtained so far has the resolution needed to "see" the fine structures of the rings, they undoubtedly are composed of relatively small particles (moonlets) that orbit the planet much like any other satellite. Radar observations indicate that most of the rings' particles are no larger than 10 meters, and the more abundant particles are perhaps as small as 10 centimeters (4 inches).

Saturn's Moons The Saturnian satellite system consists of at least 21 bodies (see chapter opening photo). All but two have nearly circular, counterclockwise orbits along Saturn's equatorial plane. The largest Saturnian moon, Titan, is larger than Mercury and is the second-largest satellite in the solar system (after Jupiter's Ganymede). It is the only satellite in the solar system known to have a substantial atmosphere. Because of its dense gaseous cover, the atmospheric pressure at the surface of Titan is about 1.5 times that at Earth's surface.

Data from *Voyager 1* revealed that Titan's atmosphere is roughly 80 percent nitrogen, with methane probably accounting for less than 6 percent. The orange color of Titan's atmosphere may result from a photochemical "smog" of hydrocarbon molecules. This planet-sized moon appears to have polar ice caps that show seasonal variations in size. Its surface, if unfrozen, would be an ocean of liquid nitrogen.

Uranus and Neptune: The Twins

If any two planets in the solar system can be considered twins, even more so than Earth and Venus, they are Uranus and Neptune. Only 1 percent different in diameter, both appear as pale greenish-blue, attributable to the methane in their atmospheres. Their structure and composition are believed to be similar. Neptune, however, is colder, because it is half again as distant from the Sun's warmth as is Uranus.

Uranus: The Sideways Planet A unique feature of Uranus is that it rotates "on its side"—its axis of rotation lies only 8 degrees from the plane of its orbit. Its rotational motion, therefore, has the appearance of rolling, rather than the toplike spinning of the other planets. (You can see the unusual orientation in Figure 15.10.) Because the axis of Uranus is inclined almost 90 degrees, the Sun is nearly overhead at one of its poles once each revolution, and then half a revolution later, it is overhead at the other. A surprise discovery in 1977 revealed that Uranus is surrounded by rings, much like those encircling Jupiter.

The first close-up views of Uranus and its satellites were transmitted to Earth by *Voyager 2* in 1986. Studies of these images revealed ten previously unknown satellites that are much smaller than the five

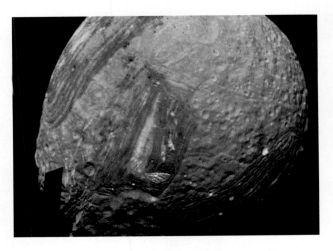

Figure 15.26 This computer-assembled mosaic of images depicts the Uranian satellite Miranda (*Voyager 2*, 1986). In addition to crater-marked areas, note Miranda's regions of folded ridges that produce curving patterns unique to this satellite. *(Courtesy of NASA Headquarters)*

large moons known from Earth-based observations. Two additional rings also were discovered, along with evidence that others may encircle the planet.

Spectacular views of the five largest moons of Uranus showed quite varied terrains. Some contain long, deep canyons and linear scars, whereas others possess large, smooth areas on otherwise crater-riddled surfaces. A spokesperson for the Jet Propulsion Laboratory described Miranda, the innermost of the five largest moons, as having a greater variety of landforms than any body yet examined in the solar system (Figure 15.26). In particular, Miranda has three vast areas with families of grooves that encircle one another, much like the lanes of a gigantic racetrack.

Neptune: The Windy Planet Even when the most powerful telescope is focused on Neptune, it appears as a bluish fuzzy disk. Until *Voyager 2*'s 1989 encounter with Neptune, astronomers knew very little about this planet, except that it had an atmosphere mostly of hydrogen, helium, and methane; that it had arcs or incomplete rings; and that it had two moons. However, *Voyager 2*'s 12-year, nearly 3-billion-mile journey has provided investigators with so much new information about Neptune and its satellites that years are needed to analyze it all.

Images from *Voyager 2* show that Neptune has a dynamic atmosphere, much like those of Jupiter and Saturn (Figure 15.27). Winds exceeding 1000 kilometers per hour (600 miles per hour) encircle the

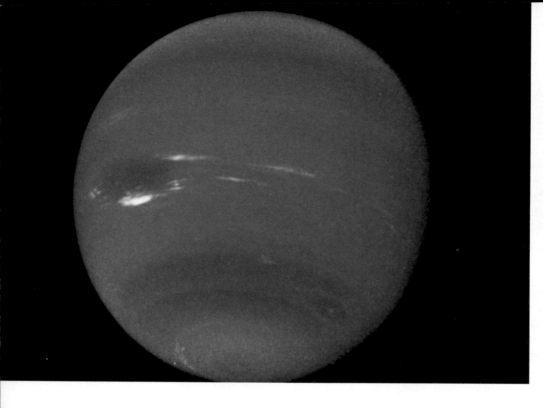

Figure 15.27 This image of Neptune shows the Great Dark Spot (left center). Also visible are bright cirruslike clouds that travel at high speed around the planet. A second oval spot is at 54° south latitude on the east limb of the planet. *(Photo courtesy of the Jet Propulsion Laboratory/NASA Headquarters)*

planet, making it one of the windiest places in the solar system. In addition, Neptune's dynamic atmosphere contains an Earth-sized blemish called the *Great Dark Spot* that is reminiscent of Jupiter's Great Red Spot. Like the Great Red Spot, this atmospheric disturbance is assumed to be a large rotating storm.

Perhaps the most surprising feature of the Neptunian atmosphere is the white, cirruslike clouds that occupy a layer about 50 kilometers above the main cloud deck. Although these clouds appear to change configurations, one group seems permanently attached to the Great Dark Spot. At the extremely low temperatures found on Neptune, frozen methane is the most likely constituent of these upper-level clouds.

Six new satellites, ranging from 50 to 400 kilometers in diameter, were discovered in the *Voyager* images. This brings the total number of known satellites for Neptune to eight. All of the newly discovered moons orbit the planet in a direction opposite that of the two larger satellites.

Voyager images also confirmed the existence of a ring system around Neptune's center. At least two narrow and two broad rings were discovered circling the planet. The outermost has three thicker arclike segments.

Triton: Neptune's Largest Moon Triton proved to be a most interesting object (Figure 15.28). Its diameter is roughly 2700 kilometers (nearly as large as Earth's Moon). Triton is the only large moon in the solar system that has a highly inclined retrograde orbit, meaning that it orbits Neptune in the direction opposite to the direction in which all the planets travel. This indicates that Triton formed independently of Neptune and was gravitationally captured by the parent planet.

Much has been learned about Triton. It has the lowest surface temperature of any body in the solar system, 38K (−391°F). Its very thin atmosphere is composed mostly of nitrogen, with a little methane. The surface of Triton apparently is largely water ice, covered with layers of solid nitrogen and methane.

Despite low surface temperatures, Triton displays a volcanic-like activity. Two active plumes were discovered that extended to an altitude of 8 kilometers and were blown downwind for more than 100 kilometers. Presumably, solar energy is absorbed more readily by the surface layers of darker methane ice. Such surface warming vaporizes some of the underlying nitrogen ice. As subsurface pressures increase, explosive eruptions eventually result.

Pluto: Planet X

Pluto lies on the fringe of the solar system, almost 40 times the distance from the Sun to Earth. It is 10,000 times too small to be visible to the unaided eye. Because of its great distance and slow orbital speed, it takes Pluto 248 years to orbit the Sun. Ever since its discovery in 1930, it has completed about one-fourth

The average temperature of Pluto is estimated at –210°C, cold enough to solidify most gases that might be present. Thus, Pluto probably has a meager atmosphere composed mainly of nitrogen. Pluto might be best described as a large, dirty iceball made up of a mixture of frozen gases with lesser amounts of rocky substances.

A recent proposal suggests that Pluto once was a satellite of Neptune and was displaced from its original orbit when it collided with a large foreign object. The discovery of a satellite around Pluto is considered evidence that this event broke the Neptunian satellite into two pieces and sent them into an elongated orbit around the Sun.

Minor Members of the Solar System

Asteroids: Microplanets

Asteroids are relatively small bodies that have been likened to "flying mountains." The largest, Ceres, is about 1000 kilometers (620 miles) in diameter, but most of the 50,000 that have been observed are only

Figure 15.28 A photomosaic of Neptune's largest moon, Triton. The large south polar cap is on the right half of the image. Seasonal ice (probably nitrogen) covers the region. Because spring in Triton's southern hemisphere extends from 1960 to the year 2000, some of the polar cap has evaporated. *(Photo courtesy of the Jet Propulsion Laboratory/NASA)*

of a revolution. Pluto's orbit is noticeably elongated (highly eccentric), causing the planet to occasionally travel inside the orbit of Neptune, where it currently resides. There is no chance that Pluto and Neptune will ever collide, because their orbits are inclined to each other and do not actually cross.

In 1978, the moon *Charon* was discovered orbiting Pluto. Because of its close proximity to the planet, the best ground-based images of Charon show it only as an elongated bulge. In 1990, the Hubble Space Telescope produced an image that clearly resolves the separation between these two icy worlds (Figure 15.29). Charon orbits Pluto once every 6.4 days at a distance of about 19,700 kilometers, or 20 times closer than our Moon is to Earth.

The discovery of Charon greatly altered earlier estimates of Pluto's size. Current data indicate that Pluto has a diameter of 2445 kilometers, about one-fifth the size of Earth, making it the smallest planet in the solar system. Charon is about 1300 kilometers across.

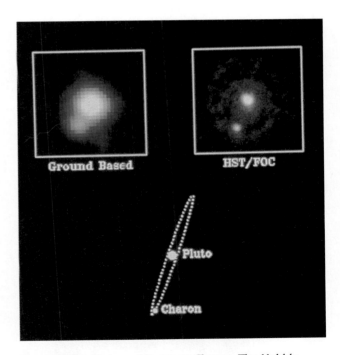

Figure 15.29 Pluto and its moon Charon. The Hubble Space Telescope produced the first image (upper right) that resolved these two icy worlds into separate objects. The image in the upper left is the best ground-based photo produced to date. *(Courtesy of NASA Headquarters)*

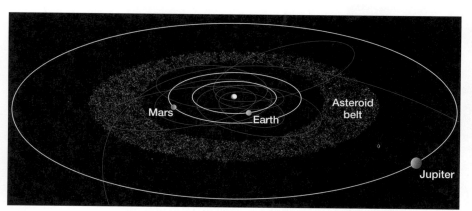

Figure 15.30 The orbits of most asteroids lie between Mars and Jupiter. Also shown are the orbits of a few known near-Earth asteroids. Perhaps a thousand or more asteroids have near-Earth orbits. Luckily, only a few dozen are thought to be larger than 1 kilometer in diameter.

about a kilometer across. The smallest asteroids are assumed to be no larger than grains of sand.

Most asteroids lie between the orbits of Mars and Jupiter and have periods ranging from three to six years (Figure 15.30). Some asteroids have very eccentric orbits and travel very near the Sun, and a few larger ones regularly pass close to Earth and the Moon. Many of the most recent impact craters on the Moon and Earth were probably caused by collisions with asteroids.

Because many asteroids have irregular shapes, planetary geologists first speculated that they might have formed from the breakup of a planet that once occupied an orbit between Mars and Jupiter (Figure 15.31). However, the total mass of the asteroids is esti-

mated to be only one-thousandth that of Earth, which itself is not a large planet. What, then might have happened to the remainder of the original planet? Others have hypothesized that several larger bodies once coexisted in close proximity and that their collisions produced numerous smaller ones. The existence of several "families" of asteroids has been used to support this explanation. However, no conclusive evidence has been found for either hypothesis.

Comets: Dirty Snowballs

Comets are among the most spectacular and unpredictable bodies in the solar system (see Figure 15.1). They have been compared to large, dirty snowballs, as they are made of frozen gases (water, ammonia, methane, carbon dioxide, and carbon monoxide) that hold together small pieces of rocky and metallic materials. Many comets travel along very elongated orbits that carry them beyond Pluto. On their return, these comets become visible again when they come within the orbit of Saturn.

When first observed, a comet appears very small. But as it approaches the Sun, solar energy begins to vaporize its frozen gases, producing a glowing head called the *coma* (Figure 15.32). The size of the coma varies greatly from one comet to another. Extremely rare ones exceed the size of the Sun, but most approximate the size of Jupiter. Within the coma, a small glowing nucleus with a diameter of only a few kilometers can sometimes be detected. As comets approach the Sun, some, but not all, develop a tail that extends for millions of kilometers. Despite the enormous size of their tails and comas, comets are relatively small members of the solar system.

The fact that the tail of a comet points away from the Sun in a slightly curved manner (Figure 15.32)

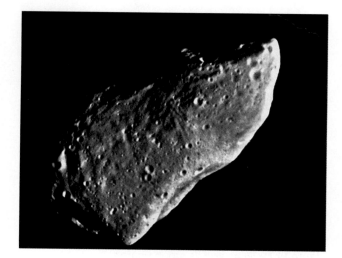

Figure 15.31 Image of asteroid 951 (Gaspra) obtained by the Jupiter-bound *Galileo* spacecraft. Like other asteroids, Gaspra is probably a collision-produced fragment of a larger body. *(Courtesy of NASA Headquarters)*

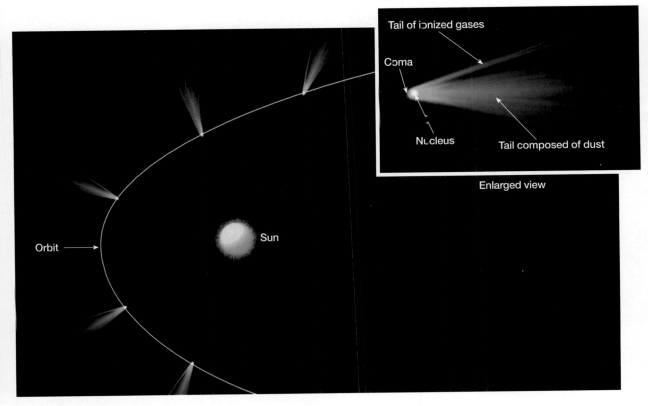

Figure 15.32 Orientation of a comet's tail as it orbits the Sun.

led early astronomers to propose that the Sun has a repulsive force that pushes the particles of the coma away, thus forming the tail. Today, two solar forces are known to contribute to this formation. One, *radiation pressure*, pushes dust particles away from the coma. The second, known as *solar wind*, is responsible for moving the ionized gases, particularly carbon monoxide. Sometimes a single tail composed of both dust and ionized gases is produced, but often two tails are observed (see Figure 15.1).

As a comet moves away from the Sun, the gases forming the coma begin to recondense, the tail disappears, and the comet once again returns to "cold storage." The material that was blown from the coma to form the tail is lost from the comet forever. Consequently, it is believed that most comets cannot survive more than a few hundred close orbits of the Sun. Once all the gases are expended, the remaining material—a swarm of unconnected metallic and stony particles—continues the orbit without a coma or a tail.

Little is known about the origin of comets. The most widely accepted proposal considers them to be members of the solar system that formed at great distances from the Sun. Accordingly, millions of comets are believed to orbit the Sun beyond Pluto with periods measured in hundreds of years. It is proposed that the gravitational effect of stars passing nearby sends some of them into the highly eccentric orbits that carry them toward the center of our solar system. Here, the gravitation of the larger planets, particularly Jupiter, alters a comet's orbit and accelerates its period of revolution. Many short-period comets of this type have been discovered. However, because they have a short life expectancy, we can be fairly certain that they are always being replaced by other long-period comets that are being deflected toward the Sun.

In July 1994 Comet Shoemaker-Levy impacted Jupiter with the force equal to 6 million megatons of energy. (A megaton is the equivalent of a million tons of the explosive, TNT.) Clearly, this was the most dramatic event in the solar system ever observed by people. Concern that a similar event might occur on Earth led NASA to establish the Near-Earth Object Search Committee to detect asteroids or comets with trajectories that might cross Earth's orbit.

Figure 15.33 Dark blemishes on Jupiter produced by the impact of fragments of Comet Shoemaker-Levy in July 1994. *(Courtesy of Hubble Space Telescope Comet Team and NASA)*

It is possible that a few small fragments of a comet may have actually impacted Earth in 1908. That year, a strong explosion flattened more than 1000 square kilometers of a remote Siberian forest.

Comet Shoemaker-Levy was discovered at California's Mount Palomar Observatory barely a year before it impacted Jupiter. Careful observation showed that it consisted of two dozen fragments. Researchers concluded that the comet had broken up during an earlier pass by Jupiter. As the larger fragments penetrated Jupiter's outer atmosphere, they produced brilliant impact flashes and debris plumes soaring thousands of kilometers above the planet. The result of these fiery impacts were dark zones in Jupiter's atmosphere that exceeded Earth in size. The largest of these dark blemishes lasted for months (Figure 15.33). Investigators are learning more about the dynamics of Jupiter's atmosphere by studying how these blemishes disperse.

Meteoroids: Visitors to Earth

You may have seen a **meteor**, popularly (but inaccurately) called a "shooting star." This streak of light, or what may look like a falling star or blazing rock, lasts for, at most, a few seconds and occurs when a small solid particle, a **meteoroid**, enters Earth's atmosphere from interplanetary space. The friction between the meteoroid and the air heats both and produces the light we see.

Although an occasional meteoroid is as large as an asteroid, most are about the size of sand grains and weigh less than 1/100 gram. Consequently, they vaporize before reaching Earth's surface. Some, called *micrometeorites*, are so tiny that their rate of fall becomes too slow to cause them to burn up, so they drift down as "space dust." Each day, thousands of meteoroids enter Earth's atmosphere. During nighttime, a half dozen or more are bright enough to be seen with the unaided eye each hour from anywhere on Earth.

Occasionally, meteor sightings increase dramatically to 60 or more per hour. These sometimes spectacular displays, called **meteor showers**, result when Earth encounters a swarm of meteoroids traveling in the same direction and at nearly the same speed as Earth. The close association of these swarms to the orbits of some short-term comets strongly suggests that they represent material lost by these comets (Table 15.2). Some swarms not associated with orbits of known comets are probably the remains of the nucleus of a long-defunct comet. The meteor showers that occur regularly each year around August 12 are believed to be the remains of the Comet 1862 III, which has a period of 110 years.

Meteoroids associated with comets are small and not known to reach the ground. Most meteoroids

Table 15.2 ■ Major meteor showers

Shower	Approximate Dates	Associated Comet
Quadrantids	January 4–6	—
Lyrids	April 20–23	Comet 1861 I
Eta Aquarids	May 3–5	Halley's comet
Delta Aquarids	July 30	—
Perseids	August 12	Comet 1862 III
Draconids	October 7–10	Comet Giacobini–Zinner
Orionids	October 20	Halley's comet
Taurids	November 3–13	Comet Encke
Andromedids	November 14	Comet Biela
Leonids	November 18	Comet 1866 I
Geminids	December 4–16	—

Figure 15.34 Meteor Crater, about 32 kilometers (20 miles) west of Winslow, Arizona. *(Photo by Michael Collier)*

large enough to survive the heated fall are thought to originate among the asteroids where chance collisions modify their orbits and send them toward Earth. Earth's gravitational force does the rest.

The remains of meteoroids, when found on Earth, are referred to as **meteorites**. A few very large meteoroids have blasted out craters on Earth's surface that are not much different in appearance from those on the lunar surface. The most famous is Meteor Crater in Arizona (Figure 15.34). This huge cavity is about 1.2 kilometers (0.75 mile) across, 170 meters (560 feet) deep, and has an upturned rim that rises 50 meters (165 feet) above the surrounding countryside. Over 30 tons of iron fragments have been found in the immediate area, but attempts to locate the main body have been unsuccessful. Judging from the amount of erosion, it appears that the impact occurred within the last 20,000 years.

Prior to Moon rocks brought back from lunar exploration, meteorites were the only samples of extraterrestrial material that could be directly examined. Meteorites are classified by their composition: (1) **irons**—mostly iron with 5 to 20 percent nickel, (2) **stony**—silicate minerals with inclusions of other minerals, and (3) **stony-irons**—mixtures. Although

stony meteorites are probably more common, most meteorites found by people are irons. This is understandable, because irons tend to withstand the impact, weather more slowly, and are much easier for a lay person to distinguish from terrestrial rocks than are stony meteorites.

Iron meteorites are probably fragments of once-molten cores of large asteroids or small planets. One rare kind of meteorite, called a *carbonaceous chondrite*, was found to contain some simple amino acids and other organic compounds, which are the basic building blocks of life. This discovery confirms similar findings in observational astronomy that indicate the existence of numerous organic compounds in the frigid realm of outer space.

If the composition of meteorites is representative of the material that makes up the Earthlike planets, as some planetary geologists suggest, the Earth must contain a much larger percentage of iron than is indicated by the surface rock. This is one of the reasons why geologists suggest that the core of Earth may be mostly iron and nickel. In addition, the dating of meteorites has indicated that our solar system has an age that certainly exceeds 4.5 billion years. This "old age" has been confirmed by data obtained from lunar samples.

The Chapter in Review

1. Early Greeks held the *geocentric* ("Earth-centered") view of the universe, believing that Earth was a sphere that stayed motionless at the center of the universe. Orbiting Earth were the Moon, Sun, and the known planets—Mercury, Venus, Mars, Jupiter, and Saturn. To the early Greeks, the stars traveled daily around Earth on a transparent, hollow *celestial sphere*. In A.D. 141, *Claudius Ptolemy* documented this geocentric view called the Ptolemaic system.

2. Modern astronomy evolved through the work of many dedicated individuals during the 1500s and 1600s. *Nicolaus Copernicus* (1473–1543) reconstructed the solar system with the Sun at the center and the planets orbiting around it, but erroneously continued to use circles to represent the orbits of planets. *Tycho Brahe*'s (1546–1601) observations were far more precise than any made previously and are his legacy to astronomy. *Johannes Kepler* (1571–1630) ushered in the new astronomy with his three laws of planetary motion. After constructing his own telescope, *Galileo Galilei* (1564–1642) made many important discoveries that supported the Copernican view of a Sun-centered solar system. *Sir Isaac Newton* (1642–1727) developed laws of motion and proved that the force of gravity, combined with the tendency of an object to move in a straight line, results in elliptical orbits for planets.

3. The planets can be arranged into two groups: the *terrestrial* (Earthlike) *planets* (Mercury, Venus, Earth, and Mars) and the *Jovian* (Jupiterlike) *planets* (Jupiter, Saturn, Uranus, and Neptune). Pluto is not included in either group. When compared to the Jovian planets, the terrestrial planets are smaller, more dense, contain proportionally more rocky material, and have slower rates of rotation.

4. The *nebular hypothesis* describes the formation of the solar system. The planets and Sun began forming about 5 billion years ago from a large cloud of dust and gases called a *nebula*. As the nebular cloud contracted, it began to rotate and flatten into a disk. Material that was gravitationally pulled toward the center became the *protosun*. Within the rotating disk, small centers formed, called *protoplanets*, sweeping up more and more of the nebular debris. Because of their high temperatures and weak gravitational fields, the inner planets (Mercury, Venus, Earth, and Mars) were unable to accumulate much of the lighter components (hydrogen, ammonia, methane, and water) of the nebula. However, because of the very cold temperatures existing far from the Sun, the fragments from which the Jovian planets formed contained a high percentage of ices—water, carbon dioxide, ammonia, and methane.

5. The lunar surface exhibits several types of features. Most *craters* were produced by the impact of rapidly moving debris (meteoroids). Bright, densely cratered highlands make up most of the lunar surface. Dark, fairly smooth lowlands are called *maria*. Maria basins are enormous impact craters that have been flooded with layer upon layer of very fluid basaltic lava. All lunar terrains are mantled with a soil-like layer of gray, unconsolidated debris, called *lunar regolith*, which has been derived from a few billion years of meteoric bombardment.

6. Closest to the Sun, Mercury is small, dense, has no atmosphere, and exhibits the greatest temperature extremes of any planet. Venus, the brightest planet in the sky, has a thick, heavy, carbon dioxide atmosphere, a surface of relatively subdued plains and inactive volcanoes, a surface atmospheric pressure 90 times that of Earth's, and surface temperatures of 475°C. Mars, the red planet, has a carbon dioxide atmosphere only 1 percent as dense as Earth's, extensive winds and dust storms, numerous inactive volcanoes, many large canyons, and several valleys of debatable origin resembling drainage patterns similar to stream valleys on Earth. *Jupiter*, the largest planet, rotates rapidly, has a banded appearance, a Great Red Spot that varies in size, a ring system, and at least 16 moons (one of which, Io, is volcanically active). *Saturn*, best known for its rings, also has a dynamic atmosphere with winds up to 930 miles per hour and "storms" similar to Jupiter's Great Red Spot. *Uranus* and *Neptune* are often called the twins because of similar structure and composition. Uranus is unique in rotating "on its side." Neptune has white, cirruslike clouds above its main cloud deck and an Earth-sized Great Dark Spot, assumed to be a large rotating storm similar to Jupiter's Great Red Spot. *Pluto*, a small frozen world with one moon, may have once been a satellite of Neptune. Pluto's elongated orbit causes it to occasionally travel inside the orbit of Neptune, but with no chance of collision.

7. The minor members of the solar system include *asteroids, comets*, and *meteoroids*. No conclusive evidence has been found to explain the origin of the asteroids. Comets are made of frozen gases with small pieces of rocky and metallic material. Many travel in very elongated orbits that carry them beyond Pluto. Meteoroids, small solid particles that travel through interplanetary space, become *meteors* when they enter Earth's atmosphere and vaporize with a flash of light. *Meteor showers* occur when Earth encounters a swarm of meteoroids, probably fragments of a comet. *Meteorites* are the remains of meteoroids. The three *types of meteorites* are iron, stony, and stony-iron.

Key Terms

asteroid (p. 401)

astronomical unit (AU) (p. 379)

Cassini division (p. 397)

celestial sphere (p. 374)

comet (p. 402)

escape velocity (p. 384)

geocentric (p. 374)

iron meteorite (p. 405)

Jovian planet (p. 382)

lunar regolith (p. 388)

maria (p. 386)

meteor (p. 404)

meteorite (p. 405)

meteoroid (p. 404)

meteor shower (p. 404)

nebular hypothesis (p. 384)

Ptolemaic system (p. 375)

retrograde motion (p. 375)

stony-iron meteorite (p. 405)

stony meteorite (p. 405)

terrestrial planet (p. 382)

Questions for Review

1. Why did the ancients think that celestial objects had some control over their lives?

2. Describe what produces the retrograde motion of Mars. What geometric arrangement did Ptolemy use to explain this motion?

3. What major change did Copernicus make in the Ptolemaic system?

4. What was Tycho Brahe's contribution to science?

5. Did Galileo invent the telescope?

6. Explain how Galileo's discovery of a rotating Sun supported the Copernican view of a Sun-centered universe.

7. Newton learned that the orbits of the planets are the result of two actions. Describe these actions.

8. Compare the rotational periods of the terrestrial and Jovian planets.

9. How fast (in kilometers per hour) does a location on the equator of Jupiter rotate? Note that Jupiter's circumference equals approximately 452,000 kilometers (280,000 miles).

10. By what criteria are the planets placed into either the Jovian or terrestrial group?

11. What are the three types of materials thought to make up the planets? How are they different? How does their distribution account for the density differences between the terrestrial and Jovian planetary groups?

12. Briefly describe the events that are thought to have led to the formation of the solar system.

13. How are the maria of the Moon thought to be similar to the Columbia Plateau?

14. What surface features does Mars have that are also common on Earth?

15. Although Mars has valleys that appear to be products of stream erosion, what fact makes it unlikely that Mars has had a water cycle like that on Earth?

16. What is the nature of Jupiter's Great Red Spot?

17. What is distinctive about Jupiter's satellite Io?

18. Why are the four outer satellites of Jupiter thought to have been captured?

19. What is unique about Saturn's satellite Titan?

20. What three bodies in the solar system exhibit volcano-like activity?

21. What do you think would happen if Earth passed through the tail of a comet?

22. Compare a meteoroid, meteor, and meteorite.

23. Why are meteorite craters more common on the Moon than on Earth, even though the Moon is much smaller?

Testing What You Have Learned

Multiple-Choice Questions

1. The geocentric model of the universe that illustrates the motion of planets using both deferents and epicycles was proposed by _____
   - **a.** Copernicus.
   - **b.** Brahe.
   - **c.** Galileo.
   - **d.** Ptolemy.
   - **e.** Kepler.

2. In his monumental work, *On the Revolution of the Heavenly Spheres*, this early astronomer correctly viewed the planets as orbiting the Sun, but incorrectly used circles to describe the orbits.
   - **a.** Copernicus
   - **b.** Brahe
   - **c.** Galileo
   - **d.** Ptolemy
   - **e.** Kepler

3. His third law relates the orbital period of planets and their distances to the Sun.
   - **a.** Copernicus
   - **b.** Brahe
   - **c.** Galileo
   - **d.** Ptolemy
   - **e.** Kepler

4. His discovery in the early 1600s of four moons orbiting Jupiter dispelled the old idea that Earth was the only center of motion in the universe.
   - **a.** Copernicus
   - **b.** Brahe
   - **c.** Galileo
   - **d.** Ptolemy
   - **e.** Kepler

5. Which of the following describes the terrestrial planets?
   - **a.** large, gaseous
   - **b.** small, rocky
   - **c.** short days, cold
   - **d.** multiple moons
   - **e.** massive, low density

6. Dark regions of layered lava flows that resemble "seas" on the moon are called _____
   - **a.** highlands.
   - **b.** meteoroids.
   - **c.** regolith.
   - **d.** terraces.
   - **e.** maria.

7. This planet, similar to Earth in size, density, and mass, is shrouded in thick clouds.
   - **a.** Mercury
   - **b.** Venus
   - **c.** Mars
   - **d.** Jupiter
   - **e.** Pluto

8. Although the carbon dioxide-rich atmosphere of this red planet is only 1 percent as dense as that of Earth, extensive dust storms may be responsible for color changes.
   - **a.** Mercury
   - **b.** Venus
   - **c.** Mars
   - **d.** Jupiter
   - **e.** Uranus

9. Until the recent discovery that Jupiter, Uranus, and Neptune also have these features, this phenomenon was thought to be unique to Saturn.
   - **a.** moons
   - **b.** spots
   - **c.** rings
   - **d.** craters
   - **e.** bands

10. Many of the most recent impact craters on the Moon and Earth were probably caused by collisions with _____
    - **a.** comets.
    - **b.** meteoroids.
    - **c.** nebulae.
    - **d.** asteroids.
    - **e.** the solar wind.

Fill-In Questions

11. The _____ view of the universe held by early Greeks had Earth, motionless at the center, being orbited by the Moon, Sun, known planets, and stars.

12. The planets fall quite nicely into two groups: the _____ planets (Mercury, Venus, Earth, and Mars) and the _____ planets (Jupiter, Saturn, Uranus, and Neptune).

13. The idea that the Sun and planets formed from the same cloud of gas and dust in interstellar space is known as the _____ hypothesis.

14. The most striking feature on Jupiter is the _____ _____ Spot in its southern hemisphere.

15. As a comet approaches the Sun, solar energy begins to vaporize the frozen gases, producing a glowing head called a _____.

True/False Questions

16. The apparent eastward motion of the planets with respect to the stars is called retrograde motion. _____

17. Sir Isaac Newton was the first to formulate and test the law of universal gravitation. _____

18. The difference between the average density of the Moon and Earth suggests that the Moon's iron core is quite large. _____

19. Mercury and Pluto are the only planets in the solar system without moons. _____

20. Although stony meteorites are probably more common, most meteorites found by people are irons. _____

Answers:

1. d; 2. a; 3. e; 4. c; 5. b; 6. e; 7. b; 8. c; 9. c; 10. d; 11. geocentric; 12. terrestrial, Jovian; 13. nebular; 14. Great Red; 15. coma; 16. F; 17. T; 18. F; 19. F; 20. T.

Web   Work

The following are informative and interesting World Wide Web sites that address topics related to those presented in this chapter:

- Lunar and Planetary Institute
  http://cass.jsc.nasa.gov/lpi.html

- National Aeronautics and Space Administration (NASA)
  http://ww0w.nasa.gov/

For direct access to these sites and other relevant Web locations, contact the *Foundations of Earth Science* WWW home page at http://www.prenhall.com/lutgens.

CHAPTER 16

Beyond the Solar System

To assist you in learning the important concepts
in this chapter, you will find it helpful to focus
on the following questions:

1. How is the principle of parallax used to measure the distance to a star?
2. What are the major intrinsic properties of stars?
3. What are the size categories of stars?
4. What are the different types of nebulae?
5. What is the most plausible model for stellar evolution? What are the stages in the life cycle of a star?
6. What are the possible final states that a star may assume after it consumes its nuclear fuel and collapses?
7. What is our own Milky Way galaxy like?
8. What are the different types of galaxies?
9. What is the Doppler shift?
10. What is the Big Bang theory for the origin of the universe?

Eagle Nebula in the constellation Serpens. *Courtesy of National Optical Astronomy Observatories*

The star Proxima Centauri is about 4.3 light-years away, roughly 100 million times farther than the Moon. Yet, other than our own Sun, it is the closest star to Earth. This fact suggests that the universe is incomprehensibly large. What is the nature of this vast cosmos beyond our solar system? Are the stars distributed randomly, or are they organized into distinct clusters? Do stars move, or are they permanently fixed features, like lights strung out against the black cloak of outer space? Does the universe extend infinitely in all directions, or is it bounded? To consider these questions, this chapter will examine the universe by taking a census of the stars—the most numerous objects in the night sky.

Properties of Stars

The Sun is the only star whose surface we can observe. Yet, a great deal is known about the universe beyond our solar system. In fact, more is known about the stars than about our outermost planet, Pluto. This knowledge hinges on the fact that stars, and even gases in the "empty" space between stars, radiate energy in all directions into space (Figure 16.1). The key to understanding the universe is to collect this radiation and unravel the secrets it holds. Astronomers have devised many ingenious methods to do just that. We will begin by examining stellar distances and some intrinsic properties of stars, including *color, brightness, mass, temperature,* and *size.*

Figure 16.1 Lagoon Nebula. It is in glowing clouds like these that gases and dust particles become concentrated into stars. *(Courtesy of National Optical Astronomy Observatories)*

Measuring Distances to the Stars

Measuring the distance to a star is very difficult. Obviously, we cannot journey to the star. Nevertheless, astronomers have developed some *indirect* methods of measuring stellar distances. The most basic of these measurements is called stellar parallax.

Recall from Chapter 15 that **stellar parallax** is the extremely slight back-and-forth shifting in the apparent position of a nearby star due to the orbital motion of Earth. To review, the principle of parallax is easy to visualize. Close one eye, and with your index finger in a vertical position, use your eye to line up your finger with some distant object. Now, without moving your finger, view the object with your other eye and notice that its position appears to have changed. The farther away you hold your finger, the less its position seems to shift.

In practice, parallax is determined by photographing a nearby star against the background of distant stars. Then, six months later, when Earth has moved halfway around its orbit, a second photograph is taken. When these photographs are compared, the position of the nearby star appears to have shifted with respect to the background stars. Figure 16.2 illustrates this shift and the parallax angle determined from it. The nearest stars have the largest parallax angles, whereas those of distant stars are too slight to measure. Recall that the medieval astronomer Tycho Brahe was unable to detect stellar parallax, leading him to reject the idea that Earth orbits the Sun.

It should be emphasized that parallax angles are *very* small. The parallax angle to the nearest star, Proxima Centauri, is less than 1 second of arc, which equals $1/_{3600}$ of a degree. To put this in perspective, fully extend your arm and raise your little finger. Your finger is roughly 1 degree wide. Try doing this on a moonlit night, covering the Moon with your finger. The moon is only about $1/_2$ degree wide. Now, imagine detecting a movement that is only $1/_{3600}$ as wide as your finger. It should be apparent why Tycho Brahe, without the aid of a telescope, was unable to observe stellar parallax.

The distances to stars are so large that the conventional units, such as kilometers or astronomical units, are often too cumbersome to use. A better unit to express stellar distance is the **light-year**, which is the distance light travels in one Earth year—about 9.5 trillion kilometers (5.8 trillion miles).

In principle, the method used to measure stellar distances is elementary and was known to the ancient Greeks. But in practice, measurements are greatly

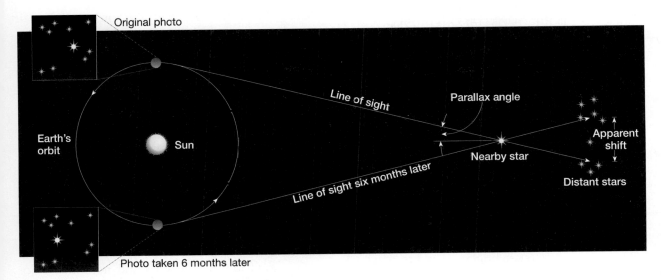

Figure 16.2 Geometry of stellar parallax. The parallax angle shown here is enormously exaggerated to illustrate the principle. Because the distance to even the nearest stars is thousands of times greater than the Earth–Sun distance, the triangles that astronomers work with are very long and narrow, making the angles that are measured very small.

complicated because of the tiny angles involved and because the Sun, and the star being measured, also have actual motion through space. The first accurate stellar parallax was not determined until 1838. Even today, parallax angles for only a few thousand of the nearest stars are known with certainty.

Most stars have such small parallax shifts that accurate measurement is not possible. Fortunately, a few other methods have been derived for estimating distances to these stars. Also, Hubble Space Telescope, which is not hindered by Earth's turbulent, light-distorting atmosphere, is expected to obtain accurate parallax distances for many additional remote stars.

Stellar Brightness

Three factors control the apparent brightness of a star as seen from Earth: *how big* it is, *how hot* it is, and *how far away* it is. The stars in the night sky are a grand assortment of sizes, temperatures, and distances, so their brightnesses vary widely.

Apparent Magnitude Stars have been classified according to their brightness ever since the second century B.C., when Hipparchus placed about 1000 of them into six categories. The measure of a star's brightness is called its **magnitude**. Some stars may appear dimmer than others only because they are farther away. Therefore, a star's brightness,

as it appears when viewed from Earth, has been termed its **apparent magnitude**.

When numbers are employed to designate relative brightness, the larger the magnitude number, the dimmer the star. *Stars that appear the brightest are of the first magnitude, while the faintest stars visible to the unaided eye are of the sixth magnitude.* With the invention of the telescope, many stars fainter than the sixth magnitude were discovered.

In the mid-1800s, a method was developed for comparing the brilliance of stars using magnitude. Just as we can compare the brightness of a 50-watt light bulb to that of a 100-watt bulb, we can compare the brightness of stars having different magnitudes. It was determined that a first-magnitude star was about 100 times brighter than a sixth-magnitude star. Therefore, on the scale that was devised, two stars that differ by 5 magnitudes have a ratio in brightness of 100 to 1. Hence, a seventh-magnitude star is 100 times brighter than a twelfth-magnitude star.

It follows, then, that the brightness ratio of two stars differing by only one magnitude is about 2.5.* Thus, a star of the first magnitude is about 2.5 times brighter than a star of the second magnitude. Table 16.1 shows some magnitude differences and the corresponding brightness ratios.

---

*Calculations: $2.512 \times 2.512 \times 2.512 \times 2.512 \times 2.512$, or 2.512 raised to the fifth power, equals 100.

Table 16.1 ■ Ratios of star brightness	
Difference in Magnitude	Brightness Ratio
0.5	1.6:1
1	2.5:1
2	6.3:1
3	16:1
4	40:1
5	100:1
10	10,000:1
20	100,000,000:1

Because some stars are brighter than first-magnitude stars, zero and negative magnitudes were introduced. The brightest star in the night sky, Sirius, has an apparent magnitude of −1.4, about 10 times brighter than a first-magnitude star. On this scale, the Sun has an apparent magnitude of −26.7. At its brightest, Venus has a magnitude of −4.3. At the other end of the scale, the 5-meter (200-inch) Hale Telescope can view stars with an apparent magnitude of 23, approximately 100 million times dimmer than stars that are visible to the unaided eye.

Absolute Magnitude   Astronomers are also interested in the "true" brightness of stars, called their **absolute magnitude**. Stars of the same luminosity, or brightness, usually do not have the same *apparent* magnitude, because their distances from us are not equal. To compare their true, or intrinsic, brightness, astronomers determine what magnitude the stars would have if it were at a standard distance of about 32.6 light-years. For example, the Sun, which has an apparent magnitude of −26.7, would, if located at a distance of 32.6 light-years, have an absolute magnitude of about 5. Thus, stars with absolute magnitudes greater than 5 (smaller numerical value) are intrinsically brighter than the Sun, but because of their distance, they appear much dimmer.

Table 16.2 lists the absolute and apparent magnitudes of some stars as well as their distances from Earth. Most stars have an absolute magnitude between −5 and 15, which puts the Sun near the middle of this range.

Stellar Color and Temperature

The next time you are outdoors on a clear night, take a good look at the stars and note their color (Figure 16.3). Some that are quite colorful can be found in the constellation Orion (see Appendix D, winter chart). Of the two brightest stars in Orion, Betelgeuse (α Orionis) is definitely red, whereas Rigel (β Orionis) appears blue.

Very hot stars with surface temperatures above 30,000 K (kelvin) emit most of their energy in the form of short-wavelength light and therefore appear blue. Red stars, on the other hand, are much cooler, generally less than 3000 K, and most of their energy is emitted as longer-wavelength red light. Stars with temperatures between 5000 and 6000 K appear yellow, like the Sun. Because color is primarily a manifestation of a star's temperature, this characteristic provides the astronomer with useful information about a star.

Figure 16.3 Time-lapse photograph of stars in the constellation Orion. These star trails show some of the various star colors. It is important to note that the eye sees color somewhat differently than does photographic film. *(Courtesy of National Optical Astronomy Observatories)*

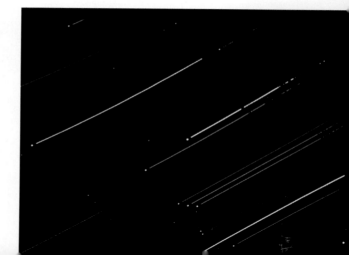

Table 16.2 ■ Distance, apparent magnitude, and absolute magnitude of some stars			
Name	Distance (Light years)	Apparent Magnitude*	Absolute Magnitude*
Sun	NA	−26.7	5.0
Alpha Centauri	4.27	0.0	4.4
Sirius	8.70	−1.4	1.5
Arcturus	36	−0.1	−0.3
Betelgeuse	520	0.8	−5.5
Deneb	1600	1.3	−6.9

*The more negative, the brighter; the more positive, the dimmer.

Binary Stars and Stellar Mass

One of the night sky's best-known constellations, the Big Dipper, appears at first glance to consist of seven stars. But individuals with good eyesight can resolve the second star in the handle of the Big Dipper (Mizar) as two stars. During the 1700s, astronomers used their new tool, the telescope, to discover numerous such star pairs. One of the stars in the pair is usually fainter than the other, and for this reason it was considered to be farther away. In other words, the stars were not considered true pairs but were thought only to lie in the same line of sight.

In the early 1800s, careful examination of numerous star pairs by William Herschel revealed that many actually orbit one another. The two stars are in fact united by their mutual gravitation. These pairs of stars, in which the members are far enough apart to be resolved telescopically, are called *visual binaries*. The idea of one star orbiting another may seem unusual, but evidence indicates that more than 50 percent of the stars in the universe occur in pairs or multiples.

Binary stars are used to determine the star property most difficult to calculate—its mass. The mass of a body can be established if it is gravitationally attached to a partner, which is the case for any binary star system. Binary stars orbit each other around a common point called the *center of mass* (Figure 16.4). For stars of equal mass, the center of mass lies exactly halfway between them. If one star is more massive than its partner, their common center will be located closer to the more massive one. Thus, if the sizes of their orbits can be observed, a determination of their individual masses can be made. You can experience this relationship on a seesaw by trying to balance a person who has a much greater mass.

For illustration, when one star has an orbit half the size (radius) of its companion, it is twice as massive. If their combined masses are equal to 3 times the mass of the Sun, then the larger will be twice as massive as the Sun, and the smaller will have a mass equal to that of the sun. Most stars have masses that range between 1/10 and 50 times the mass of our Sun.

Hertzsprung-Russell Diagram

Early in this century, Einar Hertzsprung and Henry Russell independently studied the relation between the true brightness (absolute magnitude) and temperature of stars. From this research each developed a graph, now called a **Hertzsprung-Russell diagram**

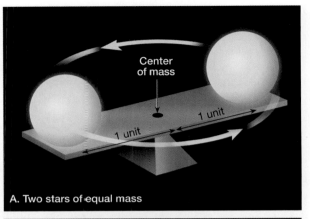

A. Two stars of equal mass

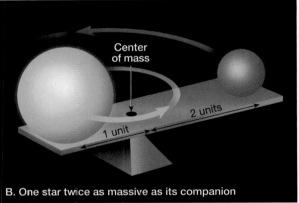

B. One star twice as massive as its companion

Figure 16.4 Binary stars orbit each other around their common center of mass. **A.** For stars of equal mass, the center of mass lies exactly halfway between them. **B.** If one star is twice as massive as its companion, it is twice as close to their common center of mass. Therefore, more massive stars have proportionately smaller orbits than do their less massive companions.

(H-R diagram), that exhibits these intrinsic stellar properties. By studying H-R diagrams, we learn a great deal about the sizes, colors, and temperatures of stars.

To produce an H-R diagram, astronomers survey a portion of the sky and plot each star according to its luminosity (brightness) and temperature (Figure 16.5). Notice that the stars in Figure 16.5 are not uniformly distributed. Rather, about 90 percent of all stars fall along a band that runs from the upper-left corner to the lower-right corner of an H-R diagram. These "ordinary" stars are called **main-sequence stars**. As shown in Figure 16.5, *the hottest main-sequence stars are intrinsically the brightest, and vice versa.*

The luminosity of the main-sequence stars is also related to their mass. The hottest (blue) stars are

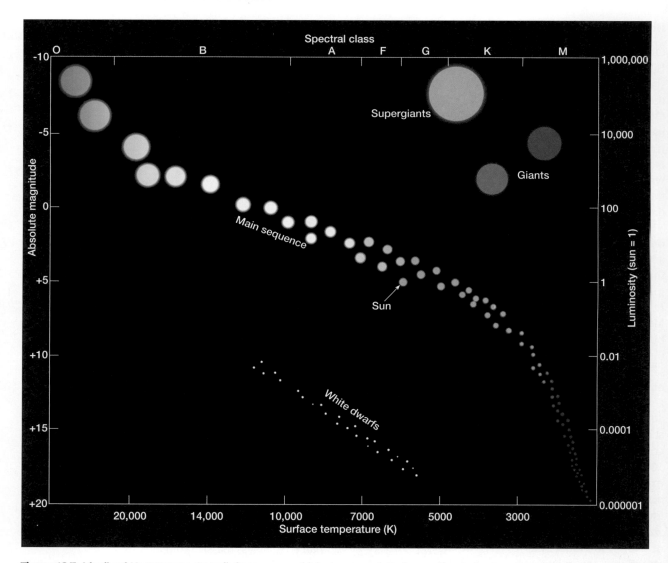

Figure 16.5 Idealized Hertzsprung-Russell diagram on which stars are plotted according to temperature and absolute magnitude.

about 50 times more massive than our Sun, whereas the coolest (red) stars are only 1/10 as massive. Therefore, on the H-R diagram, the main-sequence stars appear in a decreasing order, from *hotter, more massive* blue stars to *cooler, less massive* red stars.

Note the location of our Sun in Figure 16.5. The Sun is a yellow, main-sequence star with an absolute magnitude of about 5. Because the magnitude of a vast majority of main-sequence stars lie between –5 and 15, and because the Sun falls midway in this range, the Sun is often considered an average star. However, more main-sequence stars are cooler and less massive than our Sun.

Just as all humans do not fall into the normal size range, some stars do not fit in with the main-sequence stars. Above and to the right of the main

sequence in the H-R diagram in Figure 16.5 lies a group of very luminous stars called *giants*, or, on the basis of their color, **red giants**. The size of these giants can be estimated by comparing them with stars of known size that have the same surface temperature. We know that objects having equal surface temperatures radiate the same amount of energy per unit area. Therefore, any difference in the brightness of two stars having the same surface temperature is attributable to their relative sizes.

As an example, let us compare the Sun, which has been assigned a luminosity of 1, with another yellow star that has a luminosity of 100. Because both stars have the same surface temperature, they both radiate the same amount of energy per unit area. Therefore, for the more luminous star to be 100 times brighter than the Sun, it must have 100 times more surface area. It should be clear why stars whose plots fall in the upper-right position of an H-R diagram are called *giants*.

Some stars are so large that they are called **supergiants**. Betelgeuse, a bright-red supergiant in the constellation Orion, has a radius about 800 times that of the Sun. If this star were at the center of the solar system, it would extend beyond the orbit of Mars, and Earth would find itself inside the star! Other red giants that are easy to locate in our sky are Arcturus in the constellation Bootes and Antares in Scorpius.

In the lower-central portion of the H-R diagram, the opposite situation occurs. These stars are much fainter than main-sequence stars of the same temperature, and by using the same reasoning as before, they must be much smaller. Some probably approximate Earth in size. This group has come to be called **white dwarfs**, although not all are white.

Soon after the first H-R diagrams were developed, astronomers realized their importance in interpreting stellar evolution. Just as with living things, a star is born, ages, and dies. Owing to the fact that almost 90 percent of the stars lie on the main sequence, we can be relatively certain that stars spend most of their active years as main-sequence stars. Only a few percent are giants, and perhaps 10 percent are white dwarfs.

Next, we will examine the nature of interstellar matter, and then return to the topic of stellar evolution.

Interstellar Matter

Between existing stars is "the vacuum of space." However, it is far from being a pure vacuum, for it is populated with accumulations of dust and gases. The name applied to these concentrations of interstellar matter is **nebula** (Latin for "cloud"). If this interstellar matter is close to a very hot (blue) star it will glow and is called a **bright nebula**. The two main types of bright nebulae are known as *emission nebulae* and *reflection nebulae*.

Emission nebulae are gaseous masses that consist largely of hydrogen. They absorb *ultraviolet radiation* emitted by an embedded or nearby hot star. Because these gases are under very low pressure, they reradiate, or emit, this energy as *visible light*. This conversion of ultraviolet light to visible light is known as *fluorescence*, an effect you observe daily in fluorescent lights. A well-known emission nebula easily seen with binoculars is the sword of the hunter in the constellation Orion (Figure 16.6).

Figure 16.6 The Orion Nebula is a well-known emission nebula. Bright enough to be seen by the naked eye, the Orion Nebula is located in the sword of the hunter in the constellation of the same name. *(Courtesy of National Optical Astronomy Observatories)*

Figure 16.7 A faint blue reflection nebula, in the Pleiades star cluster, is caused by the reflection of starlight from dust in the nebula. The Pleiades star cluster, just visible to the naked eye in the constellation Taurus, is spectacular when viewed through binoculars or a small telescope. *(Palomar Observatories/California Institute of Technology (Caltech))*

Reflection nebulae, as the name implies, merely reflect the light of nearby stars (Figure 16.7). Reflection nebulae are thought to be composed of rather dense clouds of large particles called **interstellar dust**. This view is supported by the fact that atomic gases with low densities could not reflect light sufficiently to produce the glow observed.

When a dense cloud of interstellar material is not close enough to a bright star to be illuminated, it is referred to as a **dark nebula**. Exemplified by the Horsehead Nebula in Orion, dark nebulae appear as opaque objects silhouetted against a bright background (Figure 16.8). Dark nebulae can also easily be seen as starless regions—"holes in the heavens"—when viewing the Milky Way.

Although nebulae appear very dense, they actually consist of very thinly scattered matter. Because of their enormous size, however, the total mass of rarefied particles and molecules may be many times that of the Sun. Interstellar matter is of great interest to astronomers because it is from this material that stars and planets are formed.

Stellar Evolution

The idea of describing how a star is born, ages, and then dies may seem a bit presumptuous, for many of these objects have life spans that surely exceed billions of years. However, by studying stars of different ages,

astronomers have been able to piece together a plausible model for stellar evolution.

The method that was used to create this model is analogous to what an alien being, upon reaching Earth, might do to determine the developmental stages of human life. By examining a large number of humans, this stranger would be able to observe the birth of human life, the activities of children and adults, and the death of the elderly. From this information, the alien would attempt to put the stages of human development into their proper sequence. Based on the relative abundance of humans in each stage of development, it would even be possible to conclude that humans spend more of their lives as adults than as toddlers. In a similar fashion, astronomers have pieced together the story of the stars.

Simply, stars exist because of gravity. The mutual gravitational attraction of particles in a thin, gaseous cloud causes the cloud to collapse. As the cloud is squeezed to unimaginable pressures, its temperature rises, igniting its nuclear furnace, and a star is born. A star is a ball of very hot gases, caught between the opposing forces of gravity trying to contract it and thermal nuclear energy trying to expand it. Eventually, all of a star's nuclear fuel will be exhausted and gravity takes over, collapsing the stellar remnant into a small, dense body.

Star Birth

The birthplaces of stars are dark, cool, interstellar clouds, which are comparatively rich in dust and gases (Figure 16.9). In the neighborhood of the Milky Way, these gaseous clouds consist of 92 percent hydrogen, 7 percent helium, and less than 1 percent of the remaining heavier elements. By some mechanism not yet fully understood, these thin gaseous clouds become concentrated enough to begin to gravitationally contract. One proposal to explain the triggering of stellar formation is a shock wave traveling from a catastrophic explosion (supernova) of a nearby star. But, regardless of the force that initiates the concentration of interstellar matter, once it is accomplished, mutual gravitational attraction of the particles squeeze the cloud, pulling every particle toward the center. As the cloud shrinks, gravitational energy (potential energy) is converted into energy of motion, or heat energy.

The initial contraction spans a million years or so. With the passage of time, the temperature of this gaseous body slowly rises, eventually reaching a temperature sufficiently high to cause it to radiate energy

Figure 16.8 The Horsehead Nebula, a dark nebula in a region of glowing nebulosity in Orion. *(© Anglo-Australian Observatory/ Royal Observatory Edinburgh)*

from its surface in the form of long-wavelength red light. Because this large red object is not hot enough to engage in nuclear fusion, it is not yet a star. The name **protostar** is applied to these bodies.

Protostar Stage

During the protostar phase, gravitational contraction continues, slowly at first, then much more rapidly (Figure 16.9). This collapse causes the core of the developing star to heat much more intensely than the outer envelope. When the core has reached at least 10 million K, the pressure within is so great that groups of four hydrogen nuclei are fused together into single helium nuclei. Astronomers refer to this nuclear reaction as **hydrogen burning** because an enormous amount of energy is released. However, keep in mind that thermonuclear "burning" is not burning in the usual chemical sense.

Heat from hydrogen fusion causes the stellar gases to increase their motion. This in turn results in an increase in the outward gas pressure. At some point, this outward pressure exactly balances the inward force of gravity. When this balance is reached, the star becomes a stable *main-sequence star* (Figure 16.9). Stated another way, a stable main-sequence star is balanced between two forces: *gravity*, which is trying to squeeze it into a smaller sphere, and *gas pressure*, which is trying to expand it.

Main-Sequence Stage

From this point in the evolution of a main-sequence star until its death, the internal gas pressure struggles to offset the unrelenting force of gravity. Typically, hydrogen burning continues for a few billion years and provides the outward pressure required to support the star from gravitational collapse.

419

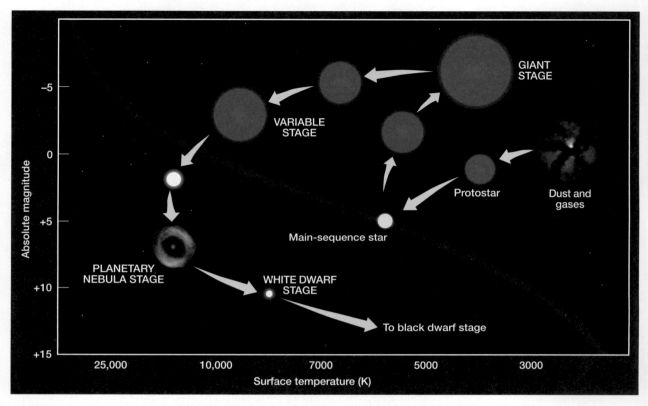

Figure 16.9 Diagram of stellar evolution on H-R diagram for a star about as massive as the Sun.

Different stars age at different rates. Hot, massive blue stars radiate energy at such an enormous rate that they substantially deplete their hydrogen fuel in only a few million years. By contrast, the very smallest (red) main-sequence stars may remain stable for hundreds of billions of years. A yellow star, such as the Sun, remains a main-sequence star for about 10 billion years. Because the solar system is about 5 billion years old, it is comforting to know that our Sun will remain stable for another 5 billion years.

An average star spends 90 percent of its life as a hydrogen-burning, main-sequence star. Once the hydrogen fuel in the star's core is depleted, it evolves rapidly and dies. However, with the exception of the least-massive (red) stars, a star can delay its death by burning other fuels and becoming a giant.

Red Giant Stage

The evolution to the red giant stage results because the zone of hydrogen burning continually migrates outward, leaving behind an inert helium core. Eventually, all of the hydrogen in the star's core is consumed. While hydrogen fusion is still progressing in the star's outer shell, no fusion is taking place in the core. Without a source of energy, the core no longer has

enough pressure to support itself against the inward force of gravity. As a result, the core begins to contract.

Although the core cannot generate nuclear energy, it does grow hotter by converting gravitational energy into heat energy. Some of this energy is radiated outward, initiating a more vigorous level of hydrogen fusion in the star's outer shell. This energy in turn heats and enormously expands the star's outer envelope, producing a giant body hundreds to thousands of times its main-sequence size (Figure 16.9).

As the star expands, its surface cools, which explains the star's reddish appearance. Eventually the star's gravitational force will stop this outward expansion. Once again, the two opposing forces, gravity and gas pressure, will be in balance, and this gaseous mass will be a stable but much larger star. Some red giants overshoot the equilibrium point and rebound like an overextended spring. Such stars continue to oscillate in size, becoming variable stars.

While the envelope of a red giant expands, the core continues to collapse and heat until it reaches 100 million K. At this incredible temperature, it is hot enough to initiate a nuclear reaction in which helium is converted to carbon. Thus, a red giant consumes both hydrogen and helium to produce energy. In stars more massive than our Sun, still other thermonuclear reactions occur that generate all the elements on the periodic table up to number 26, iron. Nuclear "burning" of elements heavier than iron requires an additional source of energy to keep the reaction progressing. Hence, these elements are not produced in ordinary stars.

Eventually, all the usable nuclear fuel in these giants will be consumed. Our Sun, for example, will spend less than a billion years as a giant, and the more massive stars will pass through this stage even more rapidly. The force of gravity will again control the star's destiny as it squeezes the star into the smallest, most dense piece of matter possible.

Burnout and Death

Most of the events of stellar evolution discussed thus far are well documented. What happens to a star after the red-giant phase is more speculative. We do know that a star, regardless of its size, must eventually exhaust all of its usable nuclear fuel and collapse in response to its immense gravitational force. With this in mind, we will now consider the final stage of stars in three different mass categories.

Death of Low-Mass Stars   Stars less than one-half the mass of the Sun (0.5 solar mass) consume their fuel at a comparatively slow rate (Figure 16.10A). Consequently these small, *cool red stars* may remain

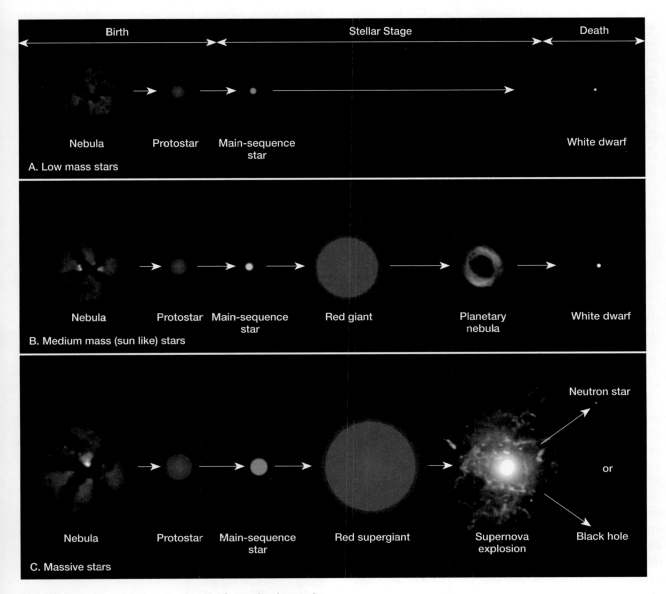

Figure 16.10 The evolutionary stages of stars having various masses.

on the main sequence for up to 100 billion years. Because the interior of a low-mass star never attains sufficiently high temperatures and pressures to fuse helium, its only energy source is hydrogen fusion. Thus, low-mass stars never evolve to become bloated red giants. Rather, they remain as stable main-sequence stars until they consume their hydrogen fuel and collapse into a hot, dense *white dwarf*. As we shall see, white dwarfs are small, compact objects unable to support nuclear burning.

Death of Medium-Mass (Sunlike) Stars Main-sequence stars with masses ranging between half that of the Sun and three times that of the Sun evolve in essentially the same way (Figure 16.10B). During their giant phase, sunlike stars fuse hydrogen and helium fuel at an accelerated rate. Once this fuel is exhausted, these stars (like low-mass stars) collapse into an Earth-sized body of great density—a white dwarf. The density of a white dwarf is as great as physics will allow short of destroying protons and electrons.

The gravitational energy supplied to a collapsing white dwarf is reflected in its high surface temperature. However, without a source of nuclear energy, a white dwarf becomes cooler and dimmer as it continually radiates its remaining thermal energy into space.

During their collapse from red giants to white dwarfs, medium-mass stars are thought to cast off their bloated outer atmosphere, creating an expanding spherical cloud of gas. The remaining hot, central white dwarf heats the gas cloud, causing it to glow. These often beautiful, gleaming spherical clouds are called **planetary nebulae**. A good example of a planetary nebula is the Helix nebula in the constellation Aquarius (Figure 16.11). This nebula appears as a ring because our line of sight through the center traverses less gaseous material than at the nebula's edge. It is, nevertheless, spherical in shape.

Death of Massive Stars In contrast to sunlike stars, which expire gracefully, stars exceeding 3 solar masses have relatively short life spans and terminate in a brilliant explosion called a **supernova** (Figure 16.10C). During a supernova event, a star becomes millions of times brighter than its prenova stage. If one of the nearest stars to Earth produced such an outburst, its brilliance would surpass that of the Sun. Supernovae are rare; none have been observed in our galaxy since the advent of the telescope, although Tycho Brahe and Galileo each recorded one about 30 years apart. An even larger supernova was recorded in A.D. 1054 by the Chinese. Today, the remnant of this great outburst is the Crab Nebula, shown in Figure 16.12.

A supernova event is thought to be triggered when a massive star consumes most of its nuclear fuel. Without a heat engine to generate the gas pressure required to balance its immense gravitational field, it collapses. This implosion is of cataclysmic proportion, resulting in a shock wave that moves out from the star's interior. This energetic shock wave destroys the star and blasts the outer shell into space, generating the supernova event.

Theoretical work predicts that during a supernova, the star's interior condenses into a very hot object, possibly no larger than 20 kilometers in diameter. These incomprehensibly dense bodies have been named *neutron stars*. Some supernovae events are thought to produce even smaller, and most intriguing, objects called *black holes*. We will consider the nature of neutron stars and black holes in the Stellar Remnants section.

H-R Diagrams and Stellar Evolution

The Hertzsprung-Russell (H-R) diagrams have been very helpful in formulating and testing models of stellar evolution. They are also useful for illustrating the changes that take place in an individual star

Figure 16.11 The Helix Nebula, the nearest planetary nebula to our solar system. A planetary nebula is the ejected outer envelope of a sunlike star that formed during the star's collapse from a red giant to a white dwarf. (© *Anglo-Australian Observatory, photography by David Malin*)

Figure 16.12 Crab Nebula in the constellation Taurus: the remains of the supernova of A.D. 1054. *(UCO/ Lick Observatory Image)*

during its life span. Figure 16.9 (p. 420) shows on an H-R diagram the evolution of a star about the size of our Sun. Keep in mind that the star does not physically move along this path, but rather that its position on the H-R diagram represents the color (temperature) and absolute magnitude (brightness) of the star at various stages in its evolution.

For example, on the H-R diagram, a protostar would be located to the right and above the main sequence (Figure 16.9). It is formed to the right

because of its relatively cool surface temperature (red color), and above because it would be more luminous than a main-sequence star of the same color, a fact attributable to its large size. Careful examination of Figure 16.9 should help you visualize the evolutionary changes experienced by a star the size of our Sun. In addition, Table 16.3 provides a summary of the evolutionary history of stars having various masses.

Stellar Remnants

Eventually, all stars consume their nuclear fuel and collapse into one of three final states—white dwarf, neutron star, or black hole. Although different in some ways, these small, compact objects are all composed of incomprehensibly dense material and all have extreme surface gravity.

White Dwarfs

White dwarfs are extremely small stars with densities greater than any known terrestrial material. It is believed that white dwarfs once were low-mass or medium-mass stars whose internal heat was able to keep these gaseous bodies from collapsing under their own gravitational force.

Although some white dwarfs are no larger than Earth, the mass of such a dwarf can equal 1.4 times that of the Sun. Thus, their densities may be a million times greater than water. A spoonful of such matter would weigh several tons. Densities this great are possible only when electrons are displaced inward from their regular orbits, around an atom's nucleus, allowing the atoms to take up less than the "normal" amount of space. Material in this state is called **degenerate matter**.

Table 16.3 ■ Summary of evolution for stars of various masses

Initial mass of interstellar cloud (Sun = 1)	Main-sequence stage	Giant phase	Evolution after giant phase	Terminal state (final mass*)
0.001	None (Planet)	No	None	Planet (0.001)
0.1	Red	No	None	White dwarf (0.1)
1–3	Yellow	Yes	Planetary nebula	White dwarf (<1.4)
6	White	Yes	Supernova	Neutron star (1.4–3)
20	Blue	Yes Supergiant	Supernova	Black hole (> 3.0)

*These mass numbers are estimates.

In degenerate matter, the atoms have been squeezed together so tightly that the electrons are displaced much nearer to the nucleus. Degenerate matter uses electrical repulsion rather than molecular motion to support itself from total collapse. Although atomic particles in degenerate matter are much closer together than in normal Earth matter, they still are not packed as tightly as possible. Stars made of matter that has an even greater density are thought to exist.

As a star contracts into a white dwarf, its surface becomes very hot, sometimes exceeding 25,000 K. Even so, without a source of energy, it can only become cooler and dimmer. Although none have been observed, the terminal stage of a white dwarf must be a small, cold, nonluminous body called a *black dwarf.*

Neutron Stars

A study of white dwarfs produced what might at first appear to be a surprising conclusion. The smallest white dwarfs are the most massive, and the largest are the least massive. The explanation for this is that a more massive star, because of its greater gravitational force, is able to squeeze itself into a smaller, more densely packed object than can a less massive star. Thus, the smallest white dwarfs were produced from the collapse of larger, more massive stars than were the larger white dwarfs.

This conclusion led to the prediction that stars smaller and more massive than white dwarfs must exist. Named **neutron stars**, these objects are thought to be the remnants of supernova events. In a white dwarf, the electrons are pushed close to the nucleus, whereas in a neutron star, the electrons are forced to combine with protons to produce neutrons (hence the name). If Earth were to collapse to the density of a neutron star, it would have a diameter equivalent to the length of a football field. A pea-sized sample of this matter would weigh 100 million tons. This is approximately the density of an atomic nucleus; thus, neutron stars can be thought of as large atomic nuclei.

During a supernova implosion, the envelope of the star is ejected (Figure 16.13), while the core collapses into a very hot neutron star about 20 kilometers (12.4 miles) in diameter. Although neutron stars have high surface temperatures, their small size would greatly limit their luminosity. Consequently, locating one visually would be extremely difficult.

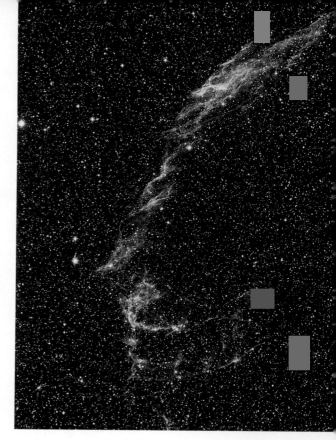

Figure 16.13 Veil Nebula in the constellation Cygnus is the remnant of an ancient supernova implosion. *(Palomar Observatories/California Institute of Technology (Caltech))*

However, theory predicts that a neutron star would have a very strong magnetic field. Further, as a star collapses, it will rotate faster, for the same reason ice skaters rotate faster as they pull in their arms. If the Sun were to collapse to the size of a neutron star, it would increase its rate of rotation from once every 25 days to nearly 1000 times per second. Radio waves generated by these rotating stars would be concentrated into two narrow zones that would align with the star's magnetic poles. Consequently, these stars would resemble a rapidly rotating beacon emitting strong radio waves. If Earth happened to be in the path of these beacons, the star would appear to blink on and off, or pulsate, as the waves swept past.

In the early 1970s, a source that radiates short pulses of radio energy, called a **pulsar** (pulsating radio source), was discovered in the Crab Nebula. Visual inspection of this radio source revealed it to be a small star centered in the nebula. The pulsar found in the Crab Nebula is undoubtedly the remains of the supernova of A.D. 1054 (see Figure 16.12). Thus, the first neutron star had been discovered.

Black Holes

Are neutron stars made of the most dense materials possible? No. During a supernova event, remnants of stars greater than 3 solar masses apparently collapse into objects even smaller and denser than neutron stars. Even though these objects would be very hot, their surface gravity would be so immense that even light could not escape the surface. Consequently, they would literally disappear from sight. These incredible bodies have appropriately been named **black holes**. Anything that moved too near a black hole would be swept in by its irresistible gravity and devoured forever.

How can astronomers find an object whose gravitational field prevents the escape of all matter and energy? One strategy is to seek evidence of matter being rapidly swept into a region of apparent nothingness. Theory predicts that as matter is pulled into a black hole, it should become very hot and emit a flood of X-rays before being engulfed. Because isolated black holes would not have a source of matter to engulf, astronomers first looked at binary star systems.

A likely candidate for a black hole is Cygnus X-1, a strong X-ray source in the constellation Cygnus. In this case, the X-ray source can be observed orbiting a supergiant companion with a period of 5.6 days. It appears that gases are pulled from this companion and spiral into the disk-shaped structure around the black hole (Figure 16.14). The result is a stream of x-rays. Because X-rays cannot penetrate our atmosphere efficiently, the existence of black holes was not confirmed until recently.

The first X-ray sources were discovered in 1971 by detectors on satellites. Cygnus X-1 is such a source.]

The Milky Way Galaxy

On a clear, moonless night away from city lights, you can see a truly marvelous sight—our own Milky Way galaxy (Figure 16.15). With his telescope, Galileo discovered that this band of light was produced by countless individual stars that the unaided eye is unable to resolve. Today, we realize that the Sun is actually a part of this vast system of stars, which number about 100 billion. The "milky" appearance of our galaxy results because the solar system is located within the flat *galactic disk*. Thus, when it is viewed from the "inside," a higher concentration of stars appears in the direction of the galactic plane than in any other direction. You can see this in the edge-on view in Figure 16.16B.

When astronomers began telescopically to survey the stars located along the plane of the Milky Way, it appeared that equal numbers lay in every direction. Could Earth actually be at the center of the galaxy? A better explanation was advanced. Imagine that the trees in an enormous forest represent the stars in the galaxy. After hiking into this forest a short distance, you look around. What you see is an equal number of trees in every direction. Are you really in the center of the forest? Not necessarily; anywhere in the forest, except at the very edge, you will seem to be in the middle.

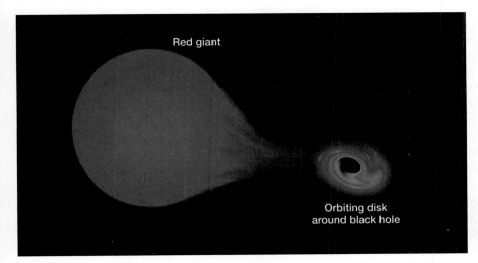

Red giant

Orbiting disk
around black hole

Figure 16.14 This illustration shows how astronomers believe a binary pair (red giant/black hole) might function.

Figure 16.15 Panorama of our galaxy, the Milky Way. Notice the dark bands caused by the presence of interstellar dark nebulae. *(Lund Observatory photograph)*

Structure of the Milky Way Galaxy

Attempts to inspect the Milky Way visually are hindered by the large quantities of interstellar matter that lie in our line of sight. Nevertheless, with the aid of radio telescopes, the gross structure of our galaxy has been determined. The Milky Way is a rather large spiral galaxy whose disk is about 100,000 light-years wide and about 10,000 light-years thick at the nucleus (Figure 16.16). As viewed from Earth, the center of the galaxy lies beyond the constellation Sagittarius.

Radio telescopes reveal the existence of at least three distinct *spiral arms*, with some showing splintering (Figure 16.17). The Sun is positioned in one of these arms about two-thirds of the way from the center, at a distance from the hub of about 30,000 light-years. The stars in the arms of the Milky Way rotate around the *galactic nucleus*, with the most outward ones moving the slowest, such that the ends of the arms appear to trail. The Sun and the arm it is in require about 200 million years for each orbit around the nucleus.

Surrounding the galactic disk is a nearly spherical *halo* made of very tenuous gas and numerous globular clusters. These star clusters do not participate in the rotating motion of the arms, but rather have their own orbits that carry them through the disk. Although some clusters are very dense, they pass among the stars of the arms with plenty of room to spare.

Galaxies

In the mid-1700s, German philosopher Immanuel Kant proposed that the telescopically visible fuzzy patches of light scattered among the stars were actually distant galaxies like the Milky Way. Kant described them as "island universes." Each galaxy, he believed, contained billions of stars and, as such, was a universe in itself. The weight of opinion, however, favored the hypothesis that they were dust and gas clouds (nebulae) within our galaxy.

This matter was not resolved until the 1920s, when American astronomer Edwin Hubble was able to locate, within one of these fuzzy patches, some unique stars that are known to be intrinsically very bright. Because these very bright stars appeared only very faintly in the telescope, Hubble believed they must lie outside the Milky Way.

This fuzzy patch, which lies over a million light-years away, was named the Great Galaxy in

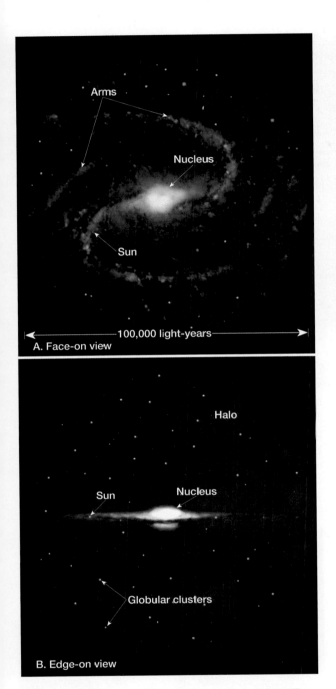

Figure 16.16 Structure of the visible portion of the Milky Way galaxy.

Figure 16.17 If the Milky Way were photographed from a distance, it might appear like the spiral galaxy NGC 2997. (© *Anglo-Australian Observatory, photography by David Malin*)

Three Types of Galaxies

From the hundreds of billions of galaxies, three basic types have been identified: spiral, elliptical, and irregular.

The Milky Way and the Great Galaxy in Andromeda are examples of fairly large **spiral galaxies** (Figure 16.19). Andromeda can be seen with the unaided eye as a fuzzy fifth-magnitude object. Typically, spiral galaxies are disk-shaped with a somewhat greater concentration of stars near their centers, but there are numerous variations. Viewed broadside, arms are often seen extending from the central nucleus and sweeping gracefully away. The outermost stars of these arms rotate most slowly, giving the galaxy the appearance of a fireworks pinwheel.

The most abundant group, making up 60 percent of the total, is the **elliptical galaxies**. These are generally smaller than spiral galaxies. Some are so much smaller, in fact, that the term *dwarf* has been applied. Because these dwarf galaxies are not visible at great distances, a survey of the sky reveals more of the conspicuous large spiral galaxies. Although most elliptical galaxies are small, the very largest known galaxies (200,000 light-years in diameter) are also elliptical. As their name implies, elliptical galaxies have an ellipsoidal shape that ranges

Andromeda (Figure 16.18). Hubble had extended the universe far beyond the limits of our imagination, to include hundreds of billions of galaxies, each containing hundreds of billions of stars. It has been said that a million galaxies are found in that portion of the sky bounded by the cup of the Big Dipper. There are more stars in the heavens than grains of sand in all the beaches on Earth.

Figure 16.18 Great Galaxy, a spiral galaxy, in the constellation Andromeda. The two bright spots to the left and right are dwarf elliptical galaxies. *(Palomar Observatories/California Institute of Technology(Caltech))*

to nearly spherical, and they lack spiral arms. The two dwarf companions of Andromeda shown in Figure 16.18 are elliptical galaxies.

Only 10 percent of the known galaxies lack symmetry and are classified as **irregular galaxies**.

The best-known irregular galaxies, the Large and Small Magellanic Clouds in the Southern Hemisphere, are easily visible with the unaided eye. Named after the explorer Ferdinand Magellan, who observed them when he circumnavigated Earth in 1520, they are our nearest galactic neighbors—only 150,000 light-years away.

One of the major differences among the galactic types is the age of the stars they contain. Irregular galaxies consist mostly of young stars, whereas elliptical galaxies contain old stars. The Milky Way and other spiral galaxies consist of both young and old stars, with the youngest located in the arms.

Galactic Clusters

Once astronomers discovered that stars were associated in groups, they set out to determine whether galaxies also were grouped or just randomly distributed throughout the universe. They found that, like stars, galaxies are grouped in **galactic clusters** (Figure 16.20). Some abundant clusters contain thousands of galaxies. Our own, called the **Local Group**, contains at least 28 galaxies. Of these, 3 are spirals, 11 are irregulars, and 14 are ellipticals. Galactic clusters also reside in huge swarms called *superclusters*. From visual observations, it appears that superclusters may be the largest entities in the universe.

Figure 16.19 Two views illustrating the idealized structure of spiral galaxies. *(Courtesy of Hansen Planetarium/U.S. Naval Observatory)*

Doppler in 1842 and is called the **Doppler effect**. The reason for the difference in pitch is that it takes time for the wave to be emitted. If the source of the wave is moving away, the beginning of the wave is emitted nearer to you than the end of the wave, effectively "stretching" the wave. This gives it a longer wavelength (Figure 16.21). The opposite is true for an approaching source.

In the case of light, when a source is moving away, its light appears redder than it actually is, because its waves appear lengthened. Objects approaching have their light waves shifted toward the blue (shorter wavelength). Therefore, the Doppler effect reveals whether Earth and another celestial body are approaching or leaving one another. In addition, the amount of shift allows us to calculate the *rate* at which the relative movement is occurring. Large Doppler shifts indicate higher velocities; smaller Doppler shifts indicate slower velocities.

Expanding Universe

One of the most important discoveries of modern astronomy was made in 1929 by Edwin Hubble. Observations completed several years earlier revealed that most galaxies have Doppler shifts toward the red end of the spectrum. Recall that red shift occurs because the light waves are "stretched," indicating that Earth and the source are moving away from each other. Hubble set out to explain the predominance of red shift.

Hubble realized that dimmer galaxies were probably farther away than brighter galaxies. Thus, he tried to determine if there is a relation between the distances to galaxies and their red shifts. Using estimated distances based on relative brightness and the observed Doppler red shifts, Hubble discovered that galaxies that exhibit the greatest red shifts are the most distant.

Figure 16.20 Numerous galaxies grouped in the constellation Hercules. *(Courtesy of National Optical Astronomy Observatories)*

Red Shifts

You probably have noticed the change in pitch of a car horn or ambulance siren as it passes by. When it is approaching, the sound seems to have a higher-than-normal pitch, and when it is moving away, the pitch sounds lower than normal. This effect, which occurs for all wave motion, including sound and light waves, was first explained by Christian

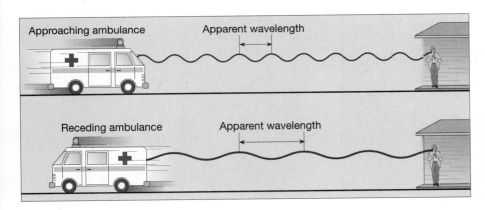

Figure 16.21 The Doppler effect, illustrating the apparent lengthening and shortening of wavelengths caused by the relative motion between a source and an observer.

A consequence of the universal red shift is that it predicts that most galaxies (except for a few nearby) are receding from us. Recall that the amount of Doppler red shift is dependent on the velocity at which the object is moving away. Greater red shifts indicate faster recessional velocities. Because more distant galaxies have greater red shifts, Hubble concluded that they must be retreating from us at greater velocities. This idea is currently called **Hubble's law** and states that galaxies are receding from us at a speed that is proportional to their distance.

Hubble was surprised at this discovery because it implied that the most distant galaxies are moving away from us many times faster than those nearby. What type of cosmological theory can explain this fact? It was soon realized that an *expanding universe* can adequately account for the observed red shifts.

To help visualize the nature of this expanding universe, we will employ a popularly used analogy. Imagine a loaf of raisin bread dough that has been set out to rise for a few hours. As the dough doubles in size, so does the distance between all of the raisins. However, the raisins that were originally farther apart traveled a greater distance in the same time span than those located closer together. We therefore conclude that, in an expanding universe, as in our analogy, those objects located farther apart recede at greater velocities.

Another feature of the expanding universe can be demonstrated using the raisin bread analogy. No matter which raisin you select, it will move away from all the other raisins. Likewise, no matter where one is located in the universe, every other galaxy (except those in the same cluster) will be receding. Edwin Hubble had indeed advanced our understanding of the universe. The Hubble Space Telescope is named in his honor.

The Big Bang

The universe—did it have a beginning? Will it have an end? Cosmologists are trying to answer these questions, and that makes them a rare breed.

First and foremost, any viable theory regarding the origin of the universe must account for the fact that all galaxies (except for the very nearest) are moving away from us. Because all galaxies appear to be moving away from Earth, is our planet in the center of the universe? Probably not, because if we are not even in the center of our own solar system, and our solar system is not even in the center of the galaxy, it seems unlikely that we could be in the center of the universe.

A more probable explanation exists: Imagine a balloon with paper-punch "dots" glued to its surface. When the balloon is inflated, each dot spreads apart from every other dot. Similarly, if the universe is expanding, every galaxy would be moving away from every other galaxy.

This concept of an expanding universe led to the widely accepted **Big Bang** theory. According to this theory, the entire universe was at one time confined to a dense, hot, supermassive ball. Then, about 20 billion years ago, a cataclysmic explosion occurred, hurling this material in all directions. The Big Bang marks the inception of the universe; all matter and space were created at this instant. The ejected masses of gas cooled and condensed, forming the stellar systems we now observe fleeing from their birthplace.

Will the stars eventually dim from view and invisible galaxies travel on forever? It has been suggested that after a certain point, perhaps 20 billion years in the future, the galaxies will slow and eventually stop their outward flight. Gravitational contraction would follow. The galaxies would then collide and coalesce, and a new fireball would be born. For this event to occur, the universe must have an average density of one atom for every cubic meter of space. But present estimates indicate that the concentration of matter in the universe is far less than this amount, so this hypothesis remains unsubstantiated.

However, this subject does not end here. It has been proposed that heretofore undetected matter exists in large quantities in the universe. For example, numerous black holes may occupy many of the voids in the universe. If this is true, the galaxies could, in fact, collapse upon themselves.

"Absence of evidence is not evidence of absence."

—Anonymous

The Chapter in Review

1. One method for determining the distance to a star is to use a measurement called *stellar parallax*, the extremely slight back-and-forth shifting in a nearby star's apparent position due to the orbital motion of Earth. *The farther away a star is, the less its parallax.* A unit used to express stellar distance is the *light-year*, which is the distance light travels in a year—about 9.5 trillion kilometers (5.8 trillion miles).

2. The intrinsic properties of stars include *brightness, color, temperature, mass,* and *size.* Three factors control the brightness of a star as seen from Earth: how big it is, how hot it is, and how far away it is. *Apparent magnitude* is how bright a star appears from Earth. *Absolute magnitude* is the "true" brightness if a star were at a standard distance of about 32.6 light-years. Color reveals a star's temperature. Very hot stars (surface temperatures above 30,000 K) appear blue; red stars are much cooler (surface temperatures generally less than 3000 K). Stars with surface temperatures between 5000 and 6000 K appear yellow, like the Sun. *Binary stars* are two stars kept together by gravitational attraction and revolving around a common center of mass. This center is used to determine the mass of the individual stars.

3. A *Hertzsprung-Russell (H-R) diagram* is constructed by plotting the absolute magnitudes and temperatures of stars on a graph. A great deal about the sizes of stars can be learned from H-R diagrams. Stars that fall in the upper-right position of an H-R diagram are called *giants*, a luminous star of large radius. *Supergiants* are very large. Very small *white dwarf* stars fall in the lower-central portion of an H-R diagram. Ninety percent of all stars, called *main-sequence stars*, fall in a band that runs from the upper-left corner to the lower-right corner of an H-R diagram.

4. New stars are born out of enormous accumulations of dust and gases, called *nebulae*, that are scattered between existing stars. A *bright nebula* glows because the matter is close to a very hot (blue) star. The main types of bright nebulae are known as *emission nebulae* (which derive their visible light from the fluorescence of ultraviolet light from a star in or near the nebula) and *reflection nebulae* (relatively dense dust clouds in interstellar space that reflect the light of nearby stars). When a nebula is not close enough to a bright star to be illuminated, it is a *dark nebula*.

5. Stars are born when their nuclear furnaces begin operating under the unimaginable pressures and temperatures in collapsing nebulae. New stars not yet hot enough for nuclear fusion are called *protostars*. When collapse causes the core of a protostar to reach a temperature of at least 10 million K, the fusion of hydrogen nuclei into helium begins in a process called *hydrogen burning*. The opposing forces acting on a star are *gravity*, which tries to contract it, and *gas pressure (thermal nuclear energy)*, which tries to expand it. When the two forces are balanced, the star becomes a stable *main-sequence star*. When the hydrogen in a star's core is consumed, its outer envelope expands enormously to form a *red giant* star, hundreds to thousands of times larger than its main-sequence size. When all the usable nuclear fuel in these giants is exhausted and gravity takes over, the stellar remnant collapses into a small dense body.

6. The *final fate of a star is determined by its mass.* Stars with less than one-half the mass of the Sun collapse into hot, dense *white dwarf* stars. Medium-mass stars (between 0.5 and 3.0 times the mass of the Sun) become red giants, collapse, and end up as white dwarf stars, often surrounded by expanding spherical clouds of glowing gas called *planetary nebulae*. Stars more than three times the mass of the Sun terminate in a brilliant explosion called a *supernova*. Supernovae events can produce small, extremely dense *neutron stars*, composed entirely of subatomic particles called neutrons; or even smaller and more dense *black holes*, objects that have such immense gravity that light cannot escape their surface.

7. The *Milky Way galaxy* is a large, disk-shaped, *spiral galaxy* about 100,000 light-years wide and about 10,000 light-years thick at the center. There are three distinct *spiral arms* of stars, with some showing splintering. The Sun is positioned in one of these arms about two-thirds of the way from the galactic center, a distance of about 30,000 light-years. Surrounding the galactic disk is a nearly spherical halo made of very tenuous gas and numerous *globular clusters* (nearly spherical groups of densely packed stars).

8. The various types of galaxies include: (1) *spiral galaxies*, which are typically disk-shaped with a somewhat greater concentration of stars near their centers and often having arms of stars extending from their central nucleus, (2) *elliptical galaxies*, the most abundant type, which have an ellipsoidal shape that ranges to nearly spherical, and lack spiral arms, and (3) *irregular galaxies*, which lack symmetry and account for only 10 percent of the known galaxies.

9. The *Doppler effect* is the apparent change in wavelength of radiation caused by the motions of the source and the observer. By applying Doppler effect to the light of galaxies, galactic motion can be determined. Most galaxies have Doppler shifts toward the red end of the spectrum, indicating increasing distance. The amount of Doppler shift is dependent on the velocity at which the object is moving. Because the most distant galaxies have the greatest red shifts, Edwin Hubble concluded in the early 1900s that they were retreating from us faster than were more nearby galaxies. It was soon realized that an *expanding universe* can adequately account for the observed red shifts.

10. The belief in the expanding universe led to the widely accepted *Big Bang theory*. According to this theory, the entire universe was at one time confined in a dense, hot, supermassive concentration. About 20 billion years ago, a cataclysmic explosion hurled this material in all directions, creating all matter and space. Eventually the ejected masses of gas cooled and condensed, forming the stellar systems we now observe fleeing from their place of origin.

Key Terms

absolute magnitude (p. 414)

apparent magnitude (p. 413)

Big Bang (p. 430)

black hole (p. 425)

bright nebula (p. 417)

dark nebula (p. 418)

degenerate matter (p. 423)

Doppler effect (p. 429)

elliptical galaxy (p. 427)

emission nebula (p. 417)

galactic cluster (p. 428)

Hertzsprung-Russell (H-R) diagram (p. 415)

Hubble's law (p. 430)

hydrogen burning (p. 419)

interstellar dust (p. 418)

irregular galaxy (p. 428)

light-year (p. 412)

Local Group (p. 428)

magnitude (p. 413)

main-sequence stars (p. 415)

nebula (p. 417)

neutron star (p. 424)

planetary nebula (p. 422)

protostar (p. 419)

pulsar (p. 424)

red giant (p. 416)

reflection nebula (p. 418)

spiral galaxy (p. 427)

stellar parallax (p. 412)

supergiant (p. 417)

supernova (p. 422)

white dwarf (p. 417)

Questions for Review

1. How far away in light-years is our nearest stellar neighbor, Proxima Centauri? Convert your answer to kilometers.

2. What is the most basic method of determining stellar distances?

3. Explain the difference between a star's apparent and absolute magnitudes. Which one is an intrinsic property of a star?

4. What is the ratio of brightness between a twelfth-magnitude and fifteenth-magnitude star?

5. What information about a star can be determined from its color?

6. What color are the hottest stars? Medium-temperature stars? Coolest stars?

7. Which property of a star can be determined from binary star systems?

8. Make a generalization relating the mass and luminosity of main-sequence stars.

9. The disk of a star cannot be resolved telescopically. Explain the method that astronomers have used to estimate the size of stars.

10. Where on an H-R diagram does a star spend most of its lifetime?

11. How does the Sun compare in size and brightness to other main-sequence stars?

12. Why is interstellar matter important to stellar evolution?

13. Compare a bright nebula and a dark nebula.

14. What element is the fuel for main-sequence stars? Red giants?

15. What causes a star to become a giant?

16. Why are less massive stars thought to age more slowly than more massive stars, even though they have much less "fuel"?

17. Enumerate the steps thought to be involved in the evolution of sunlike stars.

18. What is the final state of a low-mass (red) main-sequence star?

19. What is the final state of a medium-mass (sunlike) star?

20. How do the "lives" of the most massive stars end? What are the two possible products of this event?

21. Describe the general structure of the Milky Way.

22. Compare the three general types of galaxies.
23. Explain why astronomers consider elliptical galaxies more abundant than spiral galaxies, even though more spiral galaxies have been sighted.
24. How did Edwin Hubble determine that the Great Galaxy in Andromeda is located beyond our galaxy?
25. What evidence supports the Big Bang theory?

Testing What You Have Learned

Multiple-Choice Questions

1. The unit of stellar distance defined as the distance light travels in a year, about 9.5 trillion kilometers (5.8 trillion miles), is called a(n) _____
   a. stellar unit.
   b. megameter.
   c. astronomical unit.
   d. angstrom.
   e. none of the above.

2. The measure of a star's brightness is called its _____
   a. radiance.
   b. lumen index.
   c. magnitude.
   d. wattage.
   e. brilliance.

3. The color of a star like the Sun, with temperatures between 5000 and 6000 K, will be _____
   a. red.
   b. red-orange.
   c. blue.
   d. white.
   e. yellow.

4. Pairs of stars, in which the members are far enough apart to be resolved telescopically, are called _____
   a. white dwarfs.
   b. optical giants.
   c. telescopic stars.
   d. visual binaries.
   e. visual clusters.

5. On a Hertzsprung-Russell diagram, any differences in the brightness of two stars having the same temperature is attributable to their relative _____
   a. sizes.
   b. compositions.
   c. internal reactions.
   d. a. and b.
   e. a. and c.

6. What force is most responsible for the existence of stars?
   a. magnetic
   b. nuclear
   c. gravity
   d. electrical
   e. potential

7. The proton-proton chain reaction that is responsible for the energy of most stars involves the fusion of hydrogen nuclei together to produce _____ nuclei.
   a. oxygen
   b. carbon
   c. iron
   d. nitrogen
   e. helium

8. Stars with over three times the mass of the Sun have relatively short life spans and terminate in a brilliant explosion called a _____
   a. supernova.
   b. detonation.
   c. supergiant.
   d. fusion reaction.
   e. none of the above.

9. What type of galaxy is the Milky Way?
   a. elliptical
   b. spiral
   c. irregular
   d. barred
   e. amorphous

10. The apparent change in the wavelength of radiation caused by the relative motions of the source and the observer is called the _____ effect.
    a. Hubble
    b. Dobsonian
    c. Doppler
    d. Frequency
    e. Galactic

Fill-In Questions

11. The extremely slight back-and-forth shifting in a nearby star's position due to the orbital motion of Earth is called _____ _____.

12. An accumulation of dust and gases scattered between existing stars is called a _____.

13. A star is a ball of gases scattered between the opposing forces of _____ trying to contract it and _____ pressure trying to expand it.

14. During a supernova implosion, when the electrons of atoms are forced to combine with protons, a small, very dense object called a _____ star forms.

15. The Milky Way is part of a galactic cluster, called the _____, containing at least 28 galaxies.

True/False Questions

16. Stars with the same absolute magnitudes can have different apparent magnitudes because they are not the same size. _____
    distance

17. Binary stars can be used to determine stellar mass. _____

18. The red giant stage is the first step in stellar formation. _____

19. Black holes, whose gravitational field prevents the escape of all matter and energy, theoretically can be found by examining matter being pulled into them. ___T___

20. The predominance of red shifts which have been observed for galaxies suggests that the universe is contracting. _____

Answers:

1. e; 2. c; 3. e; 4. d; 5. a; 6. c; 7. e; 8. a; 9. a; 10. c; 11. stellar parallax; 12. nebula; 13. gravity, gas; 14. neutron; 15. Local Group; 16. F; 17. T; 18. F; 19. T; 20. F.

Web Work

The following are informative and interesting World Wide Web sites that address topics related to those presented in this chapter:

- Electronic Universe
  http://zebu.uoregon.edu/galaxy.html

- Hubble Space Telescope Photographs
  http://oposite.stsci.edu/pubinfo/pictures.html

For direct access to these sites and other relevant Web locations, contact the *Foundations of Earth Science* WWW home page at http://www.prenhall.com/lutgens.

Metric and English Units Compared

Units

1 kilometer (km)	=	1000 meters (m)
1 meter (m)	=	100 centimeters (cm)
1 centimeter (cm)	=	0.39 inch (in.)
1 mile (mi)	=	5280 feet (ft)
1 foot (ft)	=	12 inches (in.)
1 inch (in.)	=	2.54 centimeters (cm)
1 square mile (mi^2)	=	640 acres (a)
1 kilogram (kg)	=	1000 grams (g)
1 pound (lb)	=	16 ounces (oz)
1 fathom	=	6 feet (ft)

Conversions

When you want
to convert: multiply by: to find:

Length

inches	2.54	centimeters
centimeters	0.39	inches
feet	0.30	meters
meters	3.28	feet
yards	0.91	meters
meters	1.09	yards
miles	1.61	kilometers
kilometers	0.62	miles

Area

square inches	6.45	square centimeters
square centimeters	0.15	square inches
square feet	0.09	square meters
square meters	10.76	square feet
square miles	2.59	square kilometers
square kilometers	0.39	square miles

Volume

cubic inches	16.38	cubic centimeters
cubic centimeters	0.06	cubic inches
cubic feet	0.028	cubic meters
cubic meters	35.3	cubic feet
cubic miles	4.17	cubic kilometers
cubic kilometers	0.24	cubic miles
liters	1.06	quarts
liters	0.26	gallons
gallons	3.78	liters

Masses and Weights

ounces	20.33	grams
grams	0.035	ounces
pounds	0.45	kilograms
kilograms	2.205	pounds

Temperature

When you want to convert degrees Fahrenheit (°F) to degrees Celsius (°C), subtract 32 degrees and divide by 1.8.

When you want to convert degrees Celsius (°C) to degrees Fahrenheit (°F), multiply by 1.8 and add 32 degrees.

When you want to convert degrees Celsius (°C) to kelvins (K), delete the degree symbol and add 273. When you want to convert kelvins (K) to degrees Celsius (°C), add the degree symbol and subtract 273.

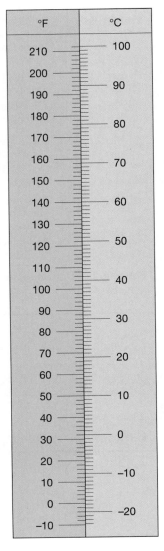

Figure A.1   A comparison of Fahrenheit and Celsius temperature scales.

Mineral Identification Key

Group I ■ Metallic Luster

Hardness	Streak	Other Diagnostic Properties	Name (Chemical Composition)
Harder than glass	Black streak	Black; magnetic; hardness = 6; specific gravity = 5.2; often granular	Magnetite (Fe_3O_4)
	Greenish-black streak	Brass yellow; hardness = 6; specific gravity = 5.2; generally an aggregate of cubic crystals	Pyrite (FeS_2)—fool's gold
	Red-brown streak	Gray or reddish brown; hardness = 5–6; specific gravity = 5; platy appearance	Hematite (Fe_2O_3)
Softer than glass	Greenish-black streak	Golden yellow; hardness = 4; specific gravity = 4.2; massive	Chalcopyrite ($CuFeS_2$)
	Gray-black streak	Silvery gray; hardness = 2.5; specific gravity = 7.6 (very heavy); good cubic cleavage	Galena (PbS)
	Yellow-brown streak	Yellow brown to dark brown; hardness variable (1–6); specific gravity = 3.5–4; often found in rounded masses; earthy appearance	Limonite ($Fe_2O_3 \bullet H_2O$)
	Gray-black streak	Black to bronze; tarnishes to purples and greens; hardness = 3; specific gravity = 5; massive	Bornite (Cu_5FeS_4)
Softer than your fingernail	Dark gray streak	Silvery gray; hardness = 1 (very soft); specific gravity = 2.2; massive to platy; writes on paper (pencil lead); feels greasy	Graphite (C)

Group II ■ Nonmetallic Luster (dark colored)

Hardness	Cleavage	Other Diagnostic Properties	Name (Chemical Composition)
Harder than glass	Cleavage present	Black to greenish black; hardness = 5–6; specific gravity = 3.4; fair cleavage, two directions at nearly 90 degrees	Augite (Ca, Mg, Fe, Al silicate)
		Black to greenish black; hardness = 5–6; specific gravity = 3.2; fair cleavage, two directions at nearly 60 degrees and 120 degrees	Hornblende (Ca, Na, Mg, Fe, OH, Al silicate)
		Red to reddish brown; hardness = 6.5-7.5; conchoidal fracture; glassy luster	Garnet (Fe, Mg, Ca, Al silicate)
	Cleavage not prominent	Gray to brown; hardness = 9; specific gravity = 4; hexagonal crystals common	Corundum (Al_2O_3)
		Dark brown to black; hardness = 7; conchoidal fracture; glassy luster	Smoky quartz (SiO_2)
		Olive green; hardness = 6.5–7; small glassy grains	Olivine ($(Mg, Fe)_2 SiO_4$)

Group II (continued) ■ Nonmetallic Luster (dark colored)

Hardness	Cleavage	Other Diagnostic Properties	Name (Chemical Composition)
Softer than glass	Cleavage present	Yellow brown to black; hardness = 4; good cleavage in six directions, light yellow streak that has the smell of sulfur	Sphalerite (ZnS)
		Dark brown to black; hardness = 2.5–3, excellent cleavage in one direction; elastic in thin sheets; black mica	Biotite (K, Mg, Fe, OH, Al silicate)
	Cleavage absent	Generally tarnished to brown or green; hardness = 2.5; specific gravity = 9, massive	Native copper (Cu)
Softer than your fingernail	Cleavage not prominent	Reddish brown; hardness = 1–5; specific gravity = 4–5; red streak; earthy appearance	Hematite (Fe_2O_3)
		Yellow brown; hardness = 1–3; specific gravity = 3.5; earthy appearance; powders easily	Limonite ($Fe_2O_3 \cdot H_2O$)

Group III ■ Nonmetallic Luster (light colored)

Hardness	Cleavage	Other Diagnostic Properties	Name (Chemical Composition)
Harder than glass	Cleavage present	Flesh colored or white to gray; hardness = 6; specific gravity=2.6; two directions of cleavage at nearly right angles	Potassium feldspar ($KAlSi_3O_8$)
			Plagioclase feldspar ($NaAlSi_3O_8$ to $CaAl_2Si_2O_8$)
	Cleavage absent	Any color; hardness = 7; specific gravity = 2.65; conchoidal fracture; glassy appearance; varieties; milky, rose, smoky, amethyst (violet)	Quartz (SiO_2)
Softer than glass	Cleavage present	White, yellowish to colorless; hardness = 3; three directions of cleavage at 75 degrees (rhombohedral); effervesces in HCl; often transparent	Calcite ($CaCO_3$)
		White to colorless; hardness = 2.5; three directions of cleavage at 90 degrees (cubic); salty taste	Halite (NaCl)
		Yellow, purple, green, colorless; hardness = 4; white streak; translucent to transparent; four directions of cleavage	Fluorite (CaF_2)
Softer than your fingernail	Cleavage present	Colorless; hardness = 2–2.5; transparent and elastic in thin sheets; excellent cleavage in one direction; light mica	Muscovite (K, OH, Al silicate)
		White to transparent, hardness = 2; when in sheets, is flexible but not elastic; varieties: selenite (transparent, three directions of cleavage); satin spar (fibrous, silky luster); alabaster (aggregate of small crystals)	Gypsum ($CaSO_4 \cdot 2H_2O$)
	Cleavage not prominent	White, pink, green; hardness = 1; forms in thin plates; soapy feel; pearly luster	Talc (Mg silicate)
		Yellow; hardness = 1–2.5	Sulfur (S)
		White; hardness = 2; smooth feel; earthy odor when moistened, has typical clay texture	Kaolinite (Hydrous Al silicate)
		Green; hardness = 2.5; fibrous; variety of serpentine	Asbestos (Mg, Al silicate)
		Pale to dark reddish brown; hardness = 1–3; dull luster; earthy; often contains spheroidal-shaped particles; not a true mineral	Bauxite (Hydrous Al oxide)

APPENDIX C
Earth's Grid System

A glance at any globe reveals a series of north-south and east-west lines that together make up Earth's grid system, a universally used scheme for locating points on Earth's surface. The north-south lines of the grid are called **meridians** and extend from pole to pole (Figure C.1). All are halves of great circles. A **great circle** is the largest possible circle that can be drawn on a globe; if a globe were sliced along one of these circles, it would be divided into two equal parts called **hemispheres**. By viewing a globe or Figure C.1, it can be seen that meridians are spaced farthest apart at the equator and converge toward the poles. The east-west lines (circles) of the grid are known as **parallels**. As their names implies, these circles are parallel to one another (Figure C.1). While all meridians are parts of great circles, all parallels are not. In fact, only one parallel, the equator, is a great circle.

Latitude and Longitude

Latitude may be defined as distance, measured in degrees, *north* and *south* of the equator. Parallels are used to show latitude. Because all points that lie along the same parallel are an identical distance from the equator, they all have the same latitude designation. The latitude of the equator is 0 degrees, while the north and south poles lie 90 degrees N and 90 degrees S, respectively.

Longitude is defined as distance, measured in degrees, *east* and *west* of the zero or prime meridian. Because all meridians are identical, the choice of a zero line is obviously arbitrary. However, the meridian that passes through the Royal Observatory at Greenwich, England, is universally accepted as the reference meridian. Thus, the longitude for any place on the globe is measured east or west from this line. Longitude can vary from 0 degrees along the prime meridian to 180 degrees, halfway around the globe.

It is important to remember that when a location is specified, directions must be given, that is, north or south latitude and east or west longitude (Figure C.2). If this is not done, more than one point on the globe is being designated. The only exceptions, of course, are places that lie along the equator, the prime meridian, or the 180-degree meridian. It should also be noted that while it is not incorrect to use fractions, a degree of latitude or longitude is usually divided into minutes and seconds. A minute (') is $\frac{1}{60}$th of a degree, and a second (") is $\frac{1}{60}$th of a minute. When locating a place on a map, the degree of exactness will depend on the scale of the map. When using a small-scale world map or globe, it may be difficult to estimate latitude and longitude to the nearest whole degree or two. On the other hand, when a large-scale map of an area is used, it is often possible to estimate latitude and longitude to the nearest minute or second.

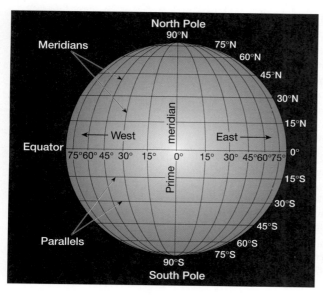

Figure C.1 Earth's grid system.

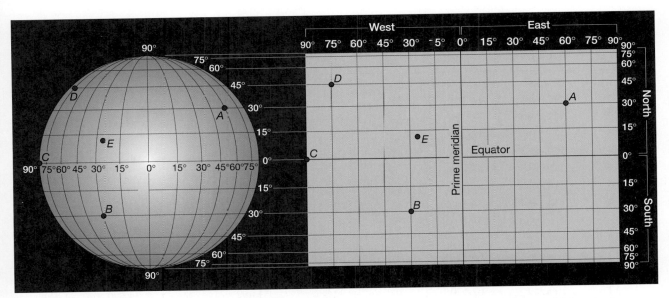

Figure C.2 Locating places using the grid system. For both diagrams: Point *A* is latitude 30 degrees N, longitude 60 degrees E; Point *B* is latitude 30 degrees S, longitude 30 degrees W; Point *C* is latitude 0 degrees, longitude 90 degrees W; Point *D* is latitude 45 degrees N, longitude 75 degrees W; Point *E* is approximately latitude 10 degrees N, longitude 25 degrees W.

Distance Measurement

The length of a degree of longitude depends on where the measurement is taken. At the equator, which is a great circle, a degree of east-west distance is equal to approximately 111 kilometers (69 miles). This figure is found by dividing Earth's circumference—40,075 kilometers (24,900 miles)—by 360. However, with an increase in latitude, the parallels become smaller, and the length of a degree of longitude diminishes (see Table C.1). Thus, at about latitude 60 degrees N and S, a degree of longitude has a value equal to about half of what it was at the equator.

As all meridians are halves of great circles, a degree of latitude is equal to about 111 kilometers (69 miles), just as a degree of longitude along the equator is. However, Earth is not a perfect sphere but is slightly flattened at the poles and bulges slightly at the equator. Because of this, there are small differences in the length of a degree of latitude.

Determining the shortest distance between two points on a globe can be done easily and fairly accurately using the "globe and string" method. It should be noted here that the arc of a great circle is the shortest distance between two points on a sphere. In order to determine the great circle distance (as well as observe the great circle route) between two places, stretch the string between the locations in question. Then, measure the length of the string along the equator (since it is a great circle with degrees marked on it) to determine the number of degrees between the two points. To calculate the distance in kilometers or miles, simply multiply the number of degrees by 111 or 69, respectively.

| °Lat. | Lenght of 1° Long. | | °Lat. | Length of 1° Long. | | °Lat. | Length of 1°Long | |
	km	miles		km	miles		km	miles
0	111.367	69.172	30	96.528	59.955	60	55.825	34.674
1	111.349	69.161	31	95.545	59.345	61	54.131	33.622
2	111.298	69.129	32	94.533	58.716	62	52.422	62.560
3	111.214	69.077	33	93.493	58.070	63	50.696	31.488
4	111.096	69.004	34	92.425	57.407	64	48.954	30.406
5	110.945	68.910	35	91.327	56.725	65	47.196	29.314
6	110.760	68.795	36	90.203	56.027	66	45.426	28.215
7	110.543	68.660	37	89.051	55.311	67	43.639	27.105
8	110.290	68.503	38	87.871	54.578	68	41.841	25.988
9	110.003	68.325	39	86.665	53.829	69	40.028	24.862
10	109.686	68.128	40	85.431	53.063	70	38.204	23.729
11	109.333	67.909	41	84.171	52.280	71	36.368	22.589
12	108.949	67.670	42	82.886	51.482	72	34.520	21.441
13	108.530	67.410	43	81.575	50.668	73	32.662	20.287
14	108.079	67.130	44	80.241	49.839	74	30.793	19.126
15	107.596	66.830	45	78.880	48.994	75	28.914	17.959
16	107.079	66.509	46	77.497	48.135	76	27.029	16.788
17	106.530	66.168	47	76.089	47.260	77	25.134	15.611
18	105.949	65.807	48	74.659	46.372	78	23.229	14.428
19	105.337	65.427	49	73.203	45.468	79	21.320	13.242
20	104.692	65.026	50	71.727	44.551	80	19.402	12.051
21	104.014	64.605	51	70.228	43.620	81	17.480	10.857
22	103.306	64.165	52	68.708	42.676	82	15.551	9.659
23	102.565	63.705	53	67.168	41.719	83	13.617	8.458
24	101.795	63.227	54	65.604	40.748	84	11.681	7.255
25	100.994	62.729	55	64.022	39.765	85	9.739	6.049
26	100.160	62.211	56	62.420	38.770	86	7.796	4.842
27	99.297	61.675	57	60.798	37.763	87	5.849	3.633
28	98.405	61.121	58	59.159	36.745	88	3.899	2.422
29	97.481	60.547	59	57.501	35.715	89	1.950	1.211
30	96.528	59.955	60	55.825	34.674	90	0	0

Glossary

Aa flow A type of lava flow that has a jagged blocky surface.

Ablation A general term for the loss of ice and snow from a glacier.

Abrasion The grinding and scraping of a rock surface by the friction and impact of rock particles carried by water, wind, or ice.

Absolute dating Determination of the number of years since the occurrence of a given geologic event.

Absolute magnitude The apparent brightness of a star if it were viewed from a distance of 10 parsecs (32.6 light-years). Used to compare the true brightness of stars.

Abyssal plain Very level area of the deep-ocean floor, usually lying at the foot of the continental rise.

Accretionary wedge A large wedge-shaped mass of sediment that accumulates in subduction zones. Here sediment is scraped from the subducting oceanic plate and accreted to the overriding crustal block.

Adiabatic temperature change Cooling or warming of air caused when air is allowed to expand or is compressed, not because heat is added or subtracted.

Advection fog A fog formed when warm, moist air is blown over a cool surface.

Aftershocks Smaller earthquakes that follow the main earthquake.

Air The mixture of gases and particles that make up Earth's atmosphere. Nitrogen and oxygen are most abundant.

Air mass A large body of air that is characterized by a sameness of temperature and humidity.

Air-mass weather The conditions experienced in an area as an air mass passes over it. Because air masses are large and fairly homogeneous, air-mass weather will be fairly constant and may last for several days.

Air pressure Force exerted by the weight of the air above.

Albedo The reflectivity of a substance, usually expressed as a percentage of the incident radiation reflected.

Alluvial fan A fan-shaped deposit of sediment formed when a stream's slope is abruptly reduced.

Alluvium Unconsolidated sediment deposited by a stream.

Alpine glacier A glacier confined to a mountain valley, which in most instances had previously been a stream valley.

Aneroid barometer An instrument for measuring air pressure that consists of evacuated metal chambers that are very sensitive to variations in air pressure.

Angular unconformity An unconformity in which the strata below dip at an angle different from that of the beds above.

Anticline A fold in sedimentary strata that resembles an arch.

Anticyclone A high-pressure center characterized by a clockwise flow of air in the Northern Hemisphere.

Apparent magnitude The brightness of a star when viewed from Earth.

Aquicludes Impermeable beds that hinder or prevent groundwater movement.

Aquifer Rock or soil through which groundwater moves easily.

Arête A narrow knifelike ridge separating two adjacent glaciated valleys.

Artesian well A well in which the water rises above the level where it was initially encountered.

Asteroids Thousands of small planet-like bodies, ranging in size from a few hundred kilometers to less than a kilometer, whose orbits lie mainly between those of Mars and Jupiter.

Asthenosphere A subdivision of the mantle situated below the lithosphere. This zone of weak material exists below a depth of about 100 kilometers and in some regions extends as deep as 700 kilometers. The rock within this zone is easily deformed.

Astronomical Unit (AU) Average distance from Earth to the sun; 1.5×10^8 km, or 93×10^6 miles.

Astronomy The scientific study of the universe; it includes the observation and interpretation of celestial bodies and phenomena.

Atmosphere The gaseous portion of a planet; the planet's envelope of air. One of the traditional subdivisions of Earth's physical environment.

Atoll A continuous or broken ring of coral reef surrounding a central lagoon.

Atom The smallest particle that exists as an element.

Atomic number The number of protons in the nucleus of an atom.

Backswamp A poorly drained area on a floodplain that results when natural levees are present.

Barograph A recording barometer.

Barometer An instrument that measures atmospheric pressure.

Barometric tendency *See* Pressure tendency.

Barred spiral A galaxy having straight arms extending from its nucleus.

Barrier island A low, elongated ridge of sand that parallels the coast.

Base level The level below which a stream cannot erode.

Basin A circular downfolded structure.

Batholith A large mass of igneous rock that formed when magma was emplaced at depth, crystallized, and subsequently exposed by erosion.

Baymouth bar A sandbar that completely crosses a bay, sealing it off from the open ocean.

Beach drift The transport of sediment in a zigzag pattern along a beach caused by the uprush of water from obliquely breaking waves.

Beach nourishment Large quantities of sand are added to the beach system to offset losses caused by wave erosion.

Bed load Sediment that is carried by a stream along the bottom of its channel.

Big Bang theory The theory that proposes that the universe originated as a single mass, which subsequently exploded.

Biogenous sediment Seafloor sediments consisting of material of marine-organic origin.

Biosphere The totality of life on Earth; the parts of the lithosphere, hydrosphere, and atmosphere in which living organisms can be found.

Black hole A massive star that has collapsed to such a small volume that its gravity prevents the escape of all radiation.

Blowout (deflation hollow) A depression excavated by the wind in easily eroded deposits.

Body waves Seismic waves that travel through Earth's interior.

Bright nebula A cloud of glowing gas excited by ultraviolet radiation from hot stars.

Caldera A large depression typically caused by collapse or ejection of the summit area of a volcano.

Capacity The total amount of sediment a stream is able to transport.

Cassini division A wide gap in the ring system of Saturn between the A ring and the B ring.

Catastrophism The concept that Earth was shaped by catastrophic events of a short-term nature.

Cavern A naturally formed underground chamber or series of chambers most commonly produced by solution activity in limestone.

Celestial sphere An imaginary hollow sphere upon which the ancients believed the stars were hung and carried around Earth.

Cenozoic era A time span on the geologic calendar beginning about 66 million years ago following the Mesozoic era.

Chemical sedimentary rock Sedimentary rock consisting of material that was precipitated from water by either inorganic or organic means.

Chinook A wind blowing down the leeward side of a mountain and warming by compression.

Cinder cone A rather small volcano built primarily of pyroclastics ejected from a single vent.

Circle of illumination The great circle that separates daylight from darkness.

Cirque An amphitheater-shaped basin at the head of a glaciated valley produced by frost wedging and plucking.

Cirrus One of three basic cloud forms; also one of the three high cloud types. They are thin, delicate ice crystal clouds often appearing as veil-like patches or thin, wispy fibers.

Cleavage The tendency of a mineral to break along planes of weak bonding.

Climate A description of aggregate weather conditions; the sum of all statistical weather information that helps describe a place or region.

Cloud A form of condensation best described as a dense concentration of suspended water droplets or tiny ice crystals.

Coarse-grained texture An igneous rock texture in which the crystals are roughly equal in size and large enough so that individual minerals can be identified with the unaided eye.

Cold front A front along which a cold air mass thrusts beneath a warmer air mass.

Color An obvious mineral characteristic that is often unreliable as a diagnostic property.

Comet A small body which generally revolves around the sun in an elongated orbit.

Competence A measure of the largest particle a stream can transport; a factor dependent on velocity.

Composite cone A volcano composed of both lava flows and pyroclastic material.

Compound A substance formed by the chemical combination of two or more elements in definite proportions and usually having properties different than those of its constituent elements.

Condensation The change of state from a gas to a liquid.

Condensation nuclei Tiny bits of particulate matter that serve as surfaces on which water vapor condenses.

Conduction The transfer of heat through matter by molecular activity. Energy is transferred through collisions from one molecule to another.

Cone of depression A cone-shaped depression in the water table immediately surrounding a well.

Conformable Layers of rock that were deposited without interruption.

Contact metamorphism Changes in rock caused by the heat from a nearby magma body.

Continental (c) air mass An air mass that forms over land; it is normally relatively dry.

Continental drift theory A theory that originally proposed that the continents are rafted about. It has essentially been replaced by the plate tectonics theory.

Continental margin That portion of the seafloor adjacent to the continents. It may include the continental shelf, continental slope, and continental rise.

Continental rise The gently sloping surface at the base of the continental slope.

Continental shelf The gently sloping submerged portion of the continental margin extending from the shoreline to the continental slope.

Continental slope The steep gradient that leads to the deep-ocean floor and marks the seaward edge of the continental shelf.

Continental volcanic arc Mountains formed in part by igneous activity associated with the subduction of oceanic lithosphere beneath a continent. Examples include the Andes and the Cascades.

Convection The transfer of heat by the movement of a mass or substance. It can take place only in fluids.

Convergence The condition that exists when the distribution of winds within a given area results in a net horizontal inflow of air into the area. Since convergence at lower levels is associated with an upward movement of air, areas of convergent winds are regions favorable to cloud formation and precipitation.

Convergent boundary A boundary in which two plates move together, causing one of the slabs of lithosphere to be consumed into the mantle as it descends beneath an overriding plate.

Coral reef Structure formed in a warm, shallow, sunlit ocean environment that consists primarily of the calcite-rich remains of corals as well as the limy secretions of algae and the hard parts of many other small organisms.

Coriolis force (effect) The deflective force of Earth's rotation on all free-moving objects, including the atmosphere and oceans. Deflection is to the right in the Northern Hemisphere and to the left in the Southern Hemisphere.

Correlation Establishing the equivalence of rocks of similar age in different areas.

Crater The depression at the summit of a volcano, or that which is produced by a meteorite impact.

Creep The slow downhill movement of soil and regolith.

Crevasse A deep crack in the brittle surface of a glacier.

Cross-bedding Structure in which relatively thin layers are inclined at an angle to the main bedding. Formed by currents of wind or water.

Cross-cutting A principle of relative dating. A rock or fault is younger than any rock (or fault) through which it cuts.

Crust The very thin outermost layer of Earth.

Crystal form The external appearance of a mineral as determined by its internal arrangement of atoms.

Crystallization The formation and growth of a crystalline solid from a liquid or gas.

Cumulus One of three basic cloud forms; also the name given one of the clouds of vertical development. Cumulus are billowy individual cloud masses that often have flat bases.

Cup anemometer An instrument used to determine wind speed.

Cutoff A short channel segment created when a river erodes through the narrow neck of land between meanders.

Cyclone A low-pressure center characterized by a counterclockwise flow of air in the Northern Hemisphere.

Dark nebula A cloud of interstellar dust that obscures the light of more distant stars and appears as an opaque curtain.

Deep-ocean trench A narrow, elongated depression on the floor of the ocean.

Deep-sea fan A cone-shaped deposit at the base of the continental slope. The sediment is transported to the fan by turbidity currents that follow submarine canyons.

Deflation The lifting and removal of loose material by wind.

Degenerate matter Incomprehensibly dense material formed when stars collapse and form a white dwarf.

Delta An accumulation of sediment formed where a stream enters a lake or ocean.

Dendritic pattern A stream system that resembles the pattern of a branching tree.

Density The weight per unit volume of a particular material.

Deposition The process by which water vapor is changed directly to a solid without passing through the liquid state.

Desalination The removal of salts and other chemicals from seawater.

Desert One of the two types of dry climate; the driest of the dry climates.

Desert pavement A layer of coarse pebbles and gravel created when wind removed the finer material.

Detrital sedimentary rock Rock formed from the accumulation of material that originated and was transported in the form of solid particles derived from both mechanical and chemical weathering.

Dew point The temperature to which air has to be cooled in order to reach saturation.

Dike A tabular-shaped intrusive igneous feature that cuts through the surrounding rock.

Dip-slip fault A fault in which the movement is parallel to the dip of the fault.

Discharge The quantity of water in a stream that passes a given point in a period of time.

Disconformity A type of unconformity in which the beds above and below are parallel.

Dissolved load That portion of a stream's load carried in solution.

Distributary A section of a stream that leaves the main flow.

Diurnal tide Tides characterized by a single high and low water height each tidal day.

Divergence The condition that exists when the distribution of winds within a given area results in a net horizontal outflow of air from the region. In divergence at lower levels the resulting deficit is compensated for by a downward movement of air from aloft; hence, areas of divergent winds are unfavorable to cloud formation and precipitation.

Divergent boundary A region where the rigid plates are moving apart, typified by the mid-oceanic ridges.

Divide An imaginary line that separates the drainage of two streams; often found along a ridge.

Dome A roughly circular upfolded structure similar to an anticline.

Doppler effect The apparent change in wavelength of radiation caused by the relative motions of the source and the observer.

Doppler radar In addition to the tasks performed by conventional radar, this new generation of weather radar can detect motion directly and hence greatly improve tornado and severe storm warnings.

Drainage basin The land area that contributes water to a stream.

Drawdown The difference in height between the bottom of a cone of depression and the original height of the water table.

Drift The general term for any glacial deposit.

Drumlin A streamlined asymmetrical hill composed of glacial till. The steep side of the hill faces the direction from which the ice advanced.

Dry adiabatic rate The rate of adiabatic cooling or warming in unsaturated air. The rate of temperature change is 1°C per 100 meters.

Dune A hill or ridge of wind-deposited sand.

Earthquake The vibration of Earth produced by the rapid release of energy.

Echo sounder An instrument used to determine the depth of water by measuring the time interval between emission of a sound signal and the return of its echo from the bottom.

Elastic rebound The sudden release of stored strain in rocks that results in movement along a fault.

Electron A negatively charged subatomic particle that has a negligible mass and is found outside an atom's nucleus.

Element A substance that cannot be decomposed into simpler substances by ordinary chemical or physical means.

Elements of weather and climate Those quantities or properties of the atmosphere that are measured regularly and that are used to express the nature of weather and climate.

Elliptical galaxy A galaxy that is round or elliptical in outline. It contains little gas and dust, no disk or spiral arms, and few hot, bright stars.

Emergent coast A coast where land that was formerly below sea level has been exposed either because of crustal uplift or a drop in sea level or both.

Emission nebula A gaseous nebula that derives its visible light from the fluorescence of ultraviolet light from a star in or near the nebula.

End moraine A ridge of till marking a former position of the front of a glacier.

Entrenched meander A meander cut into bedrock when uplifting rejuvenated a meandering stream.

Environmental lapse rate The rate of temperature decrease with increasing height in the troposphere.

Eon The largest time unit on the geologic time scale, next in order of magnitude above era.

Ephemeral stream A stream that is usually dry because it carries water only in response to specific episodes of rainfall. Most desert streams are of this type.

Epicenter The location on Earth's surface that lies directly above the forces of an earthquake.

Epoch A unit of the geologic calendar that is a subdivision of a period.

Equatorial low A belt of low pressure lying near the equator and between the subtropical highs.

Equatorial system A method of locating stellar objects much like the coordinate system used on Earth's surface.

Equinox (spring or autumnal) The time when the vertical rays of the sun are striking the equator. The length of daylight and darkness is equal at all latitudes at equinox.

Era A major division on the geologic calendar; eras are divided into shorter units called periods.

Escape velocity The initial velocity an object needs to escape from the surface of a celestial body.

Esker Sinuous ridge composed largely of sand and gravel deposited by a stream flowing in a tunnel beneath a glacier near its terminus.

Estuary A funnel-shaped inlet of the sea that formed when a rise in sea level or subsidence of land caused the mouth of a river to be flooded.

Evaporation The process of converting a liquid to a gas.

Evaporite A sedimentary rock formed of material deposited from solution by evaporation of the water.

Extrusive Igneous activity that occurs outside the crust.

Eye A zone of scattered clouds and calm averaging about 20 kilometers in diameter at the center of a hurricane.

Eye wall The doughnut-shaped area of intense cumulonimbus development and very strong winds that surrounds the eye of a hurricane.

Fault A break in a rock mass along which movement has occurred.

Fetch The distance that the wind has traveled across the open water.

Fine-grained texture A texture of igneous rocks in which the crystals are too small for individual minerals to be distinguished with the unaided eye.

Fiord A steep-sided inlet of the sea formed when a glacial trough was partially submerged.

Fissure eruption An eruption in which lava is extruded from narrow fractures or cracks in the crust.

Flood basalts Flows of basaltic lava that issue from numerous cracks or fissures and commonly cover extensive areas to thicknesses of hundreds of meters.

Floodplain The flat, low-lying portion of a stream valley subject to periodic inundation.

Focus (earthquake) The zone within Earth where rock displacement produces an earthquake.

Fog A cloud with its base at or very near Earth's surface.

Fold A bent rock layer or series of layers that were originally horizontal and subsequently deformed.

Foliated texture A texture of metamorphic rocks that gives the rock a layered appearance.

Foreshocks Small earthquakes that often precede a major earthquake.

Fossils The remains or traces of organisms preserved from the geologic past.

Fossil succession Fossil organisms succeed one another in a definite and determinable order, and any time period can be recognized by its fossil content.

Fracture Any break or rupture in rock along which no appreciable movement has taken place.

Freezing The change of state from a liquid to a solid.

Freezing nuclei Solid particles that serve as cores for the formation of ice crystals.

Front The boundary between two adjoining air masses having contrasting characteristics.

Frontal fog Fog formed when rain evaporates as it falls through a layer of cool air.

Frontal wedging Lifting of air resulting when cool air acts as a barrier over which warmer, lighter air will rise.

Galactic cluster A system of galaxies containing from several to thousands of member galaxies.

Geocentric The concept of an Earth-centered universe.

Geologic time scale The division of Earth history into blocks of time—eons, eras, periods, and epochs. The time scale was created using relative dating principles.

Geology The science that examines Earth, its form and composition, and the changes it has undergone and is undergoing.

Geostrophic wind A wind, usually above a height of 600 meters (2000 feet), that blows parallel to the isobars.

Geyser A fountain of hot water ejected periodically.

Giant (star) A luminous star of large radius.

Glacial erratic An ice-transported boulder that was not derived from bedrock near its present site.

Glacial striations Scratches and grooves on bedrock caused by glacial abrasion.

Glacial trough A mountain valley that has been widened, deepened, and straightened by a glacier.

Glacier A thick mass of ice originating on land from the compaction and recrystallization of snow that shows evidence of past or present flow.

Glassy texture A term used to describe the texture of certain igneous rocks, such as obsidian, that contain no crystals.

Glaze A coating of ice on objects formed when supercooled rain freezes on contact.

Graben A valley formed by the downward displacement of a fault-bounded block.

Gradient The slope of a stream; generally measured in feet per mile.

Greenhouse effect The transmission of short-wave solar radiation by the atmosphere coupled with the selective absorption of longer-wavelength terrestrial radiation, especially by water vapor and carbon dioxide.

Groin A short wall built at a right angle to the shore to trap moving sand.

Ground moraine An undulating layer of till deposited as the ice front retreats.

Groundwater Water in the zone of saturation.

Guyot A submerged flat-topped seamount.

Gyre The large circular surface current pattern found in each ocean.

Hail Nearly spherical ice pellets having concentric layers and formed by the successive freezing of layers of water.

Half-life The time required for one-half of the atoms of a radioactive substance to decay.

Halocline A layer of water in which there is a high rate of change in salinity in the vertical dimension.

Hanging valley A tributary valley that enters a glacial trough at a considerable height above its floor.

Hardness The resistance a mineral offers to scratching.

Heliocentric The view that the sun is at the center of the solar system.

Hertzsprung-Russell diagram A plot of stars according to their absolute magnitudes and spectral types.

Horn A pyramidlike peak formed by glacial action in three or more cirques surrounding a mountain summit.

Horst An elongated, uplifted block of crust bounded by faults.

Hot spot A concentration of heat in the mantle capable of producing magma which, in turn, extrudes onto Earth's surface. The intraplate volcanism that produced the Hawaiian Islands is one example.

Hot spring A spring in which the water is 6°–9°C (10°–15°F) warmer than the mean annual air temperature of its locality.

Hubble's Law Relates the distance to a galaxy and its velocity.

Humidity A general term referring to water vapor in the air but not to liquid droplets of fog, cloud, or rain.

Hurricane A tropical cyclonic storm having winds in excess of 119 kilometers (74 miles) per hour.

Hydrogen burning The conversion of hydrogen through fusion to form helium.

Hydrogenous sediment Seafloor sediments consisting of minerals that crystallize from seawater. An important example is manganese nodules.

Hydrologic cycle The unending circulation of Earth's water supply. The cycle is powered by energy from the sun and is characterized by continuous exchanges of water among the oceans, the atmosphere, and the continents.

Hydrosphere The water portion of our planet; one of the traditional subdivisions of Earth's physical environment.

Hygrometer An instrument designed to measure relative humidity.

Hygroscopic nuclei Condensation nuclei having a high affinity for water, such as salt particles.

Hypothesis A tentative explanation that is tested to determine if it is valid.

Ice sheet A very large, thick mass of glacial ice flowing outward in all directions from one or more accumulation centers.

Igneous rock A rock formed by the crystallization of molten magma.

Immature soil A soil lacking horizons.

Inclination of the axis The tilt of Earth's axis from the perpendicular to the plane of Earth's orbit.

Inclusion A piece of one rock unit contained within another. Inclusions are used in relative dating. The rock mass adjacent to the one containing the inclusion must have been there first in order to provide the fragment.

Index fossil A fossil that is associated with a particular span of geologic time.

Infrared Radiation with a wavelength from 0.7 to 200 micrometers.

Inner core The solid innermost layer of Earth, about 1300 kilometers (800 miles) in radius.

Interior drainage A discontinuous pattern of intermittent streams that do not flow to the ocean.

Interstellar dust Dust and gases found between stars.

Intrusive Igneous rock that formed below Earth's surface.

Iron meteorite One of the three main categories of meteorites. This group is composed largely of iron with varying amounts of nickel (5–20 percent). Most meteorite finds are irons.

Irregular galaxy A galaxy that lacks symmetry.

Isobar A line drawn on a map connecting points of equal atmospheric pressure, usually corrected to sea level.

Isotherms Lines connecting points of equal temperature.

Isotopes Varieties of the same element that have different mass numbers; their nuclei contain the same number of protons but different numbers of neutrons.

Jet stream Swift (120–240-kilometer per hour), high-altitude winds.

Jovian planet The Jupiter-like planets Jupiter, Saturn, Uranus, and Neptune. These planets have relatively low densities.

Kame A steep-sided hill composed of sand and gravel originating when sediment collected in openings in stagnant glacial ice.

Karst topography A topography consisting of numerous depressions called sinkholes.

Kettle Depressions created when blocks of ice became lodged in glacial deposits and subsequently melted.

Laccolith A massive igneous body intruded between pre-existing strata.

Land breeze A local wind blowing from land toward the water during the night in coastal areas.

Latent heat The energy absorbed or released during a change in state.

Lateral moraine A ridge of till along the sides of an alpine glacier composed primarily of debris that fell to the glacier from the valley walls.

Lava Magma that reaches Earth's surface.

Law A formal statement of the regular manner in which a natural phenomenon occurs under given conditions; e.g., the "law of superposition."

Lightning A sudden flash of light generated by the flow of electrons between oppositely charged parts of a cumulonimbus cloud or between the cloud and the ground.

Light-year The distance light travels in a year; about 6 trillion miles.

Liquefaction A phenomenon, sometimes associated with earthquakes, in which soils and other unconsolidated materials containing abundant water are turned into a fluidlike mass that is not capable of supporting buildings.

Lithification The process, generally cementation and/or compaction, of converting sediments to solid rock.

Lithosphere The rigid outer layer of Earth, including the crust and upper mantle.

Local group The cluster of 20 or so galaxies to which our galaxy belongs.

Loess Deposits of wind-blown silt, lacking visible layers, generally buff colored, and capable of maintaining a nearly vertical cliff.

Longshore current A near-shore current that flows parallel to the shore.

Lunar regolith A thin, gray layer on the surface of the moon, consisting of loosely compacted, fragmented material believed to have been formed by repeated meteoric impacts.

Luster The appearance or quality of light reflected from the surface of a mineral.

Magma A body of molten rock found at depth, including any dissolved gases and crystals.

Magnitude (earthquake) The total amount of energy released during an earthquake.

Magnitude (stellar) A number given to a celestial object to express its relative brightness.

Main-sequence stars A sequence of stars on the Hertzsprung-Russell diagram, containing the majority of stars, that runs diagonally from the upper left to the lower right.

Mantle The 2900-kilometer (1800-mile)-thick layer of Earth located below the crust.

Mantle plume A source of some intraplate basaltic magma, these structures originate at great depth and upon reaching the

crust, they spread laterally creating a localized volcanic zone called a hot spot.

Maria The Latin name for the smooth areas of the moon formerly thought to be seas.

Maritime (m) air mass An air mass that originates over the ocean. These air masses are relatively humid.

Mass number The number of neutrons and protons in the nucleus of an atom.

Mass wasting The downslope movement of rock, regolith, and soil under the direct influence of gravity.

Meander A looplike bend in the course of a stream.

Mean solar day The average time between two passages of the sun across the local celestial meridian.

Medial moraine A ridge of till formed when lateral moraines from two coalescing alpine glaciers join.

Melting The change of state from a solid to a liquid.

Mercury barometer A mercury-filled glass tube in which the height of the mercury column is a measure of air pressure.

Mesosphere The layer of the atmosphere immediately above the stratosphere and characterized by decreasing temperatures with height.

Mesozoic era A time span on the geologic calendar between the Paleozoic and Cenozoic eras—from about 245 to 66.4 million years ago.

Metamorphic rock Rocks formed by the alteration of pre-existing rock deep within Earth (but still in the solid state) by heat, pressure, and/or chemically active fluids.

Meteor The luminous phenomenon observed when a meteoroid enters Earth's atmosphere and burns up; popularly called a "shooting star."

Meteorite Any portion of a meteoroid that survives its traverse through Earth's atmosphere and strikes Earth's surface.

Meteoroid Small solid particles that have orbits in the solar system.

Meteorology The scientific study of the atmosphere and atmospheric phenomena; the study of weather and climate.

Meteor shower Many meteors appearing in the sky caused when Earth intercepts a swarm of meteoritic particles.

Middle-latitude cyclone See Wave cyclone.

Mid-ocean ridge A continuous mountainous ridge on the floor of all the major ocean basins and varying in width from 500 to 5000 kilometers (300 to 3000 miles). The rifts at the crests of these ridges represent divergent plate boundaries.

Mineral A naturally occurring, inorganic crystalline material with a unique chemical composition.

Mineral resource All discovered and undiscovered deposits of a useful mineral that can be extracted now or at some time in the future.

Mohorovičić discontinuity (Moho) The boundary separating Earth's crust from the mantle, discernible by an increase in seismic velocity.

Mohs hardness scale A series of ten minerals used as a standard in determining hardness.

Monsoon Seasonal reversal of wind direction associated with large continents, especially Asia. In winter, the wind blows from land to sea; in summer, from sea to land.

Mountain breeze The nightly downslope winds commonly encountered in mountain valleys.

Mountain building The processes such as folding, faulting, and volcanism that collectively produce mountains.

Natural levees The elevated landforms that parallel some streams and act to confine their waters, except during floodstage.

Nebula A cloud of interstellar gas and/or dust.

Nebular hypothesis The basic idea that the sun and planets formed from the same cloud of gas and dust in interstellar space.

Neutron A subatomic particle found in the nucleus of an atom. The neutron is electronically neutral and has a mass approximately that of a proton.

Neutron star A star of extremely high density composed entirely of neutrons.

Nonconformity An unconformity in which older metamorphic or intrusive igneous rocks are overlain by younger sedimentary strata.

Nonfoliated texture Metamorphic rocks that do not exhibit foliation.

Normal fault A fault in which the rock above the fault plane has moved down relative to the rock below.

Normal polarity A magnetic field that is the same as that which exists at present.

Nucleus The small heavy core of an atom that contains all of its positive charge and most of its mass.

Nuée ardente Incandescent volcanic debris buoyed up by hot gases that moves downslope in an avalanche fashion.

Oceanography The scientific study of the oceans and oceanic phenomena.

Ore Usually a useful metallic mineral that can be mined at a profit. The term is also applied to certain nonmetallic minerals such as fluorite and sulfur.

Original horizontality Layers of sediments are generally deposited in a horizontal or nearly horizontal position.

Orogenesis The processes that collectively result in the formation of mountains.

Orographic lifting Mountains acting as barriers to the flow of air force the air to ascend. The air cools adiabatically, and clouds and precipitation may result.

Outer core A layer beneath the mantle about 2200 kilometers (1364 miles) thick that has the properties of a liquid.

Outgassing The escape of gases that had been dissolved in magma.

Outwash plain A relatively flat, gentle sloping plain consisting of materials deposited by meltwater streams in front of the margin of an ice sheet.

Oxbow lake A curved lake produced when a stream cuts off a meander.

Ozone A molecule of oxygen containing three oxygen atoms.

Pahoehoe flow A lava flow with a smooth-to-ropy surface.

Paleomagnetism The natural remnant magnetism in rock bodies. The permanent magnetization acquired by rock which can be used to determine the location of the magnetic poles and the latitude of the rock at the time it became magnetized.

Paleozoic era A time span on the geologic calendar between the Precambrian and Mesozoic eras—from about 570 million to 245 million years ago.

Pangaea The proposed supercontinent which 200 million years ago began to break apart and form the present land masses.

Partial melting The process by which most igneous rocks melt. Since individual minerals have different melting points, most igneous rocks melt over a temperature range of a few hundred degrees. If the liquid is squeezed out after some melting has occurred, a melt with a higher silica content results.

Period A basic unit of the geologic calendar that is a subdivision of an era. Periods may be divided into smaller units called epochs.

Permeability A measure of a material's ability to transmit water.

Perturbation The gravitational disturbance of the orbit of one celestial body by another.

Phanerozoic eon That part of geologic time represented by rocks containing abundant fossil evidence. The eon extending from the end of the Proterozoic eon (570 million years ago) to the present.

Planetary nebula A shell of incandescent gas expanding from a star.

Plate One of numerous rigid sections of the lithosphere that moves as a unit over the material of the asthenosphere.

Plate tectonics The theory which proposes that Earth's outer shell consists of individual plates which interact in various ways and thereby produce earthquakes, volcanoes, mountains, and the crust itself.

Playa lake A temporary lake in a playa.

Pleistocene epoch An epoch of the Quaternary period beginning about 1.6 million years ago and ending about 10,000 years ago. Best known as a time of extensive continental glaciation.

Plucking (quarrying) The process by which pieces of bedrock are lifted out of place by a glacier.

Pluvial lake A lake formed during a period of increased rainfall. During the Pleistocene epoch this occurred in some nonglaciated regions during periods of ice advance elsewhere.

Polar (P) air mass A cold air mass that forms in a high-latitude source region.

Polar easterlies In the global pattern of prevailing winds, winds that blow from the polar high toward the subpolar low. These winds, however, should not be thought of as persistent winds, such as the trade winds.

Polar front The stormy frontal zone separating air masses of polar origin from air masses of tropical origin.

Polar high Anticyclones that are assumed to occupy the inner polar regions and are believed to be thermally induced, at least in part.

Polar wandering As the result of paleomagnetic studies in the 1950s, researchers proposed that either the magnetic poles migrated greatly through time or the continents had gradually shifted their positions.

Porosity The volume of open spaces in rock or soil.

Porphyritic texture An igneous texture consisting of large crystals embedded in a matrix of much smaller crystals.

Positive-feedback mechanism As used in climatic change, any effect that acts to reinforce the initial change.

Precambrian All geologic time prior to the Paleozoic era.

Precession A slow motion of Earth's axis which traces out a cone over a period of 26,000 years.

Precipitation fog Fog formed when rain evaporates as it falls through a layer of cool air.

Pressure gradient The amount of pressure change occurring over a given distance.

Pressure tendency The nature of the change in atmospheric pressure over the past several hours. It can be a useful aid in short-range weather prediction.

Prevailing wind A wind that consistently blows from one direction more than from any other.

Primary pollutants Those pollutants emitted directly from identifiable sources.

Primary (P) wave A type of seismic wave that involves alternating compression and expansion of the material through which it passes.

Proton A positively charged subatomic particle found in the nucleus of an atom.

Proton-proton chain A chain of thermonuclear reactions by which nuclei of hydrogen are built up into nuclei of helium.

Protostar A collapsing cloud of gas and dust destined to become a star.

Psychrometer A device consisting of two thermometers (wet bulb and dry bulb) that is rapidly whirled and, with the use of tables, yields the relative humidity and dew point.

Ptolemaic system An Earth-centered system of the universe.

Pulsar A variable radio source of small size that emits radio pulses in very regular periods.

Pyroclastic flow A highly heated mixture, largely of ash and pumice fragments, traveling down the flanks of a volcano or along the surface of the ground.

Pyroclastic material The volcanic rock ejected during an eruption, including ash, bombs, and blocks.

Pyroclastic texture An igneous rock texture resulting from the consolidation of individual rock fragments that are ejected during a violent eruption.

Radial pattern A system of streams running in all directions away from a central elevated structure, such as a volcano.

Radiation or electromagnetic radiation The transfer of energy (heat) through space by electromagnetic waves.

Radiation fog Fog resulting from radiation heat loss by Earth.

Radiation pressure The force exerted by electromagnetic radiation from an object such as the sun.

Radioactive decay (Radioactivity) The spontaneous decay of certain unstable atomic nuclei.

Radiometric dating The procedure of calculating the absolute ages of rocks and minerals that contain radioactive isotopes.

Rain Drops of water that fall from a cloud and have a diameter of at least 0.5 millimeter.

Rainshadow desert A dry area on the lee side of a mountain range. Many middle-latitude deserts are of this type

Rectangular pattern A drainage pattern characterized by numerous right angle bends that develops on jointed or fractured bedrock.

Red giant A large, cool star of high luminosity; a star occupying the upper right portion of the Hertzsprung-Russell diagram.

Reflecting telescope A telescope that concentrates light from distant objects by using a concave mirror.

Reflection nebula A relatively dense dust cloud in interstellar space that is illuminated by starlight.

Refracting telescope A telescope that employs a lens to bend and concentrate the light from distant objects.

Refraction The process by which the portion of a wave in shallow water slows causing the wave to bend and tend to align itself with the underwater contours.

Regional metamorphism Metamorphism associated with the large-scale mountain-building processes.

Regolith The layer of rock and mineral fragments that nearly everywhere covers Earth's land surface.

Rejuvenation A change, often caused by regional uplift, that causes the force of erosion to intensify.

Relative dating Rocks are placed in their proper sequence or order. Only the chronologic order of events is determined.

Relative humidity The ratio of the air's water vapor content to its water vapor capacity.

Reserve Already identified deposits from which minerals can be extracted profitably.

Retrograde motion The apparent westward motion of the planets with respect to the stars.

Reverse fault A fault in which the material above the fault plane moves up in relation to the material below.

Reverse polarity A magnetic field opposite to that which exists at present.

Revolution The motion of one body about another; as Earth about the sun.

Richter scale A scale of earthquake magnitude based on the motion of a seismograph.

Rift valley A region of Earth's crust along which divergence is taking place.

Rime A thin coating of ice on objects produced when supercooled fog droplets freeze on contact.

Rock A consolidated mixture of minerals.

Rock cycle A model that illustrates the origin of the three basic rock types and the interrelatedness of Earth materials and processes.

Rock flour Ground-up rock produced by the grinding effect of a glacier.

Rotation The spinning of a body, such as Earth, about its axis.

Salinity The proportion of dissolved salts to pure water, usually expressed in parts per thousand (‰).

Santa Ana The local name given a chinook wind in southern California.

Saturation The maximum possible quantity of water vapor that the air can hold at any given temperature and pressure.

Sea arch An arch formed by wave erosion when caves on opposite sides of a headland unite.

Sea breeze A local wind blowing from the sea during the afternoon in coastal areas.

Seafloor spreading The process of producing new seafloor between two diverging plates.

Seamount An isolated volcanic peak that rises at least 1000 meters (3000 feet) above the deep-ocean floor.

Sea stack An isolated mass of rock standing just offshore, produced by wave erosion of a headland.

Seawall A barrier constructed to prevent waves from reaching the area behind the wall. Its purpose is to defend property from the force of breaking waves.

Secondary (S) wave A seismic wave that involves oscillation perpendicular to the direction of propagation.

Sediment Unconsolidated particles created by the weathering and erosion of rock, by chemical precipitation from solution in water, or from the secretions of organisms and transported by water, wind, or glaciers.

Sedimentary rock Rock formed from the weathered products of pre-existing rocks that have been transported, deposited, and lithified.

Seismic sea wave A rapidly moving ocean wave generated by earthquake activity which is capable of inflicting heavy damage in coastal regions.

Seismogram The record made by a seismograph.

Seismograph An instrument that records earthquake waves.

Seismology The study of earthquakes and seismic waves.

Shadow zone The zone between 104 and 143 degrees distance from an earthquake epicenter in which direct waves do not arrive because of refraction by Earth's core.

Shield volcano A broad, gently sloping volcano built from fluid basaltic lavas.

Silicate Any one of numerous minerals that have the oxygen and silicon tetrahedron as their basic structure.

Silicon-oxygen tetrahedron A structure composed of four oxygen atoms surrounding a silicon atom that constitutes the basic building block of silicate minerals.

Sill A tabular igneous body that was intruded parallel to the layering of preexisting rock.

Sink hole A depression produced in a region where soluble rock has been removed by groundwater.

Sleet Frozen or semifrozen rain formed when raindrops freeze as they pass through a layer of cold air.

Slip face The steep, leeward slope of a sand dune; it maintains an angle of about 34 degrees.

Snow A solid form of precipitation produced by sublimation of water vapor.

Solstice (summer or winter) The time when the vertical rays of the sun are striking either the Tropic of Cancer or the Tropic of Capricorn. Solstice represents the longest or shortest day (length of daylight) of the year.

Sorting The process by which solid particles of various sizes are separated by moving water or wind. Also the degree of similarity in particle size in sediment or sedimentary rock.

Source region The area where an air mass acquires its characteristic properties of temperature and moisture.

Specific gravity The ratio of a substance's weight to the weight of an equal volume of water.

Specific humidity The weight of water vapor compared with the total weight of the air, including the water vapor.

Spiral galaxy A flattened, rotating galaxy with pinwheel-like arms of interstellar material and young stars winding out from its nucleus.

Spit An elongated ridge of sand that projects from the land into the mouth of an adjacent bay.

Spring A flow of groundwater that emerges naturally at the ground surface.

Stalactite The iciclelike structure that hangs from the ceiling of a cavern.

Stalagmite The columnlike form that grows upward from the floor of a cavern.

Steam fog Fog having the appearance of steam; produced by evaporation from a warm water surface into the cool air above.

Stellar parallax A measure of stellar distance.

Steppe One of the two types of dry climate. A marginal and more humid variant of the desert that separates it from bordering humid climates.

Stony-iron meteorite One of the three main categories of meteorites. This group, as the name implies, is a mixture of iron and silicate minerals.

Stony meteorite One of the three main categories of meteorites. Such meteorites are composed largely of silicate minerals with inclusions of other minerals.

Storm surge The abnormal rise of the sea along a shore as a result of strong winds.

Strata Parallel layers of sedimentary rock.

Stratified drift Sediments deposited by glacial meltwater.

Stratosphere The layer of the atmosphere immediately above the troposphere, characterized by increasing temperatures with height due to the concentration of ozone.

Stratus One of three basic cloud forms; also the name given one of the low clouds. They are sheets or layers that cover much or all of the sky.

Streak The color of a mineral in powdered form.

Strike-slip fault A fault along which the movement is horizontal.

Subduction zone A long, narrow zone where one lithospheric plate descends beneath another.

Sublimation The conversion of a solid directly to a gas without passing through the liquid state.

Submarine canyon A seaward extension of a valley that was cut on the continental shelf during a time when sea level was lower, or a canyon carved into the outer continental shelf, slope, and rise by turbidity currents.

Submergent coast A coast with a form that is largely the result of the partial drowning of a former land surface either because of a rise of sea level or subsidence of the crust or both.

Subpolar low Low pressure located at about the latitudes of the Arctic and Antarctic circles. In the Northern Hemisphere the low takes the form of individual oceanic cells; in the Southern Hemisphere there is a deep and continuous trough of low pressure.

Subtropical high Not a continuous belt of high pressure but rather several semipermanent, anticyclonic centers characterized by subsidence and divergence located roughly between latitudes 25 and 35 degrees.

Supergiant A very large star of high luminosity.

Supernova An exploding star that increases in brightness many thousands of times.

Superposition In any underformed sequence of sedimentary rocks, each bed is older than the layers above and younger than the layers below.

Surf A collective term for breakers; also the wave activity in the area between the shoreline and the outer limit of breakers.

Surface waves Seismic waves that travel along the outer layer of the earth.

Suspended load The fine sediment carried within the body of flowing water.

Syncline A linear downfold in sedimentary strata; the opposite of anticline.

Terrane A crustal block bounded by faults, whose geologic history is distinct from the histories of adjoining crustal blocks.

Terrestrial planet Any of the Earth-like planets including Mercury, Venus, Mars, and Earth.

Terrigenous sediment Seafloor sediment derived from weathering and erosion on land.

Texture The size, shape, and distribution of the particles that collectively constitute a rock.

Theory A well-tested and widely accepted view that explains certain observable facts.

Thermocline A layer of water in which there is a rapid change in temperature in the vertical dimension.

Thermohaline circulation Movements of ocean water caused by density differences brought about by variations in temperature and salinity.

Thermosphere The region of the atmosphere immediately above the mesosphere and characterized by increasing temperatures due to absorption of very short-wave solar energy by oxygen.

Thrust fault A low-angle reverse fault.

Thunderstorm A storm produced by a cumulonimbus cloud and always accompanied by lightning and thunder. It is of relatively short duration and usually accompanied by strong wind gusts, heavy rain, and sometimes hail.

Tidal current The alternating horizontal movement of water associated with the rise and fall of the tide.

Tidal flat A marshy or muddy area that is covered and uncovered by the rise and fall of the tide.

Tide Periodic change in the elevation of the ocean surface.

Till Unsorted sediment deposited directly by a glacier.

Tombolo A ridge of sand that connects an island to the mainland or to another island.

Tornado A small, very intense cyclonic storm with exceedingly high winds, most often produced along cold fronts in conjunction with severe thunderstorms.

Tornado warning A warning issued when a tornado has actually been sighted in an area or is indicated by radar.

Tornado watch A forecast issued for areas of about 65,000 square kilometers (25,000 square miles) indicating that conditions are such that tornadoes may develop; it is intended to alert people to the possibility of tornadoes.

Trade winds Two belts of winds that blow almost constantly from easterly directions and are located on the equatorward sides of the subtropical highs.

Transform boundary A boundary in which two plates slide past one another without creating or destroying lithosphere.

Transpiration The release of water vapor to the atmosphere by plants.

Trellis pattern A system of streams in which nearly parallel tributaries occupy valleys cut in folded strata.

Trench An elongate depression in the seafloor produced by bending of oceanic crust during subduction.

Tropical (T) air mass A warm-to-hot air mass that forms in the subtropics.

Tropical depression By international agreement, a tropical cyclone with maximum winds that do not exceed 61 kilometers (38 miles) per hour.

Tropical storm By international agreement, a tropical cyclone with maximum winds between 61 and 119 kilometers (38 and 74 miles) per hour.

Tropic of Cancer The parallel of latitude, $23^1/_2$ degrees north latitude, marking the northern limit of the sun's vertical rays.

Tropic of Capricorn The parallel of latitude, $23^1/_2$ degrees south latitude, marking the southern limit of the sun's vertical rays.

Troposphere The lowermost layer of the atmosphere. It is generally characterized by a decrease in temperature with height.

Turbidity current A downslope movement of dense, sediment-laden water created when sand and mud on the continental shelf and slope are dislodged and thrown into suspension.

Ultraviolet Radiation with a wavelength from 0.2 to 0.4 micrometer.

Unconformity A surface that represents a break in the rock record, caused by erosion or nondeposition.

Uniformitarianism The concept that the processes that have shaped Earth in the geologic past are essentially the same as those operating today.

Upslope fog Fog created when air moves up a slope and cools adiabatically.

Upwelling The rising of cold water from deeper layers to replace warmer surface water that has been moved away.

Valley breeze The daily upslope winds commonly encountered in a mountain valley.

Valley glacier See Alpine glacier.

Valley train A relatively narrow body of stratified drift deposited on a valley floor by meltwater streams that issue from a valley glacier.

Vapor pressure That part of the total atmospheric pressure attributable to water vapor content.

Vent A conduit that connects a magma chamber to a volcanic crater.

Viscosity A measure of a fluid's resistance to flow.

Visible light Radiation with a wavelength from 0.4 to 0.7 micrometer.

Volcanic island arc A group of volcanic islands formed by igneous activity associated with the subduction of oceanic lithosphere at an oceanic-oceanic convergent plate boundary.

Volcanic bomb A streamlined pyroclastic fragment ejected from a volcano while molten.

Volcanic neck An isolated, steep-sided, erosional remnant consisting of lava that once occupied the vent of a volcano.

Volcano A mountain formed of lava and/or pyroclastics.

Warm front A front along which a warm air mass overrides a retreating mass of cooler air.

Water table The upper level of the saturated zone of groundwater.

Wave-cut cliff A seawater-facing cliff along a steep shoreline formed by wave erosion at its base and mass wasting.

Wave-cut platform A bench or shelf in the bedrock at sea level, cut by wave erosion.

Wave cyclone A cyclone that forms and moves along a front. The circulation around the cyclone tends to produce the wavelike deformation of the front.

Wave height The vertical distance between the trough and crest of a wave.

Wavelength The horizontal distance separating successive crests or troughs.

Wave of oscillation A water wave in which the wave form advances as the water particles move in circular orbits.

Wave of translation The turbulent advance of water created by breaking waves.

Wave period The time interval between the passage of successive crests at a stationary point.

Wave refraction See Refraction.

Weather The state of the atmosphere at any given time.

Weathering The disintegration and decomposition of rock at or near the surface of Earth.

Well An opening bored into the zone of saturation.

Westerlies The dominant west-to-east motion of the atmosphere that characterizes the regions on the poleward side of the subtropical highs.

Wet adiabatic rate The rate of adiabatic temperature change in saturated air. The rate of temperature change is variable, but it is always less than the dry adiabatic rate.

White dwarf A star that has exhausted most or all of its nuclear fuel and has collapsed to a very small size; believed to be near its final stage of evolution.

Wind Air flowing horizontally with respect to Earth's surface.

Wind vane An instrument used to determine wind direction.

Yazoo tributary A tributary that flows parallel to the main stream because a natural levee is present.

Zone of accumulation The part of a glacier characterized by snow accumulation and ice formation. Its outer limit is the snowline.

Zone of aeration Area above the water table where openings in soil, sediment, and rock are not saturated but filled mainly with air.

Zone of saturation Zone where all open spaces in sediment and rock are completely filled with water.

Zone of wastage The part of a glacier beyond the zone of accumulation where all of the snow from the previous winter melts, as does some of the glacial ice.

Index

SYSTEM REQUIREMENTS

Platforms:

Macintosh
- PowerPC processor
- 16MB RAM
- 2X CD-ROM Drive
- System 7.1 or newer
- 16-bit color graphics (thousands of colors)
- Please see README for launch instructions

Windows
- Windows 95/98/NT4.x
- 486/33MHz or better
- 16MB RAM
- 2x CD-ROM drive
- 16-bit color graphics (thousands of colors)
- Sound Blaster 16 or compatible sound card
- Please see README for launch instructions

EarthShow CD-Rom

SITE LICENSE AGREEMENT Prentice-Hall, Inc.

YOU SHOULD CAREFULLY READ THE TERMS AND CONDITIONS BEFORE USING THE CD-ROM PACKAGE. USING THIS CD-ROM PACKAGE INDICATES YOUR ACCEPTANCE OF THESE TERMS AND CONDITIONS.

Prentice-Hall, Inc. provides this program and licenses its use. You assume responsibility for the selection of the program to achieve your intended results, and for the installation, use, and results obtained from the program. This license extends only to use of the program in the United States or countries in which the program is marketed by authorized distributors.

LICENSE GRANT

You hereby accept a nonexclusive, nontransferable, permanent license to install and use the program on an unlimited number of computers at one site. A site is a single campus or branch of an educational institution at one geographic location. You may copy the program solely for backup or archival purposes in support of your use of the program. You may not modify, translate, disassemble, decompile, or reverse engineer the program, in whole or in part.

TERM

The License is effective until terminated. Prentice-Hall, Inc. reserves the right to terminate this License automatically if any provision of the License is violated. You may terminate the License at any time. To terminate this License, you must return the program, including documentation, along with a written warranty stating that all copies in your possession have been returned or destroyed.

LIMITED WARRANTY

THE PROGRAM IS PROVIDED "AS IS" WITHOUT WARRANTY OF ANY KIND, EITHER EXPRESSED OR IMPLIED, INCLUDING, BUT NOT LIMITED TO, THE IMPLIED WARRANTIES OR MERCHANTABILITY AND FITNESS FOR A PARTICULAR PURPOSE. THE ENTIRE RISK AS TO THE QUALITY AND PERFORMANCE OF THE PROGRAM IS WITH YOU. SHOULD THE PROGRAM PROVE DEFECTIVE, YOU (AND NOT PRENTICE-HALL, INC. OR ANY AUTHORIZED DEALER) ASSUME THE ENTIRE COST OF ALL NECESSARY SERVICING, REPAIR, OR CORRECTION. NO ORAL OR WRITTEN INFORMATION OR ADVICE GIVEN BY PRENTICE-HALL, INC., ITS DEALERS, DISTRIBUTORS, OR AGENTS SHALL CREATE A WARRANTY OR INCREASE THE SCOPE OF THIS WARRANTY.

SOME STATES DO NOT ALLOW THE EXCLUSION OF IMPLIED WARRANTIES, SO THE ABOVE EXCLUSION MAY NOT APPLY TO YOU. THIS WARRANTY GIVES YOU SPECIFIC LEGAL RIGHTS AND YOU MAY ALSO HAVE OTHER LEGAL RIGHTS THAT VARY FROM STATE TO STATE.

Prentice-Hall, Inc. does not warrant that the functions contained in the program will meet your requirements or that the operation of the program will be uninterrupted or error-free.

However, Prentice-Hall, Inc. warrants the CD-ROM(s) on which the program is furnished to be free from defects in material and workmanship under normal use for a period of ninety (90) days from the date of delivery to you as evidenced by a copy of your receipt.

The program should not be relied on as the sole basis to solve a problem whose incorrect solution could result in injury to person or property. If the program is employed in such a manner, it is at the user's own risk and Prentice-Hall, Inc. explicitly disclaims all liability for such misuse.

LIMITATION OF REMEDIES

Prentice-Hall, Inc.'s entire liability and your exclusive remedy shall be:

1. the replacement of any CD-ROM not meeting Prentice-Hall, Inc.'s "LIMITED WARRANTY" and that is returned to Prentice-Hall, or

2. if Prentice-Hall is unable to deliver a replacement CD-ROM that is free of defects in materials or workmanship, you may terminate this agreement by returning the program.

IN NO EVENT WILL PRENTICE-HALL, INC. BE LIABLE TO YOU FOR ANY DAMAGES, INCLUDING ANY LOST PROFITS, LOST SAVINGS, OR OTHER INCIDENTAL OR CONSEQUENTIAL DAMAGES ARISING OUT OF THE USE OR INABILITY TO USE SUCH PROGRAM EVEN IF PRENTICE-HALL, INC. OR AN AUTHORIZED DISTRIBUTOR HAS BEEN ADVISED OF THE POSSIBILITY OF SUCH DAMAGES, OR FOR ANY CLAIM BY ANY OTHER PARTY.

SOME STATES DO NOT ALLOW FOR THE LIMITATION OR EXCLUSION OF LIABILITY FOR INCIDENTAL OR CONSEQUENTIAL DAMAGES, SO THE ABOVE LIMITATION OR EXCLUSION MAY NOT APPLY TO YOU.

GENERAL

You may not sublicense, assign, or transfer the license of the program. Any attempt to sublicense, assign or transfer any of the rights, duties, or obligations hereunder is void.

This Agreement will be governed by the laws of the State of New York.

Should you have any questions concerning this Agreement, you may contact Prentice-Hall, Inc. by writing to:

Multimedia Development
Engineering/Science/Math Division
Higher Education
Prentice-Hall, Inc.
1 Lake Street
Upper Saddle River, NJ 07458

Should you have any questions concerning technical support, you may write to:

Multimedia Production
Engineering/Science/Math Division
Higher Education
Prentice-Hall, Inc.
1 Lake Street
Upper Saddle River, NJ 07458

YOU ACKNOWLEDGE THAT YOU HAVE READ THIS AGREEMENT, UNDERSTAND IT, AND AGREE TO BE BOUND BY ITS TERMS AND CONDITIONS. YOU FURTHER AGREE THAT IT IS THE COMPLETE AND EXCLUSIVE STATEMENT OF THE AGREEMENT BETWEEN US THAT SUPERSEDES ANY PROPOSAL OR PRIOR AGREEMENT, ORAL OR WRITTEN, AND ANY OTHER COMMUNICATIONS BETWEEN US RELATING TO THE SUBJECT MATTER OF THIS AGREEMENT.